Lecture Notes in Artificial Intelligence   8346

Subseries of Lecture Notes in Computer Science

LNAI Series Editors

Randy Goebel
   University of Alberta, Edmonton, Canada
Yuzuru Tanaka
   Hokkaido University, Sapporo, Japan
Wolfgang Wahlster
   DFKI and Saarland University, Saarbrücken, Germany

LNAI Founding Series Editor

Joerg Siekmann
   DFKI and Saarland University, Saarbrücken, Germany

Hiroshi Motoda   Zhaohui Wu   Longbing Cao
Osmar Zaiane   Min Yao   Wei Wang (Eds.)

Advanced Data Mining and Applications

9th International Conference, ADMA 2013
Hangzhou, China, December 14-16, 2013
Proceedings, Part I

 Springer

Volume Editors

Hiroshi Motoda
US Air Force Office of Scientific Research, Tokyo, Japan
E-mail: motoda@ar.sanken.osaka-u.ac.jp

Zhaohui Wu
Zhejiang University, Hangzhou, China
E-mail: wzh@cs.zju.edu.cn

Longbing Cao
University of Technology, Sydney, NSW, Australia
E-mail: longbing.cao@uts.edu.au

Osmar Zaiane
University of Alberta, Edmonton, AB, Canada
E-mail: zaiane@cs.ualberta.ca

Min Yao
Zhejiang University, Hangzhou, China
E-mail: myao@zju.edu.cn

Wei Wang
Fudan University, Shanghai, China
E-mail: weiwang1@fudan.edu.cn

ISSN 0302-9743   e-ISSN 1611-3349
ISBN 978-3-642-53913-8   e-ISBN 978-3-642-53914-5
DOI 10.1007/978-3-642-53914-5
Springer Heidelberg New York Dordrecht London

Library of Congress Control Number: 2013956530
CR Subject Classification (1998): I.2, H.3, H.2.8, H.4, I.5, I.4, H.2, J.1
LNCS Sublibrary: SL 7 – Artificial Intelligence

Typesetting: Camera-ready by author, data conversion by Scientific Publishing Services, Chennai, India

Printed on acid-free paper

Springer is part of Springer Science+Business Media (www.springer.com)

Preface

It is our pleasure to welcome you to the proceedings of the 9th International Conference on Advanced Data Mining and Applications (ADMA 2013).

As the power of generating, transmitting, and collecting huge amounts of data grows continuously, information overload is an imminent problem. It generates new challenges for the data-mining research community to develop sophisticated data-mining algorithms as well as successful data-mining applications. ADMA 2013 was held in Hangzhou, China, with the purpose of promoting original research in advanced data mining and applications and providing a dedicated forum for researchers and participants to share new ideas, original research results, case studies, practical development experiences and applications in all aspects related to data mining and applications, the.

The conference attracted 222 submissions from 26 different countries and areas. All papers were peer reviewed by at least three members of the Program Committee composed of international experts in data-mining fields. The Program Committee, together with our Program Committee Co-chairs, did enormous amount of work to select papers through a rigorous review process and extensive discussion, and finally composed a diverse and exciting program including 32 full papers and 64 short papers for ADMA 2013. The ADMA 2013 program was highlighted by three keynote speeches from outstanding researchers in advanced data mining and application areas: Gary G. Yen (Oklahoma State University, USA), Xindong Wu (University of Vermont, USA), and Joshua Zhexue Huang (Shenzhen Institutes of Advanced Technology, Chinese Academy of Sciences).

We would like to thank the support of several groups, without which the organization of the ADMA 2013 would not be successful. These include sponsorship from Zhejiang University, Taizhou University, and University of Technology Sydney. We also appreciate the General Co-chairs for all their precious advice and the Organizing Committee for their dedicated organizing efforts. Finally, we express our deepest gratitude to all the authors and participants who contributed to the success of ADMA 2013.

November 2013

Hiroshi Motoda
Zhaohui Wu
Longbing Cao
Osmar Zaiane
Min Yao
Wei Wang

Organization

ADMA 2013 was organized by Zhejiang University, China

Honorary Chair

Yunhe Pan   Chinese Academy of Engineering, China

Steering Committee Chair

Xue Li   University of Queensland (UQ), Australia

General Co-chairs

Hiroshi Motoda   US Air Force Office of Scientific Research, USA
Zhaohui Wu   Zhejiang University, China
Longbing Cao   University of Technology Sydney, Australia

Program Committee Co-chairs

Osmar Zaiane   University of Alberta, Canada
Min Yao   Zhejiang University, China
Wei Wang   Fudan University, China

Organization Co-chairs

Xiaoming Zhao   Taizhou University, China
Jian Wu   Zhejiang University, China
Guandong Xu   University of Technology Sydney, Australia

Publicity Chair

Liyong Wan   Zhejiang University, China

Registration Chair

Xiaowei Xue   Zhejiang University, China

Web Master

Bin Zeng   Zhejiang University, China

Steering Committee

Kyu-Young Whang	Korea Advanced Institute of Science and Technology, Korea
Chengqi Zhang	University of Technology, Sydney, Australia
Osmar Zaiane	University of Alberta, Canada
Qiang Yang	Hong Kong University of Science and Technology, China
Jie Tang	Tsinghua University, China
Jie Cao	Nanjing University of Finance and Economics, China

Program Committee

Aixin Sun	Nanyang Technological University, Singapore
Annalisa Appice	University Aldo Moro of Bari, Italy
Atsuyoshi Nakamura	Hokkaido University, Japan
Bin Shen	Ningbo Institute of Technology, China
Bo Liu	QCIS University of Technology, Australia
Daisuke Ikeda	Kyushu University, Japan
Daisuke Kawahara	Kyoto University, Japan
Eiji Uchino	Yamaguchi University, Japan
Faizah Shaari	Polytechnic Sultan Salahudin Abddul Aziz Shah, Malaysia
Feiping Nie	University of Texas, Arlington, USA
Gang Li	Deakin University, Australia
Gongde Guo	Fujian Normal University, China
Guandong Xu	University of Technology Sydney, Australia
Guohua Liu	Yanshan University, China
Hanghang Tong	IBM T.J. Watson Research Center, USA
Haofeng Zhou	Fudan University, China
Jason Wang	New Jersey Institute of Technology, USA
Jianwen Su	UC Santa Barbara, USA
Jinjiu Li	University of Technology Sydney, Australia
Liang Chen	Zhejiang University, China
Manish Gupta	University of Illinois at Urbana-Champaign, USA
Mengchu Zhou	New Jersey Institute of Technology, USA
Mengjie Zhang	Victoria University of Wellington, New Zealand
Michael R. Lyu	The Chinese University of Hong Kong, Hong Kong, China
Michael Sheng	The University of Adelaide, Australia
Philippe Fournier-Viger	University of Moncton, Canada
Qi Wang	Xi'an Institute of Optics and Precision Mechanics of CAS, China

Sponsoring Institutions

College of Computer Science & Technology, Zhejiang University, China
College of Mathematics and information engineering, Taizhou University, China
Advanced Analytics Institute, University of Technology Sydney, Australia

Table of Contents – Part I

Sequential Data Mining

Web Mining

Image Mining

Text Mining

Social Network Mining

Classification

Table of Contents – Part II

Clustering

Association Rule Mining

Pattern Mining

Identification

Privacy Preservation

Applications

Machine Learning

Mining E-Commerce Feedback Comments
for Dimension Rating Profiles

Lishan Cui[1], Xiuzhen Zhang[1], Yan Wang[2], and Lifang Wu[3]

[1] School of Computer Science & IT, RMIT University, Melbourne, Australia
{lishan.cui,xiuzhen.zhang}@rmit.edu.au
[2] Department of Computing, Macquarie University, Sydney, Australia
yan.wang@mq.edu.au
[3] Beijing University of Technology, China
lfwu@bjut.edu.cn

Abstract. Opinion mining on regular documents like movie reviews and product reviews has been intensively studied. In this paper we focus on opinion mining on short e-commerce feedback comments. We aim to compute a comprehensive rating profile for sellers comprising of dimension ratings and weights. We propose an algorithm to mine feedback comments for dimension ratings, combining opinion mining and dependency relation analysis, a recent development in natural language processing. We formulate the problem of computing dimension weights from ratings as a factor analytic problem and propose an effective solution based on matrix factorisation. Extensive experiments on eBay and Amazon data demonstrate that our proposed algorithms can achieve accuracies of 93.1% and 89.64% respectively for identifying dimensions and ratings in feedback comments, and the weights computed can accurately reflect the amount of feedback for dimensions.

Keywords: opinion mining, typed dependency, matrix factorisation.

1 Introduction

Opinion mining, or sentiment analysis, has attracted extensive research activities [12, 15, 18, 19, 23] in recent years. See [15, 19] for a comprehensive survey of the field. Early opinion mining research has mostly focused on analysing movie reviews and product reviews [12, 18, 23]. Online social media such as blogs and microblogs have also attracted active research [6, 8, 22, 26]. E-commerce feedback comments are an important data source for opinion mining and recently attract attention in the research community [16].

On auction websites like eBay and Amazon explicit feedback ratings and textual feedback comments after transactions are kept in their feedback systems. In addition to the overall ratings of transactions, potential transactors often read the feedback comments to solicit opinions and ratings about sellers at finer granular levels. For example, a comment like *"The products were as I expected."* expresses positive opinion towards the Product dimension, whereas the comment

H. Motoda et al. (Eds.): ADMA 2013, Part I, LNAI 8346, pp. 1–12, 2013.

"Delivery was a little slow but otherwise great service. Recommend highly." expresses negative opinion towards the Delivery dimension but a positive rating to the transaction in general. By analysing the wealth of information in feedback comments we can uncover buyers' embedded opinions towards different aspects of transactions, and compute comprehensive reputation profiles for sellers.

We call the aspects of transactions *dimensions* and the aggregated opinions towards dimensions *dimension ratings*. We aim to compute a comprehensive reputation profile for sellers for e-commerce applications, comprising dimension ratings and dimension weights. We propose to compute dimension ratings from feedback comments by applying lexicon-based opinion mining techniques [2] in combination with dependency relation analysis, a tool recently developed in natural language processing (NLP) [4]. We formulate the problem of computing dimension weights from ratings as a factor analytic problem and propose a matrix factorisation technique [5] to automatically compute weights for dimensions from the sparse and noisy dimension rating matrix.

Our experiments on eBay and Amazon data show that our algorithm DR-mining can achieve accuracies of 93.1% and 89.64% for identifying dimensions and dimension ratings in comments respectively, and is more accurate than a widely used existing opinion mining approach [12]. The aggregated dimension ratings accurately reflect transactors views on dimensions. Our computed dimension weights can effectively capture the weights for different dimensions based on the amount of feedback as well as the consistency among ratings for dimensions.

2   Related Work

Recent related works include [16] and [24], where statistical topic modelling has been employed to cluster words into dimensions (also called aspects [16,24]) and to further compute aggregated ratings for dimensions. In this paper, we aim to compute comprehensive dimension rating profiles, including dimension ratings as well as dimension weights. Note that due to the unsupervised nature of topic modelling, some aspects discovered in [16] and [24] may not concur with the human understanding of important aspect for e-commerce applications. For example, "buy" is discovered as an aspect in [16] due to the frequent occurrence of phrases like "will buy", "would buy" and "buy again". But "buy" is not an aspect for e-commerce applications. In this paper, by using user specified dimensions as input, we adopt a supervised approach to assigning words to dimensions and thus can produce meaningful dimensions. The dimension ratings in [16] and [24] are computed from the overall ratings for transactions. But the dimension ratings by this approach may suffer the positive bias in e-commerce rating systems, where in average over 99% of feedback ratings are positive [21]. In this paper we compute dimension ratings by aggregating the opinions for dimensions in comments directly.

There have been other studies on analysing feedback comments in e-commerce applications [7,10,16,17]. [17] and [7] focus on sentiment classification of feedback comments. It is demonstrated that feedback comments are noisy and therefore analysing them is a challenging problem. As discussed in [17] missing aspect

Table 1. Dimension rating patterns

Dependency Relation	Patterns
adjective modifier amod(NN, JJ)	amod(price/NN, great/JJ)
	amod(postage/NN, quick/JJ)
adverbial modifier advmod(VB, RB)	advmod(shipping/VBG, fast/RB)
nominal subject nsubj(JJ, NN)	nsubj(prompt/JJ, seller/NN)
adjectival complement acomp(VB, JJ)	acomp(arrived/VBN, quick/JJ)

Note: NN = noun, VB = verb, JJ = adjective, and RB = adverb

comments are deemed negative and model built from aspect ratings are used to classify comments into positive or negative. In [10] a technique for summarising feedback comments is presented, aiming to filter out courteous comments that do not provide real feedback. In [25] a semi-supervised method is proposed to detect shilling attacks and to demonstrate its effectiveness on enhancing the product recommendations in Amazon.

Existing work on extracting features and associated opinion polarities [12, 27] from product reviews and movie reviews is similar in notion to dimension ratings in e-commerce feedback comments. Notably Hu and Liu [12] propose to extract nouns and noun phrases as candidate features and an opinion lexicon is used to identify opinion polarities on features. In [27] dependency relation analysis is further applied to extract opinions on features. In this paper we use dependency relation patterns for identifying dimension ratings in feedback comments. Experiments show that our approach significantly outperforms the Hu&Liu approach [12]. It should be noted that the Hu&Liu approach is not specifically designed for analysing short feedback comments.

Matrix factorisation is a general factor analytic technique widely applied in various areas, especially in information retrieval [5] and recommender systems [11, 14]. Our application of matrix factorisation is fundamentally different from existing approaches. While existing applications use matrices in the reduced feature space after decomposition to compute similarity among original data objects, we computes the relative weights of original data objects in the reduced feature space. Noises are removed and correlation of data objects is considered, and as a result weights computed in the new reduced feature space are more accurate description of weightings of original data objects.

3   Mining Feedback Comments for Dimension Ratings

Turney proposes to use two-word phrase patterns like *adjective-noun*, and *adverb-verb* to identify subjectivity in product reviews [23]. On the other hand, the typed dependency representation [4] is a recent NLP tool to help understand the grammatical relationships in sentences. A sentence is represented as a set of dependency relations between pairs of words in the form of (*head, dependent*), where content words are chosen as heads, and other related words depend on the heads. For the comment "*Super quick shipping.*", the adjective modifier relations *amod (shipping, super)* and *amod (shipping, quick)* indicate that *super*

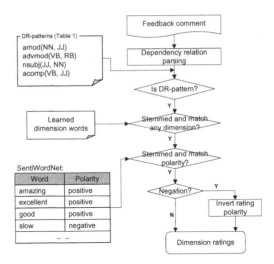

Fig. 1. The DR-mining algorithm

modifies *shipping* and *quick* modifies *shipping*. Words are also annotated with their POS tags such as noun(NN), verb (VB), adjective (JJ) and adverb (RB). Four dependency relations are found to express modifying relationship between adjectives and nouns, and adverbs and verbs. They are thus called *dimension rating patterns* (DR-patterns) and are as shown in Table 1, where head terms indicate dimensions and dependent terms indicate opinion. With our example, according to the DR-pattern *amod (NN, JJ)*, *(shipping, super)* and *(shipping, quick)* are dimension-rating pairs.

Dimension words are learned from DR-patterns annotated with dimension labels and ratings. For example the dimension-rating pairs *(shipping, super)* and *(shipping, quick)* are labelled with the Delivery with rating of +1 (positive). Given a dimension and the candidate dimension words, we apply association rule mining [1] to decide a set of dimension words for the dimension. Specifically given dimension d and a word w, w is dimension word for d if $P(d|w) \geq \alpha$, where $P(d|w)$ is called the *confidence* of w and α is a threshold ($\alpha = 50\%$ by default).

The complete dimension-rating mining (DR-mining) algorithm for identifying dimensions and associated ratings from feedback comments is shown in Fig. 1. Each comment is first analysed using the Stanford dependency relation parser. To identify dimensions in comments, the dependency relations resulted from parsing are first matched against the DR-patterns shown in Table 1. If a DR-pattern is found, first the dimension is identified using the dimension words learned from the training data. The opinion word and polarity for this DR-pattern is identified using the SentiWordNet [2] opinion lexicon. If either the head or dependent word of a DR-pattern involves the negation relation, the relevant polarity is inverted. In the process of matching dimension and opinion words, the Porter stemming algorithm [20] is used to achieve effective matching, considering the various formats for words. The opinion polarities positive and

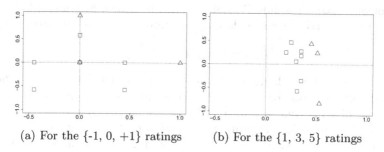

(a) For the {-1, 0, +1} ratings   (b) For the {1, 3, 5} ratings

Fig. 2. Latent component space for the rating matrix in Example 1. Comment vectors and dimension vectors are denoted as squares and triangles respectively.

negative correspond to ratings +1 and -1, and nil opinion is deemed a neutral rating of 0. DR-mining thus produces a set of dimension ratings for a comment.

The dimension ratings from comments are used to compute the *aggregated dimension ratings* by m-estimate [13]. Given a set S of ratings towards a dimension d, $S = \{v_d | v_d \in \{-1, 0, +1\}\}$, where v_d represents the positive (+1), negative(-1), or neutral (0) polarity. The aggregated rating for dimension d is:

$$r_d = \frac{|\{v_d \in S \wedge v_d > 0\}| + m * p}{|\{v_d \in S \wedge v_d \neq 0\}| + m} \tag{1}$$

In Equation 1, $\{v_d \in S \wedge v_d > 0\}$ represents positive opinions for dimension d and $\{v_d \in S \wedge v_d \neq 0\}$ represents positive or negative opinion on d. p is the prior positive opinion probability for a transaction. With the prior belief of neutral tendency we set $p = 0.5$ by default. m is a smoothing factor. From our experiments, settings of $m = 6..20$ typically give stable results, and we set $m = 6$ by default.

4 Computing Dimension Weights by Matrix Factorisation

The dimension ratings derived from feedback comments are highly noisy – comments may be from the same buyer and therefore highly correlated, and some buyers may be lenient (or harsh) raters. Viewing that there is some latent structure underlying the noisy ratings, we propose to represent dimension ratings as a rating matrix and then compute the underlying structure for the ratings using singular value decomposition (SVD) [5]. Under SVD, comment vectors and dimension vectors are projected onto vectors in the same reduced space of latent independent components, and we further compute the weights for dimensions.

4.1 Singular Value Decomposition

With SVD, latent components in the new space are ordered so as to reflect major associative patterns in the original data, and ignore the smaller and less important influences. The full SVD of a matrix is defined as follows: Let $A_{m \times n}$ denote an $m \times n$ matrix, A can be decomposed into the product of three matrices:

$$A_{m \times n} = U_{m \times m} D_{m \times n} V_{n \times n}^T$$

where U contains the orthonormal eigen vectors for AA^T, V contains the orthonormal eigen vectors for $A^T A$, and D is a diagonal matrix containing the square roots of eigen values from U (V has the same eigen value as U). Especially D contains values indicating the variance of the original data points along each latent component, ordered in decreasing level of variance. The first component represents the largest variance of the original data points. In many applications (e.g. [5]), only the first several components of SVD (typically two or three) are considered to form a reduced representation as shown below, where $k < m$ and $k < n$:

$$A_{m \times n} \approx U_{m \times k} D_{k \times k} V_{k \times n}^T$$

Example 1. The 4×10 matrix A below is an example rating matrix, representing ten comments for four dimensions.

$$A = \begin{matrix} & c_1\ c_2\ c_3 \quad c_4\ c_5\ c_6\ c_7\ c_8 \quad c_9\ c_{10} \\ & \begin{bmatrix} 1 & 0 & 0 & 1 & 0 & 0 & 0 & 1 & -1 & -1 \\ 0 & 0 & 0 & -1 & 0 & 0 & 1 & 0 & -1 & 0 \\ 0 & 0 & 1 & 0 & 0 & 0 & 0 & 0 & 0 & 0 \\ 0 & 0 & 0 & 0 & 0 & 1 & 0 & 0 & 0 & 0 \end{bmatrix} \end{matrix}$$

Applying the reduced SVD model with $k = 2$ to matrix A produces

$$A \approx UDV^T = \begin{bmatrix} 1.00 & 0.00 \\ 0.00 & 1.00 \\ 0.00 & 0.00 \\ 0.00 & 0.00 \end{bmatrix} \begin{bmatrix} 2.24 & 0 \\ 0 & 1.73 \end{bmatrix} \begin{bmatrix} 0.45 & 0.00 \\ 0.00 & 0.00 \\ 0.00 & 0.00 \\ 0.45 & -0.58 \\ 0.00 & 0.00 \\ 0.00 & 0.00 \\ 0.00 & 0.58 \\ 0.45 & 0.00 \\ -0.45 & -0.58 \\ -0.45 & 0.00 \end{bmatrix}^T$$

With $k = 2$, the comment vectors and the dimension vectors are mapped to the same latent component space of two components. As shown in Fig. 2(a), the first component is where original data points show most variance. In the latent component space, the row vectors of U are the four trust dimension vectors and the row vectors of V^T are the ten comment vectors. In Fig. 2(a), squares represent comment vectors, and triangles denote dimension vectors. Note that in Fig. 2(a) only three points are denoted for dimension vectors – the vectors for the second and third dimensions fall on the same position $(0.00, 0.00)$, which indicates that the second and third dimensions are highly correlated in the original rating matrix. Similarly only six points for comment vectors are denoted due to that some comment vectors fall on the same positions, indicating that these comment ratings are highly correlated and possibly by the same seller.

4.2   Computing Dimension Weights in the Latent Component Space

With the ratings of $\{-1, 0, +1\}$ for a rating matrix, the zeros are deemed missing values and the input matrix becomes very sparse. As a result, in addition to the high computation cost, SVD often results in latent vectors that are sparse and overlap in the latent space [3]. Indeed in Fig. 2(a) several points are at or close to the line with $x = 0$. Two comment rating vectors and two dimension rating vectors fall onto the point (0,0). To overcome this problem, and to accurately represent user ratings, we convert the ratings of $\{-1, 0, +1\}$ to a rating matrix of $\{1, 3, 5\}$ and then SVD is applied.

Example 2. The rating matrix A in Example 1 on ratings $\{-1, 0, +1\}$ is mapped to a rating matrix B on ratings $\{1, 3\ 5\}$ as follows:

$$
B = \begin{bmatrix}
5\ 3\ 3\ 5\ 3\ 3\ 3\ 3\ 5\ 1\ 1 \\
3\ 3\ 3\ 1\ 3\ 3\ 5\ 3\ 1\ 3 \\
3\ 3\ 5\ 3\ 3\ 3\ 3\ 3\ 3\ 3 \\
3\ 3\ 3\ 3\ 5\ 3\ 3\ 3\ 3
\end{bmatrix}
$$

SVD produces the following reduced decomposition for B:

$$
B \approx UDV^T = \begin{bmatrix}
0.53 & -0.83 \\
0.45 & 0.44 \\
0.51 & 0.24 \\
0.51 & 0.24
\end{bmatrix}
\begin{bmatrix}
19.88 & 0 \\
0 & 4.06
\end{bmatrix}
\begin{bmatrix}
0.35 & -0.35 \\
0.30 & 0.06 \\
0.35 & 0.18 \\
0.31 & -0.57 \\
0.30 & 0.06 \\
0.35 & 0.18 \\
0.35 & 0.28 \\
0.35 & -0.35 \\
0.20 & 0.25 \\
0.25 & 0.47
\end{bmatrix}^T
$$

In Figure 2(b), we can see that points are further away from the line where $x = 0$. There are still some vectors fall onto the same point. For example, two dimension vectors fall onto the point (0.51, 0.24). With the above reduced SVD $B_{4\times 10} \approx U_{4\times 2}D_{2\times 2}V^T_{2\times 10}$ the column vectors of $U_{4\times 2}$ are the two components in the latent space where ratings in $B_{4\times 10}$ exhibit the largest variance and are ordered decreasingly by the level of variance. Note that column vectors in U are unit vectors. Specifically, for a column vector u_j of U ($j = 1..k$), the coefficients of u_j satisfies $\sum_{d=1}^{m} u_j{}^2[d] = 1$.

The first column vector u_1 represents the latent component where the original rating matrix demonstrates the highest variance and so potentially its coefficients can indicate weights for dimensions. However u_1 itself may be unreliable as the weight for dimensions, while all k column vectors of U combined can provide a reliable estimation for dimension weights. Specifically the weight for a given dimension d is computed from k column vectors of U as follows:

$$w_d = \frac{\sum_{j=1}^{k} u_j{}^2[d]}{k} \tag{2}$$

Typically we set $k = 2$ and our experiments confirm that this setting gives reliable estimation of dimension weights while removing noises in rating matrices.

Example 3. Applying SVD to the rating matrix B with $k = 2$, the weights for four dimensions are computed from the two column vectors of U: $\boldsymbol{w}^T = [0.48\ 0.20\ 0.16\ 0.16]$ It can be seen that the third and fourth dimensions have the lowest weights, while the first dimension has the highest weight, which intuitively corresponds to the total number of positive and negative ratings and their level of consistency in ratings along these dimensions. As shown in the rating matrix A, the first dimension has a small number of positive and negative ratings, and the highest variation – three positive, two negative and five neutral ratings. The third and fourth dimensions are dominated by neutral ratings and have the highest level of consistency in ratings – one positive and nine neutral ratings.

5   Experiments

We focus on four dimensions Product, Delivery, Communication and Cost, according to the Detailed Seller Rating on eBay [1]: *item as described, shipping time, communication,* and *shipping and handling charges.* Each comment is manually annotated with dimensions and corresponding ratings of positive $(+1)$, negative (-1) or neutral (0) [2]. As a baseline we implemented an algorithm based on the Hu&Liu algorithm [12], which is simply denoted as Hu&Liu. Dimensions identified by the Hu&Liu algorithm are assigned to the four dimensions manually.

2,000 feedback comments between 31 January and 18 March 2012 for ten sellers on eBay (non-English comments removed) are used for training. For DR-mining, dimension words are learned from this training data. For the Hu&Liu algorithm, learning is conducted as follows: Nouns with a frequency of $\geq 1\%$ are deemed dimension words. Adjectives and adverbs within three words before or after dimension words are opinion words (the sentence boundary is used in [12]). SentiWordNet is used to decide the polarity of opinion words (an opinion lexicon for product reviews was used in [12] when SentiWorNet was not available). Negation words are based on the negation word list in [9], and are applied as within two words before opinion words. 2,000 comments from ten other eBay sellers and 2,000 comments (non-English comments removed) for ten Amazon sellers are then used to evaluate the performance of DR-mining and Hu&Liu.

---

[1] http://pages.ebay.com/help/feedback/contextual/
detailed-feedback-ratings.html
[2] Annotation is done by the first two authors. Agreement is reached by discussions.

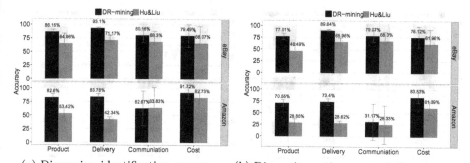

(a) Dimension identification accuracy   (b) Dimension rating computation accuracy

Fig. 3. Accuracy of DR-mining on eBay and Amazon data

Table 2. Dimension words discovered by association mining

Dimension	Words (Confidence)
Product	product (0.99), item (0.79), condition (0.98), work(verb, 0.78), quality(0.66)
Delivery	arrive (0.75), delivery (1.0), postage (1.0), receive (0.56), ship (0.98)
Communication	communication (1.0), response (1.0), email (1.0), reply (1.0)
Cost	cost (1.0), price (0.97), refund (0.71), value (1.0)
Transaction	seller (0.96), service (0.98), ebayer (1.0), transaction (1.0),
	deal (0.94), buy (0.74), business (0.97), ebay (0.89), work (noun, 0.75)

When applying SentiWordNet for opinion word polarity identification, a few words that clearly demonstrate strong polarity for e-commerce applications are annotated otherwise in SentiWordNet, due to its purpose of general applicability. As a result, ten words in SentiWordNet are re-annotated with prior polarity: *lightning (+ve), fast (+ve), prompt(+ve), safely (+ve), pretty (+ve), satisfied (+ve), scratch (-ve), squashed (-ve), late (-ve), waste (-ve),* and *cheap (+ve).*

5.1   Accuracy of the DR-Mining Algorithm

The dimension words generated by our association-mining approach with a confidence threshold of 50% are shown in Table 2. We evaluate the dimension words computed by association mining with the accuracy for dimension identification, which is the percentage of comments where a dimension is correctly identified. Fig. 3 (a) shows the average accuracies of dimension identification for DR-mining and the Hu&Liu algorithm on feedback comments of ten eBay and ten Amazon sellers for evaluation. On both datasets, DR-mining significantly outperforms the Hu&Liu approach for identifying all dimensions(Wilcoxin signed rank test, $p < 0.05$). On the eBay dataset DR-mining achieves accuracies from 79.49% on the Cost dimension to 93.1% on the Delivery dimension, whereas Hu&Liu can only reach 71.17% on the Delivery dimension. On the Amazon dataset, the accuracy of DR-mining varies from 62.67% for the Communication dimension to 91.72% for the Cost dimension, whereas the Hu&Liu accuracy varies from 42.34% on the Delivery dimension to 82.73% on the Cost dimension.

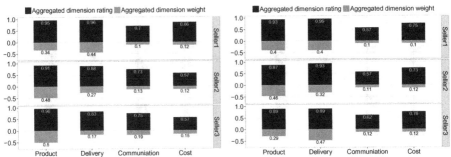

(a) Rating profiles for three eBay sellers   (b) Rating profiles for three Amazon sellers

Fig. 4. The comprehensive Rating profiles for sellers

Fig. 3 (b) compares the accuracies for computing dimension ratings of DR-mining and Hu&Liu, averaged over ten sellers from eBay and Amazon respectively. Computing dimension ratings depends on identifying dimensions first, and so generally has lower accuracy. DR-mining generally achieves consistently high accuracy. On the eBay dataset, it has accuracies of 76.12% for Cost to 89.64% for Delivery. On the Amazon dataset, it achieves accuracies from 31.17% for Communication to 83.53% for Cost. In contrast, Hu&Liu achieves reasonably good accuracy on the eBay dataset – from 46.49% for Product to 68.3% for Communication, but very modest accuracy on the Amazon dataset – from 25.33% for Communication to 61.39% for Cost. The consistently higher accuracy of DR-mining is due in part to the application of dependency relation analysis for DR-patterns to identify opinion words. The odd low accuracy with high variance of both DR-mining and Hu&Liu for Communication on the Amazon data results from the fact that comments on Communication is extremely rare – four out of ten sellers have only one comment on Communication and one false rating results in an accuracy of zero.

5.2   Dimension Rating Profiles for Sellers

Fig. 4 depicts the different dimension rating profiles for three representative eBay and Amazon sellers respectively. For each seller, the upward bars represent ratings for dimensions while the downward bars represent their weights. The figure clearly illustrates the variation of dimension ratings for each seller horizontally and those across different sellers vertically. Such comprehensive rating profiles certainly can cater to buyers' preferences for different dimensions and guide buyers in making informed decisions when choosing sellers.

In Fig. 5, the dimension weights for one eBay seller and one Amazon seller are plotted, with respect to the number of comments (Dimension weights for other sellers show similar trends and are omitted to save space). In the figure, when the number of comments is increased from ten to 190 (recall that a seller has 200 comments) the dimension weights converge at 140 and 150 comments for the eBay seller and the Amazon seller respectively. This suggests that a reasonable number

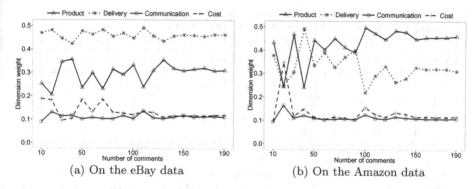

(a) On the eBay data   (b) On the Amazon data

Fig. 5. Dimension weight with respect to number of comments

of comments is sufficient to reliably estimate dimension weights. It is obvious from the figure that Communication and Cost dimensions have low weights (0.11 and 0.12 respectively). This result is consistent with our observation that there are few ratings on Communication and Cost – On the eBay data, there are on average 7.7 and 10.1 ratings (out of about 200 comments) for a seller on the two dimensions respectively whereas on the Amazon data there are 4.55 and 9.51 ratings (out of about 200 comments) for a seller respectively.

6   Conclusions

We have proposed to compute comprehensive dimension rating profiles for sellers in e-commerce applications by mining feedback comments regarding past transactions. Our dimension rating profiles comprise aggregated dimension ratings as well as their weights. We have proposed an effective algorithm for mining short feedback comments for dimension ratings. Our approach achieves significantly higher accuracy for discovering dimension ratings in comments than a commonly used opinion mining approach. We compute aggregated dimension ratings from the dimension ratings in comments. We further apply matrix factorisation techniques to compute dimension weights, which can effectively remove noises in general raw ratings mined from feedback comments.

References

1. Agrawal, R., Srikant, R.: Fast algorithms for mining association rules. In: Proc. VLDB (1994)
2. Baccianella, S., et al.: SentiWordNet 3.0: An enhanced lexical resource for sentiment analysis and opinion mining. In: Proc. LREC (2010)
3. Chandrasekaran, V., et al.: Rank-sparsity incoherence for matrix decomposition. SIAM Journal on Optimization 21(2), 572–596 (2011)
4. De Marneffe, M., et al.: Generating typed dependency parses from phrase structure parses. In: Proc. LREC, pp. 449–454 (2006)

5. Deerwester, S., et al.: Indexing by latent semantic analysis. Journal of the American Society for information Science 41(6), 391–407 (1990)
6. Fang, Y., et al.: Mining contrastive opinions on political texts using cross-perspective topic model. In: Proc. WSDM (2012)
7. Gamon, M.: Sentiment classification on customer feedback data: noisy data, large feature vectors, and the role of linguistic analysis. In: 20th Int. Conf. on Computational Linguistics (2004)
8. Go, A., et al.: Twitter sentiment classification using distant supervision. CS224N Project Report, Stanford (2009)
9. Goryachev, S., et al.: Implementation and evaluation of four different methods of negation detection. Technical report, DSG (2006)
10. Hijikata, Y., et al.: Social summarization of text feedback for online auctions and interactive presentation of the summary. Knowledge-Based Systems 20(6) (2007)
11. Hofmann, T.: Latent semantic models for collaborative filtering. ACM Transactions on Information Systems, TOIS (2004)
12. Hu, M., Liu, B.: Mining and summarizing customer reviews. In: Proc. KDD (2004)
13. Karplus, K.: Evaluating regularizers for estimating distributions of amino acids. In: Proc. Third Int. Conf. on ISMB, vol. 3 (1995)
14. Koren, Y., et al.: Matrix factorization techniques for recommender systems. Computer (2009)
15. Liu, B.: Sentiment analysis and opinion mining. Synthesis Lectures on Human Language Technologies. Morgan & Claypool Publishers (2012)
16. Lu, Y., et al.: Rated aspect summarization of short comments. In: Proc. WWW (2009)
17. O'Donovan, J., et al.: Extracting and visualizing trust relationships from online auction feedback comments. In: Proc. IJCAI, pp. 2826–2831 (2007)
18. Pang, B., Lee, L.: A sentimental education: sentiment analysis using subjectivity summarization based on minimum cuts. In: Proc. ACL (2004)
19. Pang, B., Lee, L.: Opinion mining and sentiment analysis. Found. Trends Inf. Retr. 2(1-2), 1–135 (2008)
20. Porter, M.F.: An algorithm for suffix stripping. Program: Electronic Library and Information Systems 14(3) (1980)
21. Resnick, P., et al.: Reputation Systems: Facilitating Trust in Internet Interactions. Communications of the ACM 43, 45–48 (2000)
22. Thelwall, M., et al.: Sentiment strength detection in short informal text. Journal of the American Society for information Science 61(12), 2544–2558 (2010)
23. Turney, P.D.: Thumbs up or thumbs down?: Semantic orientation applied to unsupervised classification of reviews. In: Proc. ACL, pp. 417–424 (2002)
24. Wang, H., et al.: Latent aspect rating analysis without aspect keyword supervision. In: Proc. 17th ACM SIGKDD, pp. 618–626 (2011)
25. Wu, Z., et al.: Hysad: A semi-supervised hybrid shilling attack detector for trustworthy product recommendation. In: Proc. 18th ACM SIGKDD, pp. 985–993 (2012)
26. Zhang, X., Zhou, Y., Bailey, J., Ramamohanarao, K.: Sentiment analysis by augmenting expectation maximisation with lexical knowledge. In: Wang, X.S., Cruz, I., Delis, A., Huang, G. (eds.) WISE 2012. LNCS, vol. 7651, pp. 30–43. Springer, Heidelberg (2012)
27. Zhuang, L., et al.: Movie review mining and summarization. In: Proc. CIKM (2006)

Generating Domain-Specific Sentiment Lexicons for Opinion Mining

Zaher Salah, Frans Coenen, and Davide Grossi

Department of Computer Science,
University of Liverpool, Ashton Building,
Ashton Street, L69 3BX Liverpool, United Kingdom
{zsalah,coenen,d.grossi}@liverpool.ac.uk

Abstract. Two approaches to generating domain-specific sentiment lexicons are proposed: (i) direct generation and (ii) adaptation. The first is founded on the idea of generating a dedicated lexicon directly from labelled source data. The second approach is founded on the idea of using an existing general purpose lexicon and adapting this so that it becomes a specialised lexicon with respect to some domain. The operation of the two approaches is illustrated using a political opinion mining domain and evaluated using a large corpus of labelled political speeches extracted from political debates held within the UK Houses of Commons.

Keywords: Domain-Specific Sentiment Lexicons, Opinion Mining, Sentiment Analysis.

1 Introduction

Opinion (Sentiment) mining is concerned with the use of data mining techniques to extract positive and negative opinions, attitudes or emotions, typically embedded within some form of free text, concerning some object of interest [1, 2]. Opinion mining can be conducted in a number of ways. One proposed technique is founded on the use of sentiment lexicons [3–7]. The use of sentiment lexicons for opinion mining has been shown to be a useful and promising research area [2] especially with respect to applications that require real-time response or where there is insufficient labelled data available to train an opinion mining classifier (an alternative approach to opinion mining). In the context of opinion mining, sentiment lexicons are typically used to estimate the sentiment polarity (attitude) expressed in some text by first identifying subjective words (words that convey feelings or judgement) and then "looking up" the identified words in the sentiment lexicon to obtain sentiment values. Consequently these values can be used to predict the polarity (positive or negative) of the text.

The most commonly used sentiment lexicon is the SentiWordNet 3.0 general purpose lexicon[1]. The problem with such general purpose lexicons is that they tend to not operate well with respect to specific domain corpora because of the

---

[1] sentiwordnet.isti.cnr.it

H. Motoda et al. (Eds.): ADMA 2013, Part I, LNAI 8346, pp. 13–24, 2013.

use of special purpose words (reserved words) and/or domain specific style and language. For example, given a specific domain, a collection of words and phrases may be used in a different context than their more generally accepted usage, in which case the words and phrases may reflect different sentiments than those that would be normally expected. Hence general purpose lexicons are not well suited to opinion mining in specialised domains.

One example of a specialised sentiment mining domain, and that used as a focus in this paper, is political opinion mining. To this end two approaches to generating the desired specialised sentiment lexicons are proposed: (i) direct generation and (ii) adaptation. The first is founded on the idea of generating a domain-specific lexicon directly from labelled source data using the biased occurrence of words in the source. The second approach is founded on the idea of using an existing general purpose lexicon, such as SentiWordNet 3.0, and adapting this so that it becomes a specialised lexicon with respect to some domain. Both approaches are fully described and used to build two illustrative political lexicons: (i) PoLex and (ii) PoliSentiWordNet. The effectiveness of the lexicons was tested by using then within an opinion mining setting to classify speeches taken from UK parliamentary debates.

The main contribution of the paper is the mechanism for "learning" the sentiment scores to be attached to terms using appropriately defined training data, especially the proposed ΔTF-IDF$'$ scheme. A second contribution is the mechanism for conducting the desired opinion mining using the generated lexicons.

The rest of this paper is structured as follows. Section 2 provides some background on sentiment lexicons and reviews some relevant previous work. Section 3 presents the two proposed domain specific lexicon generation approaches and Section 4 the mechanism whereby the generation mechanisms can be applied in the context of opinion mining. Some evaluation is then presented in Section 5 using UK parliamentary debate data and some concluding observations are made in Section 6.

2   Previous Work

In sentiment analysis most published research is directed at either lexicon or corpus based techniques. In lexicon-based techniques a sentiment lexicons is used to retrieve word sentiment scores after which the sentiment attitude of a word, sentence or whole document is determined by summing, averaging or counting the individual sentiment scores. In corpus-based techniques labelled corpora (exhibiting prior knowledge) are used to learn some form of classifier using established techniques such as: Naïve Bayes, Support Vector Machines (SVM), decision trees or neural networks. Corpus-based (machine learning) approaches tend to outperform lexicon-based approaches. However, the need for an appropriate training data set is often seen as a disadvantage. Learning a classifier is also a computationally expensive process in terms of time and storage space. The work described in this paper adopts a lexicon based approach and thus the remainder of this previous work section will be directed at a review of previous work in this context.

A sentiment lexicon is a lexical resource frequently used with respect to applications such as opinion mining. The lexicon is used to assign sentiment *scores* and *orientations* to single words. A sentiment score is a numeric value indicating some degree of subjectivity. Sentiment orientation is an indicator of whether a word expresses assent or dissent with respect to some object or concept. Consequently, document polarity can be judged by counting the number of assenting and dissenting words, summating their associated sentiment scores and then calculating the difference. The result represents the polarity (positive or negative) of the document.

The most commonly used generic *lexical database* (topic-independent sentiment lexicon) is the SentiWordNet 3.0 "off-the-shelf" sentiment lexicon [8]. The current version is SentiWordNet 3.0 which is based on WordNet 3.0 [3][2]. SentiWordNet 3.0 associates to each synset (set of synonyms) s of WordNet a set of three scores: $Pos(s)$ ("positivity"), $Neg(s)$ ("negativity") and $Obj(s)$ ("neutrality" or "objectivity"). The range of each score is $[0, 1]$ and for each synset s $Pos(s) + Neg(s) + Obj(s) = 1$. From the point of view of the work presented in this paper, SentiWordNet 3.0 has the key advantage, over other available lexicons (see [6] for a detailed comparison of SentiWordNet 3.0 with other popular, though manually built, lexicons) of covering the largest number of words (SentiWordNet 3.0 includes 117,659 words).

As noted above, sentiment analysis using generic sentiment lexicons is a challenging process in the context of topic-dependent domains [9]. In such cases it is desirable to use dedicated topic-dependent sentiment lexicons. The main issue with the usage of such dedicated lexicons is that they are frequently not readily available and thus have to be specially generated, a process that may be both resource intensive and error prone.

Basically, there are two techniques to generate specialised (dedicated) lexicons for specific-domain sentiment analysis: (i) creating a new dedicated lexicon or (ii) adapting an existing generic lexicon. Both techniques use labelled corpora (training data) from a specific domain. An example of the first technique (creating a new dedicated lexicon) can be found in [10] who proposed a semi-automated mechanism to extract domain-specific health and tourism words from noisy text so as to create a domain-specific lexicon. Example of the second technique (adapting an existing general lexicon) can be found in [11] and [12]. In [11] a simple algorithm is proposed to adapt a generic sentiment lexicon to a specific domain by investigating how the words from the generic lexicon are used in a specific domain in order to assign new polarities for those words. In [12] Integer Linear Programming is used to adapt a generic sentiment lexicon into a domain-specific lexicon; the method combined the relations among words and opinion expressions so as to identify the most probable polarity of lexical items (positive, negative or neutral or negator) for the given domain. There is also reported work that combines the two techniques (adapting the sentiment scores of the terms in the base lexicon and additionally appending new domain words to extend the base lexicon). For example [13] created a domain-specific

---

[2] SentiWordNet 3.0 is accessible at `sentiwordnet.isti.cnr.it`

sentiment lexicon using crowd-sourcing for assigning sentiment scores to sentiment terms and then automatically extending an initially proposed base lexicon using a bootstrapping process to add new sentiment indicators and terms. The lexicon is then customised according to some specific domain. The evaluation conducted indicated that the created lexicon outperforms a general sentiment lexicon (the General Inquirer Sentiment Lexicon). Further reported work concerned with the "dual approach" to generating domain-specific lexicons can be found in [14], [15] and [16].

Sentiment score for each term in a lexicon can be calculated either by: (i) investigating the biased occurrence of the term with respect to the class labelled documents ("positive" and "negative"), (ii) utilising the semantic, contextual or statistical relationships between terms (words) in an input domain corpus or (iii) learning a classifier to assign sentiment polarity to terms. In the context of the calculation of sentiment scores with respect to specific domains [17] proposed a method to calculate the sentiment score of each word or phrase in different domains and use these scores to quantify sentiment intensity. [9] proposed two approaches to improve the performance of polarity detection using lexical sentiment analysis with respect to social web applications, focusing on specific topics (such as sport or music). The two approaches were: (i) allowing the topic mood to determine the default polarity for false-neutral expressive text, and (ii) extending an existing generic sentiment lexicon by appending topic-specific words. The mood method slightly outperformed the lexical extension method. On the other hand it was found to be very sensitive to the mood base thus it was necessary to analyse the corpus first in order to choose an appropriate mood base relative to the corpus. Both methods require human intervention to either annotate a corpus (mood method) or to select terms (lexical extension).

3   Domain-Specific Lexicons Generation

The proposed domain-specific lexicon generation approaches are described in this Section. In both case the input is a set of n binary labelled texts (documents) $D = \{d_1, d_2, \ldots, d_n\}$. These will be parliamentary speeches in the case of the application domain used for evaluation purposes with respect to this paper. The labels are drawn from the set $\{positive, negative\}$. The output in both cases is a lexicon where each term is encoded in the form of a set of tuples:

$$\langle t_i, post_i, s_i \rangle$$

where t_i is a term that appears in the document collection D, $post_i$ is the part-of-speech tag associated with term t_i and s_i is the associated sentiment score. The proposed domain-specific lexicon generation approaches both comprise four steps: (i) part-of-speech tagging (to identify the POS tags), (ii) document pre-processing, (iii) sentiment score (s_i) and polarity ($post_i$) calculation and (iv) lexicon generation (see Figure 1). Each of these steps is described in more detail in the following four sub-sections.

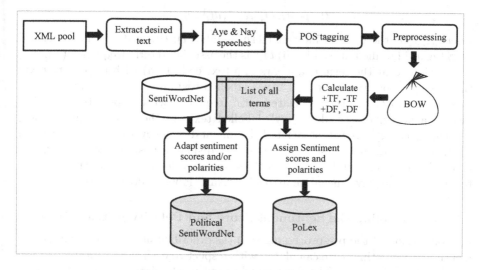

Fig. 1. Lexicon creation and adaption

3.1   Part Of Speech Tagging (POST)

With respect to opinion mining many related words (for example "suffice", "sufficiency", "sufficient" and "sufficiently") will require different sentiment scores. We can use Part Of Speech Tagging (POST) to distinguish between such groups of words. POST can also be used with respect to Word Sense Disambiguation (WSD). WSD is concerned with the process of dealing with the polysemy problem (different meanings for the same word) by discriminating the proper "semantic" sense of a word in a specific context or circumstance [18]. The first step with respect to our proposed lexicon generation approaches is therefore POS tagging. Each word in the input will be assigned a particular POS tag according to its context. With respect to the work described in this paper the TAIParse[3] POS tagger was used, the current version of which has a claimed accuracy for is 94%. At the end of the POST step we will have a list of terms $T = \{t_1, t_2, \ldots, t_m\}$ each associated with a POS tag $post_i$. Note that POST has to be performed prior to any further processing (see below) because the semantic differences between terms, required for POST, will be lost otherwise.

3.2   Preprocessing

The pre-processing phase commences with the conversion of all upper-case alphabetic characters to lower-case, this in then followed by punctuation mark and numeric digit removal. Next, given our list of terms T from the previous step (see above), we create a Bag-Of-Words (BOW) representations for all t_i in T (all the terms in the input document collection D). Each term t_i in the BOW is defined using a 6-tuple of the form:

---

[3] http://www.textanalysis.com/taiparse0_8_1.zip

$$\langle t_i, post_i, tf_i^+, tf_i^-, df_i^+, df_i^- \rangle.$$

where (i) t_i is the term of interest (term number i); (ii) $post_i$ is the associated POS tag as identified in step 1, (iii) tf_i^+ is the associated term frequency (number of occasions that the term t_i appears in a text collection) with respect to texts that display a positive attitude ("Aye" labelled texts in the case of our political speeches), (iv) tf_i^- is the associated term frequency with respect to texts that display a negative attitude ("Nay" labelled texts in the case of our political speeches), (v) df_i^+ is the associated document frequency (number of texts in which t_i appears) with respect to texts that display a positive attitude ("Aye" labelled documents) and (vi) df_i^- is the associated document frequency respect to texts that display a negative attitude ("Nay" labelled documents).

3.3 Determining the Sentiment Score and Polarity of Each Term

On completion of the pre-processing step (as described above) sentiment scores (sentiment weightings) are calculated with respect to each term contained in the generated BOW so far. The most widely used mechanism for generating term sentiment weights is the TF-IDF weighting scheme which aims to "balance out the effect of very rare and very frequent" terms in a vocabulary [19]. TF-IDF also tends to reflect the significance of each term by combining local and global term frequency [20]. The TF-IDF value W_{ij} for a term i in a text j is obtained using:

$$W_{ij} = TF - IDF = tf_{i,j} \cdot \left(log_2 \frac{n}{df_i} \right) \qquad (1)$$

where: (i) $tf_{i,j}$ is the frequency of term i in document (speeches) j (thus the local weight for the term), (ii) n is the total number of documents in the corpus (concatenated speeches in the debate), and (iii) df_i is the number of documents (speeches) containing term i (thus the global weight for the term). Alternative schemes to TF-IDF include: Term Frequency (TF), Document Frequency (DF), Term Strength (TS) and Term Contribution (TC). A disadvantage of TF-IDF, in the context of opinion mining, is that it does not reflect a term's sentiment tendency (orientation) and thus for our purposes we need either an alternative sentiment intensity weighting scheme or an alternative form of the TF-IDF scheme that takes into consideration the situation where a term t_i appears in both positive and negative documents. With respect to the latter the ΔTF-IDF provides "an intuitive general purpose technique to efficiently weight word scores" [21]. Thus ΔTF-IDF considers the biased occurrence of terms with respect to individual classes (sentiment in our case). The ΔTF-IDF value W_{ij} for a term i in a text j is obtained using:

$$W_{i,j} = \Delta TF - IDF = \left(tf_{i,j} \cdot \left(log_2 \frac{N^+}{df_i^+} \right) \right) - \left(tf_{i,j} \cdot \left(log_2 \frac{N^-}{df_i^-} \right) \right)$$

$$= tf_{i,j} \cdot \left(log_2 \frac{N^+}{df_i^+} \frac{df_i^-}{N^-} \right) \qquad (2)$$

where: (i) N^+ is the number of positive texts in the input document collection D (labelled "aye" with respect to our political opinion mining application), (ii) N^- is the number of negative texts (labelled "nay"), (iii) tf_{ij} is the term frequency for term t_i in text j, (iv) df_i^+ is the document frequency for term t_i with respect to positive texts in the input document collect D and (v) df_i^- is the document frequency for term t_i with respect to negative texts.

However, the ΔTF-IDF scheme is directed at sentiment classification of individual texts according to their ΔTF-IDF values [21]. Our research is focused on building domain-specific lexicons and thus a slightly adapted ΔTF-IDF weighting scheme is proposed so that term weightings are considered with respect to the entire document collection D and not per document. We will refer to this scheme as ΔTF-IDF'. Thus the ΔTF-IDF' value $W_{i,D}$ for a term i with respect to a document collection D is obtained using:

$$W_{i,D} = \Delta TF - IDF' = \left(tf_i^+ \cdot \left(log_2 \frac{N^+}{df_i^+} \right) \right) - \left(tf_i^- \cdot \left(log_2 \frac{N^-}{df_i^-} \right) \right) \quad (3)$$

where: tf_i^+ is the term frequency with respect to positive texts and tf_i^- is the term frequency with respect to negative texts. The advantages offered by the proposed ΔTF-IDF' scheme are that it can be used to assigning sentiment scores to each term taking into consideration term occurrences in both negative and positive texts.

Thus the ΔTF-IDF' scheme is used to determine sentiment scores for each term. On completion of step 3 each term in the BOW will comprise an 8-tuple of the form:

$$\langle t_i, post_i, tf_i^+, tf_i^-, df_i^+, df_i^-, s_i \rangle.$$

where s_i is the sentiment score associated with term i.

3.4 Lexicon Generation

In step 4 the desired domain-specific sentiment lexicon is generated. As already noted, two alternative mechanisms for generating the lexicons are proposed: (i) direct generation and (ii) adaptation. In the first case, as the name implies, we generate the lexicon directly from labelled source data following the above steps. Thus we take the generated BOW and convert this into the lexicon format described above.

In the case of the adaptation mechanism the idea is to use an existing, domain-independent, sentiment lexicon Lex, and adapt this to produce a domain-specific lexicon Lex'. More specifically the content of Lex is copied over to Lex'. The adjustment is as follows. Given a term t_i that is both in T and Lex, if the two associated sentiment scores have different polarities (thus one negative and one positive, or vice versa) we adopt the polarity from the calculated sentiment score (but not the magnitude) with respect to Lex'. Terms included in T but not in Lex are simply appended to Lex'.

4   Opinion Mining

Once we have generated a sentiment lexicon as described above we can apply these for the purpose of opinion (sentiment) mining. Given a new text which we wish to classify as expressing either a positive or a negative opinion we first apply POST so that each term in the new document is paired with a POS tag. We then lookup each term in the new text in our lexicon taking account of its associated POS tag. The overall sentiment score S for the new text is then computed using:

$$S = \sum_{i=1}^{i=z}(s_i \times \omega) \tag{4}$$

where s_i is the sentiment score $(-1.0 \leq s_i \leq 1.0)$ for term t_i in the new text (speech) obtained from a lexicon, z is the number of terms in the new text and ω is the occurrence count for term t_i in the new text. The occurrence count can be calculated according to the frequency of occurrence of trem t_i in the new text or simply as a binary value (1 for present, 0 for absent). In the following section we refer to these two techniques using the labels *TF* and *Binary* respectively. The attitude (class label) of the new text is then determined according to the overall sentiment score S_j. To this end the class label set is $\{positive, negative, objective, neutral\}$ where: (i) *positive* indicates a positive text (for the motion in the case of our political debates), (ii) *negative* indicates a negative text (against the motion), (iii) *objective* that no sentiment scores were found and (iv) neutral that the sentiment scores negate each other. In practice it was found that the last two class labels are rarely encountered.

5   Evaluation

In this section a discussion concerning the evaluation of the proposed domain-specific lexicon generation system is presented. The reported evaluation used a training set to generate the desired domain specific lexicons, using our proposed approaches, and a test set to determine their effectiveness with respect to opinion minig. To this end two parliamentary debate corpora were used: UKHCD-3 and UKHCD-4[4]. Further detail regarding these data sets is presented in Sub-section 5.1 below.

Using the training set (UKHCD-3) two lexicons were generated, PoLex and PoliSentiWordNet. The first using the direct generation mechanism and the second the adaptive mechanism (founded on SentiWordNet 3.0). PoLex comprised 170,703 terms and PoliSentiWordNet 258,353 terms. The generated lexicons were then used to assign sentiment scores to the test data to determine the attitude of the debaters (speakers). Because this attitude was known from the way that the speakers eventually voted, the predicted attitude could be compared with the known attitude. The results obtained are presented in Sub-section 5.2.

---

[4] UKHCD-1 and UKHCD-2 were two other parliamentary debate data sets used in earlier reported work by the authors [7].

5.1   Test Data

To act as a focus for the work described in this paper the UK House of Commons debates were used. Both houses in the UK parliament, the House of Commons and the House of Lords, reach their decisions by debating and then voting (with either an "Aye" or a "Nay") at the end of each debate. Proceedings of the House of Commons are published on-line (at TheyWorkForYou.com), in XML format, and thus are readily available. A second advantage offered by this collection is that the attitude of the debaters (MPs) is known, positive if a MP voted "Aye" at the end of a debate or negative if a MP voted "Nay". Thus we can evaluate the effectiveness of using our domain-specific lexicon generation approaches with respect to predicting the polarity of speeches. The authors obtained the speeches associated with the UK House of Commons debates held within the last eleven years from the TheyWorkForYou.com www site, from January 2002 to March 2013. QDAMiner4[5] was used to extract the desired textual information from the XML debate records. Speeches by MPs who did not vote were ignored, as were speeches that contained fifty words or less, as it was conjectured that no valuable sentiment attitude could be associated with these speeches. These speeches were divided into a "training" and a "testing" set referred to as the UK House of Commons Debate (UKHCD) datasets 3 and 4 (UKHCD-3 and UKHCD-4). UKHCD-3 feature speeches from debates held between January 2002 to July 2012 and UKHCD-4 debates held between August 2012 to March 2013. Some statistics concerning these two data sets are presented on Table 1.

Table 1. Dataset Statistics (UKHCD-3 and UKHCD-4)

	Number of Debates	Number of *Aye* speeches	Number of *Nay* speeches	Total Number of speeches
UKHCD-3	1101	147,559	180,566	328,125
UKHCD-4	29	1,119	949	2,068
Totals	1130	148,678	181,515	330,193

5.2   Results

One of the challenges of work on sentiment analysis is the lack of any "ground truth" data. In some cases it is possible to construct such data by hand however this still entails subjectivity and requires considerable resources (to the extent that it is not possible to construct significant benchmark data). To evaluate the proposed domain-specific lexicon generation method we compared the attitudes predicted using PoLex and PoliSentiWordNet with the known "attitude" of the speakers defined according to whether, at the end of each individual debate, they voted "Aye" or "Nay". In doing so we therefore had to assume that the speakers' attitudes during their speeches reflect how they would eventually vote. In other

---

[5] http://provalisresearch.com/products/qualitative-data-analysis-software/

Table 2. Confusion matrix

	PoLex		PoliSentiWordNet		SentiWordNet 3.0	
	TF	Binary	TF	Binary	TF	Binary
TP	893	930	870	930	857	1015
FN	226	189	249	189	262	102
TN	229	217	249	207	257	103
FP	720	732	700	742	691	844
Total	2068	2068	2068	2068	2067	2064

Table 3. Evaluation results generated from confusion matrix data given in Table 2

	PoLex TF			PoliSentiWordNet TF			SentiWordNet 3.0 TF		
	Aye	Nay	Avg.	Aye	Nay	Avg.	Aye	Nay	Avg.
Precision	0.554	0.503	0.528	0.554	0.500	0.527	0.554	0.495	0.524
Recall	0.798	0.241	0.520	0.777	0.262	0.520	0.766	0.271	0.518
F-Measure	0.654	0.326	0.490	0.647	0.344	0.496	0.643	0.350	0.497
Avg. Accuracy	54.30%			54.10%			53.90%		

	PoLex Binary			PoliSentiWordNet Binary			SentiWordNet 3.0 Binary		
	Aye	Nay	Avg.	Aye	Nay	Avg.	Aye	Nay	Avg.
Precision	0.560	0.534	0.547	0.556	0.522	0.539	0.546	0.502	0.524
Recall	0.831	0.229	0.530	0.831	0.217	0.524	0.909	0.109	0.509
F-Measure	0.669	0.320	0.495	0.666	0.307	0.486	0.682	0.179	0.430
Avg. Accuracy	55.50%			54.90%			54.20%		

words it is assumed that speakers never "change their minds" during a debate. The results are presented in tabular form in Tables 2 and 3.

Table 2 shows the confusion matrix data that results when using: (i) Polex, (ii) PoliSentiWordNet and (iii) the general purpose SentiWordNet 3.0. Note that in each case we compare the use of the Term Frequenct (*TF*) occurrence count approach with the binary (*Binary*) occurrence count approach. Table 3 shows the precision, recall, F-measure and average accuracy values obtained (using Polex, (PoliSentiWordNet and the general purpose SentiWordNet 3.0) from the confusion matrices[6] presented in Table 2. Inspection of the results presented in Table 3 indicates: (i) that there is a small improvement with respect to the average values obtained when using the domain specific lexicons and (ii) that both Polex and PolSentiWordNet produced similar results. Closer inspection

---

[6] True Positive (TP): Speaker says Aye and Machine says Aye.
False Negative (FN): Speaker says Aye and Machine says Nay.
True Negative (TN): Speaker says Nay and Machine says Nay.
False Positive (FP): Speaker says Nay and Machine says Aye.

reveals the interesting observation that all the techniques worked better with respect to predicting positive (Aye) attitudes than negative (Nay) attitudes. At present the reason for this is unclear. The suggestion is that this is because there is a wide variety of topics contained in the debates thus the domain corpus is not a pure political corpus.

6 Conclusions

In this paper we have described two approaches to generating domain-specific sentiment lexicons: (i) direct generation and (ii) adaptation. The two approaches were evaluated by generating two political sentiment lexicons (using UK House of Commons debates for training purposes). The first (Polex) was produced using the direct generation approach and the second (PoliSentiWordNet) using the adaptation approach. The effectiveness of the lexicons was evaluated by using them to predict the sentiment attitude associated with a set of test speeches (drawn from further UK House of Commons debates). The results indicated a small general improvement, with a more marked improvement with respect to positive attitude prediction, than when a general purpose sentiment lexicon is used (SentiWordNet 3.0). All approaches considered were less effective at predicting negative attitudes for reasons that are unclear thus providing a interesting avenue for further research.

References

1. Asmi, A., Ishaya, T.: A framework for automated corpus generation for semantic sentiment analysis. In: Proc. World Congress on Engineering (WCE 2012), pp. 436–444 (2012)
2. Grijzenhout, S., Jijkoun, V., Marx, M.: Opinion mining in dutch hansards. In: Proceedings of the Workshop From Text to Political Positions. Free University of Amsterdam (2010)
3. Esuli, A., Sebastiani, F.: SentiWordNet: A publicly available lexical resource for opinion mining. In: Proceedings from the International Conference on Language Resources and Evaluation, LREC (2006)
4. Denecke, K.: Are sentiwordnet scores suited for multi-domain sentiment classification? In: Fourth International Conference on Digital Information Management, ICDIM 2009, pp. 1–6 (2009)
5. Montejo-Raez, A., Martínez-Cámara, E., Martin-Valdivia, M., Ureña-López, L.: Random walk weighting over sentiwordnet for sentiment polarity detection on twitter. In: Proc 3rd Workshop on Computational Approaches to Subjectivity and Sentiment Analysis, pp. 3–10 (2012)
6. Ohana, B., Tierney, B.: Sentiment classification of reviews using sentiwordnet. In: Proceedings of the 9th IT & T Conference, Dublin Institute of Technology (2009)
7. Salah, Z., Coenen, F., Grossi, D.: Extracting debate graphs from parliamentary transcripts: A study directed at UK House of Commons debates. In: Proceedings of the Fourteenth International Conference on Artificial Intelligence and Law (ICAIL 2013), Rome, Italy, pp. 121–130 (2013)

8. Baccianella, S., Esuli, A., Sebastiani, F.: SentiWordNet 3.0: An enhanced lexical resource for sentiment analysis and opinion mining. In: Proceedings of the Seventh International Conference on Language Resources and Evaluation (LREC 2010). European Language Resources Association, ELRA (2010)
9. Thelwall, M., Buckley, K.: Topic-based sentiment analysis for the social web: The role of mood and issue-related words. Journal of the American Society for Information Science and Technology 64(8), 1608–1617 (2013)
10. Birla, V.K., Gautam, R., Shukla, V.: Retrieval and creation of domain specific lexicon from noisy text data. In: Proceedings of ASCNT-2011, CDAC, Noida, Indiaz (2011)
11. Demiroz, G., Yanikoglu, B., Tapucu, D., Saygin, Y.: Learning domain-specific polarity lexicons. In: 2012 IEEE 12th International Conference on Data Mining Workshops (ICDMW), pp. 674–679 (2012)
12. Choi, Y., Cardie, C.: Adapting a polarity lexicon using integer linear programming for domainspecific sentiment classification. In: Proceedings of the 2009 Conference on Empirical Methods in Natural Language Processing, pp. 590–598 (2009)
13. Weichselbraun, A., Gindl, S., Scharl, A.: Using games with a purpose and bootstrapping to create domain-specific sentiment lexicons. In: Proceedings of the 20th ACM International Conference on Information and Knowledge Management, CIKM 2011, pp. 1053–1060. ACM, New York (2011)
14. Qiu, G., Liu, B., Bu, J., Chen, C.: Expanding domain sentiment lexicon through double propagation. In: Proceedings of the 21st International Jont Conference on Artifical Intelligence, IJCAI 2009, pp. 1199–1204. Morgan Kaufmann Publishers Inc., San Francisco (2009)
15. Lau, R., Zhang, W., Bruza, P., Wong, K.: Learning domain-specific sentiment lexicons for predicting product sales. In: 2011 IEEE 8th International Conference on e-Business Engineering (ICEBE), pp. 131–138 (2011)
16. Ringsquandl, M., Petković, D.: Expanding opinion lexicon with domain specific opinion words using semi-supervised approach. In: BRACIS 2012 - WTI, IV International Workshop on Web and Text Intelligence (2012)
17. Zhang, J., Peng, Q.: Constructing chinese domain lexicon with improved entropy formula for sentiment analysis. In: 2012 International Conference on Information and Automation (ICIA), pp. 850–855 (2012)
18. Wilks, Y., Stevenson, M.: The grammar of sense: Using part-of-speech tags as a first step in semantic disambiguation. Natural Language Engineering 4(5), 135–143 (1998)
19. Kuhn, A., Ducasse, S., Gibra, T.: Semantic clustering: Identifying topics in source code. Information and Software Technology 49(3), 230–243 (2007)
20. Li, H., Sun, C., Wan, K.: Clustering web search results using conceptual grouping. In: Proc. 8th International Conference on Machine Learning and Cybernetics, pp. 12–15 (2009)
21. Martineau, J., Finin, T.: Delta tfidf: An improved feature space for sentiment analysis. In: Proc 3rd International ICWSM Conference, pp. 258–261 (2009)

Effective Comment Sentence Recognition
for Feature-Based Opinion Mining

Hui Song, Botian Yang, and Xiaoqiang Liu

College of Computer Science and Technology, Donghua University, Shanghai, China
{songhui,liuxq}@dhu.edu.cn, yangbotian@mail.dhu.edu.cn

Abstract. Feature-based opinion mining aims to extract fine-grained comments in product features level from the product reviews. Previous work proposed a lot of statistics-based and model-based approaches. However, the extraction result is not satisfying when these methods are actually used into an application with large useless data due to the complexity of Chinese. Through analyzing the samples hadn't been extracted correctly, we found some extracting patterns or models have been misused on the useless sentences which lead to wrong extraction. This paper focuses on improving the POS-pattern match methodology. The core idea of our approach is picking out the effective comment sentences before feature and sentiment extraction based on neural network training. Three attributes of sentences are selected to learn the classification algorithm. Experiment gives the superior parameters of the algorithm. We report the classification performance and also compare the feature extraction performance with classification process and not. The result on practical data set demonstrates the effect of this approach.

Keywords: POS-Pattern, Effective Comment Sentence, Neural Network, Products Review.

1 Introduction

With the widespread using of e-commerce, more and more people tend to shop online. Many e-commerce websites, such as Taobao, Amazon, allow customers to post their reviews freely. The user-generated comments are definitely valuable to the latecomers. Meanwhile, manufacturers take such abundant data as a positive alternative for the traditional user-feedback survey. Therefore, Opinion Mining from the magnitude of the reviews and generating summaries of the data has attracted many researchers during these years [1, 2].

Researches on feature-specific opinion mining aim at recognizing the characteristics of a product combining with emotional orientation from sentences automatically, which has been defined formally by Ding, Liu, and Yu [3]. It usually includes three topics: the extraction of product entities from reviews, the identification of opinion words that are associated with the entities, and the determination of these opinions' polarities (e.g. positive, negative, or neutral), which grouped as a feature triple.

H. Motoda et al. (Eds.): ADMA 2013, Part I, LNAI 8346, pp. 25–35, 2013.

Although different approaches have been applied to feature-level opinion mining, the mainstream basically consists of three major methodologies: rule-based technique [1], statistic-based and model-based approach [4]. These approaches generated statistic patterns or probabilistic models from training data which will be used to complete the extraction issue. Previous efforts mostly achieved not bad results in Chinese document, but made fault decision on the following sentences:

Example 1: "妈妈说效果不错。"（Mother said the moisturizing effect is good.）
Example 2: "如果能有一份小赠品就更开心了。"（A little sample will be better.）

In Example 1, the algorithms extract out "moisturizing effect" and "good" as the feature of a product and corresponding comment, what are we desired. When they go on Example 2, the result of "sample" and "better" is obviously wrong. We analyze the data set from an e-commerce website and discover that about 80% of the extraction errors are similar to sample 2. In fact, in practical data set from website, over 75% of the sentences are useless for feature triples extraction, but they distort the extraction algorithm seriously [5].

This paper mainly focuses on dealing with the above problem. We try to filter out the sentences useless for extraction to obtain effective comment sentences set before executing the extraction process. This has been generalized as a classification problem. We selected three attributes from the text of the product reviews to implement the neural networks based categorization algorithm [6]. Comparing the experiment results between adding the effective sentence recognition process and not, the precision have been improved obviously.

The remainder of the paper is organized as follows. Section 2 covers some work related to feature-level opinion mining and sentence classification. Section 3 describes the origin and the framework of our approach. In section 4, we provide the detail algorithm to recognize the effective sentence. Section 5 presents experiment's result of the sentence classification and feature triples extraction. We conclude and show directions for future work in section 6.

2   Related Work

Most approaches on Web opinion mining with the aim to discover the important information from product reviews can be classified into two major branches: document-level and feature-level. Document-level opinion mining focuses on producing an overall opinion for one document; and feature-level opinion mining aims to discover product features from sentences and identify opinion words associated with each feature.

Baccianella used POS patterns to extract larger-than-word units, some complex features, in their paper [7].Li and et al. proposed a general framework for determining unexpected sentences with the structure POS_1-POS_2-...POS_n called a class pattern model [8]. They just chose the class patterns limited to 2- and 3-length sequential patterns for experiments.

Hu and Liu [1] proposed a feature-based opinion summarization system that captured highly frequent feature words through finding association rules under a statistical framework. Popescu and Etzioni [9] improved Hu and Liu's work by removing frequent noun phrases that may not be real features. Their method can identify part-of relationship and achieve a better precision, but a small drop in recall.

The other suggested by H. Zhang, Z. Yu [10] is also based on association rules. But the difference is that they combine rules with PMI and the sentiment dictionary Hownet to extract product features and analyze the opinion orientation. Lei Zhang and Bing Liu ranked feature candidates by feature importance which was determined by two factors: feature relevance and feature frequency [11].

Recently, some researchers introduced the model-based approaches in order to increase the efficiency of the opinion. Opinion Miner [3] was built based on lexicalized Hidden Markov Models (L-HMMs) to integrate multiple important linguistic properties into an automatic learning process. Luole Qi [5] employed another model, Conditional Random Field (CRFs) model, which naturally considered arbitrary, non-independent entities without conditional independence assumption. Although they reported a satisfied precision on their experiment, but in large practical data set filled with useless data, their performance dropped seriously.

In text related classification area, Lin Zheng [12] researched the sentimental orientation of a document, proposed a supervised and semi-supervised classification method to key sentences.

The most related work is Zhongwu Zhai's [13]. They presented a semi-supervised approach to extract the comment sentences from the websites, which used a set of evaluative opinion words and a set of emotion words available to extract the evaluative sentence. However, the emotion words and the corresponding feature are hard to prepare for different domain, and in some cases, only opinion words and emotion words are not enough to distinguish the majority of sentences desired.

3   Feature Triples Extraction Approach

For feature-based sentiment analysis on customer reviews, there are three subtasks:

1. To find the contained feature in the reviews;
2. To get corresponding comment word of each sentence;
3. To obtain sentiment orientation of each feature.

So, the mining objective is to discover every triple $ft = \{e, a, o\}$ (where e: feature word, a: comment word, o: orientation) in a given Sentence s, which is called feature triple in this paper. Previous approaches can be generalized into the following steps: Pre-processing of the data set, labeling the training data; Learning the extraction or matching algorithm on the basis of the training set; Executing the extraction with validation.

3.1   Pattern Matching Method Analysis

We proposed a probability method in our former work [14]. Firstly we split the reviews in the source data set to sentences by punctuations and get word segmentation and part-of-speech (POS) tags for each. Then the typical sequential POS which called POS-patterns are extracted. We calculate the probability for each pattern identifying a subjective sentence and choose some effective patterns to create the pattern set. With the probability for each pattern, the rules for pattern matching are designed. The pattern set with rules could be applied to match the new reviews.

We select a data set with 1714 sentences of one domain (training set: 1415, validation set: 299). After training process, we extract the feature triple from both training and validation set. Each sentence handled wrongly has been examined. They are divided into five groups according to the factors resulted in error. Table 1 lists the error analysis result.

Table 1. Error Types Description

Type No.	Error Description	Training Set	Test Set	Total
1	Sentence split incorrectly without punctuation, a sentence contains two patterns	11	7	18
2	Word segmentation incorrectly due to non-standard description	9	1	10
3	Sentence with negative words	5	1	6
4	Sentence contains moral verb, not a comment	3	0	3
5	Match a certain POS-pattern, but not a comment	69	27	96

Shown as figure 1, long sentence leads to type 1 error (about 14%). Error of type 2 (about 8%) caused by the characteristics of Chinese, and word segmentation errors are almost inevitable. The problem listed in type 3 (about 4%) is supposed to handle with more semantic analysis, some work discuss it specially. Type4 and 5 (about 74%) refers to the sentences matching a certain POS-pattern, but not comments. In this paper we focus on decreasing the errors of type 1, 4 and 5, which make up about 88% of all.

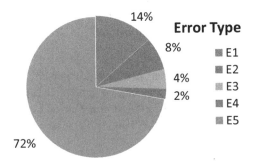

Fig. 1. Error Analysis

3.2 Improved Extraction Process

From above analysis, the algorithm mismatches some useless sentences, which lead to lower precision.

Let cs be the number of sentences correctly handled, ws be the sentences handled improperly, ls be the number of all sentences containing feature triples, rs be the number of all sentences returned by the system, then

$$\text{Precision} = \frac{cs}{cs + ws} \qquad \text{Recall} = \frac{cs}{ls} \tag{1}$$

If we can deduce the value of ws and keep ls the same, then the precision can be improved without sacrificing the recall.

From the above error analysis, type 1, 4 and 5 are the sentences labeled as useless which had been mismatched. And the rest are defined as Effective Comment Sentences (abbreviated as ECS in this paper). We try to recognize the ECS sentence before extraction process, so about 88% errors will not occur. The approach is illustrated as following:

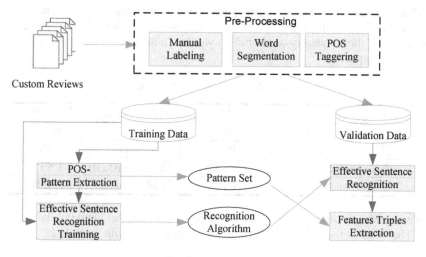

Fig. 2. Framework

Fig. 2 depicts the framework of our method. A specific product and the customer reviews of it are first given, after some preprocessing like Chinese word segmentation and POS tagging, we calculate the POS-Pattern Sets, then use it with other sentences features to train the effective sentence recognition algorithm. During the extraction process in contrast, recognition step is performed first. We only execute the pattern matching algorithm on the effective sentence set.

4   Effective Comment Sentence Extraction

In this section we present the ECS recognition algorithm in detail. The problem to choose the effective comment sentences can be generalized as a classification problem, that is, the sentences are labeled as ECS and ULS (useless sentence) in the source data set S. So the attributes that can discriminate the ECS and ULS are carefully selected and quantified. Then a supervised learning method, neural network are used to train the classification algorithm.

4.1   Attributes Selection of Effective Comment Sentence

We analysis the sample data, measure several attributes while labeling. And also the sentences extracted wrongly are examined item by item. At last, we take the following three attributes: keyword, comment pattern and the length of sentences to metric the ECS and ULS. Suppose a sentence set S, each sentence s is composed of a sequence of words

$$s = \{w_1, w_2, ..., w_m\}$$

And the ECS in S make up a set S_E. We discuss each attribute in detail.

1. Keywords

In subject sentences, author often uses some words containing strong sentiment which give explicit hint to the ECS. We calculate the words in about 2000 sentences and keep count of the frequently-used opinion words. Table 2 lists the top 10 words with the highest frequency.

Table 2. Top 10 Sentiment Words

好 (good)	21.6%	不错 (not bad)	19.3%
正品 (Favorable comment)	18.2%	快 (fast)	8%
好评 (granted)	3%	慢 (slow)	2.9%
贵 (expensive)	1.85%	满意 (satisfied)	1.83%
好用 (handy)	1.78%	滋润 (moisten)	1.45%

We consider the high-frequency words as keywords, and put them into keyword set K_W. The value of the keyword's attribute of s is quantified with formula 2:

$$f\_keyword(s) = \sum_{j=1}^{m} p_k(w_j) \times word\_appear(w_j) \qquad (2)$$

where $p_k(w_j)$ is the frequency of occurrence of word w_j in S_E, as formula 3. And $word\_appear(w_j)$ is defined as formula 4

$$p_k(w_j) = \frac{\text{the number of sentence containing word } w_j \text{ in } S_E}{\text{the sentence's number of } S_E} \qquad (3)$$

$$word\_appear(W_j) = \begin{cases} 1 & w_j \in K_W \\ 0 & w_j \notin K_W \end{cases} \qquad (4)$$

2. Comment Patterns

The typical POS-patterns indicate the semantic styles followed by most reviews. They also play a pivotal role to recognize ECS, so it is selected as one attribute. The POS-pattern set $P_T = \{pt_1, pt_2, ..., pt_n\}$ is generated in the first step (in 3.1), and the patterns are ordered according to their occurrence frequency in sentence set S. Table 2 lists the top 10 POS-patterns with the highest frequency.

Table 3. Top 10 POS-patterns

#n	8.92%	#n#d#a	8.57%
#a	8.34%	#d#a	8.31%
#v#n	4.17%	#n#d#v	3.89%
#v#d#a	3.45%	#v#v#n	3.03%
#n#a	2.78%	#d#v	2.19%

We remove two pos-patterns (#n, #a) from POS-pattern set P_T, because later experiment shows they cause most of the errors to classify the ECS with ULS, however some precision has been sacrificed.

The value of the pattern's attribute of a sentence s is quantified with formula 5:

$$f\_pattern(s) = p_m(h(s)) \times pattern\_appear(h(s)) \tag{5}$$

where function $h(s)$ gives the pattern that sentence s first matches, and the pattern pt_0 represents every POS-pattern that has been excluded from the P_T. And $p_m(pt_j)$ is the frequency of occurrence of the pt_j in S_E, as formula 6:

$$p_m(pt_j) = \frac{\text{the number of pattern } pt_j \text{ occurs in } S_E}{\text{the sentence's number of } S_E} \tag{6}$$

The function $pattern\_appear(pt_j)$ is defined as formula 7:

$$pattern\_appear(pt_j) = \begin{cases} 1 & j \neq 0 \\ 0 & j = 0 \end{cases} \tag{7}$$

3. Sentence Length

We discover that the long sentence often contains too many different POS-patterns, which cause wrong extraction. Meanwhile, such sentences are always released by author without proper punctuation.

Through calculating the length S_L of two types of sentences, we choose it as the third attribute for classification. And the value of the length attribute of a sentence s is quantified with formula 8:

$$f\_length(s) = \begin{cases} 1 & S_L \leq \gamma \\ 1 - \dfrac{(S_L - \gamma)}{S_{OL}} & S_L > \gamma \end{cases} \tag{8}$$

Where γ is the average length of ECS, and S_{OL} is the length of the longest sentences in the data set.

4.2   Neural Network Classification

In this subsection we describe the neural network classification method adopted. The model for classification is comprised of linear combinations of fixed basis functions, and the parameter values are adapted during training, which is called feed-forward neural network.

The input vector is $= \{x_1, x_2, x_3\}$, where x_i represents one of the three attribute values of a sentence s: $f\_keyword(s)$, $f\_pattern(s)$ and $f\_length(s)$. The output $y(X)$ is either 1 or 0 depending on whether the sentence is an effective comment sentence or not.

Then, the two-level network function takes the form

$$y(X, \Omega) = \sigma \left(\sum_{j=1}^{D} \omega_j^{(2)} h \left(\sum_{i=1}^{3} \omega_{ji}^{(1)} x_i + \omega_{j0}^{(1)} \right) + \omega_0^{(2)} \right) \tag{9}$$

where the set of all weight and bias parameters have been grouped together into a vector Ω , and D is the number of the basis functions of layer 2.

We choose the gradient descent method to learn parameter Ω of the function 8, minimize the error through back propagation. Though the gradient descent training doesn't convergence quickly, it reduces over-fitting dramatically.

After training, we obtain a classification function $y(X)$. For each inputting X derived from sentence s, it outputs a value y between 0 and 1. The effective sentence is recognized with the following formula

$$Type(s) = \begin{cases} ECS & y(X) \geq \delta \\ ULS & y(X) < \delta \end{cases} \tag{10}$$

where δ is a threshold set manually.

5   Experiment

We develop our approach according to an actual demanding. A dealer of cosmetics wants to mine the customer reviews to survey the popularity of a product. The method is developed and tested with the data from the dealer. We implement our approach and the algorithm proposed in work [5], and then compare the result between them.

5.1   Data Set

All the data are customers' reviews about a product of cosmetics in Chinese. The data set contains 5000 reviews. Each review is split into sentences by the punctuation automatically, and we get a source set with 19500 sentences. We divide it into two sets randomly: 14599 sentences into training set and 4001 sentences into validation one. The feature and the type (ECS or ULS) of each sentence are manually annotated by two research assistants. And there are 3584 ECS vs. 11951 ULS in training set and 1055 ECS vs. 2946 ULS in validation one. As the statistics shows, there are only about one fourth sentences are valuable for feature extraction.

5.2   Experiment Results Analysis

To get the optimization parameter of neural network, we use three-layer networks to train the data with variety training methods and different size of the hidden nodes.

The three training algorithms have been tried are the fast descent (TrainGD), Bayesian regularization (TrainBR) and Quasi-Newton (TrainBFG) Methods.

In our experiment, we evaluate the traditional precision, recall and F-score. Figure 3 gives the precision of three algorithms with different number of hidden nodes.

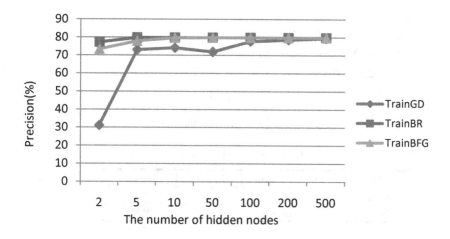

Fig. 3. The Precision of Training Algorithm

Figure 3 shows, the precision of TrainBR and TrainBFG reach the optimized value (79.36%) at the nodes number 50, and keep the same when the nodes number increases. Since the computation time of TrainBR is obviously less than that of TrainBFG, we choose the TrainBR as the training methods.   At the other hand, while the number of hidden nodes increases more than 100, precision doesn't grow. So the hidden nodes number is set to 100.

Figure 4 gives the F-score with different classification threshold δ when the algorithm selects TrainBR and the hidden nodes is100.

As the above figure shows, the total F-score reaches the high 76.8% when δ is set to 0.4. Although the F-score of ULS rises to 86.8% if we set δ to 0.7, it decreases dramatically for ECS. Our goal is to get better performance for ECS, so δ is still set to 0.4.

We implement the algorithm described in paper [5], which is a CRF-based model to extract the feature triples, and compare with the result of our approach (ECS filtering). Table 4 gives the result.

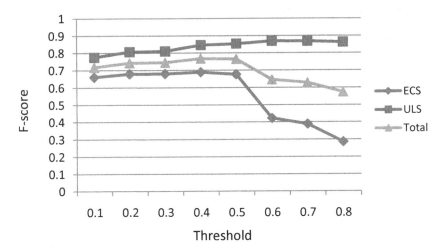

Fig. 4. F-score with different threshold δ

Table 4. The Feature Triples Extraction Result

Method	Precision	Recall	F-score
CRF-based	48.5%	46.7%	47.6%
ECS filtering	65.9%	40.2%	49.9%

Although some works reported CRF-base method achieved a high precision on their data base, but in the practical application, the data set is filled with too many garbage, the precision drops dramatically.

Our approach increases the precision about 15%, however, the recall drops about 7%, because in the ECS recognition process, some ECS have been classified to ULS. In this actual situation, dealer wants to give a report of custom feedback based on the tremendous data from the e-commerce site, the precision is more important than recall here.

6   Conclusion

In this paper, we introduce the effective comment recognition process to the feature triples extraction approach to improve the performance in practical application. The machine-learning method, neural network has been used to make a distinction between the EFS and useless sentences in the product reviews set. Many experiments have been done to find the better parameters of the algorithm from the variety candidates. The method increases about 15% precision, thought it takes extra time, it's worthy in actual application.

Acknowledgment. This work is partially supported by NSFC under grant number 61370205 and Innovation Program of Shanghai Municipal Education Commission (No. 12ZZ060).

References

1. Hu, M., Liu, B.: Mining and summarizing customer reviews. In: Proceedings of the Tenth ACM SIGKDD International Conference on Knowledge Discovery and Data Mining, KDD 2004, pp. 168–177. ACM (2004)
2. Turney, P.: Thumbs up or thumbs down? Semantic orientation applied to unsupervised classification of reviews. In: Proc of ACL, pp. 417–424. ACM, New York (2002)
3. Ding, X., Liu, B., Yu, P.S.: A holistic lexicon-based approach to opinion mining. In: Proceedings of the International Conference on Web Search and Web Data Mining, WSDM 2008, pp. 231–240. ACM (2008)
4. Jin, W., Ho, H., Srihari, R.: Opinion Miner: a novel machine learning system for web opinion mining and extraction. In: Proceedings of the 15th ACM SIGKDD International Conference on Knowledge Discovery and Data Mining, pp. 1195–1204. ACM (2009)
5. Qi, L., Chen, L.: Comparison of Model-Based Learning Methods for Feature-Level Opinion Mining. In: Proceedings of the 2011 IEEE/WIC/ACM International Conferences on Web Intelligence and Intelligent Agent Technology, WI-IAT 2011, vol. 01, pp. 265–273 (2011)
6. Su, J., Zhang, B., Xu, X.: Advances in Machine Learning Based Text Categorization. Journal of Software, 1848–1859 (2006) (in Chinese)
7. Baccianella, S., Esuli, A., Sebastiani, F.: Multi-facet Rating of Product Reviews. In: Boughanem, M., Berrut, C., Mothe, J., Soule-Dupuy, C. (eds.) ECIR 2009. LNCS, vol. 5478, pp. 461–472. Springer, Heidelberg (2009)
8. Dong (Haoyuan), L., Anne, L., Pascal, P., Mathieu, R.: Extraction of Unexpected Sentences: A Sentiment Classification Assessed Approach. J. Intelligent Data Analysis 14, 31–46 (2010)
9. Popescu, A., Etzioni, O.: Extracting product features and opinions from reviews. In: Proceedings of the Conference on Human Language Technology and Empirical Methods in Natural Language Processing, pp. 339–346. Association for Computational Linguistics (2005)
10. Zhang, H., Yu, Z., Xu, M., Shi, Y.: Feature-level sentiment analysis for Chinese product reviews. In: 3rd International Conference on Computer Research and Development, Shanghai, pp. 135–140 (2011)
11. Zhang, L., Lim, S.H., Liu, B., O'Brien-Strain, E.: Extracting and Ranking Product Features in Opinion Documents. In: Proceedings of COLING (2010)
12. Lin, Z., Tan, S., Cheng, X.: Sentiment Classification Analysis Based on Extraction of Sentiment Key Sentence. Journal of Computer Research and Development 49(11), 2376–2382 (2012) (in Chinese)
13. Zhai, Z., Liu, B., Zhang, L., Xu, H., Jia, P.: Identifying Evaluative Sentences in Online Discussions. In: The Twenty-Fifth AAAI Conference on Artificial Intelligence, pp. 933–938 (2011)
14. Song, H., Fan, Y., Liu, X.: Frequent Pattern Learning Based Feature-level Opinion Mining on Online Consumer Reviews. Advances in Information Sciences and Service Sciences 11(4), 133–141 (2012)

Exploiting Co-occurrence Opinion Words for Semi-supervised Sentiment Classification

Suke Li[1], Jinmei Hao[2], Yanbing Jiang[1], and Qi Jing[1]

[1] School of Software and Microelectronics, Peking University, China
{lisuke,jyb,jingqi}@ss.pku.edu.cn
[2] Beijing Union University

Abstract. This work proposes a semi-sentiment classification method by exploiting co-occurrence opinion words. Our method is based on the observation that opinion words with similar sentiment have high possibility to co-occur with each other. We show co-occurrence opinion words are helpful for improving sentiment classification accuracy. We employ the co-training framework to conduct semi-supervised sentiment classification. Experimental results show that our proposed method has better performance than the Self-learning SVM method.

1 Introduction

Now Web users are used to publishing product reviews in commercial Web sites such as Amazon (www.amazon.com) after they finish their shopping activities. Web reviews usually contain some objective descriptions and subjective information such as sentiment expressions about purchased products Product reviews can roughly be classified into three basic categories: positive, negative, and neural. How to automatically classify Web reviews into different sentiment categories has become a very important research problem in the opinion mining field. Sentiment classification techniques have been widely applied in business intelligence and public opinion investigation.

Machine learning methods can be used in sentiment classification. There are three categories of machine learning methods. The methods in the first category use unsupervised methods to construct sentiment classifiers. Unsupervised methods usually rely on lexical, syntax or semantic analysis to do sentiment classification. Unsupervised methods could use outer language resources to improve the classification accuracy. The second category includes traditional classification methods which must firstly label some training instances. Sometime, the labeling work needs a lot of time. Semi-supervised machine learning-based methods fall into the third category. In this case, we only need label a small number of training instances by exploiting a large number of unlabeled instances.

Our proposed method is a semi-supervised sentiment classification method. We represent training data set as symbol T, $T = \{(x_1, y_1), (x_2, y_2), ..., (x_n, y_n)\}$, where x_i presents a training instance, and $y_i \in \{-1, +1\}$. In this work we only consider two sentiment classes, so we use -1 to present the class label for negative

H. Motoda et al. (Eds.): ADMA 2013, Part I, LNAI 8346, pp. 36–47, 2013.

sentiment, $+1$ presents the class label for positive sentiment. Let unlabeled data set be $U = \{u_1, u_2, ..., u_t\}$. If our data space is in the real space $X \subseteq \mathbb{R}^m$. Then we can denote a new review as $x_i = \{x_{i_1}, x_{i_2}, ..., x_{i_m}\}$, x_i is m dimensional input vector. The task of semi-supervised classification problem is to obtain a classification function $f(x)$ to predict the polarity of a consumer review x as below equation shows.

$$f : X_{(T,U)} \longrightarrow L \qquad (1)$$

Our method is based on an observation: opinion words in the same review have high possibility to have the same sentiment polarity. For example, "great" and "good" could be in the same positive review, and they have the same positive sentiment polarity. While "great" and "dirty" are easy to be in different reviews which may have contrary sentiment polarities. Fig. 1 is an example review from Tripadvisor[1]. In Fig. 1, we can find some useful opinion words or phrases such

"3500 reasons to be at home"

◉◉◉◉◉ Reviewed July 29, 2013

My family and I stayed at this hotel for twenty-one nights between the 28th of June and the 19th of July 2013. Having read previous reviews, we were not disappointed. We experience a clean and quiet hotel with friendly and accommodating staff members accompanied by excellent amenities.

Stayed July 2013, traveled with family

◉◉◉◉◉ Value
◉◉◉◉○ Location
◉◉◉◉◉ Sleep Quality

◉◉◉◉◉ Rooms
◉◉◉◉◉ Cleanliness
◉◉◉◉◉ Service

Fig. 1. An example Web review

as "not disappointed", "clean", "friendly" and "excellent". All these opinion words have positive sentiment polarity. The research question is weather we can use the co-occurrence relationship among opinion words to boost sentiment classification. Our method is based co-training framework. In order to do semi-supervised sentiment classification, we need to construct three basic sentiment classifiers. We firstly use uni-grams to train a SVM to get the first sentiment classifier f_1. Then, we adopt the extracted opinion words from training instances to train to get another sentiment classifier f_2. The third sentiment classifier f_3 uses co-occurrence opinion words of the original extracted opinion words extracted from training and unlabeled data sets. The three sentiment classifiers iteratively co-train together to get the final sentiment classifier. Experimental results show the proposed method is effective and promising, and it outperforms the Self-learning SVM method.

---

[1] http://www.tripadvisor.com

2   Related Work

Opinion mining has three important tasks [4]:(1) the development of language resources; (2) sentiment classification; and (3) opinion summarization. Sentiment classification is one of the most tasks of opinion mining. There are many excellent publications in the field of sentiment classification. Hatzivassiloglou and McKeown [3] proposed a method to predict sentiment orientations of conjoined adjectives. Hatzivassiloglou and McKeown's work is the most early work addressing sentiment classification problem. Turney [12] proposed PMI (Pointwise Mutual Information) method to determine the sentiment orientation of words. PMI is a method of inferring sentiment orientation a word from its statistical association with a set of positive and negative words [12]. PMI is also be extended to be combined with information retrieval to form a new document sentiment classification method, namely PMI-IR [11]. Besides lexical-based methods, machine learning techniques are widely used for sentiment classification too. Pang et al. [10] employed Naive Bayes, maximum entropy and support vector machines to conduct sentiment classification on movie reviews, and experimental results proved SVM-based method was effective to do sentiment classification and it outperformed human-produced baselines. Readers who want to understand further recent research progress and future research challenge of sentiment analysis can read Pang's survey [9].

Besides various supervised machine learning methods, there are some publications focus on semi-supervised sentiment classification methods. Zhou [13] gave the semi-supervised machine learning sentiment classification method based on the active deep network. The method has high computing complexity. Li's work [5] tries to conduct sentiment classification based on co-training framework combining two views: one view is "personal view" , the other view is "impersonal view". However, in Li's work [5], text in personal view and text in impersonal view must be classified firstly.

We focus on opinion mining and have done some basic search work. Our previous work [8] also gave a spectral clustering-based semi-supervised sentiment classification method. The work [8] adopt a spectral clustering algorithm to map sentiment units of consumer reviews into new features which are extended into the original feature space. A sentiment classifier is built on the features in the original training space, and the original training features combined with the extended features are used to train the other sentiment classifier. The two basic sentiment classifiers together form the final sentiment classifier through selecting instances in the unlabeled data set into the training data set. But the work [8] firstly has high computing complexity, and it also needs to decide some parameters such as sentiment unit cluster number before sentiment classification, and it also requires much more computing resources than the proposed method in this work. Another work [6] also addresses the problem sentiment classification using subjective and objective views. We also adapt simple opinion word extract method and use these opinion words to rank product features. In the work [7], we propose a linear regression with rules-based approach to ranking product features according to their importance.

3   Proposed Approach

Web users use subjective phrases to express their opinions. It is common to find
sentiment related opinion words in product reviews. Each review usually has
one or several opinion words. Most of these opinion words indicate sentiment
polarities. We have mentioned that opinion words in the same review have high
possibility to have same sentiment polarity. For instance, in our phone data set
"good" and "great" have similar sentiment polarity. They are likely to co-occur
in reviews with positive sentiment. In the hotel data set, the opinion words
"great" and "unfriendly" also have the different sentiment polarity. They have
low possibility to co-occur in the same positive review. Of course, Web user
could express both positive comments and negative comments, because they
could satisfy with some product features but dislike some other product aspects.
Anyway, we assume the major sentiment of a review should expressed by major
sentiment words. The research problem is whether we can use opinion words to
improve sentiment classification.

Fig.2 shows the framework of our method. The co-training framework [1] is
typical semi-supervised self-boost algorithm framework. The co-training method
uses different views to train the final classifier. Our method is based on co-
training algorithm framework, hence we must construct three sentiment classifiers. The first sentiment classifier f_1 is trained using common uni-gram features
coming from training review. We extract opinion words from the training instances and the unlabeled instances, then we find the nearest neighbor opinion
words for extracted opinion words of each review. The second sentiment classifier
f_2 depends on the extracted opinion words of each review. The extracted opinion
words are used straightly as training features. The third sentiment classifier f_3
exploits the neighbor opinion words as training features. The three classifiers
co-train iteratively to get the final sentiment classifier f.

Web users like to use adjectives to describe product features and express their
opinions in their reviews. For example, in the sentence "The room is *clean* and
nice", both "clean" and "nice" adjectives have positive sentiment. So the simple
and straight way is to extract adjectives as candidate opinion words. But there
is a situation that an adjective is associated with a negative indicator. To simply
the problem, in this work, we look a negative indicator and an opinion word
together as a single opinion word. For example, in the phrase "*not good*", the
word "*good*" is an opinion word, and "*not*" is a negative indicator, and "*not
good*" is looked as a single opinion word. We extract adjectives with POS (Part-
of-Speech) labels of JJ, JJR, JJS. These negative indicators, just as in our previous work [7] [8], including "*not*", "*no*", "*donot*", "*do not*", "*didnt*", "*did not*",
"*was not*", "*wasn't*", "*isnt*", "*isn't*", "*weren't*", "*werent*", "*doesnt*", "*doesn't*",
"*hardly*", "*never*", "*neither*", and "*nor*". The context of a word window [-3,0]
means the left and right distance coverage of the word in a clause is 3 words
from the opinion word. At the same time, if there is a negative indicator in the
context window [-3,0] of an opinion word, then the negative indicator combined
with the sentiment word together constitute a sentiment unit.

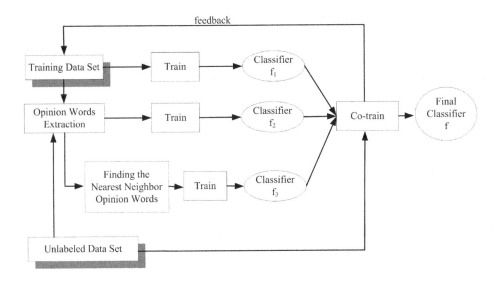

Fig. 2. Overview of our proposed method

The first sentiment classifier takes the frequencies of uni-grams extracted from the training data set as training features. In this case, we remove the tokens whose sizes are less than 3, as well as some stop words. After the preprocessing action, we use TinySVM[2] to get the first sentiment classifier f_1.

The second sentiment classifier is trained by extracted opinion words which have POS tags of "JJ","JJR" "JJS". If in the context window of an opinion word there is a negative indicator, then the negative indicator as well as the opinion word together to form a sentiment unit which is looked as a single training feature. We use frequencies of these extracted opinion words or sentiment units from training data set to get the second sentiment classifier f_2.

To get the third sentiment classifier f_3, we must get the nearest neighbor opinion word of every original opinion word of each review. In order to exploit the relationship of co-occurrence opinion words, we construct co-occurrence matrix $\mathbf{M}$ for these opinion words. $\mathbf{M}_{ij}$ is the co-occurrence frequency of opinion words o_i and o_j. If an opinion word o_i has the highest co-occurrence frequency with the opinion word o_j in o_i's co-occurrence word set coming from both the training data set and the unlabeled data set, we consider o_j is the nearest neighbor opinion word of o_i. Suppose O_T is the set of total opinion words extract from the training data set. Let O_U be the set of total opinion words extracted from the unlabeled data set. Let $O_{TU} = O_T \bigcup O_U$. For each review r_i we will get an opinion word set O_i for r_i, while $O_i \subseteq O_{TU}$. O_i could include one or several opinion words.

According to the co-occurrence matrix M, we can get the co-occurrence opinion word set for every element of O_i. We also call the co-occurrence opinion

---

[2] http://chasen.org/~taku/software/TinySVM/

word of an opinion word o_{ij} as the "neighbor opinion word" set of o_{ij}. For any element o_{ij}, $o_{ij} \in O_i$, $0 \le j < |O_i|$, we denote o_{ij}'s co-occurrence opinion word set as N_{ij}. For each original opinion word, we select one or several opinion words from N_{ij} as the training features of classifier f_2 (As Fig. 2 shows.). The co-occurrence opinion words are ranked in decreasing order according to their co-occurrence frequencies with o_{ij}. For each opinion word of a review, we can get b neighbor opinion words, where $b \ge 1$. In our experiments, we set $b = 1$, $b = 2$, and $b = 3$ respectively to conduct our experiments. If we set parameter $b = 2$, the two co-occurrence opinion words of O_{ij} with the highest co-occurrence frequencies will be selected as the nearest neighbor opinion words. If the nearest opinion words has be seen, we continue the next co-occurrence opinion word. The nearest neighbor opinion word selection algorithm is shown as Algorithm 1 which is only a part of our proposed method. The detailed steps of our proposed method is shown as Algorithm 2 which is the Co-Training SVM method using

---

Algorithm 1. Nearest Neighbor Opinion Words Selection Algorithm

Input:
    $reivew_i$, a review in our data set;
    O_i, original opinion word set of $review_i$;
    b, the limit number of the nearest opinion words of an original opinion word;
    M, co-occurrence matrix constructed from both the training data set and the unlabeled data set;

Output:
    The opinion word neighbor set N_i {Note: Opinion word set is a subset of the co-occurrence word set.};

1: Let W be the neighbor opinion word map;
2: **for** each original opinion word $o_{ij} \in O_i$ **do**
3:    $limit = 0$;
4:    Get co-occurrence opinion words and frequency of o_{ij} from co-occurrenc Matrix M;
5:    Put all o_{ij}'s distinct co-occurrence opinion words into vector V_i;
6:    Sort V_i in decreasing order according to their co-occurrence frequencies with o_{ij};
7:    **for** z=0 to sizeof(V_i) **do**
8:       **if** W.find(V_{iz}) != W.end() **then**
9:          break;
10:      **else**
11:          Put V_{iz} into $review_i$'s neighbor opinion word set N_i;
12:          Insert V_{iz} into W;
13:          $limit = limit + 1$;
14:          **if** $(limit >= b)$ **then**
15:             break;
16:         **end if**
17:      **end if**
18:    **end for**
19: **end for**
20: **return** N_i;

co-occurrence opinion words. The co-training framework [1] is often used in semi-supervised learning methods. To used the co-training framework, we must use different views found in the data set to construct corresponding classifier. We put the selected neighbor opinion words (the number is determined by the parameter b) into r_i's neighbor opinion word set N_i. Finally, we get the neighbor opinion word set N_i for O_i. We can get different number of neighbor opinion words. The Co-Training SVM method has three sentiment classifiers which select classified instances from the unlabeled data set into training space. In each iteration, every sentiment classifier only select the most 50 possible instances from the unlabeled data set, until the unlabeled data set is empty. The test features of classifier f_2, f_3 are opinion words extract from the test instances, while the test features of classifier f_1 are just uni-grams of test instances.

Algorithm 2. Co-Training SVM Method using Co-occurrence Opinion Words

Input:

Training data set $T = \{t_1, t_2, ..., t_x\}$, T includes balanced positive reviews and negative reviews;

Unlabeled data set $U = \{u_1, u_2, ..., u_y\}$;

Output:

Sentiment classifier C;

1: Extract opinion words from T and U, we get the training opinion word set O_T and the unlabeled opinion word set O_U.
2: Construct co-occurrence matrix M using co-occurrence relationship (if two opinion words are in the same review) among opinion words coming from O_T and O_U;
3: For each training instance in training data set T and unlabeled data set T , we get the nearest neighbor opinion word set N_i of its original opinion word O_i using Algorithm 1 (Nearest Neighbor Opinion Words Selection Algorithm).
4: **while** There are unlabeled instances or iterating times are not max **do**
5:    Use training data set T and SVM to get the first sentiment classifier f_1;
6:    Use original opinion words extracted from the training instances to train SVM to get the second classifier f_2;
7:    Use the nearest neighbors of extracted original opinion words from training data set and SVM to get the third sentiment classifier f_3;
8:    Use f_1 to classifier instances in U,and get positive instance set P_{f_1} and Negative instance set N_{f_1} (each iteration we take the most 50 possible classified instances);
9:    Use f_2 to classify unlabeled data set U, and get positive instance set P_{f_2} and negative instance set N_{f_2} (each iteration we take the most 50 possible classified instances);
10:    Use f_3 to classify unlabeled data set U, and get positive instance set P_{f_3} and negative instance set N_{f_3} (each iteration we take the most 50 possible classified instances);
11:    $L = L \cup P_{f_1} \cup N_{f_1} \cup P_{f_2} \cup N_{f_2} \cup P_{f_3} \cup N_{f_3}$;
12: **end while**
13: We get the final classifier $C = f_1$;
14: **return** C

4   Experiments

4.1   Experimental Data

We crawled *phone* and *laptop* reviews from **Amazon**[3] and hotel reviews from **Tripadvisor**[4] respectively. We extracted consumer reviews from these Web pages. We used **OpenNLP**[5] to conduct some shallow nature language processing to get sentences and POS tags. For each data category, we have 1000 positive reviews and 1000 negative reviews, as Table 1 shows. These data sets also studied in our previous work [8] [7] [6]. We also got the sentence number of each data set. For instance, the sentence number is and laptop reviews are segmented into 14814 sentences. The sentiment polarity of a review in a training data set is assigned according to their review ratings. A review rating is a real number ranging from 1 to 5. When the a review rating is greater than 3, then the review is a positive review; when a review rating is less than 3, the review is a negative review.

Table 1. Experimental statistics

data set	# of positive reviews	# of negative reviews	# of sentences
phone	1000	1000	28811
laptop	1000	1000	14814
hotel	1000	1000	18694

4.2   Compared Methods

Self-learning SVM Method. The earliest use of SVM (Support Vector Machine) Support vector machine [2] to determine the sentiment polarity of consumer reviews is Pang's research work [10]. It is believed SVM is one of the best text classification approach. Because our proposed method is semi-supervised sentiment classification method, we try to compare our method with Self-learning SVM method. Therefore, Self-learning SVM method is our baseline. We use the frequencies of uni-grams of reviews as the training and test features for SVM. We employ TinySVM[6] software to conduct SVM-based sentiment classification. The Self-learning SVM is a bootstrap approach to learning as Algorithm 3 shows. The method is also used in our previous work [8]. The algorithm iteratively selects the most likely correctly classified reviews which are determined by their distances to classification hyperplane, and put them into the training data set. A new sentiment classifier is built by training on the new training data set. Repeat the above steps until there are no unlabeled review that can be added to the

---

[3] http://www.amazon.com/
[4] http://www.tripadvisor.com/
[5] http://opennlp.apache.org/
[6] http://chasen.org/~taku/software/TinySVM/

Algorithm 3. Self-learning SVM Method

Input:

    Training set $T = \{t_1, t_2, ..., t_x\}$, T includes positive reviews and negative reviews;
    Unlabeled data set $U = \{u_1, u_2, ..., u_y\}$;

Output:

    Sentiment classifier C;

1: n is the number of selected reviews which are the most likely correctly classified reviews;

2: use SVM to get the initial sentiment classifier C on training data set T;

3: **while** unlabeled data set is not empty **do**

4:     Use sentiment classifier C to classify the unlabeled instances in U: get positive set P and the negative set N;

5:     if $|P| >= d$, then select the most likely correctly classified d instances from P (the set is P_d) into T, $T = T \cup P_d$, $P = P - P_d$; otherwise put all the instances in the P into T, $T = T \cup P$; {d=50 in our experiments}

6:     if $|N| >= d$, then select the most likely correctly classified d instances from N (the set is N_d) into T, $T = T \cup N_d$, $N = N - N_d$; otherwise put all the instances in the N into T, $T = T \cup N$; {d=50 in our experiments}

7:     Employ SVM to train a new sentiment classifier C on the current training data set T;

8: **end while**

9: **return** C;

---

training data set so far. When a classified review has greater distance from SVM hyperplane, the review is considered to have higher probability to be correctly classified.

4.3 Experimental Results

We have three data sets which include phone, laptop, hotel reviews respectively. We randomly sample 800 reviews from each data set to form the test data set. Each test set includes 400 positive reviews and 400 negative reviews. For each data set, we also sample 100, 200, 300, and 400 reviews for training respectively. The remaining reviews are unlabeled reviews. A training data set is a balanced training data set, for instance, 100 training instances include 50 positive instances and 50 negative instances. We compare our proposed method, namely the Co-Training SVM method, with the Self-learning SVM method as Algorithm 3 shows. We use accuracy to evaluate our proposed method.

$$accuracy = \frac{number\ of\ right\ classified\ instances}{total\ number\ of\ test\ instances} \tag{2}$$

Fig. 3 shows the experimental results of our proposed method on the phone data set. In Fig. 3, the parameter b is the limit number of the nearest opinion words of an original opinion word as Algorithm 1 shows. We can see the

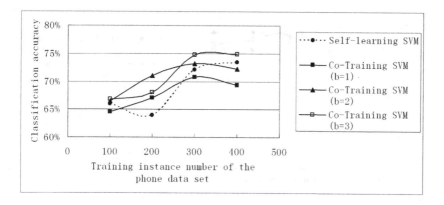

Fig. 3. Sentiment classification accuracy on the phone data set

Self-learning SVM method has the lowest accuracy when the training instance number is 200. But our proposed method doesn't have an advantage compared with the Self-learning SVM method when $b = 1$. However, when $b = 2$ and $b = 3$, the proposed method outperforms the Self-learning SVM method.

Fig. 4 shows our proposed method has better performance than the Self-learning SVM on the laptop data set in all the cases. However, the experimental results of the Co-training seems very similar.

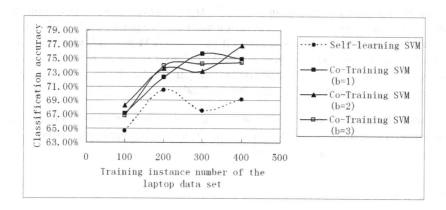

Fig. 4. Sentiment classification accuracy on the laptop data set

Fig. 5 shows our experimental results on the hotel data set. In this case, our proposed method with different b values all does better than the Self-learning SVM method. The neighbor opinion words can help sentiment classification, but increasing number of the nearest neighbor opinion words cannot improve the accuracy of sentiment classification much more.

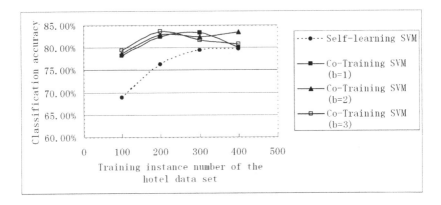

Fig. 5. Sentiment classification accuracy on the hotel data set

5   Conclusion

In this paper, we propose a novel semi-supervised sentiment classification method which is based co-training framework. Our method studies the research problem whether we can improve semi-supervised sentiment classification accuracy by using the co-occurrence relationship of opinion words extracted from training data set and the unlabeled data set. We exploit co-occurrence opinion words to improve the sentiment classification. We try to prove co-occurrence opinion words can improve sentiment classification accuracy. Experimental results show our proposed method outperform the Self-learning SVM method. Our proposed method is simple and fast, but promising. We believe there is more work to select extended opinion words from both the training and unlabeled data set. We will continue our work to find more effective methods.

Acknowledgement. We thank anonymous reviewers for their constructive comments. The paper is supported by the National Natural Science Foundation of China under Grant 61170002.

References

1. Blum, A., Mitchell, T.: Combining labeled and unlabeled data with co-training. In: Proceedings of the Eleventh Annual Conference on Computational Learning Theory, COLT 1998, pp. 92–100. ACM, New York (1998), http://doi.acm.org/10.1145/279943.279962
2. Cortes, C., Vapnik, V.: Support-vector networks. Machine Learning 20(3), 273–297 (1995), http://dblp.uni-trier.de/db/journals/ml/ml20.html#CortesV95
3. Hatzivassiloglou, V., McKeown, K.R.: Predicting the semantic orientation of adjectives. In: Proceedings of the Eighth Conference on European Chapter of the Association for Computational Linguistics, EACL 1997, pp. 174–181. Association for Computational Linguistics, Stroudsburg (1997)

4. Lee, D., Jeong, O.R., Lee, S.G.: Opinion mining of customer feedback data on the web. In: Proceedings of the 2nd International Conference on Ubiquitous Information Management and Communication, ICUIMC 2008, pp. 230–235. ACM, New York (2008)

5. Li, S., Huang, C.-R., Zhou, G., Lee, S.Y.M.: Employing personal/impersonal views in supervised and semi-supervised sentiment classification. In: Proceedings of the 48th Annual Meeting of the Association for Computational Linguistics, ACL 2010, pp. 414–423. Association for Computational Linguistics, Stroudsburg (2010)

6. Li, S.: Sentiment classification using subjective and objective views. International Journal of Computer Applications 80(7), 30–34 (2013)

7. Li, S., Guan, Z., Tang, L., Chen, Z.: Exploiting consumer reviews for product feature ranking. J. Comput. Sci. Technol. 27(3), 635–649 (2012)

8. Li, S., Hao, J.: Spectral clustering-based semi-supervised sentiment classification. In: Zhou, S., Zhang, S., Karypis, G. (eds.) ADMA 2012. LNCS, vol. 7713, pp. 271–283. Springer, Heidelberg (2012)

9. Pang, B., Lee, L.: Opinion mining and sentiment analysis. Found. Trends Inf. Retr. 2(1-2), 1–135 (2008)

10. Pang, B., Lee, L., Vaithyanathan, S.: Thumbs up?: sentiment classification using machine learning techniques. In: Proceedings of the ACL 2002 Conference on Empirical Methods in Natural Language Processing, EMNLP 2002, vol. 10, pp. 79–86. Association for Computational Linguistics, Stroudsburg (2002)

11. Turney, P.D.: Thumbs up or thumbs down?: semantic orientation applied to unsupervised classification of reviews. In: Proceedings of the 40th Annual Meeting on Association for Computational Linguistics, ACL 2002, pp. 417–424. Association for Computational Linguistics, Stroudsburg (2002)

12. Turney, P.D., Littman, M.L.: Measuring praise and criticism: Inference of semantic orientation from association. ACM Trans. Inf. Syst. 21(4), 315–346 (2003)

13. Zhou, S., Chen, Q., Wang, X.: Active deep networks for semi-supervised sentiment classification. In: Proceedings of the 23rd International Conference on Computational Linguistics: Posters, COLING 2010, pp. 1515–1523. Association for Computational Linguistics, Stroudsburg (2010)

HN-Sim: A Structural Similarity Measure over Object-Behavior Networks

Jiazhen Nian, Shanshan Wang, and Yan Zhang*

Department of Machine Intelligence, Peking University
Key Laboratory on Machine Perception, Ministry of Education
Beijing 100871, P.R. China
{nian,shanshanwang}@pku.edu.cn
zhy@cis.pku.edu.cn

Abstract. Measurement of similarity is a critical work for many applications such as text analysis, link prediction and recommendation. However, existing work stresses on content and rarely involves structural features. Even fewer methods are applicable for heterogeneous network, which is prevalent in the real world, such as bibliographic information network. To address this problem, we propose a new measurement of similarity from the perspective of the heterogeneous structure. Heterogeneous neighborhood is utilized to instantiate the topological features and categorize the related nodes in graph model. We make a comparison between our measurement and some traditional ones with the real data in DBLP[1] and Flickr[2]. Manual evaluation shows that our method outperforms the traditional ones.

Keywords: Structural similarity measurement, Heterogeneous network, Object-behavior network.

1 Introduction

Similarity measurement is quite an important and fundamental study for many practical information retrieval tasks, such as relevance search [2], clustering and even ontology generation and integration [9]. It evaluates the similarity between objects in the relation networks. To deal with the relevance search problem, content-based models are commonly used, with bag-of-words as an instance. Semantic information is learned from the words and text relation is extracted by the comparison of contents [3]. For instance, when we measure the similarity between two authors in a bibliographical information network, the co-author and conference-participation behaviors can be extracted to characterize the similarity. The method gets beyond linguistic analysis, but it still works effectively.

Though these state-of-art methods exhibit really well, some other features can be borrowed to further enhance the similarity measurement result. Take the popularization of social network as an example. Behavioral characteristics and social

---

* Corresponding author.
[1] http://www.informatik.uni-trier.de/~ley/db/
[2] http://www.flickr.com/

H. Motoda et al. (Eds.): ADMA 2013, Part I, LNAI 8346, pp. 48–59, 2013.
© Springer-Verlag Berlin Heidelberg 2013

relationship often reflect the similarity between users. To study the behavior of objects, link-based graph model, or object-behavior network are always established, where behavioral objects are the nodes and the particular behavior or relationship turns into the links between them. Hence, the analysis of object behavior is translated into the analysis of this object-behavior network. New similarity measures can be proposed based on structure analysis over object-behavior networks.

In graph model, some homogeneous structural features have been extensively studied such as common neighbors and link network [7]. However, objects and relations are always of various types in abundant networks which are called the heterogeneous information networks. In this case, conventional methods cannot work well and thus researchers exert some new ideas of similarity measurement, such as PathSim [11] and HeteSim [10].

In this paper, we tackle the structural similarity measure problem at a general level. Every object in the network is categorized in accordance with their types and attributes. Object-behavior network will be set up and nodes are linked by heterogeneous edges even some of them are not actually connected. We call these connected nods as *heterogeneous neighbors*. For a particular object, all of its heterogeneous neighbors which are connected with the same relationship constitute a *heterogeneous neighborhood*. Thus far, the measurement of similarity will be established based on these heterogeneous neighborhoods. Meanwhile, this method is more efficient and cost-saving because it merely takes into account a very small part of data.

We conduct some experiments on DBLP data set, calculating the similarity between authors. Besides, a manual evaluation on Flickr is made to compare our method with the conventional measurements.

Our contribution can be summarized as follows: we study the structural similarity beyond semantic researches and this method can be applied to measure similarity efficiently in object-behavior networks. Meanwhile, we propose heterogeneous neighborhoods in order to categorize different neighbor objects and extract the semantic information of the relationships.

The remainder of the paper is organized as follows. We review the related works in Section 2. The details of ths similairty measurement method are described in Section 3. The experimental results are shown in Section 4. Finally, we conclude our work in Section 5.

2   Related Work

The problem of similarity search between structured objects has been studied in the domain of structural pattern recognition and pattern analysis. The most common approach in previous work is based on the comparison of structure. Two objects are considered structurally equivalent if they share many common neighbors. Contrary to homogeneous networks, in heterogeneous information networks neighbors cannot be treated as the same because of the multi-type relations.

In order to utilize the information of structure in heterogeneous information network, Sun et al. advised PathSim [11] about the measurement of the similarity of same-typed objects based on symmetric paths, by considering semantics in meta-path which is constituted by different-typed objects) Shi et al. proposed another measure, called HeteSim [10]. That is based on the theory that relatedness of object pairs is defined according to the search path, which connects two objects through following a sequence of nodes.

Some similarity measurements based on link structure are effective such as SimRank [5] and P-Rank (Penetrating Rank) [13]. These methods consider that two vertices are similar if their immediate neighbors in the network are similar. The difference is that SimRank only calculates vertex similarity between same-typed vertices and only partial structural information from inlink direction is considered during similarity computation. P-Rank enriches SimRank by jointly encoding both inlink and outlink relationships into structural similarity computation. Though P-Rank has considered heterogenous relations between vertexes, only directly-linked neighbors are put into computation. Hence some semantic information is lost. Furthermore, both SimRank and P-Rank are iterative algorithms. When the network gets larger, the cost will become heavier.

3   Methodology

3.1   Notation and Definition

To formalize our method, we will first give some concepts.

Definition 1. *Network*: A network is defined as a directed graph, noted as $G = (V, E)$. V is the set of vertexes and E is that of edges. In an object-behavior network, every object is abstracted as a vertex and if two objects are related by some a particular behavior or relationship, they will be linked by an edge.

Definition 2. *Node*. A node stands for an object in an object-behavior network, noted as $v \in V$. It is represented as a triple $< info, type, \Phi >$, where $info$ and $type$ denote the information of this object; Φ is the general set of neighborhoods of node v. We refer to info, type and Φ of v with $v.info$, $v.type$ and $v.\Phi$.

Definition 3. *Heterogeneous Neighborhood*. A heterogeneous neighborhood is actually a set of nodes which describe a kind of topological structural features of an object in an object-behavior network. We formalize it as an infinite set ϕ. Each ϕ is defined as a triple $< relation, distance, \nu >$, where $relation$ is the description of this relationship and $distance$ is the length of the path from the cynosure to each node v in the node set ν.

Each type of links represents a kind of relation($\mathbb{R}$), and each relation is defined as $R \in \mathbb{R}$. Function $R(v_p) = \{v | v \in V, v \text{ and } v_p \text{ are connected by relation } R\}$. In Definition 3, if a heterogeneous neighborhood $\phi \in v_p.\Phi$ is relevant with relation R_m. For simplicity, we use subscript relation name to modify the neighborhood, e.g. ϕ_R denotes the ϕ which $relation$ is R, and it will be inferred that $\phi_{R_m}.\nu = R_m(v_p)$.

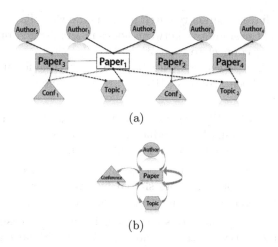

(a)

(b)

Fig. 1. A heterogeneous information network and a schematic diagram of the relations in a heterogeneous network

In this paper, distance plays a very important role. Hence, for making a clear introduction, another notation will be defined: regard to a particular node v, every $\phi \in v.\Phi$ is labeled by $\phi.distance$. Each neighborhood set Φ^k, is defined as $\{\phi_i | \phi_i \in \Phi, \ \phi_i.distance = k\}$, where $\Phi = \bigcup_{k=1}^{n} \Phi^k$.

A heterogeneous neighborhood contains same type of nodes which are connected by the same relation. In homogeneous networks, the relation is only "linked" or not. Here, relation is expressed by a sequence of related-object pairs. For instance, in Fig. 1(a), $Author_2$ is $Paper_1$'s homogeneous neighbor because there is an edge between them. $Paper_2$ is $Paper_1$'s heterogeneous neighbor for they can be connected by $(Paper_1, Author_2)$ and $(Author_2, Paper_2)$. Fig. 1(b) shows all the feasible relations in that heterogeneous information network.

Example 1. *Heterogeneous neighborhood of $Paper_1$* (in Fig. 1(a)). Node $Paper_1$: $< Paper_1, \ paper, \ \Phi >, \Phi = \bigcup_{k=1}^{3} \Phi^k. \ \Phi^1 = \{\phi_0, \ \phi_1, \ \phi_2, \ \phi_3\}.$

Thereinto $\phi_0.relation$ is *"be written"*, $\phi_1.relation$ is *"be published in"*, etc. Now we use P, A, C, T to represent paper, author, conference and terms respectively, meanwhile the relation $R_{bewritten}$ can be noted as "PA". For simplification and clearness, we use ϕ_{PA} instead of ϕ_0. Hence, $\phi_{PA}.distance$ is 1, $\phi_{PA}.\nu$ is $\{Author_1, \ Author_2\}$.

3.2 Similarity Measure Based on Heterogeneous Neighborhood

Homogenous Similarity Measure Review. The principle of establishing information network is to connect the relevant objects. For instance, two authors are related because they have once collaborated on one paper, and Hawaii is connected to CIKM in bibliographical network because CIKM was once held in Hawaii. Consequently, if two authors share so many common neighbors just like papers, they must have common research interests. So we believe that structural features fully reveal the similarity between nodes in an information network, and the similar nodes are connect to each other strongly.

A variety of similarity measurements based on structural features in homogeneous network have been proposed. Here we review some of the classical similarity measurements.

Personalized PageRank (PR): Personalized PageRank [1] is an extension for personalization of PageRank by introducing preference set P. The Personalized PageRank equation is defined as $\mathbf{v} = (1 - c)\mathbf{A}\mathbf{v} + c\mathbf{u}$.

Common neighbors (CN): Common neighbors [1] is defined in graph model as the number of common neighbors shared by two vertex v_i and v_j, namely $|\Gamma(v_i) \bigcap \Gamma(v_j)|$. A larger value means these two objects are more similar.

Jaccard's coefficient (JC): Jaccard's coefficient is another measure to evaluate the similarity between two vertex v_i and v_j, which is the normalized value of common neighbors, namely $\frac{|\Gamma(v_i) \bigcap \Gamma(v_j)|}{|\Gamma(v_i) \bigcup \Gamma(v_j)|}$.

HN-Sim Formula. However, the neighbors in different neighborhood or of diverse types always express differently. Taking DBLP as an example, an institute and a conference can be the neighbors of the same papers, but it's equivocal to tell whether they are similar.

By analyzing the neighbor ingredient, we can measure the similarity between two nodes in an information network. However, the diversity of neighbors provides different semantic information about nodes. Neighbors in different level represent different properties. In this paper, heterogeneous neighborhood is proposed to categorize the neighbors. Accordingly the similarity of two objects will be measured in each heterogeneous neighborhood separately and finally merged by an influence-based function.

As is discussed in Section 3.1, we have known that a node's heterogeneous neighbors can be denoted by sequences of object pairs. While the sequences' length can be unlimited, $|\Phi|$ will be infinite, which will lead to a big problem. On one hand, computing too many neighbors may bring in over fitting. Hence, a strategy needs to be settled to decide which neighborhoods should be considered. On the other hand, when the distance of neighborhood gets larger, the object on the neighbor level will probably be reiteration. In response to this issue, experiments and intuitions show that the length of the path used should be positively correlated to the diversity of the system and negatively correlated to the scale of the node set.

Meanwhile, since content is not utilized, neighborhoods with same distance will have the same influence in similarity measurement. These neighborhoods have different semantic meanings, however, we can average their similarity contributions in terms of their influence. Therefore the problem of synthesizing the similarity of individual heterogeneous neighborhoods is simplified as synthesizing the similarity of each neighborhood set.

The similarity between node v_1 and v_2 is defined as Formula 1.

$$HeteroSim(v_1, v_2) = \frac{1}{1+\varepsilon} \prod_{\Phi^k \in v_1.\Phi}^{k} \{ \varepsilon + \mathcal{L}[\frac{1}{|\Phi^k|} \sum_{R_i:\phi_i \in \Phi^k}^{i} \theta_{R_i}(v_1, v_2)]\}. \quad (1)$$

Where function $\mathcal{L}()$ controls the influence of each level of neighbors. When it comes to path-drive model, $\mathcal{L}()$ can be defined as a power function according to the distance:

$$\mathcal{L}[x] = [x]^\delta. \quad (2)$$

Here δ is decided by the length and node type of the heterogeneous neighborhood. Intuitively, further nodes make less influence to an object. Thus, δ should be positive and it is positively related to $\phi_{R_i}.distance$.

Smooth factor ε is employed in case of none common neighbors in some heterogeneous neighborhoods.

Function $\theta_R()$ measures the similarity between two nodes based on the particular neighborhood connected by relation R. In the experiment, Jaccard's coefficient will be used to calculate it.

$$\theta_{R_i}(v_1, v_2) = \frac{|\Gamma_{R_i}(v_1) \cap \Gamma_{R_i}(v_2)|}{|\Gamma_{R_i}(v_1) \cup \Gamma_{R_i}(v_2)|}. \quad (3)$$

Where $\Gamma_i(v)$ is the element of ϕ_i, namely $\phi_i.\nu$.

Homo-Info Adjustment. In some particular homogeneous networks, nodes quality and influence always play important roles to measure the similarity between any of the two entities. Take social network as an example, person with good reputation is always similar with other people who have the same good reputation. So in this step, we will extract some homogeneous information to adjust the heterogeneous similarity. Wang and some other researchers have already proposed several methods in which they use influence to enhance similarity measurement [12].

We extend the idea of influence ranking in homogeneous network as *HomoSim* to adjust *HN-Sim* model. The similarity between node v_1 and v_2 is defined as Formula 4.

$$Sim(v_1, v_2) = HomoSim(v_1, v_2)^\lambda \cdot HeteroSim(v_1, v_2)^{1-\lambda}. \quad (4)$$

Here *HomoSim* shows the homogeneous ranking similarity between these two nodes, and *HeteroSim* represents basic *HN-Sim*. By adjusting the weighting parameter λ, we can draw the following conclusions: (1) When $\lambda = 0$, *HN-Sim* only computes heterogeneous neighborhood based similarity; (2) When $\lambda = 1$, *HN-Sim* is reduced to homogeneous ranking similarity; (3) Setting λ between 0 to 1, it will balance the leverage between homogeneous ranking bias and the basic *HN-Sim*. It will be further discussed in Section 4.4.

HomoSim is an adjustment function to enhance the *HN-Sim*. It can be estimated as Formula 5.

$$HomoSim(v_1, v_2) = \sqrt{1 - [Rank(v_1) - Rank(v_2)]^2}. \quad (5)$$

This ranking function in Formula 3 needs to satisfy the following three properties: (i) *Global calculations*; (ii) *Rapid convergence*; (iii) *Bare content involved*. *HN-Sim* is based on local structure, therefore it has serious limitation in global view. And it should not gain much more cost by involving global calculation.

Different networks always hold different quality ranking strategies. In bibliographic network, paper, author and conference are connected to each other. In order to find the papers with the similar quality, PageRank [8] might be used. In Wikipedia[3], there are bare links between entities but entities are linked with numbers of editors. We believe that high-quality authors will edit high-quality entities, and high-quality entities are edited by high-quality authors. Therefor, in Wikipedia networks, HITS [6] will be used as the *Rank*() function. In addition, content based ranking (CRank) is also suitable. For instance, in IMDB[4] data set, the review score can be taken as the parameter of the ranking function to measure the similarity of two movies.

4   Experiments

In this Section, we present our experiments on a bibliographical information network and a social picture-sharing network.

4.1   Datasets

Bibliographical network is a typical heterogeneous information network, which is a kind of object-behavior networks. We use our method to analyse the similarity between different authors. In this paper, we use the DBLP dataset downloaded in January 2013 which contains 869,113 papers, 689,177 authors and 1,304 conferences.

Fig. 2 shows the number of neighbors in different Φ^i. The majority of the nodes have less than 10 neighbors in Φ^1, and more than 40% of the nodes have a heterogeneous neighborhood with nodes amount greater than 100. The diversity of heterogeneous neighborhood is revealed in Table 1. For instance, an author-type node will have 18 heterogeneous neighborhoods whose distance is 5.

Another dataset is the one used in [11], downloaded from Flickr. Flickr is a web site providing free and paid digital photo storage, sharing, and some other online social services, it is an image hosting and online community. This information network contains images, users, tags and groups. The dataset covers 10,000 images from 20 groups with 10,284 tags, and 664 users.

In this experiment we limit the length of the related-pair sequence as 4, that means only Φ^1, Φ^2 and Φ^3 will be calculated. In DBLP dataset, the size of neighborhood set Φ is 9. In Formula 2, δ is the exact length of the neighborhood. For example, δs of θ_{RAP} and θ_{RAPC} are 2 and 3 respectively.

---

[3] http://www.wikipedia.org/

[4] http://www.imdb.com/

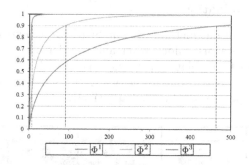

Fig. 2. Cumulative probability distribution of the scale of heterogeneous neighbors in different distances: Φ^1, Φ^2 and Φ^3

Table 1. The number of heterogeneous neighborhood types in different distances

Distance	1	2	3	4	5
Author	1	3	5	10	18
Conference	1	3	5	10	18
Paper	3	5	10	18	38

4.2 Case Study

In DBLP dataset, we calculate the similarity between the author Christos Faloutsos and others. The top 10 similar authors with the similarity marks are listed in Table 2.

We can see that *HN-Sim* satisfies self-similarity. The authors found by *HN-Sim* Model either publish similar papers or have strong connections with him and hold similar research interests. For instance, Philip S. Yu and Christos Faloutsos are both major in data mining and have the similar reputation in this area. Faloutsos and Kleinberg have collaborated on high-level papers.

Table 2. Similar authors to Christos Faloutsos

Rank	Author	Similarity
1	Christos Faloutsos	1.000
2	Philip S. Yu	0.7733
3	Jure Leskovec	0.6794
4	Joseph M. Hellerstein	0.6586
5	Yufei Tao	0.6452
6	Divesh Srivastava	0.6310
7	Jon M. Kleinberg	0.6018
8	Petros Drineas	0.5760
9	Yannis Manolopoulos	0.5525
10	Gerhard Weikum	0.5507

4.3   Result and Evaluation

To evaluate the effect of our method, we conduct another experiment on Flickr data set. Use, tag, category information and a brief description is given to each image. We extract the relation between images with tags and groups, constructing a heterogeneous information network. "I" is used to represent image, "T" and "G" represent tag and group respectively.

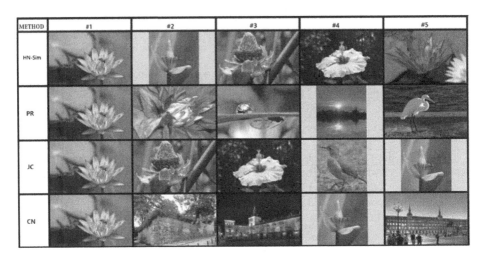

Fig. 3. Top 5 similar images in Flickr found by different methods

The relations in this object-behavior network can be combined with these three initial relations, R_{I-T}: tag is assigned to image; R_{I-G}: image is categorized as group; R_{T-G}: tag is assigned to group. To measure the similarity between two images, we limit the neighborhood distance under 3.

For another example, $\Phi^2 = \{\phi_{ITI}, \phi_{IGI}, \phi_{ITG}, \phi_{IGT}\}$. δ in Formula 2 is set as each distance of ϕ. We use our method to find the similar pictures. Common neighbors(CN), Jaccard's coefficient(JC) and P-PageRank(PR) are used as adversaries. The result is shown in Fig. 3.

To evaluate the performance of the measurement based on heterogeneous neighborhood, we calculate normalized Discounted Cumulative Goal (nDCG) [4]for each baseline method. NDCG is the most common way to evaluate the search result quality, the expression of DCG is as follows.

$$DCG_p = \sum_{i=0}^{p} \frac{2^{rel_i} - 1}{log_2(i + 1)} \tag{6}$$

Where p is the position of an image in a particular rank witch made by volunteers, and rel_i is the relevance values of image.

We invite 6 volunteers to rank 144 images by the similarity with the first image in Fig. 3. This rank is defined as ground truth. The relevance function rel in nDCG formula is defined as follows.

$$rel_i = relevance(|rank_i - goldrank_i|) \qquad (7)$$

Where $rank_i$ is the position of image i ranked by experimental method and $goldrank_i$ is that of ground truth. The values are mapped to discrete numeric (Table 3).

Table 3. Relevance mapping

	[0,1]	[2,5)	[5,10)	[10,144]
relevance	4	3	2	1

For a comparison, we normalize the DCG_p of result by dividing $IDCG_p$ which is calculated by golden standard rank. Fig. 4. shows the performance of all the methods we use. Our method achieves the best performance. Homogeneous methods with structural features are not stable and perform not well in heterogeneous information network. This validates that structural similarity is significant and it should be applied in heterogeneous way.

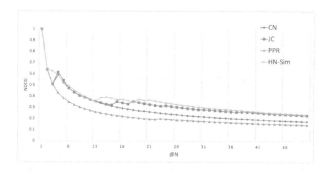

Fig. 4. nDCG of each similarity measure in Filckr dataset

4.4 Homo-Info Adjustment Discussion

Basic *HN-Sim* is based on local structure. It outperforms conventional methods in the experiments on DBLP and Flickr datasets. As is introduced in Section 3.2, basic *HN-Sim* can be enhanced by involving global computation. In this section, we integrate PageRank, Crank, and *Hetero*PageRank(PageRank in heterogeneous) to *HN-Sim*. Fig. 5 provides the the average results of different λ in terms of nDCG @ 65 similar images on the Flickr data sets.

There is little difference between all other methods at nDCG@15 while basic *HN-Sim* performs outstandingly. However we would like to point out that after nDCG@15, the integration of global information enhances the results.

To prove the effects of the incorporation of heterogeneous and homogeneous information, we sampled some movies on IMDB[5] to continue another experiment. λ is set as 0.4 and PageRank is used as the global homogeneous ranking. We calculate

---

[5] http://www.imdb.com/

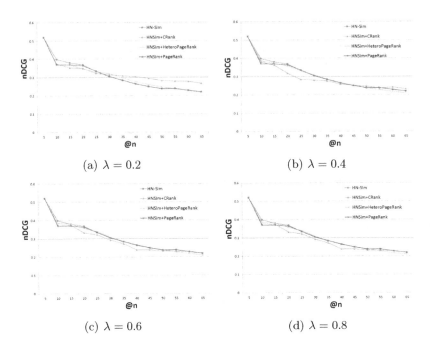

(a) $\lambda = 0.2$ (b) $\lambda = 0.4$

(c) $\lambda = 0.6$ (d) $\lambda = 0.8$

Fig. 5. The experimental results in terms of NDCG

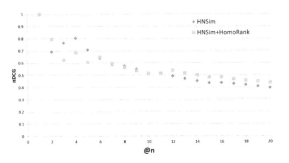

Fig. 6. nDCG of IMDB dataset

the most similar movies to Avatar[6], five volunteers were payed to rank the similar movies from a candidate set as the ground truth. The nDCG@20 is shown in Fig. 6.

We notice that in this experiment, the homogeneous information does not improve the top 10 results though it does help the overall effects.

5   Conclusions

In this paper, we propose a novel method to measure the similarity between objects from the perspective of structure. Heterogeneous neighborhood is borrowed

---

[6] http://www.imdb.com/tt499549/

to categorize heterogeneous neighbors, which contain more structural semantic information than initial neighbors. Compared with conventional methods, our *HN-Sim* measure puts more focus on topological features. The experimental results show that our method performs better.

As to future works, we will make more studies on the extraction of heterogeneous neighborhood and a formalized level-chosen pattern will be given. The aggregation of distance measure of each neighbor set may be more diversified in consideration of different data sets and the semantic information of the data.

Acknowledgements. This work was supported by NSFC with Grant No.61073081 and 61370054, and 973 Program with Grant No.2014CB340405.

References

1. Chakrabarti, S.: Dynamic personalized pagerank in entity-relation graphs. In: Proceedings of the 16th International Conference on World Wide Web, pp. 571–580. ACM (2007)
2. Hatzivassiloglou, V., Klavans, J.L., Eskin, E.: Detecting text similarity over short passages: Exploring linguistic feature combinations via machine learning. In: Proceedings of the 1999 Joint Sigdat Conference on Empirical Methods in Natural Language Processing and Very Large Corpora, pp. 203–212. Citeseer (1999)
3. Islam, A., Inkpen, D.: Semantic text similarity using corpus-based word similarity and string similarity. ACM Transactions on Knowledge Discovery from Data (TKDD) 2(2), 10 (2008)
4. Järvelin, K., Kekäläinen, J.: Cumulated gain-based evaluation of ir techniques. ACM Transactions on Information Systems (TOIS) 20(4), 422–446 (2002)
5. Jeh, G., Widom, J.: Simrank: a measure of structural-context similarity. In: Proceedings of the Eighth ACM SIGKDD International Conference on Knowledge Discovery and Data Mining, pp. 538–543. ACM (2002)
6. Kleinberg, J.M.: Authoritative sources in a hyperlinked environment. Journal of the ACM (JACM) 46(5), 604–632 (1999)
7. Liben-Nowell, D., Kleinberg, J.: The link-prediction problem for social networks. Journal of the American Society for Information Science and Technology 58(7), 1019–1031 (2007)
8. Page, L., Brin, S., Motwani, R., Winograd, T.: The pagerank citation ranking: bringing order to the web (1999)
9. Ruotsalo, T., Hyvönen, E.: A method for determining ontology-based semantic relevance. In: Wagner, R., Revell, N., Pernul, G. (eds.) DEXA 2007. LNCS, vol. 4653, pp. 680–688. Springer, Heidelberg (2007)
10. Shi, C., Kong, X., Yu, P.S., Xie, S., Wu, B.: Relevance search in heterogeneous networks. In: Proceedings of the 15th International Conference on Extending Database Technology, pp. 180–191. ACM (2012)
11. Sun, Y., Han, J., Yan, X., Yu, P.S., Wu, T.: Pathsim: Meta path-based top-k similarity search in heterogeneous information networks. In: VLDB 2011 (2011)
12. Wang, G., Hu, Q., Yu, P.S.: Influence and similarity on heterogeneous networks. In: Proceedings of the 21st ACM International Conference on Information and Knowledge Management, pp. 1462–1466. ACM (2012)
13. Zhao, P., Han, J., Sun, Y.: P-rank: a comprehensive structural similarity measure over information networks. In: Proceedings of the 18th ACM Conference on Information and Knowledge Management, pp. 553–562. ACM (2009)

Community Based User Behavior Analysis on Daily Mobile Internet Usage*

Jamal Yousaf, Juanzi Li, and Yuanchao Ma

Department of Computer Science and Technology,
Tsinghua National Laboratory for Information Science and Technology,
Tsinghua University, Beijing, 100084, China
{jamalyousaf,ljz,myc}@keg.tsinghua.edu.cn

Abstract. Laptops, handhelds and smart phones are becoming ubiquitous providing (almost) continuous Internet access and ever-increasing demand and load on supporting networks. Daily mobile user behavior analysis can facilitate personalized Web interactive systems and Internet services in the mobile environment. Though some research have already been done, there are still some problems need to be investigated. In this paper, we study the community based user behavior analysis on the daily Mobile Internet usage. What we focus on in this paper is to propose a framework which can calculate the proper number of the clusters in mobile user network. Given a mobile user Internet access dataset of one week which contains thousand of users, we firstly calculate the hourly traffic variation for the whole week. Then, we propose to use cluster coefficient and network community profile to confirm the presence of communities in mobile user network. Principal Component Analysis (PCA) is employed to capture the dominant behavioral patterns and uncover the several communities in the network. At last, we use communities/clusters to work out the various interests of the users on the timeline of the day.

Keywords: Data driven, Internet usage, Clustering, User Behavior.

1 Introduction

Wireless mobile networks are growing significantly in every aspect of our lives. Laptops, handhelds and smart phones are becoming ubiquitous providing (almost) continuous Internet access and ever-increasing demand and load on supporting networks. More and more mobile users use mobile technology to interact, create, and share content based on physical location. This change requires a deeper understanding of user behavior on the mobile Internet, and also requires uncovering the structure in the behavior of mobile internet community, inferring the relationships and recognizing the social patterns in daily user activity. This can greatly help improve the design of future mobile networks and mobile applications.

---

* The work is supported by the National Science Foundation, China.

H. Motoda et al. (Eds.): ADMA 2013, Part I, LNAI 8346, pp. 60–71, 2013.

A number of researches have been done in the analysis of mobile user behavior. Earlier work on mobility modeling tend to be trace-driven and some account for location preferences and temporal repetition [10]. More recent work [4] adopt a behavior-profile based on users' location preferences and then calculate the behavioral distance to capture the user similarity. In [1], use the location-based context data, user interaction records for extracting the latent common habits among user behavior patterns. The user profiling problem also has been addressed as a clustering problem where the objective was to group users who exhibit similar behavior [8] and co-clustering user behavior profiling was proposed [5]. Among these methods, the number of clusters is either pre-defined or flexible. This raises the problem whether we could find the proper number of clusters to facilitate the mobile user behavior analysis. This is the focus of this paper.

In this paper, we collect and process a large scale mobile network usage traces which contain thousand of users at one city of north China. We want to provide the practical technique for integrating and aggregating the data about their internet usage, and then analysis mobile users' behavior in their daily life. This addresses the following questions: **Q1)** *Could we use the implicit information like the time spent on a particular Web-page by a user, to infer the interest of the user?* **Q2)** *Could we confirm the presence of mobile user communities from the theoretical data observation and group the users that have similar navigation patterns, interests and preferences?* **Q3)** *Could we propose an effective approach for multivariate analysis and efficient technique for clustering the data into meaningful clusters?*

In this paper we propose an effective approach for multi-dimensional analysis of mobile user behavior on Internet access. At first, we examine the structural properties of mobile user network. We use clustering coefficient and network community profile to get insight of the mobile user usage network. We obtain the best community size by plotting clustering coefficient and conductance in mobile user network. Then, this community size is used to help the identification of similar user groups or communities by using scalable principal component analysis. At last, by analyzing the topics of different clusters, we work out the various interests of the users on the timeline of the day.

The key contributions described in this paper are as follows: (1) We discuss the possibility to identify the number of clusters by exploring the mobile user network for daily mobile user behavior analysis. We show how to use the coefficient and conductance to confirm the presence of the communities. (2) We use matrix factorization to reduce the data sparse problem which facilitates the extraction of the dominant trends and the internal structure of the original data. (3) By analyzing the topics of each community, we plot the characteristics of each cluster in different time of the day.

The rest of the paper is organized as follows. In section 2, we introduce the related work. In section 3, we present the initial observations about the mobile data and the network structure. In section 4, we discuss our approach for detecting the mobile communities and talk about the clustering methodology.

The global analysis with some meaningful observations and results are discussed in the section 5. Finally, section 6 concludes this paper with future research.

2   Related Work

In this paper we examine the community based user behavior on daily mobile usage. User behaviors can be seen everywhere in daily life for example in business and social life. For over century, social scientists have conducted surveys to learn about human behavior. However surveys are plagued with issues such as bias, sparsity of data and lack of continuity between discrete questionnaires. Over the last two decades there has been significant amount of research attempting to address these issues by collecting rich behavioral data. Researchers have examined previously about the mobility modeling and most work focus on using the observed user behavior characteristics to design realistic and practical mobility models [6],[9],[11] and [12].

Generally, mobility models are either synthetic or trace-driven. The most common models include random mobility generators, including random direction, random waypoint and random walk. These models are useful for simple initial evaluations, but not based on real traces and have been shown to lack important characteristics of user location preferences and periodicity [6]. Some community models [7] and [13] represent social connections based on social network theory, but do not capture mobility related preferences of similarity. Trace-based mobility models by contrast, derive their parameters from analysis of measurements and can reproduce user preferences [6].

More recent work Thakur et. al. [4] use similarity index to quantitatively compare the mobility profiles and compared similarity characteristics of the traces to existing common and community based mobility models to capture similarity. The user profiling problem has been addressed in the past as a clustering problem where the objective was to group users who exhibit similar behavior [8]. Keralapura et. al. [5] found that there exists a number of prevalent and distinct behavior patterns that persist over time, suggesting that user browsing behavior in 3G cellular networks can be captured using a small number of co-clusters.

In web usage and preference analysis, web access information, weblogs and session information for navigational history, experience and location are mainly used to 'simulate' web user behavior [14], [15] and [16]. Recently [3], explores the dynamics of Web user behaviors, such as detection of periodicities and surprises. They develop a novel learning and prediction algorithm, which can be used to enhance query suggestions, crawling policies and result ranking.

3   Data Set and Initial Findings

We first wants to analyze the daily traffic behavior, peak times of the mobile Internet usage and then engage some high-level investigation about the network structure. This will help us to further explore and get inside of the mobile Internet users network.

3.1   Mobile Data

The data we use is the mobile Web usage log (HTTP request log) of the cellular network (including 2G and 3G) of an mobile operator. It contains the traces of thousands mobile Internet users, having attributes like International Mobile Subscriber Identity[1](IMSI) number, (TCPIP and HTTP) timings and the Host address e.t.c. The dataset contains total of 1012728 users, 165055773 user traces and 107784 websites over a span of one week (Sept. 22-28, 2012). It covers the geographical range of one of the main city of China. We have filtered the records of most active users[2] and discard the records of the inactive users.

3.2   Daily Usage

We now analyze the data collected for one week and see the daily traffic spectrum at different timings of the day. Considering that consecutive daily traffic volumes well represents users' daily behavior on the mobile internet, we assess the daily traffic volume variations for the week long observation period by calculating the TCPIP connection and disconnection timings. The data is plotted by every one hour and the result is depicted in Fig. 1. According to the study of the traditional Internet traffic [17], the variation of traffic volumes in a day should match our daily life schedule.

As can be seen in Fig. 1, there is sharp drop of Traffic after 12:00 p.m, an upward traffic trend between (6:00-11:00) a.m and there are three traffic peaks every day. The peaks regularly occur between 7:00 to 11:00, between 3:30 to 5:30 and between 9:00 to 12:00 at night time. More over these peaks are not obvious on weekends. We also evaluate the number of users online in every one hour and the result is also shown in Fig. 1 as well. Clearly, there are also several peaks of the number of users, similar trend is observed in this case also. In our point of view, people tends to use the mobile Internet using the mobile phones when they do not have access to conventional wired computers or it is not convenient to do so. Moreover the times of the three peaks coincide with the times when people go to work, get off work, and ready to sleep, respectively. The wireless, small, light and portable nature of mobile terminals provides a perfect means to access the Internet at these timings.

3.3   User Network

Mobile Internet user network is a network where the vertices are mobile users who access the different websites using their mobiles at different timings of the day and the edges represent the visiting of the common website between two users. If two users are accessing the same website or URL, i.e. they have some common like between their interests.

Mobile Internet user network can be formally defined as: P is the set of all websites accessed by all the users in the dataset. For each website p, we define

---

[1] Unique identification associated with all cellular networks.
[2] Users which have atleast 20 records in whole week.

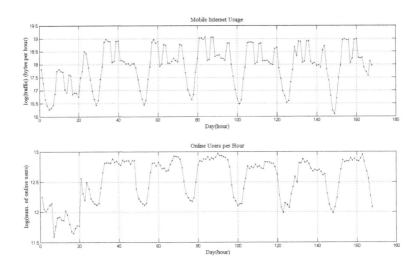

Fig. 1. Hourly Traffic variation for the whole week

the set U_p of all users who visits it. We select the set of users of website p as the subset $M_p \subseteq U_{p \in P}$. Then the mobile Internet user network graph $G = (V, E)$ is defined as follows: where the set $V = U_{p \in P} M_p$ is the vertices and the set $E = \{(u_1, u_2) | \exists p \in P : u_1, u_2 \in M_p\}$ is edges. Generally it is assumed that G is finite, connected and symmetric.

Clustering Coefficient: The Clustering coefficient is the measure of transitivity in the network especially on the social network [18]. It is observed by plotting the average clustering of the whole network as shown in the Fig. 2 with average clustering coefficient 0.7798 respectively. The value of the average clustering decreases with the increase in degree of the node in the graph. It is similar like in other real networks the average clustering decreases as the node degree increases because of the fact that low-degree nodes belong to very dense sub graph and those sub graphs are connected through the hubs.

In our analysis the decreasing trend of the clustering coefficient can also be seen as a symptom of the growing centralization of the network. This shows that a strong centralization of the network around some common interests (hubs/stars) due to visiting the same popular websites.

Network Community Profile: We have plotted the network community profile plot (NCP), which measures the quality of the best possible community in a large network, as a function of the community size. In [2] Jaewon, finds that the conductance and triad-participation-ratio consistently give the best performance in identifying ground-truth communities. In the Fig. 3, we have the two general observations. First, minima in the NCP plot, i.e., the best low-conductance cuts of a given size, correspond to communities-like sets of nodes. Second, the plot is gradually sloping downwards, meaning that smaller communities can be

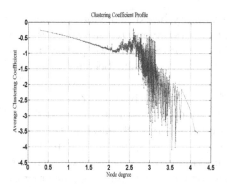
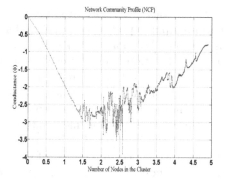

Fig. 2. Clustering Coefficient of Network   **Fig. 3.** Network Community Profile

combined into larger sets of nodes that can also be meaningfully interpreted as communities. Also it is steadily increasing for nearly its entire range and by applying the Local Spectral Algorithm, we have observed that the slope of the plot is sloping downward up to a certain size scale, which is roughly 400 nodes. It is the global minimum and according to the conductance measures, this set of nodes achieving local minima are the best communities. Above the size scale of roughly 400 nodes, the plot gradually increases over several orders of magnitude. The best communities in the entire graph are quite good (in that they have size roughly $10^{2.6}$ nodes and conductance scores less than $10^{-3.5}$).

Summary: Initial results confirms the presence of several communities in the network and the magnitude of the conductance tells us how well clusters of different sizes can be separated from the rest of the network. The clustering coefficient indicates the strong centralization of the network around some common interests due to the similar navigation patterns and preferences.

4   Mobile Communities Identification

The congregation of mobile agents with similar characteristic patterns naturally develops mobile communities in the networks. Upon reflection it should come as no surprise that these characteristics in particular also have a big impact on the overall behavior of the system. One major observation is that people demonstrate periodic reappearances and visit the same websites periodically, which in turn breeds connection among similar instances. Thus, people with similar behavioral and navigation pattern principle tie together.

4.1   Methodology

To identify the communities with the same best size as determined above, a very well-known approach is to cluster entities (e.g., users, websites) with similar characteristics. Therefore to study the community based behavior, we use their spatio-temporal preferences and preferential attachment to the websites they visited and duration (time) spent on these websites. Here we prefers to use the time duration spent by each user on the every website because it directly reflects his interests and preference in quantitative form. We devise a scalable representation of this information in the form of the activity matrix.

Each individual column corresponds to a unique website in the trace and each row represents is an n-element interest vector, where each entry in the vector represents the fraction of online time the mobile user spent on that particular website during a certain period of time. This time period can be flexibly chosen, such as an hour, a day or week, etc. Thus for t distinct websites (the fraction of time spent) and for n users, we generate an n-by-t size activity matrix. This matrix summarizes all the activity of the online mobile user during the specific period of time and now it can be used for the multivariate data analysis.

For a succinct measure of the mobile user behavior, we capture the dominant behavioral patterns by using the Principal Component Analysis (PCA) of the activity matrix. Since we already have the rough estimate of best community size and could also infers the number of best communities from it. We have used the Principal Component Analysis (PCA) for dimension reduction and then used the k-means clustering to uncovers the various clusters present in the form of communities. The PCA uses a vector space transform to reduce the dimensionality of large data sets. Here we use Singular Value Decomposition (SVD) of the activity matrix to perform the factorization, which converts high dimensional and high variable data to lower dimensional space. It will extract the dominant trends and expose the internal structure of the original data more clearly. The SVD of a given matrix A can be represented as a product of three matrices:

$$A = P \wedge P^t \tag{1}$$

Where $PP^t = P^tP = I$, where P is the orthogonal matrix whose columns are orthonormal eigen vectors, $\wedge$ is a diagonal matrix containing the square roots of eigen values of matrix A in descending order of magnitude, and the transpose of an orthogonal matrix i.e P^t. The eigen behavior vectors of the P summarize the important trends in the original matrix A.

It has been observed that SVD achieves great data reduction on the original activity matrix and more than 95% power for most of the users is captured by seven components of the association vectors. Therefore in this case we have chosen the first seven principal components for multivariate analysis. From this, we infer that user's few top websites preferences are more dominant than the remaining one.

4.2   Clustering Approach

Initially we identify a new set of orthogonal coordinate axes (called first principal axes) through the data. This is achieved by finding the direction of maximal variance through the coordinates in the t dimensional space (in the form of n x

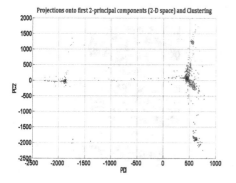

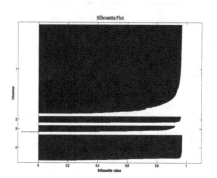

Fig. 4. Projection using 2-principal components

Fig. 5. Silhouette value plot

t activity matrix). Then we obtains a second principal coordinate (axis) which
is both orthogonal to the first principal axes, and is the next best direction
for approximating the original data (i.e. it finds the direction of second largest
variance in the data, chosen from directions which are orthogonal to the first
principal component). We now have two orthogonal principal components defining a plane which we can project our coordinates down onto this plane. After
the projecting the coordinates using the first 2-principal components, we use
k-means clustering to clusters the data points as shown in the Fig. 4. From the
figure it is clearly shown that four clusters are clearly visible, which approximately 77% of the variance in the data is accounted for by the first 2-principal
components. The silhouette plot as shown in Fig. 5, we can see that most points
in all the clusters have a large silhouette value, greater than 0.8, indicating that
those points are well-separated from neighboring clusters. However, each cluster
also contains a few points with low silhouette values, indicating that they are
nearby to points from other clusters.

Now we have added the contribution of the third principal component, and we
now have three orthogonal principal components. The data points are projected
using these principal components and again the k-means clustering is used to
clusters the data points. Now this time seven clusters are clearly visible as shown
in Fig 6. It is significant and provides more insight into the existing Mobile Internet users communities, having the common interests and preferences. Moreover
the corresponding silhouette plot shown as Fig. 7 explained that the clusters are
well separated from the neighboring clusters and some cluster also contains a
few points with low silhouette values, indicating that they are nearby to points
from other clusters. But approximately 80% of the variance in the data is accounted for by the first 3-principal components, so we have to further include
contributions of more principal components.

Finally we consider the first 7-principal components, which approximately accounted the 95% of the variance and by using again the k-means clustering to
clusters the data points to uncover the mobile internet user communities present

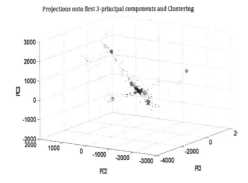

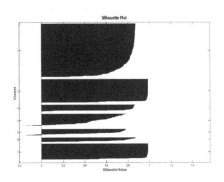

Fig. 6. Projection using 3-principal compo-   **Fig. 7.** Silhouette value plot
nents

in the network. As a result we have got the ten different clusters or communities. These communities have similar behavior like accessing the similar websites
around specific topics, interests and affiliations with the similar navigation patterns. We have considered up to 20-principal components, but did not find any
significant improvement in the results in comparison with 7-principal components.

5   Global Analysis Results

In this section we will discuss the results of ten clusters[3]. The approach confirms
the best community size approximation and able to group the users with similar
navigation pattern and the interests into the clusters. At a high level, we observe
that the Chinese community is clustered around some popular domains, which
can be considered as the nucleus of various interests and topics. A deeper look
into the clusters reveals some interesting trends and behaviors. The overall most
popular activities among the Chinese communities (which the data suggests is
the world's largest mobile Web community) are: instant messaging (chatting),
news/microblogging, web literature, social networking, search, games etc. Finally
we summarized the most frequent keywords used with the domain name and the
induced topics are as shown in the following Table 1.

A deeper look into the clusters reveals some interesting trends. Cluster 3 shows
users who mostly just utilize the mobile Internet for search via 'baidu.com' and
'uc.com' and visits the 'mobilesearch-update.com' for probably the Nokia Maps
including the directions-data updates. Cluster 10, by contrast, shows users frequently go to 'apple.com' and 'andriod.com' websites for the software updates of
their mobiles. Moreover these users are commonly interested in reading books,
jokes and other material through their mobiles. The clusters 2 and 4 have shown
that the qq, renren, pengyou and the weixin are the most popular social networking websites among the users. By doing the similar most frequent keywords

---

[3] Here we use the best community size for the possible number of clusters.

Table 1. Major related websites clustered together

No	Topic	Keywords used	Domains
1	NEWS/Micro-blog	news, video, book1, finance, forecast, learn, eng, read, mail.	sina.cn, sohu.com, chinanews.com, sinaimg.cn, soso.com, fivesky.net
2	Instant messaging, Chatting/Dating	store, photo, images, upload, info, blog, register, new, search.	qq.com, renren.com, pengyou.com, zhenai.com, wechatapp.com
3	Search	wapp, mov, wap, wapiknow, imgt4, usec, ucdesk, mynavi, wap3, uc6, ucsecc, update.	baidu.com, uc.com, soso.com, sogou.com, ucweb.com, qunar.com, youku.com, mobilesearch-update.com
4	Social Network and Games	talk, blog, 3g, api, game, ic, play, register, live, hao.	renren.com, douban.com, gamedog.cn, weibo.cn, zhenai.com, 51.com, uc.cn
5	News Media	sports, iblog, weather, travel, nba, story, content, cache, tv.	sohu.com, sina.cn, xinhuanet.com, ifeng.com, news.cn, 6xgame.com
6	Micro-blogging	api, sports, book, stock, mobile, app, log.	weibo.cn, sina.cn, qq.com, soso.com, sogou.com, gtimg.cn, baidu.com,
7	Social Search	wap, news2, music, mmsns, qzonestyle, picwap.	soso.com, 51.com, qpic.cn, umeng.com, nipic.com, baidu.com, weibo.cn
8	Videos, Music and Reviews	music, mp3, fm, audio, wireless, datamobile, fashion2, download.	douban.com, funbox.cc, 163.com, i139.cn, jiayuan.com, kaixin001.com, sohu.com
9	Service Providers Portals	api, video, pict, info, mob, wap, mbs.	monternet.com, 10086.cn, tencent.com, boc.cn, chinamobile.com, fetion.com
10	Miscellaneous	read, book, upload, info, look, download, buy, joke, iphone.	apple.com, android.com, kdbooks.cn, 91woku.com, jokeji.cn, bookbao.com

analysis as shown in Fig. 8, it has been observed that uploading the images, sharing the photos, voice messaging, chatting, information sharing and current status update etc are the popular activities done through these applications. Moreover the results shows that the most popular news reading websites among the mobile Internet users of the cluster 1 and 5 includes the sina, sohu, ifeng and the xinghuanet. The reason may be behind is that the users can view the customized news as per his interests.

Another very important observation is that some websites/domains like qq, sina, weibo etc are repeating in different clusters. For example the most popular news web-site is sina.com, which is present in the cluster 1, 5 and 6 simultaneously. Now the question is what does sina.com means in cluster 1, 5 and 6? Therefore by further investigating the sina.com domain url's in these clusters, we have seen that there are several different keywords used in combination with this domain. We have considered the position and computed the term frequency of each keyword in all the url's related to that domain. The results are as shown in the following Fig. 8. It clearly shows that users visit different pages of the sina website based on their interests and preferences. Some users in cluster 1 often use the 'news.sina.com.cn', technology, military news updates and the cluster 4 users are more frequently visit the 'blog.sina.com.cn', weather update, finance and stock news updates. This means that the users of the cluster 4 are more interested in reading and contributing the blogs and the users of the cluster 1 are also more motivated towards news application which will display the important news in proper format. Whereas the cluster 6 people are more interested in reading about the sports, games, videos, jokes and the mobile news.

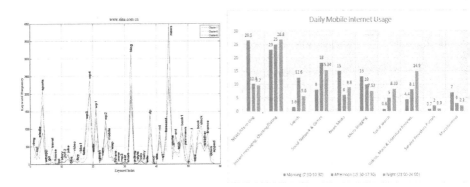

Fig. 8. Multi keywords with sina.com   **Fig. 9.** Interests with timeline of the day

Now we have used the clusters as determined above and work out the various interests of the users on the timeline of the day. From the initial observation as shown in the Fig. 1, we have concluded the three peak times of the day when the internet traffic is maximum. In the morning we have considered the time (7:30 - 10:30) a.m, in the afternoon time (15:30 - 17:30) p.m and the night time (21:00 - 24:00) p.m, the respective results are as shown in Fig. 9. It is very interesting to see that the users are interested to read the news, microblogs in the morning time, whereas listening music, watching videos and reading books are mostly done at the night time. Another very common observation is that the users use the instant messaging or dating application like the qq.com irrespective of the day and time. The figure confirms that the users login their messenger right from the morning till evening as a background activity and all the time they are connected with their social network. Mostly the users do the search related activities at the day time as compared to the other timings of the day. The social network applications like renren, douban.com e.t.c are mostly captures the day and night timings. This shows various interests and preferences of the mobile Internet users on timeline of the day.

6   Conclusion

This study is motivated by the need for a paradigm shift that is data-driven to develop realistic models. We provide a systematic method to process and analyze the largest mobile trace with millions of records of Internet usage including thousands of users. The initial results imply the evidence of strong centralization and presence of several communities in the network. Our work confirms that the magnitude of the conductance provides the good estimate of the community size, which can be used to determine the number of communities present in the network. Our method reduces the dimensions of the input, carries out the clustering of the distinct web access profiles, which identifies the communities of users having similar interests. We have shown the user's interests and preferences with the timeline of the day which is useful in defining the adaption mechanism that will target a different user interface experience in Web environments on mobile or hand held devices.

References

1. Ma, H., Cao, H., Yang, Q., Chen, E., Tian, J.: A habit mining approach for discovering similar mobile users. In: Proceedings of the 21st International Conference on World Wide Web, pp. 231–240. ACM (2012)
2. Yang, J., Leskovec, J.: Defining and evaluating network communities based on ground-truth. In: Proceedings of the ACM SIGKDD Workshop on Mining Data Semantics. ACM (2012)
3. Radinsky, K., Svore, K., Dumais, S., Teevan, J., Bocharov, A., Horvitz, E.: Modeling and predicting behavioral dynamics on the web. In: Proceedings of the 21st International Conference on WWW, pp. 599–608. ACM (2012)
4. Thakur, G.S., Helmy, A., Hsu, W.-J.: Similarity analysis and modeling in mobile societies: the missing link. In: Proceedings of the 5th ACM Workshop on Challenged Networks, pp. 13–20. ACM (2010)
5. Keralapura, R., Nucci, A., Zhang, Z.-L., Gao, L.: Profiling users in a 3g network using hourglass co-clustering. In: Proceedings of the 16th Annual International Conference on Mobile Computing and Networking, pp. 341–352 (2010)
6. Hsu, W.-J., Spyropoulos, T., Psounis, K., Helmy, A.: Modeling spatial and temporal dependencies of user mobility in wireless mobile networks. IEEE/ACM Transactions on Networking 17(5), 1564–1577 (2009)
7. Garbinato, B., Miranda, H., Rodrigues, L.: Middleware for Network Eccentric and Mobile Applications. Springer (2009)
8. Antonellis, P., Makris, C.: XML Filtering Using Dynamic Hierarchical Clustering of User Profiles. In: Bhowmick, S.S., Küng, J., Wagner, R. (eds.) DEXA 2008. LNCS, vol. 5181, pp. 537–551. Springer, Heidelberg (2008)
9. Jain, R., Lelescu, D., Balakrishnan, M., Model, T.: a model for user registration patterns based on campus WLAN data. Wirel. Netw. 13(6), 711–735 (2007)
10. Bai, F., Helmy, A.: A Survey of Mobility Modeling and Analysis in Wireless Adhoc Networks (October 2006)
11. Lelescu, D., Kozat, U.C., Jain, R., Balakrishnan, M.: Model T++: an empirical joint space-time registration model. In: Proceedings of the 7th ACM MOBIHOC, Florence, Italy. ACM (May 2006)
12. Kim, M., Kotz, D., Kim, S.: Extracting a Mobility Model from Real User Traces. In: Proceedings of the IEEE INFOCOM 2006, Barcelona, Spain (April 2006)
13. Musolesi, M., Mascolo, C.: A community based mobility model for ad hoc network research. ACM REALMAN (2006)
14. Facca, F., Lanzi, P.: Mining interesting knowledge from weblogs: a survey. Data and Knowledge Engineering 53(3), 225–241 (2005)
15. Flesca, S., Greco, S., Tagarelli, A., Zumpano, E.: Mining user preferences, page content and usage to personalize website navigation. World Wide Web, 317–345 (2005)
16. Srivastava, J., Cooley, R., Deshpande, M., Tan, P.: Web usage mining: discovery and applications of usage patterns from Web data. ACM SIGKDD, 12–23 (2000)
17. http://www.caida.org/data/realtime/passive/
18. Watts, D., Strogatz, S.: Collective dynamics of small-world networks. Nature 393, 440–442 (1998)

Tracking Drift Types in Changing Data Streams

David T.J. Huang[1], Yun Sing Koh[1], Gillian Dobbie[1], and Russel Pears[2]

[1] Department of Computer Science, University of Auckland, New Zealand
{dtjh,ykoh,gill}@cs.auckland.ac.nz
[2] School of Computing and Mathematical Sciences, AUT University, New Zealand
rpears@aut.ac.nz

Abstract. The rate of change of drift in a data stream can be of interest. It could show, for example, that a strand of bacteria is becoming more resistant to a drug, or that a machine is becoming unreliable and requires maintenance. While concept drift in data streams has been widely studied, no one has studied the rate of change in concept drift. In this paper we define three new drift types: relative abrupt drift, relative moderate drift and relative gradual drift. We propose a novel algorithm that tracks changes in drift intensity relative to previous drift points within the stream. The algorithm is based on mapping drift patterns to a Gaussian function. Our experimental results show that the algorithm is robust and achieving accuracy levels above 90%.

Keywords: Data Stream, Relative Drift Types, Gaussian Curve.

1 Introduction

A data stream is an ordered sequence of instances that are potentially unbounded in size making it impossible to process by most traditional data mining approaches. As a consequence, traditional data mining techniques that make multiple passes over data or that ignore distribution changes are not applicable to data streams. The processing of data streams generates new requirements for algorithms, such as constrained memory use, restricted learning and testing time, and one scan of incoming instances. Furthermore, due to the non-stationary nature of data streams, the target concepts tend to change over time, a phenomenon known as concept drift. Examples where real life concept drift can be observed include email spam categorization, weather prediction, monitoring systems, financial fraud detection, and evolving customer preferences.

An important characteristic of unbounded data streams is that changes in the underlying distribution can signal important changes over time. Many drift detection techniques have been proposed [4,1,6,3] to detect these changes. Most of these techniques monitor the evolution of a test statistic between two distributions: between past data in a reference window and the current window of the most recent data points [6]. Overall these drift detection techniques analyze the change in distribution within a stream. However, these techniques do not consider whether such change is a rapid sudden change over a short period of time, or a slow gradual change that has occurred over a long period of time.

H. Motoda et al. (Eds.): ADMA 2013, Part I, LNAI 8346, pp. 72–83, 2013.

These approaches operate under the assumption that drift occurs on a continuous basis, thus avoiding the complexity of trying to determine the type of drift or the rate at which drift occurs. Several classification algorithms that cope with concept drift have been proposed; however, most of the techniques specialize in detecting change but do not look at the change in detail. In this research we concentrate on categorizing the drifts that have been detected by a drift detector. We are not attempting to build a new drift detector, as current research in drift detection has solved the problem of drift detection to some extent.

The main contribution of this paper is our novel technique that analyzes drifts in data streams based on prior knowledge of the stream and categorizes the type of drift that has occurred by comparing it to previous drift points. The relative drift type assists users in making important decisions, knowing whether or not a change occurred rapidly or gradually will give a further dimension to the decision making process. An example where knowing and categorizing the type of drift can provide substantial information is bioinformatics. Consider the case where the user is concerned with antibiotic resistance. A concept drift in this case could represent that a strand of bacteria is now more resistant towards one of the commonly prescribed antibiotics, Amoxicillin. Current drift detection algorithms only identify whether a concept drift has occurred. However, only identifying that a concept drift has occurred does not provide the user with any further information regarding the regularity of the drift. For instance, the user does not know the intensity of the drift or how fast the drift occurred. If the intensity of the drift was small, then it could signal an evolutionary adaptation of the bacteria to the antibiotic which may be considered less serious; whereas, if the drift occurred with a very large intensity, this could mean that the bacteria is rapidly building up resistance to the antibiotic and could signal a serious phenomenon that needs immediate attention.

In Section 2 we discuss previous research. Section 3 describes the preliminaries and definitions of the problem we address. In Section 4 we provide a relative drift categorization technique. Section 5 presents thorough experimental evaluations of our contributed work and Section 6 concludes the paper.

2   Related Work

A drift point corresponds to a statistically significant shift in a sample of data that initially had a single homogeneous distribution. All current concept drift detection approaches formulate the problem from this perspective but the d models and algorithms differ greatly in detail. Sebastiao and Gama [8] present a concise survey on drift detection methods. They point out that methods used fall into four basic categories: Statistical Process Control (SPC), Adaptive Windowing [4,1,3], Fixed Cumulative Windowing Schemes [6] and finally other classic statistical drift detection methods such as the Page Hinkley test, and support vector machines. All of these techniques concentrate on detecting drift quickly with a minimum possible false positive rate and detection delay. There has been some research on drift types [2,10,7]. In [7] the authors introduced a metric that is an indicator of amount and speed of change between different concepts in data and in [10] the authors categorized concept drift into two scenarios which they termed: Loose Concept Drifting (LCD) and Rigorous Concept Drifting (RCD) and then proposed solutions to handle each. Even though the topic of drift types have been looked at, none of these works automatically classifies drifts into different types like our proposed technique does.

3   Preliminaries

Concept Drift Detection Problem: Let us frame the problem of drift detection and analysis more formally. Let $S_1 = (x_1, x_2, ..., x_m)$ and $S_2 = (x_{m+1}, ..., x_n)$ with $0 < m < n$ represent two samples of instances from a stream with population means μ_1 and μ_2 respectively. The drift detection problem can be expressed as testing the null hypothesis H_0 that $\mu_1 = \mu_2$, i.e. the two samples are drawn from the same distribution against the alternate hypothesis H_1 that they arrive from different distributions and $\mu_1 \neq \mu_2$. In practice the underlying data distribution is unknown and a test statistic based on sample means needs to be constructed by the drift detector. If the null hypothesis is accepted incorrectly when a change has occurred then a false negative is said to have taken place. On the other hand if the drift detector accepts H_1 when no change has occurred in the data distribution then a false positive is said to have occurred. Since the population mean of the underlying distribution is unknown, sample means need to be used to perform the above hypothesis tests. The hypothesis tests can be restated as the following. We accept hypothesis H_1 whenever $Pr(|\hat{\mu}_1 - \hat{\mu}_2|) \geq \epsilon) > \delta$, where δ lies in the interval $(0, 1)$ and is a parameter that controls the maximum allowable false positive rate, while ϵ is a function of δ and the test statistic used to model the difference between the sample means. Most drift detection approaches rely on well established test statistics for the difference between the true population and sample mean. A number of bounds exist that do not assume a particular data distribution e.g. Chernoff Inequality and Bernstein Inequality. Due to its generality, Hoeffding's Inequality has been widely used in data stream mining [3]. Hoeffding inequality [5] states that with probability at least $1 - \delta$:

$$Pr\left(\left|\frac{1}{n}\sum_{i=1}^{n} X_i - E[X]\right| > \epsilon\right) \leq 2\exp\left(-n\epsilon^2\right)$$

where $\{X_1, ..., X_n\}$ are independent random variables and $E[X]$ is the expected value or population mean. In summary, this statistical test for different distributions in two subwindows simply checks whether the observed mean in both sub-windows differs by more than the threshold ϵ .

In all drift detection algorithms a detection delay is inevitable and is consequently an important performance measure. Detection delay is the distance between c and $n^{'}$, where c is the true cut point and $n^{'}$ is the instance at which change is actually detected. Thus detection delay is determined by $\left(n^{'} - (c+1)\right)$.

Types of Drift: The fundamental problem in learning to recognize drift is how to identify, in a timely manner, whether new incoming data is inconsistent with the current concept in the stream. Because of this, several criteria are used to measure concept drift such as rate of change or speed. Speed is the inverse of the time taken for the old concept to be completely replaced by the new concept. There are three forms that exist as shown in Figure 1. Specifically they are, gradual, moderate, and abrupt drift. For a gradual drift the time step taken for

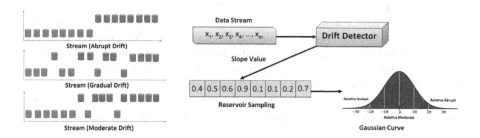

Fig. 1. Traditional Drift Types   **Fig. 2.** Overview

the change of an old concept to be replaced by a new concept is larger when compared to the other two drift types. Gradual drifts are associated with a slower rate of changes. Gradual drift refers to a transition phase where examples from two different distributions S_j and S_{j+1} are mixed. As time goes on, the probability of observing examples from S_j decreases, while examples from S_{j+1} increase. For an abrupt drift the change is immediate or sudden for the old concept to be completely replaced by a new concept. Abrupt drift occurs when at a specific moment in time the source distribution S_j is suddenly replaced by a different distribution in S_{j+1}. A moderate drift is an intermediate category between abrupt and gradual drift types.

4   Relative Drift Type Detector

In this section we present our drift categorizer. We start by giving an overview of how the categorizer works alongside some of the technical details used. Overall the categorizer works in two phases as shown in Figure 2. The data stream is first processed by passing it through a drift detector. In our experiments, we use the state-of-the-art drift detector ADWIN2 [3]. However, the categorizer is not tightly coupled to ADWIN2 and hence can be interchanged with any other drift detector. In Phase 1, when a new drift point, say x, is detected, we compute the mean rate of change of the stream segment that has arrived since the last drift point. The mean rate of change is then stored in a data repository. This process continues for each drift point detected. As the size of a data stream is infinite the number of samples in the repository will potentially be infinite in number. Thus, in order to capture a representative set of samples in a finite sized data repository, we use the reservoir sampling algorithm [9]. The reservoir sampling algorithm is an effective and elegant method of obtaining a random sample when the population size cannot be specified in advance or when the population is not bounded in size. Thus, our gathering of samples in the data stream is ideally suited to the use of reservoir sampling. In reservoir sampling, the reservoir has a maximum capacity value where all the incoming samples are stored until the capacity is reached. Newly arriving samples will replace existing ones in the reservoir with monotonically decreasing probability once the reservoir reaches maximum capacity. In Phase 2, a sample representing the mean rate of change computed in Phase 1 is fitted to a Gaussian distribution which is asymptotically approached by the set of samples already in the reservoir. We consider the critical regions defined on the Gaussian distribution through the use of a critical value γ. We will discuss the Gaussian distribution in Section 4.2 and the use of the Gaussian distribution to categorize drift in Section 4.3.

In general our technique determines the type of drift in a relative sense based on the data stream's previous history and categorizes the drift into three types.

4.1   Slope and Magnitude

In this section we define two concepts: the slope of change and the magnitude of change.

Definition 1. *(**Magnitude of Change**) The magnitude of change (ϕ) is defined as $f(t_c) - f(t_p)$ where t_c is the time step at which the current drift occurs and t_p denotes the time at which the previous drift occurs, where the function $f(t)$ defines the trajectory of the population mean over time.*

Magnitude represents the change in population mean (μ) given two time steps. It represents a change between two points but does not account for the speed of change that occurs.

Definition 2. *(**Slope of Change**) The slope of change is defined as $f'(x) = |\frac{f(t_c)-f(t_p)}{t_c-t_p}|$ with the same meaning attached to the f and t variables as in the definition of magnitude above.*

Essentially, the slope as defined above represents the mean rate of change defined over the time interval $[t_p, t_c]$. A drift with a higher slope value equates to a drift with a larger rate of change. This is characterized by being a more *abrupt* drift. On the other hand, a drift with a smaller slope value is associated with a more *gradual* drift. Consider two different drifts, the first has a magnitude of change 0.5 which occurs over 1000 steps and the second has a magnitude of change 0.5 but occurs over 5000 steps. In these two cases, they have the same magnitude value but the first drift has a slope of 0.0005 and the second drift has a slope of 0.0001. The first drift is considered more abrupt. In the next subsection we discuss the Gaussian distribution used in our approach.

4.2   Gaussian Distribution

According to the Central Limit Theorem the characteristics of the population of the means is determined from the means of an infinite number of random samples, whereby all of them are drawn from a given parent population. The Central Limit Theorem asserts that regardless of the distribution of the parent population the distribution of sample means will approximate a normal distribution as the size of samples used in the computation of the sample means increases. Thus we chose to model the distribution of slopes using a Gaussian distribution.

Our algorithm provides a relative and adaptive measure that is capable of categorizing drift points for various data distributions. The random replacement of samples in the reservoir also means that the algorithm is capable of adapting to distributional changes within the stream. If the current stream segment experiences a shift in volatility, the reservoir will react by storing a greater proportion of new (slope) samples due to the fact that a greater proportion of old samples will be flushed out.

4.3   Types of Relative Drift

From the samples in the reservoir we can fit a Gaussian probability density function. By identifying the region where the new drift point falls on the Gaussian, the drift point can be categorized as being abrupt, moderate or gradual in a relative sense.

Definition 3. (Relative Abrupt Drift) *Given a critical value γ and a Gaussian function $G(\phi)$ where ϕ is the slope variable, a drift is a relative abrupt drift when $\int G(\phi) \geq 1 - \gamma$.*

Here $\int G(\phi)$ represents the area under the curve where the slope ϕ is located. For example, given a hundred slope samples in the range $[0.4, 0.6]$ when a new drift arrives with a slope of 0.9 it will be categorized with a given confidence value γ by evaluating $\int G(\phi)$ and comparing with γ. If $\gamma = 0.05$, then the new drift point with slope value 0.9 is a relative abrupt drift because $\int G(0.9) \geq 1 - 0.05$. Figure 3 shows the critical regions of a Gaussian distribution formulated.

Definition 4. (Relative Moderate Drift) *A drift is a relative moderate drift when $\int G(\phi) < 1 - \gamma$ and $\int G(\phi) > \gamma$.*

Using the previous example with a reservoir of a hundred samples in the range between 0.4 and 0.6 (i.e. $\{0.45, 0.4, 0.5, 0.6, ..., 0.55\}$) and $\gamma = 0.05$, if we now have a new drift with slope of 0.5, this new drift is considered a relative moderate drift in the current stream because $\int G(0.5) < 1 - 0.05$ and $\int G(0.5) > 0.05$.

Definition 5. (Relative Gradual Drift) *A drift is a relative gradual drift when $\int G(\phi) \leq \gamma$.*

In the example, if we have a new drift point with slope value 0.1, this drift will be considered a relative gradual drift in this current stream because $\int G(0.1) \leq 0.05$.

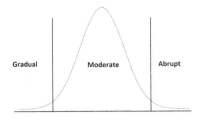

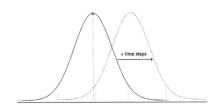

Fig. 3. Critical Regions   **Fig. 4.** Right Shift of Gaussian Curve

When the stream undergoes a distributional change, the corresponding probability density function will undergo a corresponding change. For instance, consider our example of a reservoir with a hundred slope values in the range 0.4 to 0.6 (i.e. {0.45, 0.4, 0.5, 0.6, ..., 0.55}). Here a drift with a slope value of 0.5 is classified as a relative moderate drift. Assume that after a period of time the stream has changed in distribution and the reservoir now holds a hundred samples in the range of 0.8 to 1.0 (i.e. {0.9, 0.95, 0.85, 0.95, ..., 0.85}). This corresponds to a right shift in the Gaussian curve as illustrated in Figure 4.

A right shift occurs when the mean slope of the drifts is tending towards a higher value. In the example, a drift with slope of 0.5 that would have previously been categorized as relative moderate will now be categorized as relative gradual. This is illustrated by the spot in Figure 4. Similarly, a left shift represents the fact that the slope of the drifts are tending towards a smaller value and a drift that would have been previously categorized as relative moderate will now be categorized as being relative abrupt.

One way of finding whether the shift in the Gaussian curve is significant is by keeping a record of the mean slope values stored in the reservoir across the stream and passing these stored values as a second order stream into a change detector (i.e. ADWIN2) to find changes. Another method is through the use of suitable statistical significance tests.

5   Experimental Results

In this section we describe both the data generator we proposed in Section 5.1 and the various experiments we carried out in Section 5.2.

5.1   Synthetic Data Generator

Various data generators have been used in previous data stream research. These generators, however, are unable to generate data with very specific drift magnitudes and patterns. In order to examine our approach fairly, we have created a data generator that injects drifts with controllable magnitudes and at controllable intervals.

The generator produces a raw binary data stream in several steps. The first step is drift magnitude selection. During this step, a random magnitude value between 0.0 and 1.0 is selected according to the specified distribution. The drift is generated through a stationary Bernoulli process. At each step the current mean of the stream is used as the probability in the stationary Bernoulli sequence. After each step the mean (probability) is adjusted by a slope value. In addition to generating linear drifts, we have incorporated a Gaussian perturbation of the drift. The perturbation allows the drifts to better resemble a real world drift by creating fluctuations. A Gaussian perturbation is achieved by adjusting the slope at each step based on a Gaussian probability. The formula we used for the Gaussian perturbation is: perturbed slope $= slope + (slope * \beta * \alpha)$ where β is the perturbation parameter and α is a number randomly generated from a Gaussian distribution with mean of 0 and standard deviation of 1.

5.2   Experiments

The accuracy of our algorithm is evaluated on the basis of different stream variables: parent (raw) data distribution, change detection delay classification noise, change detection delay, and fluctuation due to perturbation. In all experiments our drift categorizer ran with a runtime of less than 2 milliseconds and used a constant memory of 1086 bytes. The algorithms were coded in Java and ran on an Intel Core i5-2400 CPU @ 3.10 GHz with 8GB of RAM running Windows 7. In the fitting process, the standard normal function is used.

Table 1. Noise Study: Uniform Distribution

γ	Noise				
	0%	1%	5%	10%	20%
0.05	0.6554	0.6528	0.6321	0.6066	0.5563
0.10	0.8600	0.8512	0.8216	0.7821	0.6997
0.15	0.9486	0.9383	0.9029	0.8566	0.7621
0.20	0.8807	0.8699	0.8390	0.7991	0.7138
0.25	0.7850	0.7784	0.7508	0.7175	0.6489

Table 2. Noise Study: Uniform Distribution with ADWIN2

γ	Noise				
	0%	1%	5%	10%	20%
0.05	0.6618	0.6584	0.6388	0.6137	0.5593
0.10	0.7668	0.7604	0.7342	0.7031	0.6349
0.15	0.8111	0.8019	0.7735	0.7374	0.6646
0.20	0.7802	0.7705	0.7445	0.7119	0.6436
0.25	0.7158	0.7081	0.6860	0.6589	0.5983

Noise Study. The drift categorizer that we designed takes as input a stream of classification decisions made by a classifier. Each decision results in a binary 0 if the classification was correct and binary 1 if the decision is incorrect. Noise in the raw data submitted to the classifier will not only affect the accuracy of the classifier but will also affect the accuracy of the drift detector. We simulate the effects of noise in the raw data by setting a noise parameter in the data generator. The higher the noise level the lower the accuracy of drift categorization. Noise is introduced by randomly switching (from 0 to 1 or vice versa) a certain percentage of the labels of the data fed to the classifier. For example, a noise percentage of 1% means that 1% of the drifts in the stream have an incorrect class label.

We specifically looked at the accuracy of the algorithm under two different situations. The first situation is when we know the actual location of the drift point. The second situation corresponds to the case when we estimate the location of the drift points detected by a change detector such as ADWIN2. The drift points estimated by any change detector are always associated with some degree of detection delay. As mentioned before the drift categorizer is not coupled to a particular drift detector and any other detector can be used in place of ADWIN2. In the first situation, we are operating in an ideal environment where

Table 3. Noise Study: Gaussian Distribution

$\sigma = 0.15$				
	Noise			
γ	0%	1%	5%	10%	20%
0.05	0.9438	0.9355	0.8969	0.8533	0.7629
0.10	0.8427	0.8351	0.8052	0.7677	0.6910
0.15	0.7407	0.7355	0.7106	0.6821	0.6214
0.20	0.6417	0.6363	0.6182	0.5968	0.5501
0.25	0.5397	0.5368	0.5243	0.5107	0.4792

$\sigma = 0.20$				
	Noise			
γ	0%	1%	5%	10%	20%
0.05	**0.9690**	0.9592	0.9209	0.8742	0.7749
0.10	0.9174	0.9083	0.8722	0.8305	0.7397
0.15	0.8131	0.8061	0.7784	0.7418	0.6681
0.20	0.7128	0.7059	0.6820	0.6549	0.5958
0.25	0.6147	0.6086	0.5913	0.5726	0.5263

$\sigma = 0.25$				
	Noise			
γ	0%	1%	5%	10%	20%
0.05	0.9046	0.8982	0.8649	0.8200	0.7349
0.10	0.9588	0.9502	0.9104	0.8650	0.7716
0.15	0.8750	0.8683	0.8340	0.7944	0.7116
0.20	0.7744	0.7669	0.7414	0.7103	0.6404
0.25	0.6743	0.6704	0.6514	0.6259	0.5730

$\sigma = 0.30$				
	Noise			
γ	0%	1%	5%	10%	20%
0.05	0.8450	0.8365	0.8073	0.7651	0.6908
0.10	0.9466	0.9355	0.8994	0.8505	0.7614
0.15	0.9168	0.9039	0.8706	0.8268	0.7397
0.20	0.8208	0.8142	0.7863	0.7452	0.6754
0.25	0.7264	0.7212	0.6968	0.6647	0.6075

$\sigma = 0.35$				
	Noise			
γ	0%	1%	5%	10%	20%
0.05	0.7713	0.7641	0.7361	0.7067	0.6371
0.10	0.9135	0.9041	0.8698	0.8252	0.7362
0.15	0.9171	0.9097	0.8750	0.8302	0.7392
0.20	0.8541	0.8454	0.8149	0.7760	0.6962
0.25	0.7647	0.7572	0.7330	0.7001	0.6335

Table 4. Noise Study: Gaussian Distribution with ADWIN2

$\sigma = 0.15$				
	Noise			
γ	0%	1%	5%	10%	20%
0.05	0.8969	0.8887	0.8537	0.8121	0.7308
0.10	0.7998	0.7923	0.7620	0.7301	0.6602
0.15	0.7028	0.6982	0.6748	0.6484	0.5928
0.20	0.6057	0.6025	0.5844	0.5636	0.5236
0.25	0.5099	0.5075	0.4953	0.4830	0.4576

$\sigma = 0.20$				
	Noise			
γ	0%	1%	5%	10%	20%
0.05	0.8809	0.8724	0.8402	0.7990	0.7142
0.10	0.8178	0.8101	0.7827	0.7462	0.6698
0.15	0.7403	0.7320	0.7079	0.6787	0.6151
0.20	0.6501	0.6454	0.6280	0.6054	0.5520
0.25	0.5592	0.5565	0.5416	0.5281	0.4901

$\sigma = 0.25$				
	Noise			
γ	0%	1%	5%	10%	20%
0.05	0.8351	0.8275	0.7989	0.7603	0.6857
0.10	0.8261	0.8170	0.7883	0.7503	0.6778
0.15	0.7712	0.7662	0.7401	0.7076	0.6398
0.20	0.6953	0.6920	0.6705	0.6441	0.5865
0.25	0.6121	0.6074	0.5922	0.5714	0.5283

$\sigma = 0.30$				
	Noise			
γ	0%	1%	5%	10%	20%
0.05	0.7825	0.7760	0.7501	0.7132	0.6475
0.10	0.8089	0.8011	0.7748	0.7370	0.6662
0.15	0.7849	0.7793	0.7519	0.7170	0.6501
0.20	0.7263	0.7216	0.6979	0.6675	0.6086
0.25	0.6508	0.6461	0.6270	0.6031	0.5564

$\sigma = 0.35$				
	Noise			
γ	0%	1%	5%	10%	20%
0.05	0.7332	0.7259	0.7029	0.6710	0.6115
0.10	0.7834	0.7790	0.7500	0.7165	0.6488
0.15	0.7871	0.7816	0.7528	0.7197	0.6516
0.20	0.7462	0.7409	0.7158	0.6843	0.6231
0.25	0.6839	0.6776	0.6581	0.6309	0.5775

we assume the true drift position is known, which is normally not the case. However, the first situation allows us to assess performance of the drift categorizer without the compounded effect of detection delay. In the second situation we experiment with actual situations where detection delay is present and show the added effect of detection delay on categorization accuracy.

For all of the experiments the streams were generated with 1000 drifts and each drift occurs over an interval of 1000 time steps. To enhance reliability of results each experiment was replicated over 100 trials. The standard deviation of accuracy over the 100 trials is less than 0.0005 which shows that there is a high degree of consistency across trials. The slope of the drifts are randomly selected using the Uniform, Gaussian, and Binomial distributions. The δ value for ADWIN2 is set at 0.05. We chose to experiment with five critical values: {0.05, 0.10, 0.15, 0.20, 0.25} because they are all reasonable sizes of critical regions that users would normally set. Table 1 and Table 2 shows the results for the noise study when the slope value is generated from a uniform distribution. The ground truth for the drifts generated here is set at a value where there is a reasonable number of gradual and abrupt drifts. Specifically in these experiments, the left bound is 0.2 and right bound is 0.8. This means that drifts with slope below 0.2 is labeled gradual and drifts with slope above 0.8 is labeled abrupt.

Overall we observed that the highest accuracy of 94.85% is seen when the critical value is set at 0.15. As we tighten or loosen the critical region we observed a decrease in accuracy. In general as noise is introduced into the stream, accuracy drops as we would expect. When the experiment is performed using ADWIN2, the accuracy is lower than when we use the actual drift point. This is because of the combined effects of a detection delay associated with change detection and the possibility of a drift not being detected. Tables 3 and Table 4 display the same noise study but replicated on streams with slopes generated from a Gaussian distribution. The parameters for the Gaussian distribution is $mean(\mu) = 0.5$ with varying standard deviation (σ) of 0.15 to 0.35. The best accuracy is observed at 96.90%. A further replication of the noise study done on Binomial distribution showed similar results. Due to space limitations we have omitted the results.

Table 5. Delay Study: Uniform Distribution

γ	Delay				
	0	50	100	150	200
0.05	0.6554	0.6582	0.6654	0.6670	0.6637
0.10	0.8600	0.8478	0.8369	0.8075	0.7598
0.15	0.9486	0.9089	0.8920	0.8483	0.7890
0.20	0.8807	0.8654	0.8515	0.8218	0.7740
0.25	0.7850	0.7852	0.7786	0.7652	0.7359

Table 6. Perturbation Study: Uniform Distribution

γ	$\beta = 0$	$\beta = 1$	$\beta = 5$
0.05	0.6554	0.6562	0.6558
0.10	0.8600	0.8582	0.8585
0.15	0.9486	0.9367	0.9328
0.20	0.8807	0.8741	0.8776
0.25	0.7850	0.7797	0.7873

Perturbation Study. In this study we examine the effects of varying the perturbation variable β. A larger β represents a larger deviation from a linear increase in drift magnitude within the stream, whereas a β of 0 represents a pure linear drift pattern. The results are shown in Table 6. We observe from this study that a reasonable perturbation variable in general does not signifi

Delay Study. In a real world scenario, drift points cannot be found with exact precision as true population means are not known in a stream environment undergoing change. Thus in reality, a change detector will always be associated with some sort of detection delay. In this experiment we look at the effects of delay on the accuracy of our technique. This delay study is performed on the uniform distribution with no perturbation and noise with varied delay values.

The range of the delay values were set from 0 to 200. A delay value of 0 represents no delay. The results are shown in Table 5. Overall we observed that delay has a large effect on the accuracy of the algorithm. In general accuracy is not heavily affected when delay is within 100 but starts to decrease when delay is at 150. cantly affect the accuracy of the algorithm.

Case Study: Sensor Stream. In this study we use a real-world data stream called the Sensor Stream obtained from the Stream Data Mining Repository (http://www.cse.fau.edu/~xqzhu/stream.html). It contains a total of 2,219,803 instances each with 5 attributes and 54 classes. We use the identified drift points in our algorithm and evaluate the slope and relative drift type of the drifts found.

Table 7. Sensor Stream Summary Stats   **Table 8.** Case Study: Stream Progression

Total Drifts Found	1200
Total Drifts Categorized	1100
Relative Gradual Drifts Identified	160
Average Slope Value of Gradual Drifts	0.0005
Relative Moderate Drifts Identified	779
Average Slope Value of Moderate Drifts	0.0062
Relative Abrupt Drifts Identified	161
Average Slope Value of Abrupt Drifts	0.0274

nth drift	step of nth drift	average slope
100	186047	N/A
200	375007	0.0102
300	559615	0.0086
400	776703	0.0086
500	982335	0.0079
600	1184575	0.0080
700	1376607	0.0065
800	1545023	0.0068
900	1728447	0.0068
1000	1887071	0.0088
1100	2048223	0.0097
1200	2219039	0.0093

In this study we used a critical value $\gamma = 0.25$ and the summary statistics obtained are shown in Table 7. Overall there were 1200 total drifts found by ADWIN2 from the stream. The first 100 drifts were used to fill the original reservoir samples, therefore, the number of drifts that are actually categorized is 1100. Out of the 1100 drifts that are categorized, approximately 15% of the drifts are categorized as relative gradual, 70% of the drifts are categorized as relative moderate, and the remaining 15% categorized as relative abrupt. This distribution of drifts is relatively stable and conforms to the Gaussian curve. From the table we also observe that the average slope value for the three different drift types are distinctively different from each other with relative gradual drifts sitting at an average of 0.0005, relative moderate drifts sitting at an average of 0.0062, and relative abrupt sitting at 0.0274. Table 8 shows the shifting of the average slope value as the stream progresses. This table presents the progression of the reservoir mean sampled every 100 drifts. We observed from the table that the stream started off with an average detected slope of 0.0102, then in the midpoint of the stream, the average slope drifted down to a smaller value of 0.0065 then back up again to 0.0093. In terms of shifting of the Gaussian curve, the stream experienced a left shift when the stream progressed to its midpoint, then had a right shift from its midpoint until it reached the end.

6   Conclusion and Future Work

We have described a novel approach one pass algorithm for categorizing drifts in a data stream based on a relative measure. This is to the best our knowledge, the first algorithm that looks at drift categorization in data streams. We also present a new data generator that allows the user to generate drifts in the data stream with controllable magnitudes and distribution. In our evaluations we have shown that our categorization algorithm is accurate and robust in both ideal situations and in realistic situations with various noise and stream fluctuation characteristics. Our future work includes adapting the critical value for the Gaussian curve on-the-fly instead of having the user provide an input. By automatically adapting the critical value to produce the best accuracy of the algorithm, it saves the user the trouble of first determining a best value and also allows easier adaptation to the most recent conditions of the data stream.

References

1. Baena-García, M., del Campo-Ávila, J., Fidalgo, R., Bifet, A., Gavaldá, R., Morales-Bueno, R.: Early drift detection method. In: Fourth International Workshop on Knowledge Discovery from Data Streams (2006)
2. Bartlett, P., Ben-David, S., Kulkarni, S.: Learning changing concepts by exploiting the structure of change. Machine Learning 41(2), 153–174 (2000)
3. Bifet, A., Gavaldá, R.: Learning from time-changing data with adaptive windowing. In: SIAM International Conference on Data Mining (2007)
4. Gama, J., Medas, P., Castillo, G., Rodrigues, P.: Learning with drift detection. In: Bazzan, A.L.C., Labidi, S. (eds.) SBIA 2004. LNCS (LNAI), vol. 3171, pp. 286–295. Springer, Heidelberg (2004)
5. Hoeffding, W.: Probability inequalities for sums of bounded random variables. Journal of the American Statistical Association 58, 13–29 (1963)
6. Kifer, D., Ben-David, S., Gehrke, J.: Detecting change in data streams. In: Proceedings of the Thirtieth International Conference on VLDB, vol. 30, pp. 180–191. VLDB Endowment (2004)
7. Kosina, P., Gama, J., Sebastião, R.: Drift severity metric. In: Proceedings of the 2010 Conference on ECAI 2010: 19th European Conference on Artificial Intelligence, pp. 1119–1120. IOS Press, Amsterdam (2010)
8. Sebastião, R., Gama, J.: A study on change detection methods. In: 4th Portuguese Conf. on Artificial Intelligence, Lisbon (2009)
9. Vitter, J.S.: Random sampling with a reservoir. ACM Trans. Math. Softw. 11(1), 37–57 (1985)
10. Zhang, P., Zhu, X., Shi, Y.: Categorizing and mining concept drifting data streams. In: Proceedings of the 14th ACM SIGKDD International Conference on Knowledge Discovery and Data Mining, KDD 2008, pp. 812–820. ACM, New York (2008)

Continuously Extracting High-Quality Representative Set from Massive Data Streams[*]

Xiaokang Ji[1,2], Xiuli Ma[1,2,**], Ting Huang[1,2], and Shiwei Tang[1,2]

[1] Key Laboratory of Machine Perception(Peking University), Ministry of Education
[2] School of Electronics Engineering and Computer Science, Peking University,
Beijing, 100871, China
{jixk,maxl}@cis.pku.edu.cn, {TingHuang,tsw}@pku.edu.cn

Abstract. In many large-scale real-time monitoring applications, hundreds or thousands of streams should be continuously monitored. To ease the monitoring, a small set of representatives can be extracted to represent all the streams. To get a high-quality representative set, not only representativeness but also its stability should be guaranteed. In this paper, we propose a method to continuously extract high-quality representative set from massive streams. First, we cluster streams based on core clustering model. The tightness of core set, which means any two streams in core set are highly correlated, ensures high representativeness of representative set; second, we use topological relationship to force each cluster to be connected in the network where streams are generated from. Because streams in one cluster are driven by similar underlying mechanisms, so the representative set becomes much more stable. By utilizing the tightness of core sets, we can get representative set immediately. Moreover, with local optimization strategies, our method can adjust core clusters very efficiently, which enables real-time response. Experiments on real applications illustrate that our method is efficient and produces high-quality representative set.

Keywords: Data streams, Representative set, Online clustering.

1 Introduction

Nowadays, monitoring large-scale networks such as drinking water distribution networks [13] and traffic networks [8] has been paid much attention. In these applications, multiple data sources (e.g., sensors) are deployed in networks to continuously collect data, resulting in massive co-evolving streams. Online analysis on trends of streams is widely conducted in pattern discovery and anomaly detection [10]. Thus, it is necessary for surveillance managers to capture an accurate picture of current trends.

In most cases, monitoring every trend is redundant because many streams vary at different numerical levels but with similar trends. To reduce monitoring cost, we can

---

[*] This work is supported by the National Natural Science Foundation of China under Grant No.61103025.
[**] Corresponding author.

H. Motoda et al. (Eds.): ADMA 2013, Part I, LNAI 8346, pp. 84–96, 2013.

first cluster streams based on the similarities of their trends and then select a set of streams as representative set from clusters. More specifically, we define it as a minimal set of streams that can represent all trends.

In this paper, we focus on mining representative set with high quality. The quality of representative set is measured in two aspects: representativeness and stability. Traditional methods only make efforts to reduce the dissimilarity between the representatives and those streams they represent, so as to achieve high representativeness. However, in streaming condition, the stability of representative set should also be kept at a high level. With high stability, the representative set is able to provide a stable overview and support incremental analysis. To the best of our knowledge, there is no existing work that is capable of guaranteeing both high representativeness and stability.

Continuously extracting high-quality representative set faces several challenges: (1) high representativeness of representative set should be ensured; (2) Stability of representative set should be kept at a high level; (3) Real-time response is required in streaming condition. Next, we introduce our method to deal with above challenges.

— How to enhance the representativeness of representative set?

The correlation among streams is frequently utilized to cluster streams, which puts the similar trends together and separates dissimilar ones apart. Methods adopted to cluster streams have a significant impact on the representativeness of representative set. Traditional methods cannot ensure that each pair of streams in one cluster has high correlation. This shortcoming lowers the representativeness of representative set. To solve this problem, we adopt core cluster model [5]. Each core cluster contains a core set and a boundary set. In the core set, every pair of streams are correlated. We denote that as tightness. For each stream in boundary set, it should be correlated to at least one stream in the core set. By utilizing the tightness of core set, we can restrict the maximum dissimilarity between two streams in one cluster strongly.

— How to ensure the stability of representative set?

Traditionally, one stream is able to represent another stream if the correlation between them satisfies a given threshold. However, there exists one hidden trouble: the correlation between some representatives and those streams they represent may vary greatly, so the representative set alters frequently with time. Fortunately, we find the correlation between two topologically nearby streams in the network is much steadier due to the similar underlying mechanisms their trends are driven by. This enlightens us to cluster streams based on both correlation and topology. Thus, we further develop core cluster model into topology-constrained core clustering model. By mapping each core cluster to a connected region in the network, the correlation among streams in the same cluster becomes much more stable, so does the representative set.

— How to obtain representative set with high quality in real time?

Considering we can extract the representative set from core clusters very quickly by utilizing the tightness of core sets, so the respond time mostly depends on how fast streams are clustered. Performing initial clustering algorithm at every timestamp may cause a waste of time if clusters remain almost the same. Thus, online adjustment on clusters becomes necessary to obtain representative set more efficiently.

The main challenge of online adjustment is maintaining the characteristics of core clustering model in a limited time, especially the tightness of core sets. As new data arrives, core sets may not satisfy tightness any more due to the variation of correlation. The variation between contiguous timestamps is mostly tiny, which enables us to get new core sets by adjusting old ones. Because all the streams in one core set can be represented by just one of them, so we can make the best use of the tightness to reduce the size of representative set if each core set is maximum. However, mining maximum core set is an equivalent problem of mining maximum clique [3][4], which is known as a NP-hard problem. To meet real-time requirement, we target mining maximal core sets. In our method, we first shrink core sets to be tight again then expand them to maximal. For efficiency's sake, local optimization strategies are adopted to determine the order of removing streams from core set or adding streams into core set. To be specific, we construct profiles for each stream in the process of adjustment and the attributes in profiles will provide the basis for local optimization strategies.

Consequently, the main contributions of our work can be summarized as follows:

- We raise a problem of continuously extracting high-quality representative set.
- We design a topology-constrained core clustering model. Representative set extracted from core clusters can achieve excellent representativeness and stability.
- Local optimization strategies are adopted to adjust core clusters, so as to maintain characteristic of core clustering model and guarantee efficiency.

The rest of this paper is organized as follows. Section 2 discusses related work. Section 3 introduces definitions and main problem. Section 4 presents our method in details. In Section 5, we present performance study. Finally we conclude in Section 6.

2   Related Work

There are plenty of work aiming at summarizing massive data streams. Pattern discovery in multiple streams or time series continues to attract high interest. SPIRIT [10] monitors key trends by calculating hidden variables incrementally. Dynammo [6] focuses on summarization of coevolving sequences with missing values. Most of these methods use mathematical transformation tools such as PCA to discover patterns. However, their results are not intuitive enough to represent and explain, especially in the complex networks. For example, if multiple demand patterns exist in a drinking water network, many hidden variables will be discovered by SPIRIT, so the results will become very hard to interpret to users. In contrast, representative set provides a more intuitive way for users to capture trends of all streams.

A framework of identifying representatives out of massive streams is proposed in [7]. Although core clustering model is also adopted in the method, we still have three advantages: First, we consider topological relationship of streams to enhance the stability of representative set; second, the tightness of core sets is maintained during the online adjustment; third, our representative set aims to capture all trends of streams instead of identifying key trends. This is because when a stream just turns abnormal, it is precisely the stream that cannot be represented by other streams, so the information of anomalous events may be missed if only key trends are focused.

Many methods are capable of clustering streams. A new clustering method based on micro-clusters is proposed in [2]. Some work develop traditional clustering algorithms such as K-means [1] and K-medoids [11], so we can use them in online clustering. COMET-CORE [14] is a framework of continuously clustering streams by events which are defined as the marked changes of the streams. Similar to our method, it computes correlation between streams and continuously clusters them with time. However, in the above methods, pairs of streams in one cluster may have small similarities, which means high representativeness cannot be ensured. What's more, as a result of the ever-changing correlation between representatives and those streams they represent, they cannot provide a stable overview of current trends in networks.

3   Overview

In this section, we first state how to apply the two kinds of relationship among streams in our method: topology and correlation. Then we give the definition of our clustering model based on them. Finally, we state our main problem.

Definition 1. *(TOPOLOGICAL GRAPH) For a large-scale network, we summarize its topological information by an undirected graph $G <V, E>$, where V is the set of vertexes and E is the edge set. Each vertex in V corresponds to a data source in the network and an edge exists where two data source are adjacent.*

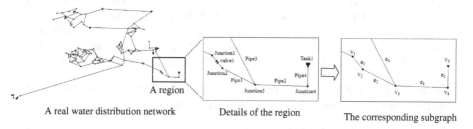

A real water distribution network   Details of the region   The corresponding subgraph

Fig. 1. Mapping a water distribution network to a topological graph

Fig.1 is an illustration of mapping a real water distribution network to a topological graph. Details of a region in the network is displayed here. Specifically, we first map junctions, tanks or reservoirs, where sensors are deployed, to vertexes in the topological graph. Then, we map pipes, valves or pumps that connect two sensors to edges. In our method, the introduced graph of each cluster from topological graph is kept connected. Next, we show how to determine whether or not two streams are correlated.

Here, we formally define all streams as $/S_1, S_2, ..., S_n/$ which are generated from corresponding data source set $/v_1, v_2 ..., v_n /$ in V. For simplicity sake, we assume v_i and S_i are the same in this paper. Each stream is denoted as $S_i [1, 2, ..., k]$, where k is the length of sliding window and $S_i[t]$ denotes the t-th value of S_i. Pearson correlation coefficient is able to measure the similarity of two streams' trend, so we use it to calculate the correlation. To fit in streaming condition, we adopt weighted Pearson correlation coefficient, which gives recently arrived data more weight than aged data.

Definition 2. *(WEIGHTED CORRELATION COEFFICIENT) Given a weight function w(t) and w(t) is a monotonically nondecreasing function of timestamp t, the weighted correlation coefficient between S_i and S_j is:*

$$wcorr(S_i, S_j) = \frac{(\sum_t w(t)S_i[t]S_j[t] - \frac{\sum_t w(t)S_i[t]\sum_t w(t)S_j[t]}{\sum_t w(t)})}{\sqrt{A(S_i) \times A(S_j)}}, \text{ where } A(S_x) = \sum_t w(t)(S_x[t])^2 - \frac{(\sum_t w(t)S_x[t])^2}{\sum_t w(t)}$$

Comparing this coefficient with a specified threshold, we can determine whether two streams are correlated. Next, we introduce our clustering model based on the correlation and topological relationship among streams.

Definition 3. *(TOPOLOGY-CONSTRAINED CORE CLUSTERING MODEL) Given a topological graph G=<V, E>, the core clustering results of V satisfied:*

$$Clusters = \{(C_1, B_1), (C_2, B_2), ..., (C_n, B_n)\}$$

where (C_i, B_i) is a core cluster in the form of core set and boundary set. Streams in core set and boundary are core members and boundary members. Given a user-specified threshold ε, C_i and B_i satisfy:
(1) $\forall c_i, c_j \in C_i$, $wcorr(c_i, c_j) \geq \varepsilon$; (2) $\forall b_i \in B_i$, $\exists c_i \in C_i$, $wcorr(c_i, c_j) \geq \varepsilon$; (3) the introduced graph of $C_i \cup B_i$ in G is connected; (4)C_i is maximal to satisfy the above principles.

Our proposed clustering model guarantees the high quality of representative set. Before we define the representative set, *ε-represent* is introduced to describe relationship between a representative and its represented streams. Finally, our main problem can be stated as follows: at every new timestamp, how to get the high-quality representative set from clusters in real time. Notations applied later are listed in Table 1.

Definition 4. *(ε-REPRESENT) Given a threshold $\varepsilon \in [0,1]$ and two streams S_i and S_j, S_i ε-represents S_j if $wcorr(S_i, S_j) \geq \varepsilon$ and they are in the same cluster.*

Definition 5. *(REPRESENTATIVE SET) Denote all streams as $\{S_1, S_2, ..., S_n\}$, representative set is a minimal set of streams that can ε-represent all the streams.*

Table 1. Description of notations

Symbol	Description	Formula
$neighborSet(v, N)$	streams in N that are adjacent to v in G<V,E>	$\{n \mid <v, n> \in E, n \in N\}$
$corrSet(v, N)$	streams in N that are correlated to v	$\{n \mid wcorr(v, n) \geq \delta, n \in N\}$
$corrDegree(v, N)$	the number of streams in corrSet(v,N)	$card(corrSet(v, N))$

4   Methodology

The framework of our methodology includes three parts: (1) initial core clustering; (2) extracting representative set; (3) online adjustment on clusters. First, we initially cluster streams based on topology-constrained core clustering model. Then, we show how to efficiently extract representative set from core clusters. In the final part, we explain our method to adjust core clusters online.

4.1   Initial Core Clustering

Original algorithm of core clustering is designed for high-dimensional data in [5]. Unlike high-dimensional data, streams cannot be all stored in advance and clustered offline. As data continuously arrives, core clusters should be updated immediately. To supply a foundation for online adjustment, an initial clustering result that satisfies the topology-constrained core clustering model is needed first.

Different from the algorithm in [5], we use topological relationship to ensure that each cluster corresponds to a connected region in the network. In our proposed algorithm, one core cluster is generated at a time. In each time, we divide the algorithm into two segments. In the first segment, we generate the core set. At the beginning, we initialize the core set with a stream based on a local optimization policy which requires that it must be correlated to the maximum number of streams in the current streams set. Then, we keep adding streams into core set which are adjacent to at least one core member and, meanwhile, are correlated to all core members based on the same local optimization policy. When no more streams can be added into core set, the core set is generated completely. In the second segment, we assign streams into boundary set if they are both adjacent and correlated to at least one core member. We keep on investigating each stream that is adjacent to newly joined boundary members. If it is correlated to at least one core member, we assign it into boundary set. When the loop stops, the boundary set is generated. Then we remove streams in the new core cluster from the current streams set and get back to the start to generate a new core cluster. The complexity of this algorithm is $O(n^2)$.

4.2   Extracting Representative Set

Now, we discuss how to extract representative set from core clusters. At the beginning, we obtain representatives for each core cluster. Then, we gather representatives of all core clusters together to form the final representative set. The pseudo-code for extracting representative set is listed in Algorithm 1.

Finding minimum representatives for one cluster is an equivalent problem of finding a minimum dominating set for a graph, which it is known as a NP-hard problem. For efficiency's sake, we set our goal as finding minimal representatives instead of minimum solution. By utilizing the following two special properties of the topology-constrained core clustering model, we can reach the goal more efficiently: (1) any core member can represent all the other core members in the same core set; (2) any boundary member can be represented by at least one core member. Based on these characteristics, we first elect the core member which is correlated to the maximum number of boundary member as a representative. Then we remove streams that can be represented by this representative from boundary set. We repeat these operations until the boundary set becomes empty, then representatives of one core cluster is generated successfully. Apparently, they are minimal to represent all streams in the cluster.

4.3   Online Adjustment on Clusters

As new data arrives, core clusters should be adjusted accordingly. The minimum requirement of online adjustment is preserving the characteristics of our clustering model so as to keep the quality of representative set at a high level. After the adjustment, we use algorithm 1 to get the up-to-date representative set.

As Fig.2 shows, the main strategy of online adjustment can be divided into four steps. The first two steps aim to adjust core sets: first, we shrink core sets to maintain the tightness and then we expand core sets to keep them maximal. In order to ensure efficiency, local optimization strategies are applied in the two steps of adjusting core sets. To support local optimization strategies, a profile for each stream is introduced.

Definition 7. *(PROFILE) The profile of a stream v in core set C is constructed with three attributes: {Cost,Gain,Energy} and they are defined as:*

$$Cost(v,C) = card(\{c \mid c \in C \wedge wcorr(v,c) \geq \delta\})$$
$$Gain(v,C) = card(\{< n,c > \mid n \in neighborSet(v, V - C) \wedge c \in C \wedge wcorr(v,n) \geq \delta\})$$
$$Energy(v,C) = card(\{n \mid n \in neighborSet(v, V - C) \wedge \forall c \in C, wcorr(c,n) \geq \delta\}) .$$

$Cost(v,C)$ represents the price of removing v from C, which is the number of core members that are correlated to it. $Gain(v,C)$ provides a bridge connecting two steps, which refers to the potential profit of retaining v inside C and is measured by the number of correlated pairs between C and its neighbor set. As defined in Table 1, a stream's neighbor set contains every stream adjacent to it in the topological graph. $Energy(v,C)$ denotes the direct profit of v if we add it into the C, which is the number of streams in its neighbor set that are correlated to all core members.

Algorithm 1. ExtractRepresentSet	**Algorithm 3.**ExpandingCoreSet
Input : $Clusters_t = \{(C_1, B_1), ..., (C_n, B_n)\}$	**Input** : core set C_{t-1}, $G = <V, E>$
Output : Rp_t //the representative set	**Output** : core set C_t
1 *For* every (C_i, B_i) in $Clusters_t$	1 Initialize $canSet = neighborSet(C_{t-1}, V)$;
2 *While* $!isEmpty(B_i)$ *do*	2 *For* every c_i in C_{t-1} and v_k in $canSet$
3 find c in C_i that satisfies:	3 *If* $wcorr(c_i, v_k) < \delta$
4 $corrDegree(c, B_i)$ is minimum;	4 $canSet = canSet - \{v_k\}$;
5 $Rp_t = Rp_t \cup \{c\}$;	5 *End if*
6 $C_i = C_i - \{c\}$;	6 *End for* //get original candidate set
7 $B_i = B_i - corrSet(c, B_i)$;	7 *While* $!isEmpty(canSet)$ *do*
8 *End while*;	8 *For* every v_k in $canSet$
9 *End for* ;	9 $Energy(v', C_{t-1}) = max\{Energy(v_k, C_{t-1})\}$;
	10 *End for*
Algorithm 2. ShrinkingCoreSet	11 $C_{t-1} = C_{t-1} \cup \{v'\}$; //expand the core set
Input : core set C_{t-1}, $G = <V, E>$	12 *For* every v_k in $canSet$
Output : core set C_t	13 *If* $corrDegree(c_i, v_k) < \delta$ or $v_k = v'$
1 *While* $!isCoreSet(C_{t-1})$ *do*	14 $canSet = canSet - \{v_k\}$;
2 $minCostSet = \varnothing$;	15 *End if*
3 *For* every c_i in C_{t-1}	16 *End for*
4 Find c_i with minimum $Cost(c_i, C_{t-1})$	17 *For* every n_k in $neighborSet(v', V - C_{t-1})$
5 $minCostSet = minCostSet \cup \{c_i\}$;	18 *If* n_k is correlated with $C_{t-1} \cup \{v'\}$
6 *End for*	19 $canSet = canSet \cup \{n_k\}$;
7 *For* every c_k in $minCostSet$	20 *End if* //update candidate set
8 $Gain(c', C_{t-1}) = min\{Gain(c_k, C_{t-1})\}$;	21 *End for*
9 *End for*	22 *End while*;
10 $C_{t-1} = C_{t-1} - \{c'\}$;	23 $C_t = C_{t-1}$;
11 *End while*;	
12 $C_t = C_{t-1}$;	

4.3.1 Shrinking Core Sets

In this step, we first attempt to shrink every core set by removing some streams so as to keep core sets tight as before. Because the variation of correlation is very small, so most pairs of streams in core sets still remain correlated (Fig.2 (a)). Therefore, only a few streams need to be removed out. Determining the order of removing streams is the key problem here. Methods aiming at achieving tightness by removing the minimum number of streams have one main shortcoming: they usually have to search the optimal solution in a space through backtrackings, which lowers the efficiency.

To deal with that problem, we use a local optimization strategy to iteratively determine which stream should be removed from core set. In each iteration of shrinking a core set, we first check if the current core set satisfies tightness. If not, we get the set of streams with the minimum *Cost*. The size of this set is usually larger than one due to the symmetry of correlation. Therefore, we remove the stream which supports expanding core set at the lowest level. Due to topological constraints, when we look for a stream to add into core set, we must start with the adjacent streams of core members. Thus, the more correlated pairs between one stream's neighbor set and core set, the more chance we can expand the core set with its neighbors later. Thus, we remove the stream with the minimum *Gain* in that set. Algorithm 2 shows the details.

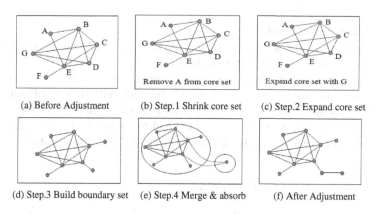

(a) Before Adjustment   (b) Step.1 Shrink core set   (c) Step.2 Expand core set

(d) Step.3 Build boundary set   (e) Step.4 Merge & absorb   (f) After Adjustment

Fig. 2. Online adjustment on core cluster. Core member: Blue; Boundary members: Orange. Streams connected with an edge are correlated at current timestamp.

4.3.2 Expanding Core Sets

After shrinking every core set, we get shrunken core sets which is tight but may not be maximal (Fig.2 (b)). In this step, we try to expand core sets until they become maximal. Same problem exists here. If we target at adding the maximum number of streams into core set, the efficiency of the algorithm cannot be guaranteed.

Local optimization strategy to determine the order of adding streams into core set here is supported by *Energy*. More *Energy* one stream possess, more streams can be added into core set after we expand core set with it. First, we go through every stream in the neighbor set of core set and add it into an original candidate set if it is correlated to all core members. Then we expand core set with the stream in candidate set with maximum *Energy*. After that, we update candidate set and core set according to the newly joined stream: (1) remove streams that are not correlated to it from candidate set; (2) add streams in its neighbor set that are correlated to all core members into core set; (3) add neighbors of those new core members into candidate set if they are correlated to all core members. Because the *Energy* of newly joined stream is maximum, so we can add the maximum streams into core set in step 2. Algorithm 3 shows the details of expanding core sets. The algorithm will finish when candidate set becomes empty, which means the current core set becomes maximal (Fig.2 (c)).

4.3.3   Building Boundary Sets

After all core sets are generated, we assign the rest of streams into boundary sets. Unsettled streams adjacent to core set will be managed first and we keep investigating unsettled streams adjacent to the current core members and boundary members until no more streams can be added in (Fig.2 (d)). It should be noted that not every unsettled stream can be assigned into an existing boundary set due to constraints of correlation and topology. Therefore, for every existing connected component consist of streams that cannot be assigned in $G<V,E>$, we do initial core clustering on them and combine the clustering results with the other clusters.

4.3.4   Merge and Absorb

At last, we finish online adjustment by seeking if two adjacent clusters can be merged or one cluster can absorb another in order to keep a certain degree of differentiation between adjacent clusters. *Merge* happens if core sets of two clusters are adjacent in topological graph and the union set of them is tight as well. *Absorb* will only occur if one isolated cluster can be added into another adjacent cluster's boundary set (Fig.2 (e)). Specifically, we adopt a bottom-up method to combine adjacent clusters which ensures that no more adjacent clusters can be combined after the adjustment.

5   Experimental Evaluations

In this section, we evaluate the proposed method, which is abbreviated as CERS, in real applications. Section 5.1 introduces the datasets. Section 5.2 studies the effectiveness, including representativeness and stability. Efficiency is analyzed in Section 5.3. We compare CERS with two methods that can be used for extracting representative set: (1) K-medoids and (2) fast K-medoids proposed in [11]. Fast K-medoids transforms K-medoids into a much more efficient form at the cost of local optimum solution. The value of k of these two methods is synchronously assigned the size of our up-to-date representative set, then we extract k centers of their clusters as representative sets to compare with ours. By unifying sizes of representative sets, we can focus on the quality of them at same starting line.

The following experiments are conducted on Windows 7 Professional operating system equipped with an Intel Pentium 3.1GHz processor and 4 GB of RAM.

5.1   Dataset

Monitoring streams in drinking water network is a typical large-scale real-time monitoring application. We used three networks of different scales in the real world as
Fig.3 shows. Network in Fig.3 (a) is a real water distribution system referred to in
BWSN [9], and networks in Fig.3 (b) and Fig.3 (c) are provided by the center for
water systems at the University of Exeter [15]. Datasets of these networks was generated by EPANET 2.0 [12] that accurately simulates the hydraulic and chemical phenomena within drinking water distribution systems. In our experiments, we monitor
the chlorine concentration level at every node in the above three networks for 5760
timestamps during 20 days (one time tick every five minutes).

Fig. 3. (a) Network1 of 129 nodes   (b) Network2 of 511 nodes   (c) Network3 of 920 nodes

5.2   Effectiveness

In this section, we first evaluate the representativeness of representative set in
two criteria: unidirectional Hausdorff distance and coverage rate. Hausdorff distance
denotes the maximum distance from the represented streams to representative set.
As in our situation, the distance is equal to subtracting the correlation coefficient
from 1. Coverage rate indicates the percentage of streams that can be ε-represented
by the representative set. The value of ε in the experiments is deliberately designated
as 0.8.

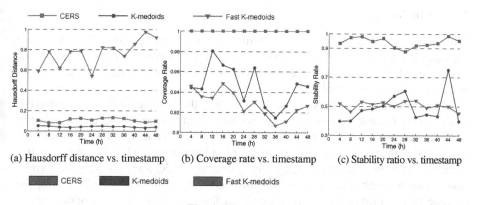

(a) Hausdorff distance vs. timestamp   (b) Coverage rate vs. timestamp   (c) Stability ratio vs. timestamp

Fig. 4. Effectiveness

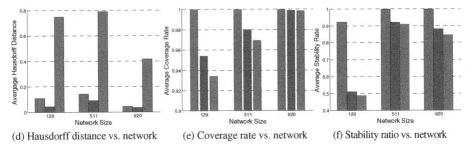

(d) Hausdorff distance vs. network   (e) Coverage rate vs. network   (f) Stability ratio vs. network

Fig. 4. (*Continued.*)

Fig.4 (a-c) shows the results versus timestamps and Fig.4 (d-f) shows the results versus networks. As shown in Fig.4 (a), the Hausdorff distance of our method is much smaller than Fast K-medoids and a bit larger than K-medoids which provides the optimum solution at the cost of high computation complexity. In Fig.4 (b), CERS maintains the coverage rate at 100% and surpasses the other two methods during a whole cycle of two days. For each network, we calculate the average Hausdorff distance and coverage rate in a cycle of 48 hours. As can be seen from Fig.4 (d), the Hausdorff distance of CERS is very close to the optimal result in each network and our coverage rate is kept optimal as the size of network grows (Fig.4 (e)).

Then, we evaluate the stability of our representative set by stability ratio, which denotes the percentage of representatives at current timestamp that also exist in the representative set of last timestamp. As can be seen from Fig.4 (c), the stability of representative set in our method is much better than in the other two methods. During the most timestamps in 48 hours, our stability rate is up to ninety percent. Moreover, the stability rate of our method becomes even higher when the scale of network increases. In Fig.4 (f), our stability rate almost reaches 1.0 in network2 and network3.

5.3   Efficiency

We compare efficiency of our method with two methods: one is called Basic here, and the other is Fast K-medoids. K-medoids is not used as a contrast because its runtime is too much larger than ours. The Basic algorithm extracts representative set through performing initial core clustering at every timestamp.

If the number of clusters is fixed as k, then the complexity of our method is $O(n^2/k)$. Fast K-medoids can reduce the complexity of K-medoids which is $O(k(n-k)^2)$ to $O(nk)$. As proved in Fig.5 (a), our method CERS is much more efficient than Basic. Compared with Fast K-medoids, our method is a little slower. The main reason of this is that our clustering model is a pair-wise clustering model and considers the correlation between much more pair of streams than Fast K-medoids. In Fig.5 (b), the runtime of our method grows very slowly as the scale of network increases, which indicates our method also has good scalability.

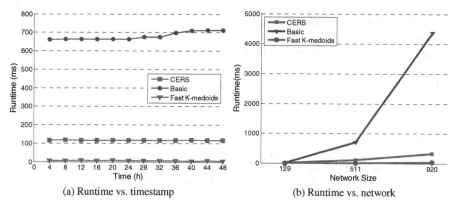

(a) Runtime vs. timestamp   (b) Runtime vs. network

Fig. 5. Efficiency

6   Conclusion

In this paper, we focused on continuously extracting high-quality representative set from massive streams that are generated from large-scale networks. We believe a stream should only be represented by another if the similarity between their trends is stably above a given threshold. Based on that, we redefined the representative and designed a topology-constrained core clustering model to cluster streams. Another contribution of our work is the framework that adjusts core clusters based on local optimization strategies and preserve the properties of our clustering model in the meantime, which enables us to respond the query of representative set in real time.

In the future, we look forward to discover the evolving patterns of representative set for events detection in large-scale real-time applications.

References

1. Ackermann, M.R., Märtens, M., Raupach, C., et al.: StreamKM++: A Clustering Algorithm for Data Streams. Journal of Experimental Algorithmics (JEA) 17(1), 2–4 (2012)
2. Aggarwal, C.C., Han, J., Wang, J., et al.: A Framework for Clustering Evolving Data Streams. In: VLDB (2003)
3. Cheng, J., Ke, Y., Chu, S., et al.: Efficient Core Decomposition in Massive Networks. In: ICDE (2011)
4. Cheng, J., Zhu, L., Ke, Y., et al.: Fast Algorithms for Maximal Clique Enumeration with Limited Memory. In: SIGKDD (2012)
5. Jiang, L., Yang, D., Tang, S., Ma, X., Zhang, D.: A Core Clustering Approach for Cube Slice. Journal of Computer Research and Development, 359–365 (2006)
6. Li, L., McCann, J., Pollard, N., Faloutsos, C.: DynaMMO: Mining and Summarization of Coevolving Sequences with missing values. In: SIGKDD (2009)
7. Li, Q., Ma, X., Tang, S., Xie, S.: Continuously Identifying Representatives out of Massive Streams. In: Tang, J., King, I., Chen, L., Wang, J. (eds.) ADMA 2011, Part I. LNCS, vol. 7120, pp. 229–242. Springer, Heidelberg (2011)

8. Liu, W., Zheng, Y., Chawla, S.: Discovering Spatio-temporal Causal Interactions in Traffic Data Streams. In: SIGKDD (2011)

9. Ostfeld, A., Uber, J.G., Salomons, E.: Battle of water sensor networks: A Design Challenge for Engineers and Algorithms. In: WDSA (2006)

10. Papadimitriou, S., Sun, J., Faloutsos, C.: Streaming Pattern Discovery in Multiple Timeseries. In: VLDB (2005)

11. Park, H.S., Jun, C.H.: A Simple and Fast Algorithm for K-medoids Clustering. Expert Systems with Applications 36(2), 3336–3341 (2009)

12. Rossman, L.A.: EPANET2 user's manual. National Risk Management Research Laboratory: U.S. Environmental Protection Agency (2000)

13. Xiao, H., Ma, X., Tang, S.: Continuous Summarization of Co-evolving Data in Large Water Distribution Network. In: WAIM (2010)

14. Yeh, M., Dai, B., Chen, M.: Clustering over Multiple Evolving Streams by Events and Correlations. TKDE 19(10), 1349–1362 (2007)

15. The Centre for Water Systems (CWS) at the University of Exeter, http://centres.exeter.ac.uk/cws/

Change Itemset Mining in Data Streams

Minmin Zhang, Gillian Dobbie, and Yun Sing Koh

Department of Computer Science,
The University of Auckland, New Zealand
mzha106@aucklanduni.ac.nz, {gill,ykoh}@cs.auckland.ac.nz

Abstract. Data stream mining is becoming very important in many application areas, such as the stock market. A data stream consists of an ordered sequence of instances and because there are usually a large number of instances along with limited computing and storage capabilities, algorithms that read the data only once are preferred. There has been some research that focuses on finding when a concept has changed, given some knowledge about the previous instances in the data stream, but little on determining the characteristics of that change. In this paper we concentrate on finding the characteristics of the changes that occur, using frequent itemset mining techniques. We propose an approach, which combines both heuristic and statistical approaches, that analyses changes that have occurred within a stream at itemset level and identify three types of change: extension, reduction, and support fluctuation. We evaluate our algorithm using both synthetic and real world datasets.

Keywords: Change Mining, Data Stream, Frequent Itemset, Hoeffding Bound.

1 Introduction

Conventional data mining methods observe a dataset and learn local models based on that dataset. Nowadays applications are processing a vast range of non-stationary data that are arriving at a rapid rate in the form of data streams. Traditional static mining methods to discover patterns are not suitable for rapidly changing and non-stationary data streams. A new challenge that arises with data streams is identifying how patterns in data streams evolve over time. Consequently, there is a new area of research that examines how a discovered pattern in a data stream may have been modified, deleted, or augmented over time.

This particular challenge is related to the field of change mining in data streams, which is a sub area of higher order mining[1]. Its objective is the discover changes in the patterns that describe an evolving stream over a succession of temporally ordered data sets known as windows. One aspect of change mining is detecting changes in frequent patterns in a data stream. Many of the approaches in this area rely on the availability of a strong assumption of the underlying model, which is inadequate when we are trying to detect change. Furthermore, a vast number of patterns are often generated at different points within a data stream. Thus we need an automated approach to discover the changes, as verifying changes in the patterns by hand is not feasible.

Based on previous research, there are three different levels where we can detect change: item level, itemset level, and pattern/rule level. Most common change detection techniques, currently either look at changes at the item level, which is known as

H. Motoda et al. (Eds.): ADMA 2013, Part I, LNAI 8346, pp. 97–108, 2013.

drift detection, or changes at the pattern level. In our research we look at changes at the intermediate level, the itemset level. Patterns are generated from valid itemsets in association rule mining.

The ability to find changes within itemsets is very useful in many application domains. In a retail environment a change can be represented by a change in the purchasing behaviors of customers. An example of this is that the retail behavior of customers may change with new trends *e.g.* increased sales of a series of mobile phones and their accessories in the market may indicate a new emerging trend. Another example is that there may be a decline in sales *e.g.* sunblock lotions and sandals sales may decline as we head into winter. Thus by only considering itemsets which have changed in some manner to form patterns, we are reducing the overhead of generating and then pruning out rules which we are not interested in. In this paper, we propose a novel algorithm to find changes in frequent itemsets in a data stream. In our algorithm we look at three types of changed itemsets: extension, reduction, and fluctuation of support.

The paper is organized as follows. In Section 2 we look at previous work in the area of change mining using association rules. In Section 3 we present basic concepts and definitions for change itemset mining and in Section 4 we introduce our approach. Section 5 describes the experimental results. Section 6 conducts the contribution. Finally, Section 7 concludes the paper.

2   Related Work

Research on how to analyse pattern histories for interesting changes has yielded three directions that differ considerably in terms of the techniques that were employed: template matching, statistical testing, and heuristics. The first approach is based on template matching of frequent patterns [2]. Trends were represented as shape queries, depicting the support and confidence histories of association patterns. The user can specify several shapes, which he regards as interesting, and the histories are then matched against them. There have been a number of papers based on this approach [3,4].

The second approach uses statistical testing mechanisms for characterising change detection. Liu et al. [5] used statistical techniques, like the χ^2-test, to determine the stability of each rule over time. Using this technique, additional information about how a rule changes, if it is semi-stable, stable or exhibits a trend, can be provided to assist a user in judging a rules value. Rules are then ranked within each class according to an interestingness measure. The idea to use previous patterns as a key to examine its interestingness has received attention from other researchers, such as Boettcher et al. [6], who proposed a framework which employs statistical tests and additionally accounts for temporal redundancies.

The last approach employs certain heuristics to determine a change within a pattern. Framework pattern monitor (PAM) [7] aims to monitor rule histories in order to detect interesting short- and long-term changes for instance in the context of email traffic monitoring [7], web usage analysis [8], and retail behaviour [9,10]. In contrast to the other approaches discussed in this section, the monitor treats a history as a dynamic object which is extended permanently after every new data mining session. Early detection of long-term changes is a key issue in this work.

All of these techniques, focus on finding changes in patterns. In our paper we look at finding changes within itemsets which are used to generate patterns.

3 Preliminary

In this section we introduce basic concepts and definitions. Mining frequent itemsets using sliding windows over data streams is described using the following concepts:

Definition 1. Let $\mathcal{I} = \{i_1, i_2, ...i_m\}$ be a set of distinct literals, called items. A subset of items is denoted as an itemset. An itemset with k items is called a k-itemset. A transaction $t = (TID, X)$ is a tuple where TID is a transaction-id and X is an itemset. A data stream S is a sequence of transactions, $S = \{t_1, t_2, ...\}$. We denote $|S|$ as the number of transactions in the data stream S.

Definition 2. The frequency of an itemset X is the number of transactions in the data stream S that contain X. The support value of an itemset X is the frequency of X divided by the total number of transactions in the data stream S, i.e., $sup(X) = \frac{|\{t|t \in S, X \in t\}|}{|S|}$. An itemset is frequent if it satisfies the support threshold (minsup). We denote I as the set of frequent itemsets, when $I = \{X|sup(X) \geq minsup\}$.

Definition 3. A sliding window W_k over a data stream is a block of the last N elements of the stream, where k is the window identifier. There are two variants of sliding windows based on whether N is fixed (fixed-sized sliding windows) or variable (variable-sized sliding windows). Fixed-sized windows are constrained to perform the insertions and deletions in pairs, except in the beginning when exactly N elements are inserted without a deletion, whereas variable-sized windows have no size constraint.

The frequency of an itemset X is the number of transactions in the current sliding window that contain X. The support value of an itemset X is the frequency of X divided by the total number of transactions in the current sliding window W_k, $sup_{W_k}(X) = \frac{|\{t|t \in W_k, X \in t\}|}{|W_k|}$. An association rule or pattern is an implication $X \rightarrow Y$ such that $X \cup Y \subseteq I$ and $X \cap Y = \emptyset$. X is the antecedent and Y is the consequent of the rule. The support of $X \rightarrow Y$ in a particular sliding window W_k is the proportion of transactions in W_k that contains $X \cup Y$. The confidence (conf) of $X \rightarrow Y$ is the proportion of transactions in W_k containing X that also contain Y. For a rule to be considered the $conf(X \rightarrow Y) \geq minconf$, where $minconf$ is a user defined threshold.

The classical change pattern mining technique has three phases: itemset generation phase, rule mining phase and change detection phase. In the initial phase, all the valid frequent patterns are determined based on the support, and then passed into the subsequent rule mining phase. The changed patterns are then evaluated to determine the changes that have occurred. We reason that by only passing itemsets which have changed into the pattern mining algorithm, we reduce the overhead of the generate then prune mechanism.

4   Change Itemset Mining

In our approach we compare two different windows in a data stream. If there is a difference between the sets of itemsets found in one window compared to the other window, we consider that there is a change in the itemsets generated between the two windows. The difference may constitute a modification to existing itemsets in the previous window as compared to the recent window, or a significant change in the support of the itemsets. Figure 1 shows an example of two windows W_k and $W_{k'}$. Here k and k'

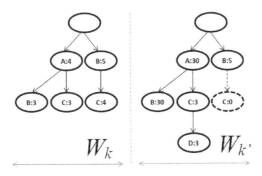

Fig. 1. Example of change between two windows

represent the index where the windows occur in the data stream, where $k < k'$. In this example two trees are built based on the transactions within the window. Current techniques that deal with data streams are based on a tree structure, this is also the structure we adopt in our technique. Most of these trees such as the DS-Tree[11] and SPO-Tree [12] store data in descending frequency order, which means that there is a compact representation of its current state. Each branch in the tree can represent an itemset, if it passes the minimum support threshold requirement. In this example, only items that fulfill the minsup threshold are shown. In the figure each item is represented by a solid node and each node contains its item and a frequency counter. A node which has a dashed outline represents a deleted node. By comparing the trees, we notice that there are several changes that have occured between these two windows. There is a new item appearing within a current itemset. In W_k we notice an itemset $\{A, C\}$ and in W_k we notice that an additional item has been appended to the itemset extending the item $\{A, C\}$ to $\{A, C, D\}$. We also notice that an itemset that was present has been removed. In W_k, we notice that and item C was removed from the itemset $\{B, C\}$ that was valid in window W_k. An itemset may appear in both windows but the support levels for the itemset may be significantly different. For example, $\{A, B\}$ has support levels that are vastly different between the two windows.

In the following section we define the kinds of changes to itemsets that we recognise, and the algorithm that detects change.

4.1   Types of Changed Itemsets

In this section we discuss the three kinds of changes that our algorithm recognises: extension, reduction, and support fluctuation itemsets.

Definition 4 (Extension Set). Given two sliding windows W_k and $W_{k'}$, if an itemset X exists in W_k and itemset X' exists in $W_{k'}$ and $X' \supset X$, then X' is an extension of X.

For example, given we detect an itemset $\{A, B\}$ in window W_1 and we detect an itemset $\{A, B, C\}$ in window W_2, and $\{A, B, C\}$ has not been generated for W_1 then $\{A, B, C\}$ is an extension of $\{A, B\}$.

Here we can deduce that given an extension itemset X, and two windows W_k and $W_{k'}$:

$$\frac{sup_{W_{k'}}(X')}{sup_{W_k}(X')} = \infty$$

Intuitively this follows similar characteristics of emerging patterns [13].

Definition 5 (Reduction Set). Given two sliding windows W_k and $W_{k'}$, if an itemset X exists in W_k and itemset X' exists in $W_{k'}$ and $X' \subset X$, then X' is a reduction of X.

For example, given we detect an itemset $\{A, B, C\}$ in window W_1 and we detect an itemset $\{A, B\}$ in window W_2, and $\{A, B, C\}$ has not been generated in window W_2 then $\{A, B\}$ is a reduction of $\{A, B, C\}$.

Here we can deduce that given a reduction of an itemset X, and two windows W_k and $W_{k'}$:

$$\frac{sup_{W_{k'}}(X)}{sup_{W_k}(X)} = 0$$

To ensure that we are only detecting itemsets with significant changes, in both the extension and reduction set we also apply a further threshold:

$$|sup_{W_k}(X) - sup_{W_{k'}}(X')| > \epsilon.$$

Most change detection approaches rely on well established test statistics for the difference between the true population and sample mean. A number of bounds exist that do not assume a particular data distribution e.g. Chernoff Inequality and Bernstein Inequality. Due to its generality, Hoeffding's Inequality has been widely used [14], and thus is used in our research. Hoeffding Inequality states that with probability at least $1 - \delta$:

$$Pr\left(\left|\frac{1}{n}\sum_{i=1}^{n} X_i - E[X]\right| > \epsilon\right) \leq 2\exp\left(-n\epsilon^2\right)$$

where $\{X_1, ..., X_n\}$ are independent random variables and $E[X]$ is the expected value or population mean. In summary, this statistical test for different distributions in two windows simply checks whether the observed mean in both windows differs by more than the threshold ϵ.

The Hoeffding bound states that, with probability $1 - \delta$, the true mean of a random variable, r, is at least $\bar{r} - \epsilon$ when the mean is estimated over N samples, where

$$\epsilon = \sqrt{\frac{R^2 ln(1/\delta)}{2N}} \qquad (1)$$

R is the range of r. In our case the variable r has a range, R of 1, and the number of samples N is the window size. Thus for any value of $|sup_{W_k}(X) - sup_{W_{k'}}(X')| > \epsilon$, our assumption of no change is violated and we consider there to be a support fluctuation set.

Definition 6 (Support Fluctuation Set). Given two sliding windows W_k and $W_{k'}$, if an itemset X exists in W_k and X exists in $W_{k'}$ and there is sufficient change in the support of the itemset where $|sup_{W_k}(X) - sup_{W_{k'}}(X)| > \epsilon$ then X is a support fluctuation set. We calculate ϵ using Equation 1.

For example, suppose we detect an itemset $\{A, B, C\}$ in window W_1 and we detect an itemset $\{A, B, C\}$ in window W_2, where the support for itemset $\{A, B, C\}$ in window W_1 is $sup_{W_1}(\{A, B, C\}) = 0.1$ and support for $\{A, B, C\}$ in window W_2 is $sup_{W_2}(\{A, B, C\}) = 0.9$. If the difference of the support values is greater than the ϵ value then $\{A, B, C\}$ is considered a fluctuation set.

4.2 Our Algorithm

In this section we describe the algorithm that detects changes in itemsets. The algorithm is outlined in Algorithm 1.

The windows can either be completely exclusive, or have partial overlaps in its transaction. Figure 2 and 3 show examples of these scenarios. We control the amount of

Fig. 2. Overlapping   **Fig. 3.** Exclusive

overlap with the interval parameter. This is currently a user defined threshold but in the future a more robust input mechanism could be incorporated, for example using an item level drift detector to signal when higher level change itemset mining should occur. When a transaction t is processed, it is inserted into a frequent tree for the data stream[11]. This is a compact data structure that stores all valid itemsets, however the tree can only store up to a certain number of most recent transactions. This is limited by the sliding window size, N. In our experiments, we use two windows which are exclusive.

In our algorithm, we restrict the minimum support value that is used for cutting the large itemsets off. Potentially, the length of extension, reduction and support fluctuation itemsets are bounded by the size of the itemset. Furthermore, the possibility of having an itemset with a large number of items is very small.

Algorithm 1. Change Itemset Detector

Input: S Data Stream, N sliding window size, $interval$ between change itemset detection
Output: $C_{1,2,3}^m$ set of changed itemset.

```
1:  count ← 0
2:  j ← 0
3:  while hasNext(t_i) do
4:      ADDELEMENTTREE(t, W_c, N)
5:      count ← count + 1
6:      if count == interval then
7:          for each i ∈ {I_c|I_c ∩ W_c, sup_{W_c}(I) ≥ minsup} do
8:              Check for Extension Set
9:              if (sup_{W_c}(i))/(sup_{W_p}(i)) = ∞ and |sup_{W_c}(i) − sup_{W_p}(i)| > ε then
10:                  add i to C_1^j
11:             end if
12:             Check for Reduction Set
13:             if (sup_{W_c}(i))/(sup_{W_p}(i)) = 0 and |sup_{W_c}(i) − sup_{W_p}(i)| > ε then
14:                  add i to C_2^j
15:             end if
16:             Check for Support Fluctuation Set
17:             if |sup_{W_c}(i) − sup_{W_p}(i)| > ε then
18:                  add i to C_3^j
19:             end if
20:         end for
21:         W_p ← W_c
22:         count ← 0
23:         j ← j + 1
24:     end if
25: end while
```

---

5   Experimental Results

We have conducted various experiments on both real-world and synthetic datasets to test the efficiency and efficacy of our proposed technique. The algorithms are all coded in Microsoft Visual C++ and run on a Intel Core i5-2500 CPU @ 3.30Ghz with 4GB of RAM running Windows 7x64. In all experiments, runtime excludes I/O costs. In order to evaluate the performance of our change itemset mining technique in relation to the effects of dataset changes, we modified the IBM synthetic data generator to inject changed itemsets into the generated datasets. In all the experiments we set the minimum support threshold (minsup) at 0.10.

We ran three sets of experiments. The first set was to determine the accuracy of the results. The second set was to evaluate the runtime. Lastly, we ran the experiments on real world datasets. We compared our technique, which used Hoeffding bound against the heuristic of having a fixed user defined threshold γ to signal a change. In this instance the change is signaled, if the difference of support for an itemset between two windows is higher than a user defined γ threshold, *i.e.* $|sup_{W_k}(X) - sup_{W_{k'}}(X)| > \gamma$.

5.1   Accuracy

In these experiments we generated three sets of datasets by varying the number of itemsets, window size, and the length of the itemset. In each of the datasets we set the same

Table 1. Average Precision for All Types of Change Itemsets

No of Itemsets	Hoeffding Bound δ			Fixed Threshold γ		
	0.0001	0.001	0.01	0.1	0.01	0.001
10	1.00	1.00	1.00	0.69	0.91	0.91
15	1.00	1.00	1.00	0.02	0.49	0.87
20	1.00	1.00	1.00	0.02	0.40	0.85
25	0.99	0.99	0.99	0.02	0.55	0.88
30	1.00	1.00	1.00	0.01	0.40	0.85
Itemset Length	0.0001	0.001	0.01	0.1	0.01	0.001
4	1.00	1.00	1.00	0.00	0.92	0.92
6	1.00	1.00	1.00	0.00	0.92	0.92
8	1.00	1.00	1.00	0.00	0.92	0.92
10	1.00	1.00	1.00	0.00	0.03	0.10
12	1.00	1.00	1.00	0.01	0.40	0.85
Window Size	0.0001	0.001	0.01	0.1	0.01	0.001
1000	0.92	0.92	0.92	0.01	0.36	0.71
5000	0.98	0.98	0.98	0.01	0.42	0.82
10000	0.99	0.99	0.99	0.01	0.42	0.84
15000	0.99	0.99	0.99	0.01	0.42	0.85
20000	1.00	1.00	1.00	0.01	0.41	0.85
25000	1.00	1.00	1.00	0.01	0.40	0.85

Table 2. Average Recall for All Types of Change Itemsets

No of Itemsets	Hoeffding Bound δ			Fixed Threshold γ		
	0.0001	0.001	0.01	0.1	0.01	0.001
10	1.00	1.00	1.00	0.69	0.91	0.91
15	0.99	0.99	0.99	0.02	0.49	0.87
20	0.99	0.99	0.99	0.02	0.40	0.85
25	0.99	0.99	0.99	0.02	0.55	0.88
30	0.99	0.99	0.99	0.01	0.40	0.85
Itemset Length	0.0001	0.001	0.01	0.1	0.01	0.001
4	1.00	1.00	1.00	0.00	0.92	0.92
6	1.00	1.00	1.00	0.00	0.92	0.92
8	1.00	1.00	1.00	0.00	0.92	0.92
10	0.99	0.99	0.99	0.01	0.03	0.10
12	0.99	0.99	0.99	0.01	0.40	0.85
Window Size	0.0001	0.001	0.01	0.1	0.01	0.001
1000	0.85	0.85	0.85	0.01	0.36	0.71
5000	0.97	0.97	0.97	0.01	0.42	0.82
10000	0.99	0.99	0.99	0.01	0.42	0.84
15000	1.00	1.00	1.00	0.01	0.42	0.85
20000	1.00	1.00	1.00	0.01	0.41	0.85
25000	1.00	1.00	1.00	0.01	0.40	0.85

number of changed itemsets between two windows. We repeated all the experiments ten times and showed the average results.

In the first set of experiments, we varied the number of itemsets I from 10 to 30 possible itemsets, we kept the size of the window N at 25K, the itemset length $|I|$ at 12 and the number of items at 1000. In the second set of experiments we varied the itemset length $|I|$ from 4 to 12, we kept the number of itemsets I at 30 and the window size at 25K, and the number of items at 1000. In the third set of experiments, we varied the window size W from 1K to 25K, we kept the number of itemsets I at 30, the itemset length $|I|$ at 12, and the number of items at 1000. The δ value for the Hoeffding bound was set at either: 0.0001, 0.001, or 0.01. The γ value for the fixed threshold was set at 0.1, 0.01, and 0.001. We varied the γ range so that it provided enough detail on the effects of using a threshold without considering the underlying distribution as a statistical bound would. A small γ value means that the difference of support of itemsets between the two windows can be relatively small for it to be considered a support fluctuation set.

To test whether our algorithm could detect changes accurately we use the precision and recall measures. Precision is the proportion of change itemsets results found that are true positives. Table 1 shows the average precision when we vary either the number of itemsets, itemset length, or window size. We note that the error rate (standard deviation) for the results is close to 0. We also note that we obtain precision values in the range of 0.92 to 1.00, which is better than just setting a fixed threshold.

Recall measures the proportion of actual positives which are correctly identified. Table 2 shows the average recall when we vary either the number of itemsets, itemset length, or window size. We note that the error rate (standard deviation) for the results is close to 0.

Table 3 shows the breakdown of recall rates for support fluctuation set. Due to space constraints, we concentrate on this particular itemset type. The range of recall values using a Hoeffding Bound is from 0.85 to 1.00.

Table 3. True Positive Rates (Recall) for the Support Fluctuation Set

No of Itemsets	Hoeffding Bound δ			Fixed Threshold γ		
	0.0001	0.001	0.01	0.1	0.01	0.001
10	1.00	1.00	1.00	0.10	0.10	0.10
15	0.99	0.99	0.99	0.13	0.27	0.16
20	0.99	0.99	0.99	0.15	0.38	0.18
25	0.99	0.99	0.99	0.12	0.21	0.13
30	0.99	0.99	0.99	0.15	0.38	0.18
Itemset Length	0.0001	0.001	0.01	0.1	0.01	0.001
4	1.00	1.00	1.00	0.00	0.09	0.09
6	1.00	1.00	1.00	0.00	0.09	0.09
8	1.00	1.00	1.00	0.00	0.09	0.09
10	0.99	0.99	0.99	0.01	0.10	0.10
12	0.99	0.99	0.99	0.15	0.38	0.18
Window Size	0.0001	0.001	0.01	0.1	0.01	0.001
1000	0.85	0.85	0.85	0.28	0.78	0.40
5000	0.97	0.97	0.97	0.17	0.42	0.21
10000	0.99	0.99	0.99	0.16	0.37	0.18
15000	1.00	1.00	1.00	0.15	0.36	0.18
20000	1.00	1.00	1.00	0.15	0.37	0.18
25000	1.00	1.00	1.00	0.15	0.38	0.18

5.2 Execution Time

Table 4 shows the average runtime when we vary either the number of itemsets, itemset length, or window size. We only consider the time taken to detect the changes within the windows, we do not consider the I/O cost or the time to build the pattern tree. We note that the execution time of our approach is still faster than setting a fixed threshold. We highlight this fact because we do need to perform additional threshold calculations, as compared to just setting a fixed threshold.

Table 4. Average Runtime in Seconds

No of Itemsets	Hoeffding Bound δ			Fixed Threshold γ		
	0.0001	0.001	0.01	0.1	0.01	0.001
10	0.84	0.83	0.87	1.09	0.83	0.83
15	259.68	259.05	233.46	279.67	269.17	269.37
20	254.46	253.33	228.06	280.36	265.70	265.25
25	264.17	263.00	236.81	289.29	273.78	273.77
30	255.48	253.72	227.89	271.62	265.66	265.30
Itemset Length	0.0001	0.001	0.01	0.1	0.01	0.001
4	0.83	0.82	0.83	0.54	0.82	0.83
6	0.83	0.80	0.81	0.55	0.82	0.84
8	0.86	0.80	0.82	0.56	0.83	0.83
10	20.63	20.45	18.35	19.81	20.97	20.51
12	255.48	253.72	227.89	271.62	265.66	265.30
Window Size	0.0001	0.001	0.01	0.1	0.01	0.001
1000	202.71	202.12	190.04	202.54	210.19	209.26
5000	247.99	247.43	227.64	265.06	257.08	257.37
10000	253.29	251.97	229.90	293.97	263.76	263.83
15000	255.14	254.67	230.33	276.84	264.75	265.06
20000	254.63	253.31	228.31	273.17	264.82	264.67
25000	255.48	253.72	227.89	271.62	265.66	265.30

5.3 Real World Datasets

This section presents the results of comparing our change itemset mining technique on real world datasets from the FIMI repository (http://fimi.ua.ac.be/). We used the Kosarak dataset and the Retail dataset for these experiments.

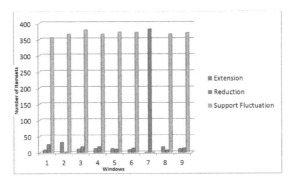

Fig. 4. Hoeffding Bound - Kosarak

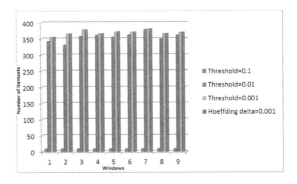

Fig. 5. Support Fluctuation Set - Kosarak

With the Kosarak dataset, we divided the dataset into ten exclusive windows. Each window is represented by 100,000 transactions. Figure 4 presents the results obtained using our approach with $\delta = 0.001$. Figure 5 presents the comparison between fixed threshold and Hoeffding bound, which is shown in Equation 1, for the support fluctuation set. In these experiments we used a γ value of 0.1, 0.01, and 0.001 for the fixed threshold. However, in the experiments, we noticed that there is no variation in the results when Hoeffding bound delta value varied from 0.1 to 0.0001. The Hoeffding bound algorithm is considered more stable and efficient than the fixed threshold approach.

With the Retail dataset, we divided the dataset into nine exclusive windows. Each window is represented by 10,000 transactions. Figure 6 presents the results obtained using our approach with $\delta = 0.001$. Figure 7 presents the comparison between fixed threshold and Hoeffding bound for the support fluctuation set. We note that for both sets of experiments the results from our technique using a Hoeffding bound δ value of 0.0001, 0.001, and 0.01 are the same.

From the experiments, we show how our approach works on real world datasets. We also notice that if we used a user defined γ value, the number of changed itemsets detected would vary based on the threshold used.

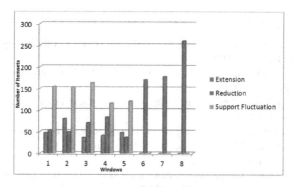

Fig. 6. Hoeffding Bound - Retail

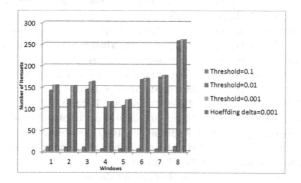

Fig. 7. Support Fluctuation Set - Retail

6 Contribution

The contributions of our research are divided into two parts. The first contribution our research makes is to formally characterise the type of changes that can occur within itemsets. The second contribution of our research is a novel detection technique for change detection at the itemset level. The difference between this approach and traditional change mining techniques is that we identify change at a structural level, *i.e.* changes within the combination of items, instead of purely relying on support.

7 Conclusions and Future Work

In this paper we presented a new technique to find changed itemsets for a data stream environment. We described three different types of possible changes within itemsets. We tested our results using both synthetic and real world datasets. We show that we are able to detect itemsets which have changed with a high accuracy while maintaining a fast runtime. In the future we will look at the impact of using different statistical bounds, such as the Bernstein bound. We will also look at incorporating drift detection to detect item level change, before analysing itemset changes.

References

1. Roddick, J.F., Spiliopoulou, M., Lister, D., Ceglar, A.: Higher order mining. SIGKDD Explor. Newsl. 10(1), 5–17 (2008)
2. Agrawal, R., Psaila, G.: Active data mining. In: Fayyad, U.M., Uthurusamy, R. (eds.) KDD, pp. 3–8. AAAI Press (1995)
3. Au, W.H., Chan, K.C.C.: Mining changes in association rules: a fuzzy approach. Fuzzy Sets Syst. 149(1), 87–104 (2005)
4. Steinbrecher, M., Kruse, R.: Identifying temporal trajectories of association rules with fuzzy descriptions. In: Annual Meeting of the North American Fuzzy Information Processing Society, NAFIPS 2008, pp. 1–6 (2008)
5. Liu, B., Ma, Y., Lee, R.: Analyzing the interestingness of association rules from the temporal dimension. In: Proceedings of the 2001 IEEE International Conference on Data Mining. ICDM 2001, pp. 377–384. IEEE Computer Society, Washington, DC (2001)
6. Böttcher, M., Höppner, F., Spiliopoulou, M.: On exploiting the power of time in data mining. SIGKDD Explor. Newsl. 10(2), 3–11 (2008)
7. Baron, S., Spiliopoulou, M., Günther, O.: Efficient monitoring of patterns in data mining environments. In: Kalinichenko, L.A., Manthey, R., Thalheim, B., Wloka, U. (eds.) ADBIS 2003. LNCS, vol. 2798, pp. 253–265. Springer, Heidelberg (2003)
8. Baron, S., Spiliopoulou, M.: Monitoring the evolution of web usage patterns. In: Berendt, B., Hotho, A., Mladenič, D., van Someren, M., Spiliopoulou, M., Stumme, G. (eds.) EWMF 2003. LNCS (LNAI), vol. 3209, pp. 181–200. Springer, Heidelberg (2004)
9. Song, H.S., Kim, J.K., Kim, S.H.: Mining the change of customer behavior in an internet shopping mall. Expert Systems with Applications 21(3), 157–168 (2001)
10. Chen, M.C., Chiu, A.L., Chang, H.H.: Mining changes in customer behavior in retail marketing. Expert Syst. Appl. 28(4), 773–781 (2005)
11. Leung, C.S., Khan, Q.: Dstree: A tree structure for the mining of frequent sets from data streams. In: Sixth International Conference on Data Mining, ICDM 2006, pp. 928–932 (2006)
12. Koh, Y.S., Dobbie, G.: Spo-tree: Efficient single pass ordered incremental pattern mining. In: Cuzzocrea, A., Dayal, U. (eds.) DaWaK 2011. LNCS, vol. 6862, pp. 265–276. Springer, Heidelberg (2011)
13. Bailey, J., Manoukian, T., Ramamohanarao, K.: Fast algorithms for mining emerging patterns. In: Elomaa, T., Mannila, H., Toivonen, H. (eds.) PKDD 2002. LNCS (LNAI), vol. 2431, pp. 39–50. Springer, Heidelberg (2002)
14. Hoeffding, W.: Probability inequalities for sums of bounded random variables. Journal of the American Statistical Association 58, 13–29 (1963)

TKS: Efficient Mining of Top-K Sequential Patterns

Philippe Fournier-Viger[1], Antonio Gomariz[2], Ted Gueniche[1],
Espérance Mwamikazi[1], and Rincy Thomas[3]

[1] Department of Computer Science, University of Moncton, Canada
[2] Dept. of Information and Communication Engineering, University of Murcia, Spain
[3] Department of Computer Science, SCT, Bhopal, India
{philippe.fournier-viger,etg8697,eem7706}@umoncton.ca,
rinc_thomas@rediffmail.com, agomariz@um.es

Abstract. Sequential pattern mining is a well-studied data mining task with wide applications. However, fine-tuning the *minsup* parameter of sequential pattern mining algorithms to generate enough patterns is difficult and time-consuming. To address this issue, the task of top-*k* sequential pattern mining has been defined, where *k* is the number of sequential patterns to be found, and is set by the user. In this paper, we present an efficient algorithm for this problem named TKS (*Top-K Sequential pattern mining*). TKS utilizes a vertical bitmap database representation, a novel data structure named PMAP (*Precedence Map*) and several efficient strategies to prune the search space. An extensive experimental study on real datasets shows that TKS outperforms TSP, the current state-of-the-art algorithm for top-*k* sequential pattern mining by more than an order of magnitude in execution time and memory.

Keywords: top-k, sequential pattern, sequence database, pattern mining.

1   Introduction

Various methods have been proposed for mining temporal patterns in sequence databases such as mining repetitive patterns, trends and sequential patterns [8]. Among them, sequential pattern mining is probably the most popular set of techniques. Given a user-defined threshold *minsup* and a set of sequences, it consists of discovering all subsequences common to more than *minsup* sequences [1]. It is a well-studied data mining problem with wide applications such as the analysis of web click-streams, program executions, medical data, biological data and e-learning data [5, 8]. Although many studies have been done on designing sequential pattern mining algorithms [1, 2, 3, 8], an important problem is how the user should choose the *minsup* threshold to generate a desired amount of patterns. This problem is important because in practice, users have limited resources (time and storage space) for analyzing the results and thus are often only interested in discovering a certain amount of patterns, and fine-tuning the *minsup* parameter is time-consuming. Depending on the choice of the *minsup* threshold, algorithms can become very slow and generate an extremely large amount of results or generate none or too few results, omitting valuable information. To address this problem, it was proposed to redefine the problem of mining sequential patterns as the problem of mining the top-*k*

H. Motoda et al. (Eds.): ADMA 2013, Part I, LNAI 8346, pp. 109–120, 2013.
© Springer-Verlag Berlin Heidelberg 2013

sequential patterns, where k is the number of sequential patterns to be found and is set by the user. The current best algorithm for this problem is TSP [4]. However, in our experimental study, (cf. section 4), we found that it does not perform well on dense datasets. Therefore, an important research question is could we develop a more efficient algorithm for top-*k* sequential pattern mining than TSP? In this paper, we address this research question by proposing a novel algorithm named TKS (Top-K Sequential pattern mining). TKS is an efficient top-*k* algorithm for sequential pattern mining. It uses the same vertical database representation and basic candidate generation procedure as SPAM [3]. Moreover, TKS incorporates several efficient strategies to prune the search space and rely on a novel data structure named PMAP (Precedence Map) for avoiding costly bit vector intersection operations. An extensive experimental study with five real datasets shows that (1) TKS outperforms the state-of-the-art algorithm (TSP) by more than an order of magnitude in terms of execution time and memory usage. Moreover, we found that TKS has excellent performance on dense datasets.

The rest of the paper is organized as follows. Section 2 formally defines the problem of sequential pattern mining and top-*k* sequential pattern mining, and presents related work. Section 3 describes the TKS algorithm. Section 4 presents the experimental study. Finally, Section 5 presents the conclusion and discusses future work.

2   Problem Definition and Related Work

The problem of sequential pattern mining was proposed by Agrawal and Srikant [1] and is defined as follows. A *sequence database SDB* is a set of sequences $S = \{s_1, s_2..., s_s\}$ and a set of items $I = \{i_1, i_2, ..., i_m\}$ occurring in these sequences. An *item* is a symbolic value. An *itemset* $I = \{i_1, i_2, ..., i_m\}$ is an unordered set of distinct items. For example, the itemset $\{a, b, c\}$ represents the sets of items a, b and c. A *sequence* is an ordered list of itemsets $s = \langle I_1, I_2, ..., I_n \rangle$ such that $I_k \subseteq I$ for all $1 \leq k \leq n$. For example, consider the sequence database *SDB* depicted in Figure 1.a. It contains four sequences having respectively the *sequences ids* (SIDs) 1, 2, 3 and 4. In this example, each single letter represents an item. Items between curly brackets represent an itemset. For instance, the first sequence $\langle \{a, b\}, \{c\}, \{f\}, \{g\}, \{e\} \rangle$ indicates that items a and b occurred at the same time, were followed successively by c, f, g and lastly e. A sequence $s_a = \langle A_1, A_2, ..., A_n \rangle$ is said to be *contained in* another sequence $s_b = \langle B_1, B_2,..., B_m \rangle$ if and only if there exists integers $1 \leq i_1 < i_2 < ... < i_n \leq m$ such that $A_1 \subseteq B_{i1}$, $A_2 \subseteq B_{i2}$, ..., $A_n \subseteq B_{in}$ (denoted as $s_a \sqsubseteq s_b$). The *support of a subsequence* s_a in a sequence database *SDB* is defined as the number of sequences $s \in S$ such that $s_a \sqsubseteq s$ and is denoted by $sup(s_a)$. The *problem of mining sequential patterns* in a sequence database *SDB* is to find all frequent sequential patterns, i.e. each subsequence s_a such that $sup(s_a) \geq minsup$ for a threshold *minsup* set by the user. For example, Figure 1.b shows six of the 29 sequential patterns found in the database of Figure 1.a for *minsup* = 2. Several algorithms have been proposed for the problem sequential pattern mining such as PrefixSpan [2], SPAM [3], GSP and SPADE [9].

Problem Definition. To address the difficulty of setting *minsup,* the problem of sequential pattern mining was redefined as the *problem of top-k sequential pattern mining* [4]. It is to discover a set L containing k sequential patterns in a sequence database

SDB such that for each pattern $s_a \in L$, there does not exist a sequential pattern $s_b \notin L$ | $sup(s_b) > sup(s_a)$. For example, for the database of Figure 1.a and $k = 10$, the top-k sequential patterns are $\langle\{g\}\rangle$, $\langle\{a\},\{f\}\rangle$, $\langle\{a\}\rangle$, $\langle\{b\}, \{e\}\rangle$, $\langle\{b\}, \{g\}\rangle$, $\langle\{a\}, \{e\}\rangle$ and$\langle\{e\}\rangle$ with a support of 3, and $\langle\{b\}, \{f\}\rangle$, $\langle\{b\}\rangle$ and $\langle\{f\}\rangle$, with a support of 4. The definition of this problem is analogous to the definition of other top-k problems in the field of pattern mining such as top-k frequent itemset mining [1], top-k association rule mining [6] and top-k sequential rule mining.

The current state-of-the-art algorithm for top-k sequential pattern mining is TSP [4]. Two versions of TSP have been proposed for respectively mining (1) top-k sequential patterns and (2) top-k closed sequential patterns. In this paper, we are addressing the first case. Extending our algorithm to the second case will be considered in future work. The TSP algorithm is based on PrefixSpan [2]. TSP first generates frequent sequential patterns containing a single item. Then it recursively extends each pattern s by (1) projecting the database by s, (2) scanning the resulting projected database to identify items that appear more than *minsup* times after s, and (3) append these items to s. The main benefit of this projection-based approach is that it only considers patterns appearing in the database unlike "generate-and-test" algorithms [1, 4]. However, the downside of this approach is that projecting/scanning databases repeatedly is costly, and that cost becomes huge for dense databases where multiples projections have to be performed (c.f. experimental study presented in Section 4). Given this limitation, an important research challenge is to define an algorithm that would be more efficient than TSP and that would perform well on dense datasets.

SID	Sequences
1	$\langle\{a, b\},\{c\},\{f, g\},\{g\},\{e\}\rangle$
2	$\langle\{a, d\},\{c\},\{b\},\{a, b, e, f\}\rangle$
3	$\langle\{a\},\{b\},\{f\},\{e\}\rangle$
4	$\langle\{b\},\{f, g\}\rangle$

ID	Pattern	Supp.
p1	$\langle\{a\},\{f\}\rangle$	3
p2	$\langle\{a\},\{c\}\{f\}\rangle$	2
p3	$\langle\{b\},\{f,g\}\rangle$	2
p4	$\langle\{g\},\{e\}\rangle$	2
p5	$\langle\{c\},\{f\}\rangle$	2
p6...	$\langle\{b\}\rangle$	4

Fig. 1. A sequence database (left) and some sequential patterns found (right)

3   The TKS Algorithm

We address this research challenge by proposing a novel algorithm named TKS. TKS employs the vertical database representation and basic candidate-generation procedure of SPAM [2]. Furthermore, it also includes several efficient strategies to discover top-k sequential pattern efficiently. The next subsection reviews important concepts of the SPAM algorithm. Then, the following subsection describes the TKS algorithm.

3.1   The Database Representation and Candidate Generation Procedure

The *vertical database representation* [3] used in TKS is defined as follows. Let *SDB* be a sequence database containing q items and m sequences, where $size(i)$ denotes the number of itemsets in the i-th sequence of *SDB*. The *vertical database representation* $V(SDB)$ of *SDB* is defined as a set of q bit vectors of size $\sum_{i=1}^{m} size(i)$, such that each item x has a corresponding bit vector $bv(x)$. For each bit vector, the j-th bit represents the p-th itemset of the t-th sequence of SDB, such that $\sum_{i=1}^{\min(0,t-1)} size(i) < j < \sum_{i=1}^{t} size(i)$ and $p = j - \sum_{i=1}^{\min(0,t-1)} size(i)$. Each bit of a bit vector $bv(x)$ is set to 1 if and only if x appears in the itemset represented by this bit, otherwise, it is set to 0. For example, the left part of Table 1 shows the bit vectors constructed for each item from the database of Figure 1.

Table 1. The vertical representation (left) and PMAP data structure (right)

item	bit vector
a	100001001100000
b	100000011010010
c	010000100000000
d	000001000000000
e	000010001000100
f	001000001001001
g	001100000000001

item	pairs of type <item, support>
a	<a,1, s> <b,2, s>, <b, 2, i>, <c, 2, s>, <d,1, i>, <e,2, s>, <e,1, i>, <f,2, s>, <f,1, i>, <g,1, s>
b	<a,1,s>, <b,1,s>,<c, 1,s>, <e,2,s>, <e,1,i>, <f, 4,s>, <f, 1,i>, <g,2,s>
c	<a,1,s>, <b,1,s>, <e,2,s>, <f,2,s>, <g,1,s>
d	<a,1,s>, <b,1,s>,<c,1,s>, <e,1,s>, <f,1,s>
e	<f,1,i>
f	<e,2,s>, <g,2,i>
g	<e,1,s>

The procedure for candidate generation [3] is presented in Figure 2. It takes as parameter a sequence database SDB and the *minsup* threshold. It first scans SDB once to construct $V(SDB)$. At the same time, it counts the support of each single item. Then, for each frequent item s, it calls the procedure "SEARCH". This procedure outputs the pattern $\langle\{s\}\rangle$ and recursively explore candidate patterns starting with the prefix $\langle\{s\}\rangle$. The SEARCH procedure (cf. Figure 3) takes as parameters a sequential pattern *pat* and two sets of items to be appended to *pat* to generate candidates. The first set S_n represents items to be appended to *pat* by *s-extension*. The result of the *s-extension* of a sequential pattern $\langle I_1, I_2, ..., I_n \rangle$ with an item x is $\langle I_1, I_2, ..., I_n, \{x\}\rangle$ [3]. The second set S_n represents items to be appended to *pat* by *i-extension*. The result of the *i-extension* of a sequential pattern $\langle I_1, I_2, ..., I_n \rangle$ with an item x is $\langle I_1, I_2, ..., I_n \cup \{x\}\rangle$ [3]. For each candidate *pat'* generated by extension, SPAM calculates the candidate's bit vector $bv(pat')$ by performing a modified logical AND (see [3] for details) of the bit vectors associated to *pat* and the appended item. The support of the candidate is calculated without scanning SDB by counting the number of bits set to 1 representing distinct sequences in $bv(pat')$ [3]. If the pattern *pat'* is frequent, it is then used in a recursive call to SEARCH to generate patterns starting with the prefix *pat'*. Note that in the recursive call, only items that resulted in a frequent pattern by extension of *pat* will be considered for extending *pat'* (see [3] for justification). Moreover, note that infrequent patterns are not extended by the SEARCH procedure because of the Apriori property (any infrequent sequential pattern cannot be extended to form a frequent pattern) [3]. The candidate generation procedure is very efficient in dense datasets because performing the AND operation for calculating the support does not require scanning the original database unlike the projection-based approach of TSP, which in the worst case performs a database projection for each item appended to a pattern. However, there is a potential downside to the candidate generation procedure of SPAM. It is that it can generate candidates not occurring in the database. Therefore, it is not obvious that building a top-k algorithm based on this procedure would result in an efficient algorithm.

SPAM(*SDB*, *minsup*)
1.   Scan *SDB* to create *V(SDB)* and identify S_{init}, the list of frequent items.
2.   **FOR** each item s ∈ S_{init},
3.   **SEARCH**($\langle s \rangle$), S_{init}, the set of items from S_{init} that are lexically larger than *s*, *minsup*).

Fig. 2. The SPAM algorithm

SEARCH(*pat*, S_n, I_n, *minsup*)
1.   Output pattern *pat*.
2. $S_{temp} := I_{temp} := \emptyset$
3.   **FOR** each item *j* ∈ S_n,
4.   **IF** the s-extension of *pat* is frequent **THEN** $S_{temp} := S_{temp} \cup \{i\}$.
5.   **FOR** each item *j*∈ S_{temp},
6.   **SEARCH**(the s-extension of *pat* with *j*, S_{temp}, elements in S_{temp} greater than *j*, *minsup*).
7.   **FOR** each item *j* ∈ I_n,
8.   **IF** the i-extension of *pat* is frequent **THEN** $I_{temp} := I_{temp} \cup \{i\}$.
9.   **FOR** each item *j* ∈ I_{temp},
10.   **SEARCH**(i-extension of *pat* with *j*, S_{temp}, all elements in I_{temp} greater than *j*, *minsup*).

Fig. 3. The candidate generation procedure

3.2   The TKS Algorithm

We now present our novel top-*k* sequential pattern mining algorithm named TKS. It takes as parameters a sequence database *SDB* and *k*. It outputs the set of top-*k* sequential patterns contained in *SDB*.

Strategy 1. Raising Support Threshold. The basic idea of TKS is to modify the main procedure of the SPAM algorithm to transform it in a top-*k* algorithm. This is done as follows. To find the top-*k* sequential patterns, TKS first sets an internal *minsup* variable to 0. Then, TKS starts searching for sequential patterns by applying the candidate generation procedure. As soon as a pattern is found, it is added to a list of patterns *L* ordered by the support. This list is used to maintain the top-*k* patterns found until now. Once *k* valid patterns are found, the internal *minsup* variable is raised to the support of the pattern with the lowest support in *L*. Raising the *minsup* value is used to prune the search space when searching for more patterns. Thereafter, each time a frequent pattern is found, the pattern is inserted in *L*, the patterns in *L* not respecting *minsup* anymore are removed from *L*, and *minsup* is raised to the value of the least interesting pattern in *L*. TKS continues searching for more patterns until no pattern can be generated, which means that it has found the top-*k* sequential patterns. It can be easily seen that this algorithm is correct and complete given that the candidate generation procedure of SPAM is. However, in our test, an algorithm simply incorporating Strategy 1 does not have good performance.

TKS(*SDB*, *k*)
1. $R := \emptyset$. $L := \emptyset$. *minsup* := 0.
2. Scan *SDB* to create $V(SDB)$.
3. Let S_{init} be the list of items in $V(SDB)$.
4. **FOR** each item $s \in S_{init}$, **IF** *s* is frequent according to bv(*s*) **THEN**
5.   **SAVE**(*s*, *L*, *k*, *minsup*).
6. $R := R \cup \{<s, S_{init}$, items from S_{init} that are lexically larger than *s*>\}.
7. **WHILE** $\exists <r, S_1, S_2> \in R$ AND sup(*r*) $\geq$ *minsup* **DO**
8.   Select the tuple $<r, S_1, S_2>$ having the pattern *r* with the highest support in *R*.
9.   **SEARCH**(*r*, S_1, S_2, *L*, *R*, *k*, *minsup*).
10.   **REMOVE** $<r, S_1, S_2>$ from R.
11.   **REMOVE** from *R* all tuples $<r, S_1, S_2> \in R$ | sup(*r*) < *minsup*.
12. **END WHILE**
13. **RETURN** *L*.

Fig. 4. The TKS algorithm

SEARCH(*pat*, S_n, I_n, *L*, *R*, *k*, *minsup*)
1. $S_{temp} := I_{temp} := \emptyset$
2. **FOR** each item $j \in S_n$,
3.   **IF** the s-extension of *pat* is frequent **THEN** $S_{temp} := S_{temp} \cup \{i\}$.
4. **FOR** each item $j \in S_{temp}$,
5.   **SAVE**(s-extension of *pat* with *j*, *L*, *k*, *minsup*).
6. $R := R \cup \{<$ s-extension of *pat* with *j*, S_{temp}, all elements in S_{temp} greater than *j*>\}.
7. **FOR** each item $j \in I_n$,
8.   **IF** the i-extension of *pat* is frequent **THEN** $I_{temp} := I_{temp} \cup \{i\}$.
9. **FOR** each item $j \in I_{temp}$,
10.   **SAVE**(i-extension of *pat* with *j*, *L*, *k*, *minsup*).
11. $R := R \cup \{<$ the s-extension of *pat* with *j*, S_{temp}, all elements in I_{temp} greater than *j* >\}.

Fig. 5. The modified candidate generation procedure

SAVE(*r*, *L*, *k*, *minsup*)
1. $L := L \cup \{r\}$.
2. **IF** |*L*| > *k* **THEN**
3.   **IF** sup(*r*) > *minsup* **THEN**
4.   **WHILE** |*L*| > *k* AND $\exists s \in L$ | sup(*s*) = *minsup*, **REMOVE** *s* from *L*.
5.   **END IF**
6.   Set *minsup* to the lowest support of patterns in *L*.
7. **END IF**

Fig. 6. The SAVE procedure

Strategy 2. Extending the Most Promising Patterns. To improve the performance of TKS, we have added a second strategy. It is to try to generate the most promising sequential patterns first. The rationale of this strategy is that if patterns with high support are found earlier, it allows TKS to raise its internal *minsup* variable faster, and thus to prune a larger part of the search space. To implement this strategy, TKS uses an internal variable *R* to maintain at any time the set of patterns that can be extended to generate candidates. TKS then always extends the pattern having the highest support first.

The pseudocode of the TKS version incorporating Strategy 1 and Strategy 2 is shown in Figure 4. The algorithm first initializes the variables R and L as the empty set, and *minsup* to 0 (line *1*). Then, *SDB* is scanned to create $V(SDB)$ (line *2*). At the same time, a list of all items in *SDB* is created (S_{init}) (line *3*). For each item s, its support is calculated based on its bit vector $bv(s)$ in $V(SDB)$. If the item is frequent, the SAVE procedure is called with $\langle s \rangle$ and L as parameters to record $\langle s \rangle$ in L (line *4* and *5*). Moreover, the tuple $<s, S_{init,}$ items from S_{init} that are lexically larger than $s>$ is saved into R to indicate that $\langle s \rangle$ can be extended to generate candidates (line *6*). After that, a WHILE loop is performed. It recursively selects the tuple representing the pattern r with the highest support in R such that $sup(r) \geq minsup$ (line *7* and *8*). Then the algorithm uses the tuple to generate patterns by using the SEARCH procedure depicted in Figure 5 (line *9*). After that, the tuple is removed from R (line *10*), as well as all tuples for patterns that have become infrequent (line *11*). The idea of the WHILE loop is to always extend the rule having the highest support first because it is more likely to generate rules having a high support and thus to allow to raise *minsup* more quickly for pruning the search space. The loop terminates when there is no more pattern in R with a support higher than *minsup*. At this moment, the set L contains the top-k sequential patterns (line *13*).

The SAVE procedure is shown in Figure 6. Its role is to raise *minsup* and update the list L when a new frequent pattern r is found. The first step of SAVE is to add the pattern r to L (line *1*). Then, if L contains more than k patterns and the support is higher than *minsup*, patterns from L that have exactly the support equal to *minsup* can be removed until only k rules are kept (line *3 to 5*). Finally, *minsup* is raised to the support of the rule in L having the lowest support. (line *6*). By this simple scheme, the top-k rules found are maintained in L.

Note that to improve the performance of TKS, in our implementation, sets L and R are implemented with data structures supporting efficient insertion, deletion and finding the smallest/largest element. In our implementation, we used a Fibonacci heap for L and R. It has an amortized time cost of $O(1)$ for insertion and obtaining the minimum/maximum, and $O(log(n))$ for deletion [9].

Strategy 3. Discarding Infrequent Items in Candidate Generation. This strategy improves TKS' execution time by reducing the number of bit vector intersections performed by the SEARCH procedure. The motivation behind this strategy is that we found that a major cost of candidate generation is performing bit vector intersections because bit vectors can be very long for large datasets. This strategy is implemented in two phases. First, TKS records in a hash table the list of items that become infrequent when *minsup* is raised by the algorithm. This is performed in line 6 of the SAVE procedure by replacing "**REMOVE** s from L." by "**REMOVE** s from L and **IF** s contains a single item **THEN** register it in the hash map of discarded items". Note that it is not necessary to record infrequent items discovered during the creation of $V(SDB)$ because those items are not considered for pattern extension.

Second, each time that the SEARCH procedure considers extending a sequential pattern *pat* with an item x (by *s-extension* or *i-extension*), the item is skipped if it is contained in the hash table. Skipping infrequent items allows avoiding performing costly bit vector intersections for these items. Integrating this strategy does not affect the output of the algorithm because appending an infrequent item to a sequential pattern cannot generate a frequent sequential pattern.

Strategy 4. Candidate Pruning with Precedence Map. We have integrated a second strategy in TKS to reduce more aggressively the number of bit vector intersections. This strategy requires building a novel structure that we name *Precedence Map (PMAP)*. This structure is built with a single database scan over *SDB*. The PMAP structure indicates for each item i, a list of triples of the form $<j, m, x>$ where m is an integer representing the number of sequences where j appears after i in *SDB* by x-extension ($x \in \{i, s\}$). Formally, an item i is said to *appear after an item j* by s-extension in a sequence $\langle A_1, A_2, ..., A_n \rangle$ if $j \in A_x$ and $i \in A_y$ for integers x and y such that $1 \le x < y \le n$. An item i is said to *appear after an item j* by i-extension in a sequence $\langle A_1, A_2, ..., A_n \rangle$ if $i, j \in A_x$ for an integer x such that $1 \le x \le n$ and j is lexicographically greater than i. For example, the PMAP structure built for the sequence database of Figure 1 is shown in right part of Table 1. In this example, the item f is associated with the pair $<e, 2, s>$ because e appears after f by s-extension in two sequences. Moreover, f is associated with the pair $<g, 2, i>$ because g appears after f by *i-extension* in two sequences. To implement PMAP, we first considered using two matrix (one for *i-extensions* and one for *s-extension*). However, for sparse datasets, several entries would be empty, thus potentially wasting large amount of memory. For this reason, we instead implemented PMAP as a hash table of hash sets. Another key implementation decision for PMAP is when the structure should be built. Intuitively, one could think that constructing PMAP should be done during the first database scan at the same time as $V(SDB)$ is constructed. However, to reduce the size of PMAP, it is better to build it in a second database scan so that infrequent items can be excluded from PMAP during its construction. The PMAP structure is used in the SEARCH procedure, which we modified as follows. Let a sequential pattern *pat* being considered for *s-extension* (equivalently *i-extension*) with an item x. If there exists an item a in *pat* associated to an entry $<x, m, s>$ in PMAP (equivalently $<x, m, i>$) and $m < minsup$, then the pattern resulting from the extension of *pat* with x will be infrequent and thus the bit vector intersection of x with *pat* does not need to be done. It can be easily seen that this pruning strategy does not affect the algorithm output, since if an item x does not appear more than *minsup* times after an item y from a pattern *pat*, any pattern containing y followed by x will be infrequent. Furthermore, x can be removed from S_{temp} (equivalently I_{temp}).

Optimizations of the Bit Vector Representation. Beside the novel strategies that we have introduced, optimizations can be done to optimize the bit vector representation and operations. For example, bit vectors can be compressed if they contain contiguous zeros and it is possible to remember the first and last positions of bits set to 1 in each bit vector to reduce the cost of intersection.   These optimizations are not discussed in more details here due to the space limitation.

Correctness and Completeness of the Algorithm. Since SPAM is correct and complete and Strategy 2, 3 and 4 have no influence on the output (only parts of the search space that lead to infrequent patterns is pruned), it can be concluded that TKS is correct and complete.

4   Experimental Study

We performed multiple experiments to assess the performance of the TKS algorithm. Experiments were performed on a computer with a third generation Core i5 processor running Windows 7 and 1 GB of free RAM. We compared the performance of TKS with TSP, the state-of-the-art algorithm for top-k sequential pattern mining. All algorithms were implemented in Java. The source code of all algorithms and datasets can be downloaded as part of the SPMF data mining framework (http://goo.gl/hDtdt). All memory measurements were done using the Java API. Experiments were carried on five real-life datasets having varied characteristics and representing four different types of data (web click stream, text from books, sign language utterances and protein sequences). Those datasets are *FIFA*, *Leviathan*, *Bible*, *Sign* and *Snake*. Table 2 summarizes their characteristics.

Table 2. Datasets' Characteristics

dataset	sequence count	distinct item count	avg. seq. length (items)	type of data
Leviathan	5834	9025	33.81 (std= 18.6)	book
Bible	36369	13905	21.64 (std = 12.2)	book
Sign	730	267	51.99 (std = 12.3)	sign language utterances
Snake	163	20	60 (std = 0.59)	protein sequences
FIFA	20450	2990	34.74 (std = 24.08)	web click stream

Experiment 1. Influence of the k Parameter. We first ran TKS and TSP on each dataset while varying k from 200 to 3000 (typical values for a top-k pattern mining algorithm) to assess the influence of k on the execution time and the memory usage of the algorithms. Results for $k = 1000$, 2000 and 3000 are shown in Table 3. As it can be seen in this table, TKS largely outperforms TSP on all datasets in terms of execution time and memory usage. TKS can be more than an order of magnitude faster than TSP and use up to an order of magnitude less memory than TSP. Note that no result are given for TSP for the *FIFA* dataset when $k = 2000$ and $k = 3000$ because it run out of memory. Figure 2 shows detailed results for the *Bible* and *Snake* dataset. From this figure, it can be seen that TKS has better scalability than TSP with respect to k (detailed results are similar for other datasets and not shown due to page limitation). From this experiment, we can also observe that TKS performs very well on dense datasets. For example, on the very dense dataset *Snake*, TKS uses 13 times less memory than TSP and is about 25 times faster for $k = 3000$.

Table 3. Results for k = 1000, 2000 and 3000

Dataset	Algorithm	Execution Time (s)			Maximum Memory Usage (MB)		
		k=1000	k=2000	k=3000	k=1000	k=2000	k=3000
Leviathan	**TKS**	**10**	**23**	**38**	**302**	**424**	**569**
	TSP	103	191	569	663	856	974
Bible	**TKS**	**16**	**43**	**65**	**321**	**531**	**658**
	TSP	88	580	227	601	792	957
Sign	**TKS**	**0.5**	**0.8**	**1.2**	**46**	**92**	**134**
	TSP	4.8	7.6	9.1	353	368	383
Snake	**TKS**	**1.1**	**1.63**	**1.8**	**19**	**38**	**44**
	TSP	19	33	55	446	595	747
FIFA	**TKS**	**15**	**34**	**95**	**436**	**663**	**796**
	TSP	182	O.O.M.	O.O.M.	979	O.O.M.	O.O.M.

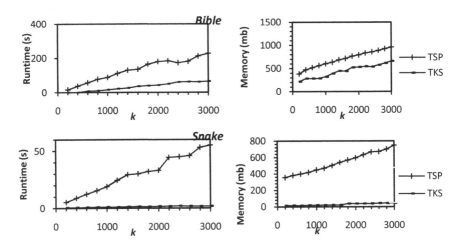

Fig. 7. Results of varying k for the *Bible* and *Snake* datasets

Experiment 2. Influence of the Strategies. We next evaluated the benefit of using strategies for reducing the number of bit vector intersections in TKS. To do this, we compared TKS with a version of TKS without Strategy 4 (TKS W4) and without both Strategy 3 and Strategy 4 (TKS W3W4). We varied k from 200 to 3000 and measured the execution time and number of bit vector intersection performed by each version of TKS. For example, the results for the *Sign* dataset are shown in Figure 8. Results for other datasets are similar and are not shown due to space limitation. As it can be seen on the left side of Figure 8, TKS outperforms TKS W4 and TKS W3W4 in execution time by a wide margin. Moreover, as it can be seen on the right side of Figure 8, Strategy 3 and Strategy 4 are effective strategies that greatly reduce the number of bit vector intersections. Note that we also considered the case of removing without Strategy 2. However, the resulting algorithm would not terminate on most datasets. For this reason, results are not shown.

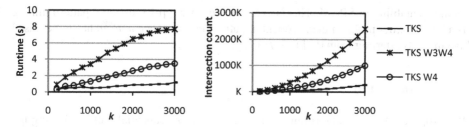

Fig. 8. Influence of the strategies for the *Sign* dataset

Experiment 3. Influence of the Number of Sequences. We also ran TKS and TSP on the five datasets while varying the number of sequences in each dataset to assess the scalability of TKS and TSP. For this experiment, we used k=1000, and varied the database size from 10% to 100 % of the sequences in each dataset. Results are shown in Figure 9 for the *Leviathan* dataset, which provides representative results. We found that both TKS and TSP shown excellent scalability.

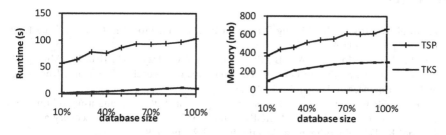

Fig. 9. Influence of the number of sequences for the *Leviathan* dataset

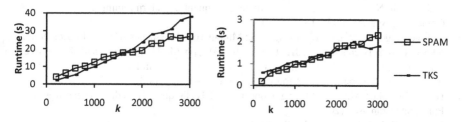

Fig. 10. Comparison of SPAM and TKS runtime for *Leviathan* (left) and *Snake* (right)

Experiment 4. Performance Comparison with SPAM. To further assess the efficiency of TKS, we compared its performance with SPAM for the scenario where the user would tune SPAM with the optimal minimum support threshold to generate k patterns (which is very hard for a user, in practice). The results for the *Leviathan* and *Snake* dataset are shown in Figure 10. Results for other datasets are not shown due to space limitation but are similar. We conclude from the results that the execution time of TKS is very close to that of SPAM for the optimal support and that TKS shows similar scalability to SPAM. This is excellent because top-*k* sequential pattern mining is a much harder problem than sequential pattern mining since *minsup* has to be raised dynamically, starting from 0   [4, 6, 7, 10],

5   Conclusion

We proposed TKS, an algorithm to discover the *top-k* sequential patterns having the highest support, where *k* is set by the user. To generate patterns, TKS relies on a set of efficient strategies and optimizations that enhance its performance. An extensive experimental study show that TKS (1) outperforms TSP, the state-of the art algorithm by more than an order of magnitude in terms of execution time and memory usage, (2) has better scalability with respect to *k* and (3) has a very low performance overhead compared to SPAM. The source code of TKS as well as all the datasets and algorithms used in the experiment can be downloaded from http://goo.gl/hDtdt.

References

1. Agrawal, R., Srikant, R.: Mining Sequential Patterns. In: Proc. Int. Conf. on Data Engineering, pp. 3–14 (1995)
2. Pei, J., Han, J., Mortazavi-Asl, B., Pinto, H., Chen, Q., Dayal, U., Hsu, M.: Mining Sequential Patterns by Pattern-Growth: The PrefixSpan Approach. IEEE Trans. Knowledge and Data Engineering 16(10), 1–17 (2001)
3. Ayres, J., Flannick, J., Gehrke, J., Yiu, T.: Sequential PAttern mining using a bitmap representation. In: Proc. 8th ACM SIGKDD Int. Conf. on Knowledge Discovery and Data Mining (KDD 2002), Edmonton, Alberta, July 23-26, pp. 429–435 (2002)
4. Tzvetkov, P., Yan, X., Han, J.: TSP: Mining Top-k Closed Sequential Patterns. Knowledge and Information Systems 7(4), 438–457 (2005)
5. Mabroukeh, N.R., Ezeife, C.I.: A taxonomy of sequential pattern mining algorithms. ACM Computing Surveys 43(1), 1–41 (2010)
6. Fournier-Viger, P., Wu, C.-W., Tseng, V.S.: Mining Top-K Association Rules. In: Kosseim, L., Inkpen, D. (eds.) Canadian AI 2012. LNCS, vol. 7310, pp. 61–73. Springer, Heidelberg (2012)
7. Kun Ta, C., Huang, J.-L., Chen, M.-S.: Mining Top-k Frequent Patterns in the Presence of the Memory Constraint. VLDB Journal 17(5), 1321–1344 (2008)
8. Han, J., Kamber, M.: Data Mining: Concepts and Techniques, 2nd edn. Morgan Kaufmann Publ., San Francisco (2006)
9. Cormen, T.H., Leiserson, C.E., Rivest, R., Stein, C.: Introduction to Algorithms, 3rd edn. MIT Press, Cambridge (2009)
10. Fournier-Viger, P., Tseng, V.S.: Mining Top-K Sequential Rules. In: Tang, J., King, I., Chen, L., Wang, J. (eds.) ADMA 2011, Part II. LNCS, vol. 7121, pp. 180–194. Springer, Heidelberg (2011)

When Optimization Is Just an Illusion

Muhammad Marwan Muhammad Fuad

Forskningsparken 3, Institutt for kjemi, NorStruct
The University of Tromsø, The Arctic University of Norway
NO-9037 Tromsø, Norway
mfu008@post.uit.no

Abstract. Bio-inspired optimization algorithms have been successfully applied to solve many problems in engineering, science, and economics. In computer science bio-inspired optimization has different applications in different domains such as software engineering, networks, data mining, and many others. However, some applications may not be appropriate or even correct. In this paper we study this phenomenon through a particular method which applies the genetic algorithms on a time series classification task to set the weights of the similarity measures used in a combination that is used to classify the time series. The weights are supposed to be obtained by applying an optimization process that gives optimal classification accuracy. We show in this work, through examples, discussions, remarks, explanations, and experiments, that the aforementioned method of optimization is not correct and that completely randomly-chosen weights for the similarity measures can give the same classification accuracy.

Keywords: Bio-inspired Optimization, Genetic Algorithms, Similarity Measures, Time Series Data Mining.

1   Introduction

Optimization is a ubiquitous problem that has a broad range of applications in engineering, economics, and others. In computer science optimization has different applications in software engineering, networking, data mining and other domains. Optimization can be defined as the action of finding the best-suited solution of a problem subject to given constraints. These constraints can be in the boundaries of the parameters controlling the optimization problem, or in the function to be optimized. Optimization problems can be classified according to whether they are: discrete/ continuous/hybrid, constrained/unconstrained, single objective/multiobjective, unimodal (one extreme point) /multimodal (several extreme points).

Formally, an optimization task can be defined as follows: Let $\vec{X} = \left[x_1, x_2, ..., x_{nbp} \right]$ be the candidate solution to the problem for which we are searching an optimal solution. Given a function $f : U \subseteq \mathbf{R}^{nbp} \rightarrow \mathbf{R}$ (nbp is the number of parameters), find the solution $\vec{X^*} = \left[x_1^*, x_2^*, ..., x_{nbp}^* \right]$ which satisfies $f\left(\vec{X^*} \right) \leq f\left(\vec{X} \right), \forall \vec{X} \in U$. The function f is called the fitness function, the objective function, or the cost function.

H. Motoda et al. (Eds.): ADMA 2013, Part I, LNAI 8346, pp. 121–132, 2013.
© Springer-Verlag Berlin Heidelberg 2013

While fitness functions can sometimes be expressed analytically using mathematical formulas, in many cases there is no mathematical formula to express this function.

Optimization algorithms can be classified in several ways, one of which is whether they are *single solution –based* algorithms; these use one solution and modify it to get the best solution. The other category is *population-based* algorithms; these use several solutions which exchange information to get the best solution.

Optimization problems can be handled using deterministic algorithms or probabilistic ones. *Metaheuristics* are general approximate optimization algorithms which are applicable to a wide range of optimization problems. Metaheuristics are usually applied when the search space is so large, or when the number of parameters of the optimization problem is very high, or, and this is particularly important, when the relationship between the fitness function and the parameters is not clear.

There are several paradigms to handle optimization problems, one of which is *bio-inspired*, also called *nature-inspired*, optimization algorithms. These optimization algorithms are inspired by natural phenomena or by the collective intelligence of natural agents.

Bio-inspired computation can be classified into two main families; the first is *Evolutionary Algorithms* (EA). This family is probably the largest family of bio-inspired algorithms. EA are population-based algorithms that use the mechanisms of Darwinian evolution such as selection, crossover and mutation.

The *Genetic Algorithm* (GA) is the main member of EA. GA is an optimization and search technique based on the principles of genetics and natural selection [8]. GA has the following elements: a population of individuals, selection according to fitness, crossover to produce new offspring, and random mutation of new offspring [11].

In the following we present a description of the simple, classical GA. The first step of GA is defining the problem variables and the fitness function. A particular configuration of variables produces a certain value of the fitness function and the objective of GA is to find the configuration that gives the "best" value of the fitness function. GA starts with a collection of individuals, also called *chromosomes*, each of which represents a possible solution to the problem at hand. This collection of randomly chosen chromosomes constitutes a population whose size *popSize* is chosen by the algorithm designer. This step is called *initialization*. In real-valued encoding GA a candidate solution is represented as a real-valued vector in which the dimension of the chromosomes is equal to the dimension of the solution vectors [1]. This dimension is denoted by *nbp*. The fitness function of each chromosome is evaluated. The next step is *selection*, which determines which chromosomes are fit enough to survive and possibly produce offspring. This is decided according to the fitness function of the chromosome. The percentage of chromosomes selected for mating is denoted by *sRate*. *Crossover* is the next step in which offspring of two parents are produced to enrich the population with fitter chromosomes. *Mutation*, which is a random alteration of a certain percentage *mRate* of chromosomes, is the other mechanism which enables the GA to examine unexplored regions in the search space.

Now that a new generation is formed, the fitting function of the offspring is calculated and the above procedures repeat for a number of generations *nGen* or until a stopping criterion terminates the algorithm.   □

Data mining is a branch of computer science that handles several tasks, most of which demand extensive computing. As with other fields of research, different papers have proposed applying bio-inspired optimization to data mining tasks [12], [13], [14], [15], [16]. Most of these tasks, however, include non-analytical fitness functions and the relationship between the parameters and fitness function is vague. Under certain circumstances this may result in "pseudo optimization", that is; an optimization algorithm that seems to be functioning normally, but the outcome is unintuitive, or not different from what a completely random choice of values for the parameters can yield.

   We will show in this paper the consequences of an inappropriate application of bio-inspired optimization on a particular data mining problem, and we try to explain the reasons for the unintuitive results obtained. In Section 2 we present this application we are referring to, in Section 3 we give our remarks on this application, we show through a counter example in Section 4 that the results of the application, which are supposed to be the outcome of an optimization process, can be obtained through a random solution of the problem, and we try to explain in Section 5 why the application we are referring to did not work, we conclude the paper in Section 6.

2   On Using the Genetic Algorithms to Combine Similarity Measures of Time Series in a Classification Task

A *time series* is a collection of observations at intervals of time points. These observations are measurements of a particular phenomenon. Formally, an n-dimensional time series S is an ordered collection:

$$S = \{(t_1, v_1), (t_2, v_2), ..., (t_n, v_n)\} \tag{1}$$

where $t_1 < t_2 < ... < t_n$, and where v_i are the values of the observed phenomenon at time points t_i.

   Time series data mining handles several tasks such as classification, clustering, similarity search, motif discovery, anomaly detection, and others.

   The goal of classification, one of the main tasks of data mining, is to assign an unknown object to one out of a given number of classes, or categories [10]. One of the most popular classification techniques of time series is *Nearest-Neighbor Classification* (NNC). In NNC the time series-query is classified according to the majority of its nearest neighbors [10].

   A Classification task is based on another, fundamental task of data mining, which is the *similarity search* problem. In this problem a pattern or a *query* is given and the similarity search task is to retrieve the data objects in the database that are "close" to that query according to some semantics that quantify this closeness. This closeness or similarity is quantified using a principal concept which is the *similarity measure* or its more rigorous concept; the *distance metric*.

There are many similarity measures or distance metrics (in this paper we will use these two terms interchangeably) in the field of time series data mining. The state-of-the-art distance measures [5] include the *Minkowski distance* (L_p) mainly the *Euclidian distance* (L_2) and the *Manhattan distance* (L_1), *Dynamic Time Warping* (DTW), the *Longest Common Subsequence* (LCSS), the *Edit Distance with Real Penalty* (ERP), the *Edit Distance on Real Sequences* (EDR), *Dissimilarity Distance* (DISSIM), *Similarity Search based on Threshold Queries* (TQ), *Spatial Assembling Distance* (SpADe), and *Sequence Weighted Alignment* (Swale).

In [2] the authors propose utilizing a similarity function defined as a weighted combination of several metrics. A similar idea was proposed in [3] where the authors present a retrieval method based on a weighted combination of feature vectors.

The method we are discussing in this work [6] is called "Combination of Similarity Measures for Time Series Classification Using Genetic Algorithms". We will refer to this method from now on as CSM-GA. The authors of CSM-GA propose the same idea of a weighted combination of similarity measures. As we can see, the idea is not new, although they describe their method as "novel". They mention however that the closest work to theirs is that of [18], which also uses a weighted combination of distances on a text classification task, and where the optimized weights are also computed by applying the genetic algorithms, so we are not sure what the novelty of CSM-GA is.

It is important to mention here that in our paper we are not going to discuss the validity of [18], since text classification is not our field of expertise (besides the application details are not presented in [18]), neither will we discuss the similarity measures proposed in that paper. We will not discuss either the correctness of applying the genetic algorithms, or any other optimization technique, to determine the weights of the similarity measures in a combination of similarity measures as a method to enhance the retrieval process in any kind of multimedia search in general. We are only discussing the validity of CSM-GA as described in [6].

As mentioned earlier, the idea of CSM-GA is to use a weighted combination of similarity measures for time series, where the weights are determined using the genetic algorithms (GA), and the authors of CSM-GA apply it on a classification task. This is expressed mathematically as:

$$S_{new} = \sum_{i=1}^{n} \omega_i . s_i \qquad (2)$$

where S_{new} is the new similarity measure, ($s_1, s_2, ..., s_n$) are the n combined similarity measures, ω_i are the associated weights, and where $0 \le \omega_i \le 1$.

CSM-GA uses a classical GA (described in Section 1) as an optimization technique. The fitness function to be optimized is the classification accuracy.

As indicated earlier, the objective of CSM-GA is to find the optimal values of ω_i which yield the highest classification accuracy.

The authors of CSM-GA use 8 similarity measures in the combination of similarity measures of relation (2). The similarity measures used in CSM-GA are the following:

i- Euclidean Distance (L_2): defined between time series p and q as:

$$s_1(p,q) = \sqrt{\sum_{i=1}^{n}(p_i - q_i)^2}$$

ii- Manhattan Distance (L_1): defined as:

$$s_2(p,q) = \left|\sum_{i=1}^{n}(p_i - q_i)\right|$$

iii- Maximum Distance (L_∞):

$$s_3(p,q) = max(|p_1 - q_1|, |p_2 - q_2|, ..., |p_n - q_n|)$$

iv- Mean Dissimilarity:

$$s_4(p,q) = \frac{1}{n}\sum_{i=1}^{n} disim(p_i, q_i), \quad \text{where: } disim(p_i, q_i) = \frac{|p_i - q_i|}{|p_i| - |q_i|}$$

v- Root Mean Square Dissimilarity:

$$s_5(p,q) = \sqrt{\frac{1}{n}\sum_{i=1}^{n} disim(p_i, q_i)^2}$$

vi- Peak Dissimilarity:

$$s_6(p,q) = \frac{1}{n}\sum_{i=1}^{n} peakdisim(p_i, q_i), \quad \text{where: } peakdisim(p_i, q_i) = \frac{|p_i - q_i|}{2\,max(|p_i|, |q_i|)}$$

vii- Cosine Distance:

$$s_7(p,q) = 1 - cos(\theta), \quad \text{where: } cos(\theta) = \frac{p.q}{\|p\|\|q\|}$$

viii- Dynamic Time Warping Distance (DTW): DTW is an algorithm to find the optimal path through a matrix of points representing possible time alignments between two time series $p=\{p_1, p_2, ..., p_n\}$ and $q=\{q_1, q_2, ..., q_m\}$. The optimal alignment can be efficiently calculated via dynamic programming [7]. The dynamic time warping between the two time series is defined as:

$$s_8 = DTW(i,j) = d(i,j) + min\begin{cases} DTW(i, j-1) \\ DTW(i-1, j) \\ DTW(i-1, j-1) \end{cases}, \text{where } 1 \leq i \leq n, 1 \leq j \leq m \quad \square$$

The authors of CSM-GA test their method on a *1*-NN classification task of time series data. In *k*-NN each time series is assigned a class label. Then a "leave one out" prediction mechanism is applied to each time series in turn; i.e. the class label of the chosen time series is predicted to be the class label of its nearest neighbor, defined based on

Table 1. The datasets used in the experiments of CSM-GA

Dataset	Size of Training Set	Size of Validation Set	Size of Test Set
Control Chart	180	120	300
Coffee	18	10	28
Beef	18	12	30
OliveOil	18	12	30
Lightning2	40	20	61
Lightning7	43	27	73
Trace	62	38	100
ECG200	67	33	100

the tested distance function. If the prediction is correct, then it is a hit; otherwise, it is a miss.

The classification error rate is defined as the ratio of the number of misses to the total number of time series [4]. *1*-NN is a special case of *k*-NN, where *k=1*.

The data sets chosen in the experimental part of CSM-GA are obtained from [9]. This archive contains 47 different datasets, where each dataset is divided into a training set, on which the algorithm for the problem at hand is trained, and a testing set, to which the outcome of the training set is applied [17]. To test CSM-GA the authors apply a different protocol where they split the original training datasets in [9] further into what they call *training set* and *validation set* (in other words, training set + validation set in CSM-GA = training set in [9]), so they have three sets in their experiments as they report in Table 1 taken from their paper (we omit two columns which show the number of classes in the datasets and the size of the time series).

The authors of CSM-GA do not explain the reason for choosing this protocol, neither do they explain the exact role of their training set and validation set. However, the authors report the results they obtain after running the genetic algorithms for 10 generations on the datasets presented in Table 1 to obtain the optimal weights ω_i of the distance measures s_i in relation (2). We show these optimal weights in Table 2 (which is also taken from their paper).

Then the weights in Table 2 are combined to form the new similarity measure which, as the authors explain in their paper, is used to classify the test data. The results are shown in Table 3, which also contains the results of applying each similarity measure separately, also on a *1*-NN classification task, which the authors report for comparison reasons to show the merits of using a combination of similarity measures. We present Table 3 exactly as it was presented in CSM-GA.

Table 2. Weights assigned to each similarity measure after 10 generations of CSM-GA

Dataset	S_1	S_2	S_3	S_4	S_5	S_6	S_7	S_8
Control Chart	0.72	0.29	0.33	0.18	0.12	0.61	0.31	0.82
Coffee	0.74	0.9	0.9	0.1	0.03	0.03	0.06	0.70
Beef	0.95	0.09	0	0.48	0	0.62	0.58	0.73
OliveOil	0.7	0	0.79	0	0	0	0.58	0.67
Lightning2	0.95	0.38	0	0.78	0.42	0.81	0.49	0.59
Lightning7	0	0	0	0.23	0.84	0.56	0	0.39
Trace	0.62	0.08	0.28	0.39	0.14	0.47	0.23	0.98
ECG200	0.052	0	0.21	0	0	0.98	0.90	0

Table 3. Comparison of classification accuracy using CSM-GA and other similarity measures

Dataset	Size	CSM-GA mean ± std	L$_2$	L$_1$	L$_\infty$	disim	root disim	peak disim	cosine	DTW
Control Chart	600	99.07± 0.37	88	88	81.33	58	53	77	80.67	99.33
Coffee	56	87.50± 2.06	75	53.57	89.28	75	75	75	53.57	82.14
Beef	60	54.45± 1.92	53.33	50	53.33	46.67	50	46.67	20	50
OliveOil	121	82.67± 4.37	86.67	80.33	83.33	63.33	60	63.33	16.67	86.67
Lightning2	121	87.54± 1.47	74.2	81.9	68.85	55.75	50.81	83.60	63.93	85.25
Lightning7	143	69.28± 2.97	71.23	71.23	45.21	34.24	28.76	61.64	53.42	72.6
Trace	200	100.0± 0.00	76	100	69	65	57	75	53	100
ECG200	200	90.00± 1.15	88	89	87	79	79	91	81	77

The authors of CSM-GA do not explain how they obtained the standard deviation. The size of the data sets in Table 3, which is the sum of the training sets, the validation sets, and the test sets, would suggest that they tested the weights on the training sets+ validation sets+ testing sets all mixed together.

The authors of CSM-GA conclude that the results obtained by their approach are "considerably better" and that their method "is guaranteed to yield better results".

3   Legitimate Remarks

In the following we present a few remarks on CSM-GA as proposed and implemented in [6]:

1- Some of the results presented in the experiments of CSM-GA are very unintuitive; e.g. L$_\infty$ gave better classification accuracy on (Coffee) than the combination itself. This means that after going through the costly optimization process, CSM-GA proposed an output which is not even as good as using a single similarity measure (whose complexity is low) and which does not require any optimization process. This was also the case with DTW on (Chart Control), L$_1$, L$_\infty$, and DTW on (OliveOil) , L$_1$, L$_2$ and DTW on (Lightning7) , L$_1$ and DTW on (Trace) , *peakdism* on (ECG200). In fact for all the datasets reported in Table 3 we see that one or even several similarity measures give very close, or even better, classification accuracy than the combination itself, which requires a costly optimization process. To give an idea, the total time of training the two smallest datasets in Table 1 for 10 generations (run on Intel Core 2 Duo CPU with 3G memory) was 29 hours 43 minutes (Beef) and 6 hours 23 minutes (Coffee), yet the datasets used in the experiments of CSM-GA are among the smallest datasets in [9]. As we can see, the optimization process is very time consuming, may be even impossible to implement for some datasets in [9] whose size is of the magnitude of 1800 (for comparison, the sizes of Beef and Coffee are 30 and 28, respectively). Yet this optimization process proposed in CSM-GA is of very little benefit, if any. These surprising remarks make the motivation of applying CSM-GA seriously questionable.

2- The choice of some of the similarity measures in the experimental part of CSM-GA seems redundant. Some of these distances are of the same family (L$_1$, L$_2$, L$_\infty$), and (*dism, rootdisim*). They are even related numerically (L$_1 \geq$ L$_2 \geq$ L$_\infty$), (*dism* $\leq$, *rootdisim*). Strange enough, although the authors present in the background of their paper the same similarity measures we mentioned in Section 2 as the state-of-the-art similarity measures, they use only 3 of these measures in their experiments.

3- The outcome of the training phase in some cases is strange and unexplainable; for instance, Table 3 shows that the weight of s_3 on dataset (Beef) is 0, which means that this similarity measure is completely useless in classifying (Beef), yet Table 3 shows that the performance of s_3 as a standalone similarity measure on (Beef) is not worse than any other similarity measure of the others (on the contrary, it seems to be the best). This is also the case with s_2, s_4, s_5, s_6 on (OliveOil) and s_1, s_2, s_3, s_7 on (Lighting7) , and this was particularly strange with s_2 on (Trace) whose value on the training set was very small; 0.08 whereas its performance as a standalone similarity measure is completely accurate (100%). In fact, this result itself refutes the basis of the whole method CSM-GA; if s_2 (which is L_1) gives and accuracy of 100% on (Trace) then why do we have to go through all this long and complicated optimization process to get the same results (which are not even intuitive since the solution obtained gives a very small weight to this distance, almost 0)?

4- The most astonishing remark to us was the abnormal existence of all these zeros in Table 2, and in 10 generations only. This point needs particular attention; in genetic algorithms the only possibility to produce a certain value is either during initialization or through mutation (or a mixture of both). As for initialization, CSM-GA uses a population of 10 chromosomes each with 8 parameters (the similarity measures) we find it very hard to understand how a random generator of real numbers between 0 and 1 (with 2 or 3 decimal digits) would generate the value zero 15 times out of 80 (=10x8) . As for mutation, again, this is also a mystery; the authors of CSM-GA do not state the mutation rate they used, they only mention that it is small. However, most genetic algorithms use a mutation rate of 0.2 [8], so the number of mutations of CSM-GA is 10 (=population size) x 8 (=number of parameters) x 10 (=number of generations) x 0.2 (=mutation rat) =160 random real numbers between 0 and 1 (with 2 or 3 decimal digits). Again we can not understand how 0 was generated 15 times. This remark strongly suggests that there was an error in the optimization process of CSM-GA.

4   A Counter Example

In this section we will show that the conclusion of CSM-GA; that there is an optimal combination of weighted similarity measures which gives an optimal classification accuracy of a time series 1-NN classification task, is unfounded, and that any, completely randomly-chosen combination of the weighted similarity measures presented in CSM-GA on the same datasets can give very close classification accuracy.

In our example we will conduct a 1-NN classification task on the same time series presented in Section 2, using the same datasets, and the same similarity measures, except that the weights of the similarity measures we use are chosen randomly and without any optimization process.

To prove that the weights we choose are random and do not result from any hidden optimization process we choose weights resulting from a known irrational number; the number $\pi=3.14159265358979323846....$ Since in the experiments of CSM-GA there were 8 weights, each between 0 and 1, with two digits after the decimal point, we will remove the decimal point of π and assign each two successive digits, after dividing them by 100 (to scale them in the range [0, 1]; the range of ω_i), to the successive similarity measures, so the weights will be: $\omega_1=0.31$, $\omega_2=0.41$, $\omega_3=0.59$, $\omega_4=0.26$, $\omega_5=0.53$, $\omega_6=0.58$, $\omega_7=0.97$, $\omega_8=0.93$. As we can see, the weights do not result from any optimization process whatsoever. Then we construct a combination of the similarity measures according to relation (2) with the above weights, so we get:

$$S_\pi = 0.31s_1 + 0.41s_2 + 0.59s_3 + 0.26s_4 + 0.53s_5 + 0.58s_6 + 0.97s_7 + 0.93s_8$$

We also wanted to test another randomly chosen number, so we took the square root of 2 ; $\sqrt{2} = 1.4142135623730950488...$, which is also an irrational number, and we proceeded in the same manner as we did with π, i.e.: we remove the decimal point and we assign each two successive digits, after dividing them by 100, to the successive similarity measures in relation (2), so we get the following weights in this case; $\omega_1=0.14$, $\omega_2=0.14$, $\omega_3=0.21$, $\omega_4=0.35$, $\omega_5=0.62$, $\omega_6=0.37$, $\omega_7=0.30$, $\omega_8=0.95$, and the combination in this case is:

$$S_{\sqrt{2}} = 0.14s_1 + 0.14s_2 + 0.21s_3 + 0.35s_4 + 0.62s_5 + 0.37s_6 + 0.30s_7 + 0.95s_8$$

Now we test S_π and $S_{\sqrt{2}}$ on the same datasets on which CSM-GA was tested. Since in our experiment we are not applying any optimization process we do not have any training phase (which is used in CSM-GA to get the optimal weights), so we apply S_π and $S_{\sqrt{2}}$ directly on the test sets. In Table 4 we present the results we obtained (together with those of CSM-GA for comparison). As we can see, the results are very close in general and we do not see any significant difference in classification accuracy between CSM-GA, which is supposed to be the outcome of an optimization process, and two other combinations with weights chosen completely at random. In other words, any combination of weights could produce the same classification accuracy. We experimented with other randomly chosen combinations of weights and we got similar results, and we urge the reader to test any combination of weights (between 0 and 1) and he/she will draw the same conclusion that we did that there is no optimal combination of weights.

Table 4. Comparison of the classification accuracy using CSM-GA and two randomly chosen combinations of weights w_i

Dataset	CSM-GA mean ± std	S_π	$S_{\sqrt{2}}$
Control Chart	99.07± 0.37	96	98.33
Coffee	87.50± 2.06	92.85	89.28
Beef	54.45± 1.92	56.66	53.33
OliveOil	82.67± 4.37	83.33	80
Lightning2	87.54± 1.47	83.61	88.53
Lightning7	69.28± 2.97	80.82	76.71
Trace	100.0± 0.00	96	95
ECG200	90.00± 1.15	90	90

It is also worth mentioning that [9] urges all those who use that archive to test on all the datasets to avoid "cherry picking" (presenting results that work well/ better with certain datasets). The authors of CSM-GA do not mention why they did not test their method on the other datasets in [9], or why they chose these datasets in particular. However, in this example we meant to test our random combinations on these datasets in particular because these are the ones on which CSM-GA was tested.

It is very important to mention here that all the results, remarks and the example we present in this paper are reproducible; the datasets are, as mentioned before, available at [9] where the reader can also find the code for the classification task. As for the 8 similarity measures used (Section 2) they are very easy to code. Besides, many scientific programming forums on the internet have the codes for these similarity measures, so all our results are easily verifiable.

5   What Went Wrong with CSM-GA

In this section we try to answer the question of why CSM-GA did not work. We were able to find two major errors in CSM-GA

1-The most serious error in CSM-GA is that the similarity measures it uses were not normalized before being used in relation (2). These similarity measures have very different numeric values so they should be normalized. To give a rough estimation, we present in Table 5 the distances between the first and the second time series of the test datasets used in the experiments of CSM-GA using each of the 8 similarity measures used with CSM-GA. We believe Table 5 clearly explains the whole point; since all the weights w_i are in the interval [0,1] then for a certain value of the weight the combination in relation (2) will be dominated by the similarity measures with the highest numeric values. We see in (Coffee) for instance that the value of DTW is 6.3×10^6 times larger than that of the Cosine Distance, so the influence of DTW on S_{new} in (2) will be far much stronger than that of the Cosine Distance in classifying (Coffee), so normalization is crucial.

Table 5. A rough estimation of the numeric values of the similarity measures used in CSM-GA

Dataset	L_2	L_1	L_∞	disim	root disim	peak disim	cosine	DTW
Control Chart	9.79	59.17	2.94	0.64	0.8	0.48	0.81	21.16
Coffee	187.58	2932.7	18.77	0.23	0.48	0.19	0.001	10567
Beef	0.93	13.8	0.14	0.53	0.73	0.38	1.05	50
OliveOil	0.07	0.84	0.02	0.007	0.08	0.007	0.000	0.004
Lightning2	32.85	358.24	18.36	0.75	0.87	0.50	0.85	294.06
Lightning7	16.55	206.18	8.88	0.59	0.77	0.43	0.43	98.28
Trace	27.57	389.97	5.08	0.76	0.87	0.56	1.39	437.4
ECG200	3.90	28.48	1.53	0.36	0.60	0.26	0.08	6.34

Since the numeric values of L_1, L_2, L_∞, DTW are far much larger than those of the other 4 similarity measures, and since, as we can see from Table 3, the performance of L_1, L_2, L_∞, DTW as standalone similarity measures is better than the other 4 similarity measures, we can easily understand the reason behind what it seems as the "optimal" classification accuracy of S_{new}. It is simply due to the fact that S_{new} is dominated by similarity measures whose performance is superior, and not as a result of an optimization process. This point explains the findings of our example in Section 4 that any combination will likely give similar classification accuracy as that of CSM-GA.

2- The I-NN classification task of time series has certain characteristics related to its nature. We will explain this through the following example; the size of (Beef) is 30, thus the number of mismatches on this dataset (whatever algorithm is used) will be $NrMis= \{0,1,2,...,30\}$, thus the I-NN classification accuracy (which is the objective function of CSM-GA), whatever algorithm is used, will take one of these values: $ClassAcc=\{30/30,29/30,28/30,...,0/30\}$. This set $ClassAcc$ is finite (and even more it has a small cardinality), while the values of the weights w_i belong to an infinite set (the interval [0, 1]), yet any combination of weights of similarity measures will yield I-NN classification accuracy that belongs to the finite (and small) set $ClassAcc$, this means that an infinite number of different combinations will give the same classification accuracy. In optimization language, such an objective function is highly multimodal, so the assumption that there is an optimal combination of similarity measures that gives maximum$I$-NN classification accuracy is incorrect and the search for such an optimal combination is meaningless (and the example we presented in Section 4 leads to the same conclusion).

6   Conclusion

There are far too many "new" bio-inspired optimization applications claiming their superior performance. But a deep investigation of these techniques may show that they are incorrect. Our purpose in this paper was by no means to refute a particular method or to show its incorrectness, but it was rather to shed light on certain cases where the careless, inappropriate, or erroneous application of bio-inspired algorithms may give false and even unintuitive results. While the widespread success of bio-inspired algorithms encourages many researchers to apply these algorithms extensively, it is important to be careful to choose the right application that takes into account the problem at hand, in addition to the correct implementation of the bio-inspired algorithm, especially that these algorithms require employing intensive computing resources, so we have to make sure that the results obtained will be better than those yielded by a random solution of the problem.

References

1. Affenzeller, M., Winkler, S., Wagner, S., Beham, A.: Genetic Algorithms and Genetic Programming Modern Concepts and Practical Applications. Chapman and Hall/CRC (2009)
2. Bustos, B., Skopal, T.: Dynamic Similarity Search in Multi-metric Spaces. In: Proceedings of the ACM Multimedia, MIR Workshop, pp. 137–146. ACM Press, New York (2006)

3. Bustos, B., Keim, D.A., Saupe, D., Schreck, T., Vrani'c, D.: Automatic Selection and Combination of Descriptors for Effective 3D Similarity Search. In: Proceedings of the IEEE International Workshop on Multimedia Content-based Analysis and Retrieval, pp. 514–521. IEEE Computer Society (2004)
4. Chen, L., Ng, R.: On the Marriage of Lp-Norm and Edit Distance. In: Proceedings of 30th International Conference on Very Large Data Base, Toronto, Canada (August 2004)
5. Ding, H., Trajcevski, G., Scheuermann, P., Wang, X., Keogh, E.: Querying and Mining of Time Series Data: Experimental Comparison of Representations and Distance Measures. In: Proc of the 34th VLDB (2008)
6. Dohare, D., Devi, V.S.: Combination of Similarity Measures for Time Series Classification using Genetic Algorithms. Congress on Evolutionary Computation, CEC (2011)
7. Guo, A.Y., Siegelmann, H.: Time-warped Longest Common Subsequence Algorithm for Music Retrieval. In: Proc. ISMIR (2004)
8. Haupt, R.L., Haupt, S.E.: Practical Genetic Algorithms with CD-ROM. Wiley-Interscience (2004)
9. Keogh, E., Zhu, Q., Hu, B.: Hao. Y., Xi, X., Wei, L. and Ratanamahatana, C.A.: The UCR Time Series Classification/Clustering, http://www.cs.ucr.edu/ ~eamonn/time_series_data/
10. Last, M., Kandel, A., Bunke, H. (eds.): Data Mining in Time Series Databases. World Scientific (2004)
11. Mitchell, M.: An Introduction to Genetic Algorithms. MIT Press, Cambridge (1996)
12. Muhammad Fuad, M.M.: ABC-SG: A New Artificial Bee Colony Algorithm-Based Distance of Sequential Data Using Sigma Grams. In: The Tenth Australasian Data Mining Conference, AusDM 2012, Sydney, Australia, December 5-7 (2012)
13. Muhammad Fuad, M.M.: Genetic Algorithms-Based Symbolic Aggregate Approximation. In: Cuzzocrea, A., Dayal, U. (eds.) DaWaK 2012. LNCS, vol. 7448, pp. 105–116. Springer, Heidelberg (2012)
14. Muhammad Fuad, M.M.: Particle Swarm Optimization of Information-Content Weighting of Symbolic Aggregate Approximation. In: Zhou, S., Zhang, S., Karypis, G. (eds.) ADMA 2012. LNCS (LNAI), vol. 7713, pp. 443–455. Springer, Heidelberg (2012)
15. Muhammad Fuad, M.M.: Towards Normalizing the Edit Distance Using a Genetic Algorithms–Based Scheme. In: Zhou, S., Zhang, S., Karypis, G. (eds.) ADMA 2012. LNCS (LNAI), vol. 7713, pp. 477–487. Springer, Heidelberg (2012)
16. Muhammad Fuad, M.M.: Using Differential Evolution to Set Weights to Segments with Different Information Content in the Piecewise Aggregate Approximation. In: KES 2012. Frontiers of Artificial Intelligence and Applications (FAIA). IOS Press (2012)
17. Xiao, Y., Chen, D.Y., Ye, X.L.: Hu: Entropy-Based Symbolic Representation for Time Series Classification. Fuzzy Systems and Knowledge Discovery. In: Fourth International Conference on In Fuzzy Systems and Knowledge Discovery (2007)
18. Yamada, T., Yamashita, K., Ishii, N., Iwata, K.: Text Classification by Combining Different Distance Functions with Weights. In: International Conference on Software Engineering, Artificial Intelligence, Networking and Parallel/Distributed Computing, and International Workshop on Self-Assembling Wireless Networks (2006)

Accurate and Fast Dynamic Time Warping

Hailin Li and Libin Yang

College of Business Administration, Huaqiao University, Quanzhou 362021, China
hailin@mail.dlut.edu.cn, ylib1982@163.com

Abstract. Dynamic time warping (DTW) is widely used to measure similarity between two time series by finding an optimal warping path. However, its quadratic time and space complexity are not suitable for large time series datasets. To overcome the issues, we propose a modified version of dynamic time warping, which not only retains the accuracy of DTW but also finds the optimal warping path faster. In the proposed method, a threshold value used to narrow the warping path scope can be preset automatically, thereby resulting in a new method without any parameters. The optimal warping path is found by a backward strategy with reduced scope which is opposite to the forward strategy of DTW. The experimental results demonstrate that besides the same accuracy, the proposed dynamic time warping is faster than DTW, which shows that our method is an improved version of the original one.

Keywords: Dynamic time warping, Time series, Similarity measure, Data mining, Computational complexity.

1 Introduction

In the field of time series data mining, distance function or similarity measure is often used to describe relationships between two time series, such as Euclidean distance [1, 2], edit distance [3], K-L distance [4] and dynamic time warping (DTW) [5–9]. Euclidean distance is one of the most popular functions used to fast measure time series with same length. Its time and space complexity are linear to the length of time series. It often combines with time series representations [10–12] to improve the results. However, Euclidean distance is sensitive to abnormal points and helpless to measure time series with different lengths.

Dynamic time warping (DTW) [6] is another popular method to measure similarity between two time series. It is not only robust to the abnormal points but also suitable for time series with different length. DTW finds an optimal alignment between two time series to measure the similarity. In this way, the points with same shape in each time series can be mapped. However, since DTW must search the best warping path in a cumulated matrix, it has to cost the quadratic time and space complexity, which causes to be not suitable for the long time series. Thereby, a new DTW with the same accuracy to the original one and less time consumption is required.

In this paper, we propose a modified version of DTW, which we call accurate and fast dynamic time warping (AF_DTW). The framework of AF_DTW consists

H. Motoda et al. (Eds.): ADMA 2013, Part I, LNAI 8346, pp. 133–144, 2013.
© Springer-Verlag Berlin Heidelberg 2013

of three parts. The first is a backward strategy used to find the optimal warping path, which is opposite to the forward strategy of DTW. The second is the process of reduced scope and the third is a choice method of threshold value which can be preset to narrow the warping path scope. Especially, the backward strategy and the way of reduced scope run concurrently with one another. The contribution of the backward strategy and threshold intervening is to discard a remaining scope when the current cost of the warping path is negative. In this way, AF_DTW finds the optimal warping path in a reduced scope and its accuracy is the same to DTW.

The remainder of the paper is organized as follows. In section 2, we give the background and related work. The framework of the improved version of DTW is presented in section 3. Some experiments are performed on serval time series datasets in section 4. In the last section the conclusions and future work are discussed.

2   Background and Related Work

Given two time series $Q = \{q_1, q_2, \cdots, q_m\}$ and $C = \{c_1, c_2, \cdots, c_n\}$, where m and n respectively represent the length of Q and C, Euclidean distance [1,13,14] is used to measure the similarity between the two time series when $m = n$. Each pair of points with same time-stamp in the two time series is used to calculate the distance. When the shapes of the two time series are considered, DTW is a good choice to measure time series. Especially, when the lengths of the two time series are different, i.e., $m \neq n$, DTW is often used instead of Euclidean distance.

DTW [2,5,6] minimizes the distance between two time series by constructing an optimal warping path P which often makes the points with same shapes map to each other. A warping path can be denoted as $P = \{p_1, p_2, \cdots, p_K\}$ and p_i means the mapping information about the time of two points (q_i and c_j) respectively deriving from the two time series, i.e., $p_k = [i, j]$, where $K \in [max(m,n), m + n - 1]$ represents the length of the path, $i \in [1, m]$ and $j \in [1, n]$. $d(p_k)$ denotes the distance between two points q_i and c_j, i.e., $d(p_k) = d(i, j) = (q_i - c_j)^2$. At the same time, the warping path much satisfy at least three constraints, such as boundary conditions, continuity and monotonicity [5].

There are many such path existed in the mapping sets, but we only need an optimal one with the minimal warping cost, i.e.

$$DTW(Q, C) = \min_P \sum_{k=1}^{K} d(p_k).$$

(1)

Generally, the best warping path can be found by using dynamic programming which defines the cumulative distance $R(i, j)$ as the distance $d(i, j)$ adding the minimum of the cumulative distance of the three adjacent elements, i.e.

$$R(i,j) = d(i,j) + \min \begin{cases} R(i,j-1) \\ R(i-1,j-1) \\ R(i-1,j) \end{cases}, \tag{2}$$

where $R(0,0) = 0$, $R(i,0) = R(0,j) = \infty$. We often call the cumulative distance R as a cost matrix.

For example, if we have two time series Q and C, i.e., $Q = [-0.6, 4.6, -6.4, -22.3, 0.9, 0.4, 0.3, 9, 12]$ and $C = [0.1, 5.3, -5.8, -22.1, 1.5, 0.7, 0.4, 9.2]$ The best warping path can be calculated by starting from ($i = 1$ and $j = 1$) to ($i = 9$ and $j = 8$). There are 10 path elements constructing the best warping path, that is, $P = \{p_1, p_2, \cdots, p_{10}\}$. Thereby, the minimum distance of DTW to measure the similarity between two time series Q and C is $DTW(Q,C) = R(9,8) = 9.67$.

DTW is widely applied to the field of time series data mining [15], speech recognition [16] and other disciplines [2], such as medicine, meteorology, gesture recognition and finance. However, the quadratic time and space complexity ($O(nm)$) of DTW constrain its performance. So far, there are some methods [18,19] used to speed up the calculation of DTW. Two of the most commonly used methods is the Sakoe-Chuba Band [17] and the Itakura Parallelogram [16] that limit the number of cells in the cost matrix R and reduce the path scope for searching the suboptimal warping path. However, The performance of DTW using the two methods depends on a constant factor. Moreover, it often cannot retrieve the best warping path which is often out of the path scope. FTW [20] is often applied to similarity search in time series and can faster retrieve the results than the original DTW. However, it uses lower bounding functions to reduce time cost when they are applied to similarity search. So FTW is preferable to similarity search and indexing rather than clustering and classification. Similarly, FastDTW [21] is proven to be a method with linear time and space complexity, but its accuracy depends on a factor which is hard to decide and the returned distance value is a result approximating to the minimum one. At the same time, the larger the factor is, the more the calculation of time and space is cost. Thereby, a technique with less time and space consumption and without any factors is required, which is the main topic and contribution of our paper.

3 Accurate and Fast Dynamic Time Warping

Accurate and fast dynamic time warping (AF_DTW) is an improved version of DTW, which includes three parts. The first one is a backward strategy to construct the main idea of AF_DTW. The second one is the introduction of a way to reduce scope in the cost matrix. The last one is a choice method of a threshold value used to reduce the scope.

3.1 Backward Strategy

DTW starting from $(i,j) = (1,1)$ to $(i,j) = (m,n)$ is a forward strategy based algorithm. The cost matrix R is constructed cumulatively from $(1,1)$ to (m,n).

Each element in the cost matrix R is larger than the three bottom left adjacent elements. In contrast to DTW, AF_DTW uses a backward strategy to construct the algorithm. It means that AF_DTW starts from (m, n) to $(1, 1)$ and each element in cost matrix R is less than the three top right adjacent ones.

In DTW a current element $R(i, j)$ in cost matrix is the distance $d(i, j)$ adding the minimum of the cumulative distance of the three bottom left adjacent elements. Contrary, in AF_DTW the element $R'(i, j)$ is the maximum of three top right adjacent elements subtracting the distance $d(i, j)$, i.e.,

$$R'(i,j) = \max \begin{cases} R'(i, j+1) \\ R'(i+1, j+1) \; - d(i,j), \\ R'(i+1, j) \end{cases} \tag{3}$$

where $i = m, m-1, \cdots, 1$, $j = n, n-1, \cdots, 1$, $R'(m+1, n+1) = 0$, $R'(i, n+1) = R'(m+1, j) = -\infty$.

In this way, a warping path $P' = \{p'_1, p'_2, \cdots, p'_K\}$, which is a contiguous set of distance matrix elements that defines a mapping between Q and C, can be constructed as DTW does. The warping path also must be subject to three constraints including boundary conditions, continuity and monotonicity.

The dynamic programming is also used to find the best warping path with a maximum warping cost, i.e.,

$$BS\_DTW(Q, C) = \max_{P'} \sum_{k=1}^{K} d(p'_k). \tag{4}$$

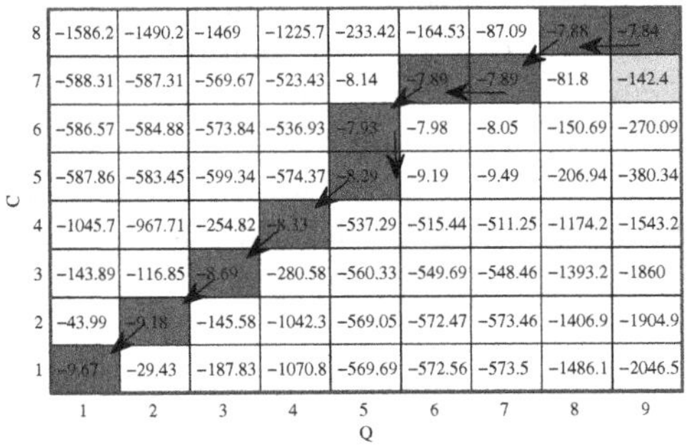

8	−1586.2	−1490.2	−1469	−1225.7	−233.42	−164.53	−87.09	−7.88	−7.84
7	−588.31	−587.31	−569.67	−523.43	−8.14	−7.89	−7.85	−81.8	−142.4
6	−586.57	−584.88	−573.84	−536.93	−7.93	−7.98	−8.05	−150.69	−270.09
5	−587.86	−583.45	−599.34	−574.37	−8.29	−9.19	−9.49	−206.94	−380.34
4	−1045.7	−967.71	−254.82	−8.33	−537.29	−515.44	−511.25	−1174.2	−1543.2
3	−143.89	−116.85	−8.69	−280.58	−560.33	−549.69	−548.46	−1393.2	−1860
2	−43.99	−9.18	−145.58	−1042.3	−569.05	−572.47	−573.46	−1406.9	−1904.9
1	−9.67	−29.43	−187.83	−1070.8	−569.69	−572.56	−573.5	−1486.1	−2046.5
	1	2	3	4	5	6	7	8	9

c (vertical axis), Q (horizontal axis)

Fig. 1. The cost matrix is constructed by the backward strategy and the best warping path can be found

Using the above example, our new method starts at $(m, n) = (9, 8)$ and ends at $(1, 1)$. The warping cost can be calculated by the backward strategy as shown in Fig. 1. The red cells denote the elements of the best warping path. It is easy to discover that the element is $R'(1, 1) = -9.67$, which is the minimum value of the best warping path in red cells. It also means that $BS\_DTW(Q, C) = R'(1, 1) = -9.67$. The arrows show the direction of our strategy.

Comparing the result of DTW with that of BS_DTW, we found that their absolute values are equal, i.e., $|BS\_DTW(Q, C)| = DTW(Q, C) = 9.67$. BS_DTW has the same accuracy of DTW. Moreover, the time and space complexity of BS_DTW are the same to that of DTW and equal to $O(mn)$. So the next subsection is proposing a method to reduce the time and space consumption when running AF_DTW based on BS_DTW.

3.2 Reduced Scope

In this section, a method is proposed to obtained the reduced scope in which the best warping path exists. It is inspired by the the Sakoe-Chuba Band [17] and the Itakura Parallelogram [16] which limit the cells in the cost matrix.

In BS_DTW, all the elements in the cost matrix R' are negative as shown in Fig. 1. The reason is that BS_DTW depends on the initial value of $R'(m+1, n+1)$. If $R'(m + 1, n + 1)$ is initially set to be 0, all the elements in R' are negative. If we set $R'(m + 1, n + 1)$ to be a positive value, then some of the elements are positive. Moreover, these positive elements are adjacent. As shown in Fig. 2, the red cells denote the best warping path which is obtained by BS_DTW according to different initial values of $R'(m + 1, n + 1)$. The positive cells are in gray color. Since some cells in red color are positive, they are also a part of the ones in gray color.

If we set $R'(m + 1, n + 1)$ to be equal to 9.5 (i.e., $R'(m + 1, n + 1) = 9.5$) and run BS_DTW, some elements in R' are positive as shown in Fig. 2(a). Moreover, it is obvious that the best warping path only appears in the scope of the gray cells. As shown in Fig. 2(b), the best warping path is surrounded by the gray cells when $R'(m + 1, n + 1)$ is set to be 200. In Fig. 2(a), except for the cell $(1, 1)$, the remaining red cells are also surrounded by the positive cells. Thereby, if $R'(m + 1, n + 1)$ is big enough, all the elements of the best warping path can be surrounded by the positive cells filled in gray color. We regard these positive cells as reduced scope.

For simplicity, we denote $R'(m + 1, n + 1)$ to be a threshold value θ, i.e., $\theta = R'(m+1, n+1)$. In addition, we know that the minimum distance between the two time series Q and C is equal to 9.67 in subsection 3.1. This same result can also be obtained by our method, i.e.,

$$AF\_DTW(Q, C, \theta) = \theta - R'(1, 1), \qquad (5)$$

AF_DTW is the new method based on BS_DTW. Fig. 2 shows that if $\theta = 9.5$, then $AF\_DTW(Q, C, \theta) = 9.5 - (-0.17) = 9.67$. If $\theta = 200$, then $AF\_DTW(Q, C, \theta) = 200 - 190.33 = 9.67$. So AF_DTW retains the same accuracy of DTW.

The above analysis tells us that the best warping path always exists in the reduced scope with regards to a special value of θ. So we only force AF_DTW to find the best warping path in the scope of the positive cells when θ is enough large. In this way, the number of the positive cells is less than that of all cells in the original cost matrix so that the time and space consumption depending on the number of the positive cells can be reduced. Therefore, AF_DTW costs less time and space than DTW and retrieves the same accurate result.

(a)R'(m+1,n+1)=9.5

	1	2	3	4	5	6	7	8	9
8	-1576.7	-1480.7	-1459.5	-1216.2	-223.92	-155.03	-77.59	1.62	1.66
7	-578.81	-577.81	-560.17	-513.93	1.36	1.61	1.61	-72.3	-132.9
6	-577.07	-575.38	-564.34	-527.43	1.57	1.52	1.45	-141.19	-260.59
5	-578.36	-573.95	-589.84	-564.87	1.21	0.31	0.01	-197.44	-370.84
4	-1036.2	-958.21	-245.32	1.17	-527.79	-505.94	-501.75	-1164.7	-1533.7
3	-134.39	-107.35	0.81	-271.08	-550.83	-540.19	-538.96	-1383.7	-1850.5
2	-34.49	0.32	-136.08	-1032.8	-559.55	-562.97	-563.96	-1397.4	-1895.4
1	-0.17	-19.93	-178.33	-1061.3	-560.19	-563.06	-564	-1476.6	-2037

(b)R'(m+1,n+1)=200

	1	2	3	4	5	6	7	8	9
8	-1386.2	-1290.2	-1269	-1025.7	-33.42	35.47	112.91	192.12	192.16
7	-388.31	-387.31	-369.67	-323.43	191.86	192.11	192.11	118.2	57.6
6	-386.57	-384.88	-373.84	-336.93	192.07	192.02	191.95	49.31	-70.09
5	-387.86	-383.45	-399.34	-374.37	191.71	190.81	190.51	-6.94	-180.34
4	-845.7	-767.71	-54.82	191.67	-337.29	-315.44	-311.25	-974.15	-1343.2
3	56.11	83.15	191.81	-80.58	-360.33	-349.69	-348.46	-1193.2	-1660
2	156.01	190.82	54.42	-842.34	-369.05	-372.47	-373.46	-1206.9	-1704.9
1	199.33	170.57	12.17	-870.81	-369.69	-372.56	-373.5	-1286.1	-1846.5

Fig. 2. Some positive elements exist in the cost matrix and the best warping path is surrounded by gray cells

Why should we choose an enough large θ? In order to let the best warping path be surrounded by the positive cells, θ must be larger than the minimum distance value, i.e., $\theta \geq DTW(Q, C)$. Otherwise, the positive cells cannot surround the best warping path as shown in Fig. 2(a). The cell (1,1) is an element of the best warping path, but it is out of the scope of the positive cells. Actually, the value of cell (1,1) is smallest in the red cells. If cell (1,1) is negative, then at least one element of the best warping path is out of the scope of the positive cells. Thereby, to make cell (1,1) be positive, we must let $\theta \geq DTW(Q, C)$, which is also inferred by Eq. (5).

After we choose a suitable θ, we take the backward strategy to construct the reduced scope in the cost matrix. If the value of current cell in the cost matrix is negative, then let it be zero. If the three top right adjacent cells are zero, then the current cell is set to be zero and the current iteration is broken. In this way, AF_DTW goes on calculating other cells whose values are positive. Finally, AF_DTW can obtain the reduced scope and retrieve the minimum distance between time series Q and C.

Different θ produces different number of non-white cells. At the same time, large difference between two values of θ does not crazily influence the changed number of not-white cells. In other words, θ and $\theta + \eta$ (where η may be very large) may retrieve the same number of non-white cells in the cost matrix, which means that AF_DTW using a large θ may be speeded up because of the reduced scope. Thereby, a suitable θ is important for AF_DTW. In next subsection, we will address this problem.

3.3   Choice of Threshold Value

We know that the bigger θ is, the larger the scope will be, and the more time and space are consumed. It means that the smallest scope is obtained when $\theta = DTW(Q, C)$. However, these minimum distance value is unknown and need us to compute. Thereby, we should find some value of θ close to $DTW(Q, C)$.

According to the Sakoe-Chuba Band and the Itakura Parallelogram, they all search the suboptimal warping path by limiting the scope along the opposite diagonal of the cost matrix, which inspires us to initialize θ. In other words, we regard one of the opposite diagonal warping paths as the initial warping path to assign θ a relative small value. The initial value of θ is the sum of elements in the initial warping path, i.e., $\theta = \sum_{k=1}^{K} d(p_k)$.

However, if the length of two time series used to measure the similarity is equal (i.e., $m = n$), then Euclidean distance between the two time series is the initial value of θ. We call this special path the Euclidean path. Actually, when $m = n$, the best warping path in opposite diagonal scope will be generalized into Euclidean path. Thereby, if $m = n$, then

$$\theta = \sum_{k=1}^{L}(q_k - c_k)^2, \qquad L = m = n. \tag{6}$$

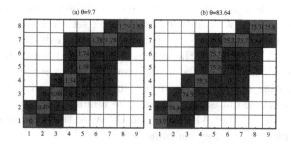

Fig. 3. The reduced scope produced by θ deriving from the initial warping path is the same to that produced by the minimum distance value $\theta = DTW(Q, C)$

In the above case, if $\theta = 83.64$ which is obtained by our method, then the reduced scope is identical to the optimal one which is produced by $\theta = DTW(Q, C) = 9.67$ as shown in Fig. 3. Thereby, our method to choose a suitable value of θ is feasible. At the same time, we must point out that any ways to produce a small θ which is bigger than $DTW(Q, C)$ can be used to reduce the search scope so as to speed up AF_DTW. Of course, the time and space complexity of the ways used to obtain a suitable θ must be linear to the length of time series. In our method, since the opposite diagonal scope is linear to the length of the two time series, the time and space complexity of the proposed method to decide the threshold value θ are linear to the length of time series, which also can be inferred by Euclidean path.

4   Experiments

Three experimental subsections about the performance of the proposed method AF_DTW are given. In the first subsection, we testify that the method about the choice of threshold θ is feasible and is used to improve the efficiency of DTW. In the second subsection, an experiment shows that the difference of time consumed by DTW and AF_DTW is more obvious with regards to the increased length of time series. The last experiments demonstrate that in contrast to DTW, the proposed AF_DTW not only has the same accuracy but also has faster calculation.

4.1   Different Thresholds Based Comparison

It is well know that the efficiency of AF_DTW depends on a reduced scope which further depends on a choice of threshold θ. In other words, the efficiency of AF_DTW indirectly depends on the choice of threshold θ. To testify that the method about the choice of threshold θ is feasible and useful, we take the following experiments on different time series datasets.

We choose the well-known UCR time series datasets [22] whose ID can be obtained in table 1. Each kind of time series dataset is consist of a training set and a testing set. Their length and size are also different from each other. For each kind of time series, we combine the training set with the testing set and regard them as a whole dataset. For a special value of threshold θ, every adjacent time series are used to compute the similarity (distance) between them and we regard the averaged time consumed as the result of the experiment with regards to the θ. Take Adiac dataset for example, for each value of θ, the number of time series in the whole dataset is 781 and the number of the pairs to compute similarity measure is 780. Let AF_DTW and DTW do this experiment. The final result is the averaged time they consumed.

In addition, the method about the choice of threshold θ runs in advanced and returns the initial value of θ for each pair of time series. We denote the initial value as a standard value θ'. At the same time, we also calculate the minimum distance value v between the pair of time series using the original DTW, i.e., $v = DTW(Q, C)$. Let l (step length) be $l = (\theta' - v)/3$. The values of θ chosen to compute the similarity by AF_DTW are $[v, v+l, v+2l, \theta', \theta'+l, \theta'+2l, \theta'+3l]$ or $[\theta'-3l, \theta'-2l, \theta'-l, \theta', \theta'+l, \theta'+2l, \theta'+3l]$ which is also identified as $[1,2,3,4,5,6,7]$ shown in the label of axis X in the Fig. 4 for simplicity.

The results of experiments in Fig. 4 shows that for each θ, the proposed AF_DTW consumes much less time than DTW. Moreover, the result marked by a cycle in each subplot is the averaged time consumed by AF_DTW according to the choice method of threshold as shown in algorithm 3. The result marked by the cycle tells us that the proposed choice method of threshold makes AF_DTW be faster than DTW and the time consumption of AF_DTW with the standard value θ' is close to the optimal one which is consumed by AF_DTW with the minimum distance value v. At the same time, the ascending trend of time consumed by AF_DTW tells us that the bigger θ is, the more time AF_DTW consumes.

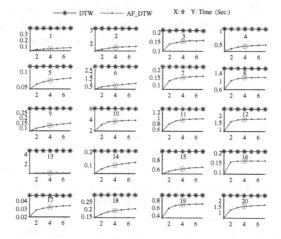

Fig. 4. The time consumption comparisons between AF_DTW and DTW according different θ

4.2   Different Length Based on Comparison

In this subsection, time series with different length are used to compare the efficiencies of AF_DTW and DTW. A long stock time series of length 2119415 is used [23] and we segment it seven groups of subsequences according to the length $L = [32, 64, 128, 256, 512, 1024, 2048]$. Each group has 50 subsequences. In each group we use AF_DTW and DTW to measure the similarity between each adjacent subsequences. In other words, each group has 49 pairs of subsequences used to measure the similarity. For each group, the result of the experiment is the averaged time.

Fig. 5 shows the result of comparison between AF_DTW and DTW according to different length of time series. It is easy to find that the longer the time

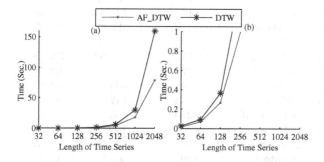

Fig. 5. The time consumption comparison between AF_DTW and DTW according different length of time series. (a) shows the result in the range $[0, 160]$ of axis Y; (b) shows the result in the range $[0,1]$ of axis Y.

series is, the more obvious the difference of time consumption between the two methods will be. It means that when the time series is much long, in contrast to DTW, AF_DTW used to measure the similarity is more efficient.

4.3   Classification

Although the previous work [16, 17, 20, 21] can fast run the variants of dynamic time warping, they retrieve the approximated results instead of the accurate one. It means that in most cases they can find the best warping path as the original DTW does. In order to experimentally shows the accuracy and efficiency of AF_DTW, we do the classification experiments on the UCR time series [22]. We use the nearest neighbor classification to classify the time series. In every time series dataset, each time series in the testing set using the classification is respectively measured by AF_DTW and DTW to find the most similar one in training set. If the label of the most similar object in the training set is not same to that of the one used to classify in the testing set, then we regard it as an error classification. Finally, we record the number of the error classifications and consider the error ratio as the accuracy of the algorithm. At the same time, for each time series dataset, we record the CPU time consumption and denote it as T. The averaged time consumption $\bar{T}$ (sec.) is used to show the efficiency of the algorithm.

Table 1. Experiment results on the UCR datasets

No.	Name	DTW & AF_DTW	DTW($\bar{T}$)	AF_DTW($\bar{T}$)
1	Adiac	0.396	168.57	25.16
2	Beef	0.5	101.39	54.11
3	CBF	0.003	6.92	4.89
4	Coffee	0.179	32.64	13.01
5	ECG200	0.23	12.99	8.08
6	FISH	0.167	563.86	121.44
7	FaceAll	0.192	124.64	94.66
8	FaceFour	0.170	39.20	29.00
9	Gun	0.093	14.65	6.10
10	Lighting2	0.131	365.60	232.28
11	Lighting7	0.274	94.34	69.30
12	OSULeaf	0.409	503.58	352.86
13	OliveOil	0.133	142.70	6.37
14	Swedish Leaf	0.210	105.85	52.83
15	Trace	0	100.56	64.48
16	2Patterns	0	212.27	162.47
17	Control	0.007	13.91	10.36
18	Wafer	0.02	201.71	194.33
19	50Words	0.310	429.12	316.75
20	Yoga	0.164	752.99	459.80

The final results of the experiments are shown in table 1. It is easy to discovery that the accuracy of classification using AF_DTW and DTW is identical, which means that the proposed method AF_DTW can retain the same accuracy as the original DTW does. At the same time, the averaged time consumption by AF_DTW is less than that of DTW, which means that AF_DTW is faster than DTW for time series classification. It also easily know that AF_DTW can be more efficient than DTW. Especially in some datasets, such as No.1, No.6 and No.13, AF_DTW is obviously more efficient than DTW.

5   Conclusions

The proposed method, accurate and fast dynamic time warping (AF_DTW), is an improved version of the original DTW. Comparing to DTW, it uses backward strategy to find the best warping path. At the same time, a new method using the backward strategy reduces the search scope in the cost matrix so that not only the accuracy of AF_DTW is the same to that of DTW but also the computation performance is more efficient. In addition, the choice method of threshold we proposed makes AF_DTW be a new version of DTW without any parameters. Actually, any values of the threshold greater than the minimum distance are able to reduce the search scope, which makes AF_DTW be faster than DTW.

Although the computation performance of AF_DTW is more efficient than DTW, the speeding-up degree indirectly depends on a choice of threshold. The proposed choice method of threshold can retrieve a good result to speed up the computation. However, other better choice methods of threshold to narrow the search scope may be existed, which will let AF_DTW be more faster. So finding a more efficient choice method of threshold is one of the most important tasks in the future.

Acknowledgements. This work has been partly supported by the National Natural Science Foundation of China (61300139) and the Society and Science Planning Projects in Fujian (2013C018).

References

1. Faloutsos, C., Ranganathan, M., Manolopoulos, Y.: Fast subsequence matching in time series databases. In: Proceedings of the ACM SIGMOD International Conference on Management of Data, pp. 419–429 (1994)
2. Chu, S., Keogh, E., Hart, D., Pazzani, M.: Iterative deepening dynamic time warping for time series. In: Proceedings of the Second SIAM International Conference on Data Mining, pp. 195–212 (2002)
3. Chen, L., Ng, R.: On the marriage of Lp-norm and edit distance. In: Proceedings of the 30th International Conference on Very Large Databases, pp. 792–801 (2004)
4. Burnham, K.P., Anderson, D.R.: Kullback-Leibler information as a basis for strong inference in ecological studies. Wildlife Research 28(1), 111–119 (2001)
5. Keogh, E., Pazzani, M.J.: Derivative dynamic time warping. In: Proceedings of The First SIAM International Conference on Data Mining, pp. 1–11 (2001)

6. Keogh, E.: Exact indexing of dynamic time warping. In: Processings of the 28th International Conference on Very large Data Bases, pp. 358–380 (2005)

7. Lemire, D.: Faster retrieval with a two-pass dynamic-time-warping lower bound. Pattern Recog. 42(9), 2169–2180 (2009)

8. Li, H., Guo, C., Yang, L.: Similarity measure based on piecewise linear approximation and derivative dynamic time warping for time series mining. Expert Systems with Applications 38(12), 14732–14743 (2011)

9. Petitjean, F., Ketterlin, A., Gancarski, P.: A global averaging method for dynamic time warping, with applications to clustering. Pattern Recogn. 44(3), 678–693 (2011)

10. Lai, C.P., Chung, P.C., Tseng, V.S.: A novel two-level clustering method for time series data analysis. Expert Systems with Applications 37(9), 6319–6326 (2010)

11. Guo, C., Li, H., Pan, D.: An improved piecewise aggregate approximation based on statistical features for time series mining. In: Bi, Y., Williams, M.-A. (eds.) KSEM 2010. LNCS, vol. 6291, pp. 234–244. Springer, Heidelberg (2010)

12. Li, H., Guo, C.: Piecewise cloud approximation for time series mining. Knowledge-Based Systems 24(4), 492–500 (2011)

13. Vlachos, M., Gunopoulos, D., Kollios, G.: Discovering similar multidimensional trajectories. In: Proceedings of the 18th International Conference on Data Engineering, pp. 673–684 (2005)

14. Lin, J., Keogh, E., Lonardi, S., Chiu, B.: A symbolic representation of time series with implications for streaming algorithms. In: Proceedings of the 8th ACM SIGMOD Workshop on Research Issues in Data Mining and Knowledge Discovery, pp. 2–11 (2003)

15. Muscillo, R., Conforto, S., Schmid, S.: Classification of motor activities through derivative dynamic time warping applied on accelerometer aata. In: Proceedings of the 29th Annual International Conference of Engineering in Medicine and Biomedicine, pp. 4930–4933 (2007)

16. Itakura, F.: Minimum prediction residual principle applied to speech recognition. IEEE Transactions on Acoustous, Speech, and Signal Process 23(1), 52–72 (1975)

17. Sakoe, H., Chiba, S.: Dynamic programming algorithm optimization for spoken word recognition. IEEE Transactions on Acoustous, Speech, and Signal Process 26(1), 43–49 (1978)

18. Keogh, E., Pazzani, M.: Scaling up dynamic time warping for datamining applications. In: Proceedings of the 6th ACM SIGKDD International Conference on Knowledge Discovery and Data Mining, pp. 285–289 (2000)

19. Zhou, M., Wong, M.H.: A segment-wise time warping method for time scaling searching. Information Sciences 173(1), 227–254 (2005)

20. Sakurai, Y., Yoshikawa, M., Faloutsos, C.: FTW: fast similarity search under the time warping distance. In: Proceedings of the 2005 Principles of Database Systems, pp. 326–337 (2005)

21. Salvador, S., Chan, P.: FastDTW: toward accurate dynamic time warping in linear time and space. Intelligence Data Analysis 11(5), 561–580 (2007)

22. Keogh, E., Xi, X., Li, W.: Welcome to the UCR time series classification/clustering (2007), http://www.cs.ucr.edu/~eamonn/time_series_data/

23. Stock datasets (2005), http://www.cs.ucr.edu/~wli/FilteringData/stock.zip

Online Detecting Spreading Events with the Spatio-temporal Relationship in Water Distribution Networks[*]

Ting Huang[1,2], Xiuli Ma[1,2,**], Xiaokang Ji[1,2], and Shiwei Tang[1,2]

[1] School of Electronics Engineering and Computer Science, Peking University
[2] Key Laboratory of Machine Perception (Ministry of Education), Peking University,
Beijing, China, 100871
TingHuang@pku.edu.cn, {maxl,jixk,tsw}@cis.pku.edu.cn

Abstract. In a water distribution network, massive streams come from multiple sensors concurrently. In this paper, we focus on detecting abnormal events spreading among streams in real time. The event is defined as a combination of multiple outliers caused by one same mechanism and once it breaks out, it will spread out in networks. Detecting these spreading events timely is an important and urgent problem both in research community and for public health. To the best of our knowledge, few methods for discovering abnormal spreading events in networks are proposed. In this paper, we propose an online method based on the spatial and temporal relationship among the streams. Firstly we utilize Bayesian Network to model the spatial relationship among the streams, and a succinct data structure to model the temporal relationship within a stream. Then we select some nodes as seeds to monitor and avoid monitoring all sensor streams, thus improving the response speed during detection. The effectiveness and strength of our method is validated by experiments on a real water distribution network.

Keywords: Spatio-temporal relationship, event detection, Bayesian Network.

1 Introduction

With the development of technology, sensors can be deployed in water distribution networks to monitor various substances in water such as chlorine [1] in real-time. And massive streams flow out of sensors continuously. All the sensor measurements need to be analyzed in real time in order to detect abnormal events, especially those caused by deliberate contaminant intrusions.

In this paper we are interested in detecting the event spreading in the network instead of the traditional problem of outlier detection [2-5]. The spreading event involves multiple streams and is a cascading of several outliers, which sequentially occur and spatially spread. Once it breaks out somewhere, the event will spread widely and

---

[*] This work is supported by National Natural Science Foundation of China under Grant No. 61103025.
[**] Corresponding author.

H. Motoda et al. (Eds.): ADMA 2013, Part I, LNAI 8346, pp. 145–156, 2013.
© Springer-Verlag Berlin Heidelberg 2013

brings much more impact and greater destruction. However, many previous works focus on simply finding outliers in a single time-series [2-5], or finding outlier regions from multiple sources [11-12]. In contrast, our proposed method is able to identify multiple outliers as one event and track the event's spreading path dynamically, instead of detecting outliers from multiple sources independently. It's much more interesting, complex and significant.

Because of massive data streams continuously coming, there are lots of challenges to do online detection of spreading events. First of all, we can't collect all information and then find events in an offline way. Secondly, algorithms for event detection should be efficient, with high accuracy and low false alarms in monitoring applications. In this paper we design a framework to address the above challenges and propose solutions to the problem of detecting abnormal events with spreading property.

Our proposed method utilizes both spatial and temporal relationship to achieve reliable results with high accuracy and low false alarm rate. However, modeling spatio-temporal network data itself is a complex and challenging problem, since the data often exhibit nonlinearity and heterogeneity. There are two fundamental challenges. One is to model dependency in both space and time and another is to fully accommodate the topology (edges and directions) of the network. Bayesian Network (BN) is able to incorporate the knowledge from different sources and encode the uncertainty relationship among data, which is a good choice for modeling spatial relationship. To the best of our knowledge, we are the first to use a BN to model the spatial relationship for a complex network. Meanwhile, we exploit the changing behavior of each stream and a succinct data structure to represent the temporal relationship.

Besides modeling spatio-temporal relationship, we use a greedy method to select some important nodes as seeds for online monitoring, instead of monitoring all nodes, thus improving the response speed. The last step is online detecting spreading events. The biggest challenge is how to identify multiple outliers as one event. Because the event has spreading property, we believe that the possible outliers may be downstream nodes. So after identifying outliers among seeds, we will expand the monitoring scope to the downstream nodes of the outliers and trace from them.

More specifically, the contributions we make in this paper are:

- We identify the problem of dynamically detecting spreading events in real time, which is a cascading of several outliers caused by the one same mechanism.
- We construct a BN with domain knowledge to encode the spatial and causal relationships among sensor streams, and put forward a data structure to succinctly and effectively model the temporal relationship within a single stream.
- Extensive experiments show that the framework proposed here can detect spreading events accurately and timely with much smaller monitoring and analysis effort.

The rest of the paper is structured as following. Section 2 discusses the related work. In Section 3 we introduce the overall framework of our model, including preliminary concepts and notations that we use in this paper. In Section 4 we propose our method for detecting abnormal events in detail. Experiments and their analysis are reported in Section 5. We conclude in Section 6 with directions for future work.

2   Related Work

Various research works have been reported to deal with finding outliers or surprising patterns in a single time series [2-5], which is relatively simpler than anomaly detection among multiple streams. For example, a framework for finding anomalous events in a time-series was proposed in [3], which is based on a time-varying Poisson process model. And the Canary software [2] compares the current value with the predicted value to detect the changes in a stream. However the event we detect involves multiple streams and we identify multiple outliers as an event by the causal relationship among multiple outliers, which is more challenging.

The problem of outlier detection in sensor networks has been studied previously. For example, Burdakis and Deligiannakis [13] proposed an algorithm for detecting outliers in sensor networks, based on the geometric approach.The work in reference [14] is about outlier detection in wireless sensor networks. They built a BN for each sensor node to capture the relationship between the attributes of sensor node as well as the history readings and neighbor readings, while we built a BN for the whole network mainly for capturing the spatial relationship among sensor streams. What's more, we are interested in detecting an event spreading in the network not outliers.

There also have been some efforts on detecting anomalous regions from distributed network recently [11-12]. Franke at al. [11] presented a system that gives every sensor an outlier degree and constructs outlier regions with outlier degree threshold. The work in [12] proposed an algorithm to detect outliers and then construct outlier trees to uncover causal relationships among these outliers. However, both of them first detect outliers independently from each stream and then construct anomalous regions for outliers, while our method not only proposes the anomaly regions in real-time but also tracks the event's spreading path, which is quite different from previous work.

Table 1. Description of notation

Symbol	Description
$G =< V, E >$	Topological structure of the water distribution network
S	The set of sensor nodes in network
S_i^t	Observation from sensor i at timestamp t
M^i	Temporal state transition matrix of stream i
$STO_{i,t}$	Spatio-temporal outlier occurring at sensor i at timestamp t

3   Preliminary Concepts

In this section, we introduce our notations and definitions of the proposed model. First of all, we assume the network infrastructure as following:

In a water distribution network, for example in Fig. 1, there are many nodes at pipe joints. We assume that the topological structure of the network is represented as a graph $G = <V, E>$, where V is the node set and E is the edge set, and we define degree of a node as the number of total edges of that node. Our goal is to find abnormal events in networks. A set of sensors are deployed on eachnode to monitor the water quality. Each sensor has a unique number, from 1 to s. Sensors can sense a variety of attributes, such as the residual chlorine, PH, temperature and so on. A study [1] has shown that free chlorine responding to most contaminants. So we select monitoring the chlorine attribute for detecting events, and we denote the observed value from sensor i at timestamp t as S_i^t.

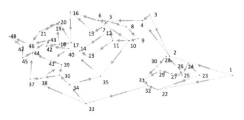

Fig. 1. An example of water distribution network

We model the complexity of the network based no discretizing the continuous observations to several equal intervals called state. BN is constructed on the basis of states to improve the efficiency and fully use the inference machinery of BN. And with states we can depict temporal behavior of a stream better.

Definition 1. *(State) A state is corresponding to an equal interval of continuous observations, and we assume there are total* n *states labeled from 1 to* n *for a range of values. Especially, the formal definition is as following:*

$$sI = \lceil (x \cdot n) / range \rceil$$

where x *is a continuous observation value to be discretized,* $range$ *is the size of the range of the observation values, and* sI *is the state label corresponding to* x.

We put forward a data structure called temporal state transition matrix based on state to succinctly and effectively describe the temporal behavior of a single stream.

Definition 2. *(Temporal state transition matrix) A temporal state transition matrix (tstMatrix)* M^i *defines how the states of two neighboring timestamps from sensor* i *transform. Especially, the formal definition is as following:*

$$M_{ij}^i = \begin{cases} 1 & \text{if state } i \text{ from sensor } i \text{ is followed by state } j \text{ from the same sensor } i \\ 0 & \text{otherwise} \end{cases}$$

In order to clarify the problem in this paper, we give the definition of outlier and event.

Definition 3. *(Spatio-Temporal outlier) A spatio-temporal outlier* $STO_{i,t} =< S_i, t >$ *is an observation value* S_i^t *from sensor i of timestamp t that deviates from expectation based on its spatial neighbors and its temporal neighbors.*

The definition of event is based on spatio-temporal outlier. The formal definition is as following, which defines the spreading property of an event.

Definition 4. *(Event) An event* $E = \{STO_{i_1,t_1}, STO_{i_2,t_2}, \cdots, STO_{i_\theta,t_\theta}\}$ *is defined as* θ *spatio-temporal outliers, satisfying 1)* $t_j \le t_k$ *when* $j \le k \le \theta$; *2) sensor* i_j *can be linked to* i_k ($j < k$)*by one edge directly or through one or more of sensors* i_x *where* $j < x < k$. *The parameter* θ *is a threshold value defined by users.*

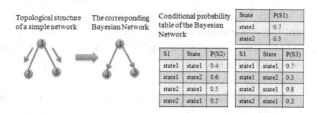

Fig. 2. View of a simple sensor network and its Bayesian Network

We give an example here to illustrate how we use BN. Assume there is a simple distribution network deployed with three sensors shown in Fig. 2. The direction labeled by the arrow means the flow direction along the water pipe. There exists causal relationship between upstream node and downstream node, so the structure of BN is just like the topological structure. Assume that there are total two states and the range of data is $[0,1]$. According to the definition of state, the state label for values in $[0,0.5)$ is 1 denoted as *state*1, while the state label for $[0.5,1]$ is 2 denoted as *state*2. The detail of training will be introduced in the Section 4. After training phase, we can learn the conditional probability table of each sensor shown in Fig. 2. For example, in the table of sensor 2 (table at the left side), the second row means the conditional probability $P(S_2 = state1 \mid S_1 = state1) = 0.4$. In the inference phase, if the value of S_1 at some timestamp is 0.34, i.e., locating at *state*1, the value of S_2 will more likely locate at *state*2 and the value of S_3 will more likely locate at *state*1 according to the given conditional probability table. If the real value of S_2 and S_3 does not belong to that state, we think that the real value deviates from the learnt spatial relationship.

Fig. 3. The overall framework of our method for detecting events

4   Methodology

The overall framework of our proposed method is illustrated in Fig. 3. Our goal is to online detecting spreading events accurately and efficiently. By modeling spatial and temporal relationship of sensor streams, we can understand the nature of data and prepare for the event detection. Also selecting seeds from all sensor nodes is able to improve the response time. The details are described in the following section.

4.1   Model Spatio-temporal Relationship

Before online detecting events, we first model the spatio-temporal relationship from normal historical data. For a large network, it's hard to simultaneously model spatial and temporal relationship because of the complexity of the data. Furthermore, how to efficiently represent the relationship is still an open problem. In this paper, we separately model spatial and temporal relationship. We creatively use conditional probability to represent the spatial relationship and innovatively use temporal state transition matrix for the temporal relationship.

Assume that the historical data are time series $(S_1, S_2, \cdots, S_s)$ from timestamp 1 to τ. We first develop a BN for the whole network G to model causal and spatial relationship between sensor stream and its neighboring sensor streams.

In the following, we will explain how to build the structure and learn the parameters of the BN. Assume that a distribution network is as shown in Fig. 1. The structure of the network and the flow directions known a priori provide a natural and rational structure of BN. We regard each sensor as a node of the BN and the flow direction on each pipe as the direction of each edge. So we get the structure, that is, the directed acyclic graph of the BN.

A case of the training data for learning the parameters of the BN consists of an s-tuple $(state_1^t, state_2^t, \cdots, state_s^t)$, where $state_i^t$ represents the state of S_i^t. We can generate lots of training data cases from the historical data set, and then we use the maximum likelihood parameter estimation algorithm [7] to learn the parameters of the BN.

After knowing how to build the BN, we now introduce how to model the temporal relationship of the data. We use the temporal state transition matrix to describe the temporal behavior of a stream from a sensor, which is not only accurate but also saving space storage. For each data stream we compute a temporal state transition matrix. We first initialize the matrix to a $n \cdot n$ matrix, and then we get the corresponding states slc and sln for two observations of neighboring timestamps. At last, we set $M_{slc,sln}^i$ to 1, which represents the trend of the data stream. The pseudo-code for building the tstMatrix is listed in Algorithm 1.

In this step, we learn the BN for the whole network and the temporal state transition matrix for each stream, which is the knowledge base to event detection and is critical for ensuring fast response.

Algorithm 1. BuildtstMatrix(historical data $(S_1, S_2, \cdots, S_s)$, total states number n)

---

1   for $i = 1$ to s do
2   initialize M^i to $n \cdot n$ Matrix
3   for $j = 1$ to $\tau - 1$ do
4   slc =computeState(S_i^j)
5   sIn =computeState(S_i^{j+1})
6           $M^i_{slc, sIn} = 1$
7   end for
8   end for

Subroutine: computeState(an observation value x)
9   initialize $range$ to 1, sI to -1
10 $sI = \lceil (x \cdot n) / range \rceil$
11 return sI

Output: tstMatrix for each sensor stream

4.2   Event Detection

4.2.1   Select Seeds

Modeling the spatio-temporal relationship is for estimating expected values when online detecting events, while selecting seeds from node set V is for improving the efficiency and response speed in the real-time environment.

Selecting optimal seeds has many different known strategies. We use a greedy algorithm to choose the node with the largest degree at the moment. We first try to select a node with largest degree, because nodes with larger degree have larger influence relatively. We then remove those selected nodes and neighbors of the selected nodes from all nodes. We iterate the above operations till no nodes left to select. The pseudo-code for selecting seeds is listed in the Algorithm 2.

Algorithm 2. SelectSeeds(topological structure of the water distribution network G, number of seeds k)

---

1   initialize $nodeSet$ to all the nodes in G
2   initialize $seedSet$ to *null* set, *counter* to 0
3   while($nodeSet$ is not *null*)
4   {
5   select the node with the largest degree from $nodeSet$ and add to $seedSet$
6   remove the selected node and its neighbors from $nodeSet$
7   *counter* ++
8   if k is not *null* and *counter* > k //user defined the number of seeds
9   break
10   end if
11   }

Output: $seedSet$

After selecting seeds, we just keep monitoring several streams from seeds instead of all streams for event detection. Moreover, our strategy for selecting seeds can guarantee that unselected nodes can find at least one seed from its neighbors. So it can minimize the loss as best as possible when the source of event locates at unselected nodes, because the event will spread to seed nodes as fast as possible.

The number of seeds can be either specified by the user or determined by the algorithm. It's easy to realize that the more the number of seeds is selected, the faster the detection of an event is.

4.2.2   Track Spreading Events

Detecting an event and finding its diffusion process is a challenging problem, since it's hard to identify multiple outliers as an event caused by one same mechanism. Although work in [12] has proposed an algorithm to build outlier tree from discovered outliers, it has to firstly keep discovering all outliers and then construct outlier tree, and the overall time complexity of construction process at each timestamp is upper bounded by $O(k^2)$, where k is the number of outliers in a timestamp. However, our method just keeps monitoring streams from seed nodes at every timestamp and traces from one outlier to its neighborhood to find how it spreads, which is easy to scale up.

More specifically, we keep monitoring data from seeds at every timestamp to analyze whether a spatio-temporal outlier appears among these seeds. If it does not, our method will wait to monitor the data of the next timestamp. Otherwise we will generate a candidate set for each abnormal node, and put the children nodes of the abnormal node to the corresponding candidate set. In the following timestamps, we will expand the scope of monitoring, including seeds and nodes in candidate sets. Once a node in a candidate set is detected as a spatio-temporal outlier, the children nodes of that node will be added to the candidate set. When the number of spatio-temporal outliers from one abnormal seed is up to the threshold θ, we think that an event has happened to that seed and a warning is given. What's more, by tracking the spatio-temporal outliers, we can find how the event happens and how it spreads.

So how do we determine an $STO_{i,t}$? if the corresponding state sI deviates from both the learnt spatial relationship and temporal relationship, an observed value S_i^t is judged as an $STO_{i,t}$. First, we use the inference ability of BN built previously to estimate the expected state sE of the target node based on the knowledge of other seeds' values. And then we compare the expected state sE with sI to determine if S_i^t deviates from the learnt spatial relationship. After that we have to check whether S_i^t goes against the trend of the stream S_i. Assume that sP is the state of S_i^{t-1}. We then check whether sP changing to sI has appeared in the temporal state transition matrix to determine if S_i^t deviates from the learnt temporal relationship. The pseudocode for detecting events is listed in the Algorithm 3.

Algorithm 3. EventDetect(observations at timestamp t $(S_1^t, S_2^t, \cdots, S_s^t)$)

---

1   static variables: *seedSet*, *candidateSet*, *counter*
2   for *sn* in *seedSet*
3   if STODetect(sensor label *sn*, observation S_{sn}^t) is *true*
4   *counter*++ and add children of *sn* to *true*
5   end if
6   end for
7   for *cn* in *candidateSet*
8   if STODetect(sensor label *cn*, observation S_{cn}^t) is true
9   *counter*++ and add children of *cn* to *candidateSet*
10   end if
11 end for
12 if *counter* larger than a *threshold*
13   a warning is given
14 end if

Subroutine: STODetect(sensor label i, observation S_i^t)

---

17 sI =computeState(S_i^t)
18 sE is the result of BN inference based on other observations
19 sP is the state of sensor i at timestamp $t-1$
20 if $sI \mathrel{!=} sE$ && $M_{sP,sI}^i = 0$
21   return *false*
22 end if
23 return *true*

---

5   Experiments

In this section, we will evaluate the results on the event detection of our method using a real water distribution network shown in the Fig. 1. We will first introduce the dataset and then by comparing with the state-of-the-art Canary, we present our experiments.

We use the EPANET 2.0 [9] to simulate the hydraulic and the chemical phenomena in the network in Fig. 1, which has 48 sensors totally and 17 of them are selected as seeds by our proposed algorithm. In our experiments, we monitor the chlorine concentration level for 5760 timestamp during 20 days (one time tick per 5 minutes). We also built a contaminant simulator using Java and EPANET toolkit [9] to simulate events in the network. In the contaminant simulator, after specifying the parameters for initial concentration of a contaminant, location, start time, and length of time for a contaminant injection, EPANET can simulate the propagation of contaminants and the reaction of chlorine and contaminant in the real time.

Our method is based on the BN, so we do our first experiment on a test set about 1000 cases to validate the reliability of the inference of our built BN. For every test case we use the accuracy ratio p / s to evaluate the inference performance, where s is the number of total nodes' state to be estimated and p is the number of nodes' state that are accurately predicted. The result is shown in Fig.4. Test case number is illustrated on the X-axis and the accuracy ratio is illustrated on the Y-axis. From the figure we can see that the average accuracy ratio is 0.85 that shows the high reliability of the inference.

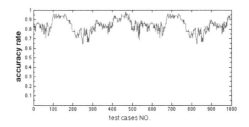

Fig. 4. The Bayesian Network inference performance

We do our next experiment to validate the effectiveness of our method comparing with the Canary software. We use two performance metrics to measure the effectiveness. The first one is the average delay time, which is the average delay in the time of identifying events from the truly beginning of events. And the second one is the sensitivity, which is the proportion of all events to be detected. And a higher sensitivity a method has, a lower missed detection rate it has.

We use Canary to keep monitoring 6 sensors in Network 1. The location of these 6 sensors is approximated optimal solutions proposed by the BWSN (the Battle of Water Sensor Networks) [10]. We use LPCF algorithm of Canary, with a window size of 576, and with a threshold value 0f 0.72. In the case of our method, we set the number of states to 9 and the threshold θ to 2, that is, when there are at least 2 neighboring spatio-temporal outliers happened sequentially we think that the spreading phenomenon has appeared and an event occurs.

Table 2. Comparison of different methods

Metrics Algorithm	Average delay time	Sensitivity
Canary	4.20h	0.50
Our method on random node	3.74h	0.76
Our method on all non-seed nodes	8.37h	0.63

We do two experiments here. The first one is to simulate 21 contaminant events on random nodes at random timestamp for injection length of 20 hours. The second one is to simulate 27 contaminant events on all non-seed nodes at random timestamp. The results for the different experiments are illustrated in table 2. The first one shows our method has a lower average delay time and a higher sensitivity than the Canary software.The second one proves that our method works well in detecting events not starting from seeds, although it takes longer to detect these events because all events start from non-seed nodes. The sensitivity is 0.63, which is still higher than the sensitivity of Canary. In general, it shows that although we just keep monitoring seeds, we can still effectively detect events starting from non-seed nodes.

The Fig.5 shows how the average delay time changes with the variation of the threshold θ. From the result we can get two conclusions. Firstly, if the system gives a warning when it detects the first spatio-temporal outlier, the delay time will be much shorter. Secondly, the average delay time will be longer with the increasing of θ, because the events need more time to spread out further.

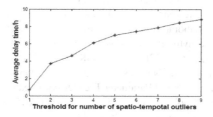

Fig. 5. Average delay time with variation of the threshold θ

Finally, we illustrate the efficiency and the scalability of our online method by the CPU time. The every running time is the average value of 40 experiments' CPU time. As shown in Fig.6, the running time is grows approximately linear when increasing the network size, and the response of our method is very quick. Therefore, we have the reason to believe that our method is scalable to streaming environments.

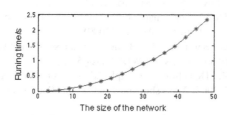

Fig. 6. Running time with variation of the size of the network

6   Conclusions

In this paper, we propose an efficient method for online detecting abnormal events in the context of multiple streams. By exploring the spatial and temporal relationship among sensor streams, we not only find those events with the spreading nature but also trace how the event spreads. We also provide experiments on chlorine concentration from a real water distribution network. The experiments illustrate that our method is of low delay time and high sensitivity and that it is easy to scale up.

Moreover, we look forward to maintaining the parameters of Bayesian Network and the temporal state transition matrixes to update the spatial and temporal relationship. So the system will keep the information latest, and it may detect the spreading events much more accurately and timely.

References

1. Hall, J., Zaffiro, A.D., Marx, R.B., Kefauver, P.C., Krishnan, E.R., Herrmann, J.G.: Online water quality parameters as indicators of distribution system contamination. Journal American Water Works Association 99(1), 66–77 (2007)
2. Hart, D.B., Klise, K.A., McKenna, S.A., Wilson, M.P.: CANARY User's Manual Version4.1. Sandia National Laboratories. U.S. Environmental Protection Agency (2009)
3. Ihler, A., Hutchins, J., Smyth, P.: Adaptive event detection with time-varying poisson processes. In: Proceedings of the 12th ACM SIGKDD Conference on Knowledge Discovery and Data Mining, New York, NY, USA, pp. 207–216 (2006)
4. Luo, W., Gallagher, M.: Faster and Parameter-Free Discord Search in Quasi-Periodic Time Series. In: Huang, J.Z., Cao, L., Srivastava, J. (eds.) PAKDD 2011, Part II. LNCS, vol. 6635, pp. 135–148. Springer, Heidelberg (2011)
5. Keogh, E., Lonardi, S., Yuan-chi Chiu, B.: Finding surprising pat-terns in a time series database in linear time and space. In: Proceedings of the Eighth ACM SIGKDD International Conference on Knowledge Discovery and Data Mining, Edmonton, Alberta, Canada, July 23-26 (2002)
6. Hawkins, D.M.: Identification of Outliers, Monographs on Applied Probability & Statistics. Chapman and Hall, London (1980)
7. https://code.google.com/p/bnt/
8. Uusitalo, L.: Advantages and challenges of Bayesian networks in environmental modelling. Ecological Modelling 203(3-4), 312–318 (2007)
9. Rossman, L.A.: EPANET2 user's manual. National Risk Management Re-search Laboratory: U.S. Environmental Protection Agency (2000)
10. Ostfeld, A., Uber, J.G., Salomons, E.: Battle of water sensor networks: A de-sign challenge for engineers and algorithms. In: WDSA (2006)
11. Franke, C., Gertz, M.: Detection and Exploration of Outlier Regions in Sensor Data Streams. In: SSTDM 2008: Workshop on Spatial and Spatiotemporal Data Min-ing at IEEE ICDM (2008)
12. Liu, W., Zheng, Y., Chawla, S., Yuan, J.: Discovering Spatio-Temporal Causal Interactions in Traffic Data Streams. In: 17th ACM SIGKOD Conference on Knowl-edge Discovery and Data Mining. ACM, San Diego (2011)
13. Burdakis, S., Deligiannakis, A.: Detecting Outliers in Sensor Networks using the Geometric Approach. In: Proc. of the International Conference on Data Engineering (2012)
14. Janakiram, D., Mallikarjuna, A., Reddy, V., Kumar, P.: Outlier Detection in Wireless Sensor Networks using Bayesian Belief Networks. In: Proc. IEEE Comsware (2006)

MLSP: Mining Hierarchically-Closed Multi-Level Sequential Patterns

Michal Šebek, Martin Hlosta, Jaroslav Zendulka, and Tomáš Hruška

Faculty of Information Technology, IT4Innovations Centre of Excellence,
Brno University of Technology, Božetěchova 2, Brno, Czech Republic
{isebek,ihlosta,zendulka,hruska}@fit.vutbr.cz

Abstract. The problem of mining sequential patterns has been widely studied and many efficient algorithms used to solve this problem have been published. In some cases, there can be implicitly or explicitly defined taxonomies (hierarchies) over input items (e.g. product categories in a e-shop or sub-domains in the DNS system). However, how to deal with taxonomies in sequential pattern mining is marginally discussed. In this paper, we formulate the problem of mining hierarchically-closed multi-level sequential patterns and demonstrate its usefulness. The MLSP algorithm based on the on-demand generalization that outperforms other similar algorithms for mining multi-level sequential patterns is presented here.

Keywords: closed sequential pattern mining, taxonomy, generalization, GSP, MLSP.

1 Introduction

Mining sequential patterns is interesting and a long studied research problem in the data mining community. It is used for many applications such as in the analysis of customer purchase patterns, web log data or biology sequences. Given a sequence database, the goal is to find sequential patterns that occur frequently in this database (i.e. more often than a given user defined threshold). Taking the market basket as an application example, the sequential pattern $\langle PC\_midtower\ ink\_printer \rangle$ can be discovered. This means that many people buy a midtower PC and then also buy an ink printer. The problem was first presented by Agrawal and Srikant in [1] and since then, many algorithms have been published trying to solve this problem.

One or more taxonomies of items can be often stored in the database. These taxonomies can be utilized in order to find patterns on which items of sequences can be on different levels of hierarchy. It allows one to find patterns which would not be retrieved without defined taxonomies. Our motivation to deal with taxonomies is primary to analyse some important facts over internet domains taxonomy (top-level domain, second-level domain, etc.), but due to privacy issues, we demonstrate our examples on customer purchase analysis. Following the previous sequence pattern example, in addition to $\langle PC\_midtower\ ink\_printer \rangle$ the pattern

H. Motoda et al. (Eds.): ADMA 2013, Part I, LNAI 8346, pp. 157–168, 2013.

$\langle PC\ printer \rangle$ can be found by replacing the items by items on a higher hierarchical level. Unfortunately, the amount of such patterns can grow enormously, but many of the resulting patterns can be considered as unuseful. For example, the pattern $\langle PC\ printer \rangle$ does not bring any new information if the number of its occurrence in the database is the same as for $\langle PC\_midtower\ ink\_printer \rangle$.We define hierarchically-closed sequential patterns which are the most useful for the analyst. In addition, the patterns can be divided into *multi-level* and *level-crossing* [2]. In *multi-level* (known also as *intra-level*) patterns, all items in one pattern are at the same level of hierarchy, whereas in *level-crossing* (known also as *inter-level*), the level can be different. The amount of retrieved level-crossing patterns is often greater than the number of multi-level patterns, as is the search space. This leads to a longer execution time of algorithms that mine level-crossing patterns. In contrast to the amount of algorithms for mining sequential patterns, very few algorithms tackled this problem with taxonomies.

Our contribution in this paper can be summarized as: formal definition of the mining hierarchically-closed multi-level sequential patterns problem, the MLSP algorithm for solving the problem and the experimental comparison of the algorithm with existing approaches.

The paper is organized as follows. In the next section, the mathematical basics for mining sequential patterns are described. In Section 3, the related work in mining sequential patterns is discussed. The section is finalized by an analysis of using taxonomies in this research area. In Section 4, terminology for multi-level sequential pattern mining is introduced. Our algorithm MLSP for mining sequential patterns is presented in Section 5. In Section 6, the algorithm is compared with techniques presented in other papers. Conclusions and future work are presented in Section 7.

2   Preliminaries

In this section the problem of mining sequential patterns is formalized.

Definition 1. *(Itemset) Let $I = \{i_1, i_2, i_3, \ldots, i_k\}$ be a nonempty finite set of items. Then an* itemset *is a nonempty subset of I.*

Definition 2. *(Sequence) A sequence is an ordered list of itemsets. A sequence s is denoted by $\langle s_1 s_2 s_3 \ldots s_n \rangle$, where s_j for $1 \leq j \leq n$ is an itemset. s_j is also called an* element *of the sequence. The length of a sequence is defined as the number of instances of items in the sequence. A sequence of length l is called an l-sequence. The sequence $\alpha = \langle a_1 a_2 \ldots a_n \rangle$ is a* subsequence *of the sequence $\beta = \langle b_1 b_2 \ldots b_m \rangle$ where $n \leq m$ if there exist integers $1 \leq j_1 < j_2 < \cdots < j_n \leq m$ such that $a_1 \subseteq b_{j_1}, a_2 \subseteq b_{j_2}, \ldots, a_n \subseteq b_{j_n}$. We denote it $\alpha \sqsubseteq \beta$ and β is a* supersequence *of α .*

Example 1. In the following examples, we use a well established convention from sequence pattern mining for denoting elements of sequences [3]. An element that has m items is denoted as $(i_1 i_2 \ldots i_m)$ where $m \geq 1$. If an element contains only one item, the braces are omitted for brevity.

Given $I = \{a, b, c, d, e\}$, an example of an itemset is $\{a, b, d\}$ and an example of a sequence is $\langle a(ab)(ce)d(cde) \rangle$. This sequence has 5 elements and 9 items. Its length is equal to 9 and it is called a *9-sequence*.

Definition 3. *(Sequence database) A sequence database D is a set of tuples $\langle SID, s \rangle$, where SID is a sequence identifier and s is a sequence. The support of a sequence s_1 is defined as the number of sequences in D containing a subsequence s_1. Formally stated, the support of a sequence s_1 is $support(s_1) = |\{\langle SID, s \rangle | (\langle SID, s \rangle \in D) \wedge (s_1 \sqsubseteq s)\}|$.*

Definition 4. *(Sequence Pattern, Mining Sequential Patterns) Given sequence database $\mathcal{D}$ and minimum support threshold min_supp, a frequent sequence is such a sequence s whose $support(s) \geq min\_sup$. A frequent sequence is called a* sequence pattern. *For a given sequence database D and a minimal support min_supp, the goal of mining sequential patterns is to find all frequent sequences in $\mathcal{D}$.*

In the problem of mining multi-level sequential patterns, taxonomies exist over items in D and are defined as follows.

Definition 5. *(Taxonomy of Items) The taxonomy structure of an item set V is a rooted tree $T = (V, E)$ with a root $r \in V$. In the context of the tree. we refer to V as a set of nodes representing items. For each node v in a tree, let $UP(v)$ be the simple unique path from v to r. If $UP(v)$ has exactly k edges then the* level *of v is k for $k \geq 0$. The level of the root is 0. The height of a taxonomy is the greatest level in the tree. The* parent *of $v \neq r$, formally $parent(v)$, is the neighbour of v on $UP(v)$, and for each node $v \in V, v \neq r$ there exists a set of its* ancestors *defined as:*

$$ancestors(v) = \{x | x \in UP(v), x \neq v\}. \tag{1}$$

The parent of r and the ancestors of r are not defined. If v is the parent of u then u is a child *of v. A* leaf *is a node having no child [4].*

In every taxonomy structure there exists a is-a *relation which is defined as follows:*

$$is - a : V \times V := \equiv \{(a, b) | b \in ancestors(a)\}. \tag{2}$$

Let $\iota = \{I_1, \ldots, I_m\}$ be a partition of a nonempty finite set of items I. Then a set of taxonomy structures of items I is a nonempty set of taxonomy structures $\tau = \{T_1, \ldots, T_m\}$ corresponding to ι such that $T_i = (V_i, E_i)$ where $I_i \in \iota$ for $1 \leq i \leq m$. It means that each item $i \in I$ appears in exactly one taxonomy structure $T_i \in \tau$. It should be noted that we do not require that items need to be only leaf nodes. In addition we refer to any ancestor of a node representing a item x a generalized item *of the item x.*

Example 2. We use a running example based on taxonomies in Figure 1. In the terms defined above, the parent of *black* is *Ink printer*; ancestors is the set $\{$*Ink printer, printer*$\}$ and the level of *black* is 2. In the following examples, some node names will be shorted (e.g. *LCD monitor* to *LCD*).

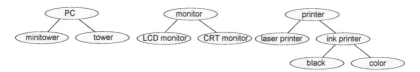

Fig. 1. Three taxonomies for running example

3   Related Work

Sequential pattern mining is a temporal extension of frequent pattern mining [5], [6]. Basic algorithms for mining sequential patterns are AprioriAll [1] and GSP [7] which are based on the generating and pruning candidates sequences. Other representatives of this category are the SPADE [8] algorithm using a vertical format of the database and SPAM [9] using a bitmap structure as an optimization. The second family of algorithms is based on the pattern growth. It is represented by the well-known fast algorithm PrefixSpan [10] or the BIDE algorithm [11] for mining closed sequential patterns.

The GSP algorithm is important for this work and, therefore, it will be described in more detail. The algorithm works iteratively. In each iteration it makes a pass over the sequence database:

1. Initially, the support of items is counted in the first pass over the sequence database and those having it higher than a minimal support *min_sup* are inserted in the resulting set L_1 containing *frequent 1-sequences*.
2. Then the following steps are executed iteratively in $1 < k \leq n$ iterations until no k-sequence is generated:
   (a) The *Candidate generation* runs in Join and Prune steps. In the *Join step*, a *candidate set* C_k is generated from sequential patterns in L_{k-1}. A pair of sequences $s_1, s_2 \in L_{k-1}$ can be joined if sub-sequences, generated by omitting of the first item of s_1 and the last item of s_2, are the same. Then the candidate k-sequence is formed by adding the last item of the s_2 at the and of the sequence s_1 as:
      i. the last new element containing one item x if x was in a separate element in s_2;
      ii. as a next item of the last element in s_1 otherwise.
      iii. When joining $x \in L_1$ with $y \in L_1$, both sequences $< (y)(x) >$ and $< (yx) >$ are generated as candidate sequences.
      The *Prune step* removes candidates with any non-frequent subsequence.
   (b) In the *Counting candidates step*, the database is passed and the support of each candidate sequence is counted. Candidates with a support greater than min_supp are added into the set L_k of sequential patterns.
3. The result sequential patterns set is $\bigcup_{k=1}^{n} L_k$.

Multi-level frequent patterns (itemsets) and association rules were described by Han and Fu in [2]. They proposed a family of algorithms for mining multi-level association rules based on Apriori which processes each level separately.

It allows us to use a different minimum support threshold for different levels of taxonomies. However, mining multi-level sequential patterns have not been deeply studied. The straightforward idea on how to mine level-crossing sequential patterns using GSP was presented in [7]. The authors proposed to use an extended-sequence database where each item of a sequence of origin database $\mathcal{D}$ is extended by all ancestors within its element. The disadvantage is that many redundant sequences are generated by this method.

Plantevit et al. in [12] firstly describe the idea of mining the most specific sequences in order to avoid redundancy. The presented algorithm performs a generalization only in the first step over single items in order to create maf-subsequences (maximally atomic frequent 1-sequences). However, the algorithm does not perform the generalization the during generation of their superse-quences.

The information theory concept was used for solving the problem of mining level-crossing sequential patterns in [13]. The proposed hGSP algorithm does not reveal a complete set of level-crossing sequential patterns, but only the most specific sequential patterns. The algorithm works on the bottom-up principle and performs a generalization only if the candidate sequence should be removed. It was demonstrated that mining level-crossing sequential patterns have an extremely large search space.

In this paper, we focus on the benefits of multi-level against the level-crossing sequential pattern mining algorithms.

4   Mining Multi-Level Sequential Patterns

The search space of mining level-crossing sequential patterns problem (studied in [13]) can be reduced by the *constraint* where levels of all items of frequent sequence are equal to a constant l_s (firstly used in [2]). The constraint could lead to a substantial simplification of the mining process. The constraint brings a compromise between the execution time of the algorithm and the specificity of sequential patterns for a user. The experimental verification is discussed in Section 6.

Example 3. The difference between the complexity of level-crossing and multi-level sequential patterns mining is shown in Figure 2. In the level-crossing way, the sequence ¡tower (LCD, laser)¿ has 6 ancestors, whereas in the multi-level way it has only one ancestor sequence.

A constrained element (ML-element) and a constrained sequence (ML-se-quence) are derived from definitions of element and sequence as follows.

Definition 6. (ML-element, ML-sequence) *Let $l \in N$ be a level of items in a taxonomy $T \in \tau$. Then an ML-sequence is an ordered list of itemsets $s_{ML} = \langle s_1 s_2 s_3 \ldots s_n \rangle$ such that the level of all items of the itemsets is equal to l. The itemset of the ML-sequence is called an ML-element. The length, subsequence and supersequence of an ML-sequence is defined analogously to the ones in Definition 2.*

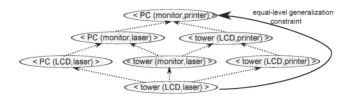

Fig. 2. Level-crossing and constrained multi-level generalization combinations

Example 4. For a better understanding of defined notions, we return to our running example and Figure 2 using taxonomies from Figure 1. From all the sequences here, only the most specific $\langle tower(LCD, laser)\rangle$ and the most generic $\langle PC(monitor, printer)\rangle$ are ML-sequences. In contrast, $\langle PC(LCD, laser)\rangle$ is not a ML-sequence because the level of item PC is 0 and the level of the items in the element $(LCD, laser)$ is 1. The levels of items in the other sequences are different .

The existence of the taxonomy over items allows us to introduce ML variants of a parent and ancestors.

Definition 7. *(ML-element parent) Given an ML-element $e = \{i_1, i_2, \ldots, i_n\}$, an ML-element parent of the ML-element e is an element whose all items are replaced by their parents. This is defined as*

$$parent_{el}(e) = \{parent(i_k)|1 \le k \le n \wedge i_k \in e\}. \tag{3}$$

Definition 8. *(ML-sequence parent, ML-sequence ancestors) Given an ML-sequence $s = \langle e_1 e_2 \ldots e_n\rangle$, where e_k is ML-element on position k, the ML-sequence parent of s is an ML-sequence such that all ML-elements of s are replaced by their ML-element parents. Formally,*

$$parent_{seq}(s) = \langle f_1 f_2 \ldots f_n\rangle, f_k \in parent_{el}(e_k), 1 \le k \le n. \tag{4}$$

For a given set of taxonomies τ, a root ML-sequence *is an ML-sequence consisting of ML-elements with items corresponding to root nodes of taxonomies. The ML-sequence parent of a root ML-sequence is not defined. Based on the definition of the ML-sequence parent, the* ML-sequence ancestors *of an ML-sequence s, $ancestors_{seq}(s)$ is defined recursively as follows:*

$$ancestors_{seq}(s) = M_i, \text{ if } M_{i+1} = M_i, \text{where} \tag{5}$$
$$M_0 = \{parent_{seq}(s)\}$$
$$M_{i+1} = M_i \cup \{parent_{seq}(x) \mid x \in M_i\} \text{ for } i \ge 0.$$

Example 5. The element $(monitor, printer)$ is an ML-element parent of the $(LCD, laser)$ element because all of its items are replaced by their parents. The ML-sequence $\langle PC(monitor, printer)\rangle$ is the ML-sequence parent of the sequence $\langle tower(LCD, laser)\rangle$ and it is also the only member of its ML-ancestors set because $\langle tower(LCD, laser)\rangle$ does not have any ML-sequence parent.

The *generalized support gen_supp* is based on the definition of support in Definition 2. It is only necessary to redefine the subset relation for two elements to deal with taxonomies according to Definition 9.

Definition 9. *(The generalized support)* A generalized subset relation $\subseteq_g$ *is defined as*

$$e_1 \subseteq_g e_2 \Leftrightarrow \forall i \in e_1 : i \in e_2 \vee$$
$$\exists j \in e_2 : i \in ancestors(j). \tag{6}$$

A sequence $\alpha = \langle a_1 a_2 \ldots a_n \rangle$ *is a* generalized subsequence *of a sequence* $\beta = \langle b_1 b_2 \ldots b_m \rangle$ *if there exists integers* $1 \leq j_1 < j_2 < \cdots < j_n \leq m$ *such that* $a_1 \subseteq_g b_{j_1}, a_2 \subseteq_g b_{j_2}, \ldots, a_n \subseteq_g b_{j_n}$. *We denote* $\alpha \sqsubseteq_g \beta$. *Then, the* the generalized support *of a sequence* s_{ML} *is*

$$gen\_supp(s_{ML}) = |\{\langle SID, s \rangle | (\langle SID, s \rangle \in D) \wedge (s_{ML} \sqsubseteq_g s)\}|.$$

Recall the term *closed* in *closed sequential pattern mining*. The *closed* means that if a sequence s and a supersequence of s has the same support, then the result set will contain only a supersequence of s. In this case, any omitted subsequence can be derived from the result set. In contrast, the problem discussed in the paper allows us to modify sequence in two dimensions – a sequence length (the same with *closed sequential pattern mining*) and a level of sequence (new for ML-sequences). We will discuss closed only in the taxonomy dimension.

5   The MLSP Algorithm

In this section, the problem of mining hierarchically-closed multi-level sequential patterns is specified and the effective **MLSP** *(Multi-Level Sequential Patterns)* algorithm for mining these patterns is presented.

Definition 10. *(Mining hierarchically-closed multi-level sequential patterns)* *The* set of hierarchically-closed ML-sequences *is such a set of ML-sequences which* does not *contain any ML-sequence* s *and ML-sequence ancestor of* s *with equal generalized support. Then, the problem of **mining hierarchically-closed multi-level sequential patterns** (hereinafter* ML-sequential patterns*) for a given input sequence database* D *and minimal generalized support threshold* min_supp *is to find a set* L_{ML} *of all ML-sequences over* D *such that*

$$L_{ML} = \{s_{ML} \sqsubseteq s | \langle SID, s \rangle \in D \wedge gen\_supp(s_{ML}) \geq min\_supp \tag{7}$$
$$\wedge \not\exists s_x [s_x \sqsubseteq_g s \wedge gen\_supp(s_x) \geq min\_supp$$
$$\wedge gen\_supp(s_x) = gen\_supp(s_{ML})$$
$$\wedge s_{ML} \in ancestor_{seq}(s_x)]\}.$$

The proposed algorithm MLSP is based on the candidate generation principle (adapted from the GSP, see Section 3) combined with the on-demand generalization. The algorithm runs in two steps *(candidate generation* and *counting)*.

Counting candidates step modification. The idea is that if the candidate ML-sequence should be removed, the generalization of the ML-sequence is performed. The generalization of the ML-sequence means to find an ancestor with the greatest level of the sequence which satisfies a minimal support threshold. The on-demand *bottom-up* generalization procedure GETFIRSTFREQUENTANCESTOR() is shown in Algorithm 1.

Algorithm 1. Method GetFirstFrequentAncestor()

1: **procedure** GETFIRSTFREQUENTANCESTOR(s, min_supp)
2:     **repeat**
3:         **if** $gen\_supp(s) \geq min\_supp$ **then**
4:             **return** s
5:         **end if**
6:         $s \leftarrow parent_{seq}(s)$
7:     **until** s is *root sequence*
8:     **return** *null*
9: **end procedure**

Example 6. Counting step generalization. If none of the candidate ML-sequences $\langle minitower\ CRT \rangle$ and $\langle minitower\ LCD \rangle$ is frequent, there can exist the ancestor of both ML-sequences $\langle PC\ monitor \rangle$ which could be frequent and important.

Nevertheless, this modification of the counting step does not guarantee the algorithm's completeness. Also, the generalization has to be performed for *join step*. Recall that the GSP allows to join a pair of sequences if the common subsequence is the same (see Section 3 for details). In difference, the MLSP algorithm firstly tries to find the common ancestor of the candidate ML-subsequences in a *bottom-up way*. If the common ML-subsequence exists, then the generalized ML-sequences are joined to the new candidate ML-sequence, otherwise, no candidate is generated.The modified procedure for candidate generation is shown in Algorithm 2.

Example 7. Join step generalization. The pair of 2-ML-sequences $\langle minitower\ CRT \rangle$ and $\langle LCD\ laser\_printer \rangle$ cannot be joined in GSP, because of items CRT and LCD. However, they have a common parent *monitor*. Therefore, the items should be generalized to the common subsequence ¡*monitor*¿ and the sequences are joined to the 3-ML-sequence ¡*PC, monitor, printer*¿ by MLSP.

Finally, we summarize the complete MLSP algorithm in Algorithm 3. In the first phase, the algorithm counts the generalized support for all items of the database D and their parents and 1-sequences are formed as candidate ML-sequences C_1. Then, the ML-sequences for each candidate 1-sequence with sufficient support are generated as ML-sequential patterns or the generalization is performed.

Algorithm 2. Method GenerateCandidateMLSequences()

1: **procedure** GENERATECANDIDATEMLSEQUENCES(L_{k-1}, k)
2:     $C_k = \emptyset$
3:     **for all** $s_1, s_2 \in L_{k-1}$ **do**
4:         **if** ML-subsequences wrt. GSP join rule of s_1 and s_2 are equal
            or can be generalized to common subsequence/s **then**
5:             Add ML-sequence joined from s_1, s_2 on the greatest level into C_k.
6:         **end if**
7:     **end for**
8:     **return** C_k
9: **end procedure**

The next phases run iteratively while some ML-sequential patterns are generated in the previous phase. In each phase, candidate ML-sequences are generated by the generalized join procedure (join step). In the counting step, the algorithm makes a pass over the database D and the generalized supports for the generated candidate ML-sequences are counted. It is effective to count also the generalized supports for the all the ancestors of candidate ML-sequences, because it will be be used by the following the generalization procedure. The generalization tries to find the frequent ML-sequence with the greatest level for each candidate ML-sequence. Because there can exist candidate ML-sequences that are not hierarchically-closed, it is necessary to verify that does not exist any child of the candidate ML-sequence with the same generalized support.

Algorithm 3. The MLSP algorithm

1: **procedure** MLSP(D, min_supp)
2:     $k \leftarrow 1$ ▷ First phase.
3:     Count gen_supp for all items in D ▷ Count also for ancestors.
4:     $C_1 \leftarrow$ Add all 1-ML-sequences for candidates c created from all items in DB
5:     Process sequences $c \in C_1$ by procedure GETFIRSTFREQUENTANCESTOR(C)
        and add resulting sequences into L_1
6:     **while** $L_k = \emptyset$ **do**   ▷ Next iterative phases.
7:         $k \leftarrow k + 1$
8:         $C_k \leftarrow$ GENERATECANDIDATEMLSEQUENCES(L_{k-1}, k)
9:         Count gen_supp for all candidate sequences in C_k and their ancestors
10:        Process items $c \in C_k$ by procedure GETFIRSTFREQUENTANCESTOR(C)
            and add resulting hierarchically-closed sequences into L_k
11:    **end while**
12:    **return** $\bigcup_{i=1}^{k} L_i$
13: **end procedure**

Algorithm Heuristic. The algorithm often performs "is a subsequence" test (e.g. for the generalized support counting). This test can be optimized to the linear time-complexity if a suitable complete ordering exists over items. The simple ordering is not usable for MLSP because the subsequence uses the *generalized subset relation* $\sqsubseteq_g$. Therefore, MLSP uses two step ordering: 1) firstly it sorts taxonomies lexicographically by their roots and 2) items in one taxonomy are sorted in the post-order walk. The first rule provides for a grouping of items in elements by a taxonomy; the second rule guarantees that it is possible to check for an ideal mapping to ancestors in the linear time complexity (the lexicographical ordering cannot be used because of the generalization which changes the order).

6 Experiments

We performed experiments on a PC in the configuration CPU i5 3.3GHz, 8GB RAM, OS MS Windows 7. Because there has not been any algorithm published for the mining multi-level sequential patterns, we compared results of our algorithm to the GSP. Authors of the GSP recommended using the GSP over an extended database [7] for mining sequential patterns with taxonomies. In extended database, each item of the original database $\mathcal{D}$ is replaced by a set of all its ancestors within its element. Both GSP and MLSP algorithms were implemented in C# on .NET platform. We also evaluated the level-crossing mining algorithm hGSP [13] but it did not finished for large datasets. The experiments were executed over synthetic datasets generated by a random generator similar to [1] modified to generate sequences with items on different levels of taxonomies.

The first experiment compares scalability of GSP and MLSP. The parameters of the synthetic datasets are: $|\mathcal{D}|$ = *from 100 000 to 1 000 000, avg. sequence length = 4, sequential patterns count = 5, avg. sequential pattern length = 3, avg. support of sequential patterns=5%, taxonomies count = 0.1% of $\mathcal{D}$, avg. count of taxonomies levels=4.* The comparison of execution times is shown in Fig. 3a. The *MLSP* algorithm is about *5-10 times* faster than the *GSP* on all sizes of datasets. For both GSP and MLSP, the execution time grows approximately linearly with the size of the dataset (it corresponds to results presented by Agrawal et. al. in [1]). The main reason for higher speed is a smaller set of candidates (see the Fig. 3b). It results in a much lower number of sequence comparisons when computing support. Moreover, the GSP algorithm generates a higher number of sequential patterns in contrast to MLSP, because sequential patterns generated by GSP are not hierarchically-closed. Note that the numbers of candidate patterns and sequential patterns were summarized over all phases.

The second experiment shows the dependency of the execution time and of the number of sequential patterns on the minimum support threshold. In this case, all experiments were executed over the same dataset and only the minimum support threshold parameter was changed. It used the smallest dataset from the previous experiment where $|\mathcal{D}|$ = *100 000* sequences. The minimum support threshold was set to *1%, 3%, 5% and 7%*. The results are shown in Fig. 4a and Fig. 4b. Total execution time decreases with a greater minimum support threshold. Also, the number of sequential patterns decreases rapidly. In all cases, the MLSP algorithm is much faster than GSP and produces a more readable result set.

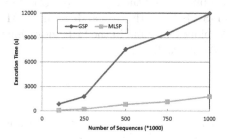 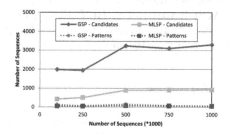

(a) Execution times of algorithms.   (b) Number of cadidates and patterns.

Fig. 3. Scalability experiment of GSP and MLSP algorithms

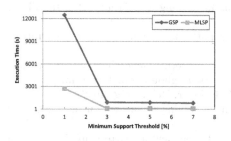 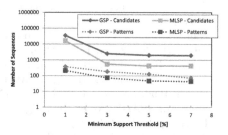

(a) Execution times of algorithms.   (b) Number of cadidates and patterns.

Fig. 4. Dependency of the execution time and of the number of sequential patterns on the minimum support threshold

7   Conclusion

In this paper was discussed the important problem of the multi-level sequential pattern mining. The term hierarchically-closed sequential pattern mining which reduces the number of output multi-level sequential patterns was introduced. Then we proposed the MLSP algorithm for mining hierarchically-closed multi-level sequential patterns. We experimentally verified that MLSP generates much smaller result set of sequential patterns without the loss of useful information for the user and the mining process is a much more effective compared to GSP when the extended database is used.

In future research, there is a challenge to develop an algorithm that does not generate candidates. We also plan to research the problem of mining multi-level sequential patterns for data streams.

Acknowledgement. This work has been supported by the research programme TAČR TA01010858, BUT FIT grant FIT-S-11-2, by the research plan MSM 0021630528, and the European Regional Development Fund in the IT4Innovations Centre of Excellence project (CZ.1.05/1.1.00/02.0070).

References

1. Agrawal, R., Srikant, R.: Mining sequential patterns. In: Proc. of the Eleventh International Conference on Data Engineering, pp. 3–14 (March 1995)
2. Han, J., Fu, A.: Mining multiple-level association rules in large databases. IEEE Trans. on Knowledge and Data Engineering 11(5), 798–805 (1999)
3. Han, J., Kamber, M.: Data mining: concepts and techniques. The Morgan Kaufmann series in data management systems. Elsevier (2006)
4. Nakano, S.I.: Efficient generation of plane trees. Inf. Process. Lett. 84(3), 167–172 (2002)
5. Agrawal, R., Srikant, R.: Fast algorithms for mining association rules. In: VLDB 1994, pp. 487–499 (1994)
6. Han, J., Pei, J., Yin, Y.: Mining frequent patterns without candidate generation. SIGMOD Rec. 29(2), 1–12 (2000)
7. Srikant, R., Agrawal, R.: Mining sequential patterns: Generalizations and performance improvements. In: Apers, P.M.G., Bouzeghoub, M., Gardarin, G. (eds.) EDBT 1996. LNCS, vol. 1057, pp. 1–17. Springer, Heidelberg (1996)
8. Zaki, M.: SPADE: An efficient algorithm for mining frequent sequences. Machine Learning 42, 31–60 (2001)
9. Ayres, J., Flannick, J., Gehrke, J., Yiu, T.: Sequential pattern mining using a bitmap representation. In: Proc. of the 8th ACM SIGKDD Int. Conf. on Knowledge Discovery and Data Mining, pp. 429–435. ACM, New York (2002)
10. Pei, J., Han, J., Mortazavi-Asl, B., Wang, J., Pinto, H., Chen, Q., Dayal, U., Hsu, M.C.: Mining sequential patterns by pattern-growth: the prefixspan approach. IEEE Trans. on Knowledge and Data Engineering 16(11), 1424–1440 (2004)
11. Wang, J., Han, J.: BIDE: efficient mining of frequent closed sequences. In: Proc. of 20th International Conference on Data Engineering, pp. 79–90 (2004)
12. Plantevit, M., Laurent, A., Laurent, D., Teisseire, M., Choong, Y.W.: Mining multidimensional and multilevel sequential patterns. ACM Trans. Knowl. Discov. Data 4(1), 4:1–4:37 (2010)
13. Šebek, M., Hlosta, M., Kupčík, J., Zendulka, J., Hruška, T.: Multi-level sequence mining based on gsp. Acta Electrotechnica et Informatica (2), 31–38 (2012)

Mining Maximal Sequential Patterns without Candidate Maintenance

Philippe Fournier-Viger[1], Cheng-Wei Wu[2], and Vincent S. Tseng[2]

[1] Department of Computer Science, University of Moncton, Canada
[2] Dep. of Computer Science and Information Engineering, National Cheng Kung University
philippe.fournier-viger@umoncton.ca,
silvemoonfox@hotmail.com, tsengsm@mail.ncku.edu.tw

Abstract. Sequential pattern mining is an important data mining task with wide applications. However, it may present too many sequential patterns to users, which degrades the performance of the mining task in terms of execution time and memory requirement, and makes it difficult for users to comprehend the results. The problem becomes worse when dealing with dense or long sequences. As a solution, several studies were performed on mining maximal sequential patterns. However, previous algorithms are not memory efficient since they need to maintain a large amount of intermediate candidates in main memory during the mining process. To address these problems, we present a both time and memory efficient algorithm to efficiently mine maximal sequential patterns, named MaxSP (Maximal Sequential Pattern miner), which computes all maximal sequential patterns without storing intermediate candidates in main memory. Experimental results on real datasets show that MaxSP serves as an efficient solution for mining maximal sequential patterns.

Keywords: sequences, sequential pattern mining, compact representation, maximal sequential patterns.

1 Introduction

Mining useful patterns in sequential data is a challenging task in data mining. Many studies have been proposed for mining interesting patterns in sequence databases [1, 2, 3] *Sequential pattern mining* is probably the most popular research topic among them. A *sub-sequence* is called *sequential pattern* or *frequent sequence* if it frequently appears in a sequence database, and its frequency is no less than a user-specified *minimum support threshold minsup*. Sequential pattern mining plays an important role in data mining and is essential to a wide range of applications such as the *analysis of web click-streams, program executions, medical data, biological data* and *e-learning data* [1, 2, 3]. Several algorithms have been proposed for sequential pattern mining such as SPAM [4], SPADE [5] and PrefixSpan [6]. However, a drawback of these algorithms is that they may present too many sequential patterns to users. A very large number of sequential patterns makes it difficult for users to analyze results to gain insightful knowledge. It may also cause the algorithms to become inefficient in

H. Motoda et al. (Eds.): ADMA 2013, Part I, LNAI 8346, pp. 169–180, 2013.
© Springer-Verlag Berlin Heidelberg 2013

terms of time and memory because the more sequential patterns the algorithms produce, the more resources they consume. The problem becomes worse when the database contains long sequential patterns. For example, consider a sequence database containing a sequential pattern having 20 distinct items. A sequential pattern mining algorithm will present the sequential pattern as well as its 2^{20-1} subsequences to users. This will most likely make the algorithm fail to terminate in reasonable time and run out of storage space. For example, the well-known sequential pattern mining algorithm PrefixSpan would have to perform 2^{20} database projection operations to produce the results.

To reduce the computational cost of the mining task and present fewer but representative patterns to users, many studies focused on developing concise representations of sequential patterns. One of the representations that has been proposed is *closed sequential patterns* [7, 8, 9]. A closed sequential pattern is a sequential pattern that is not strictly included in another pattern having the same frequency. Several approaches have been proposed for mining closed sequential patterns in sequence databases such as BIDE [7], CloSpan [8] and ClaSP [9]. Although these algorithms mines a compact set of sequential patterns, the set of closed patterns is still too large for dense databases or database containing long sequences.

To address this problem, it was proposed to mine *maximal sequential patterns* [10, 11, 12, 13, 14]. A maximal sequential pattern is a closed pattern that is not strictly included in another closed pattern. The set of maximal sequential patterns is thus generally a very small subset of the set of (closed) sequential patterns. It is widely recognized that mining maximal patterns can be faster than mining all (closed) patterns. Besides, the set of maximal sequential patterns is representative because all sequential patterns can be derived from it. Furthermore, the exact frequency of the sequential patterns can be obtained with a single database pass. This method thus provides an alternative solution to find all sequential patterns when the traditional algorithms cannot successfully mine (closed) sequential patterns in the databases.

Maximal sequential pattern mining is important and has been adopted for many applications such as discovering frequent longest common subsequences in a text, analysis of DNA sequences, data compression and web log mining [10].

Although maximal sequential pattern mining is desirable and useful in many applications, it is still a challenging data mining task that has not been deeply explored. Only few algorithms have been proposed for efficiently mining maximal sequential patterns. MSPX [12] is an approximate algorithm and therefore it provides an incomplete set of maximal patterns to the user, and thus may omit important information. DIMASP algorithm [10] is designed for the special case where sequences are strings (no more than an item can appear at the same time) and where no pair of contiguous items appears more than once in each sequence. AprioriAdjust algorithm [13] is an apriori-like algorithm, which may suffer from the drawbacks of the candidate generation-and-test paradigm. In other word, it may generate a large number of candidate patterns that do not appear in the input database and require to scan the original database many times. MSPX [12] and MFSPAN [14] algorithms need to maintain a large amount of intermediate candidates in main memory during the mining process. Although the above algorithms are pioneers for maximal sequential pattern mining, they are not memory efficient since they need to maintain a large amount of intermediate candidates in main memory during the mining process [10, 11, 12, 13, 14].

To address the above issues, we propose a both time and memory efficient algorithm, named *MaxSP* (*Maximal Sequential Pattern miner*), to efficiently mine maximal sequential patterns in sequence databases. The proposed algorithm is developed for the general case of a sequence database rather than strings and it can capture the complete set of maximal sequential patterns with only two database scans. Moreover, it discovers all maximal sequential patterns neither producing redundant candidates nor storing intermediate candidates_in main memory. Whenever a maximal pattern is discovered by MaxSP, it can be outputted immediately. We performed an experimental study with five real-life datasets to evaluate the performance of MaxSP. We compared its performance with the BIDE algorithm [8], one of the current best algorithms for mining closed sequential patterns without storing intermediate candidates in memory. Results show that MaxSP outperforms BIDE in terms of execution and memory consumption and that the set of maximal patterns is much smaller than the set of closed patterns.

The rest of the paper is organized as follows. Section 2 formally defines the problem of maximal sequential pattern mining and its relationship to sequential pattern mining. Section 3 describes the MaxSP algorithm. Section 4 presents the experimental study. Finally, Section 5 presents the conclusion and future work.

2   Problem Definition

The problem of sequential pattern mining was proposed by Agrawal and Srikant [1]. A *sequence database SDB* is a set of sequences $S = \{s_1, s_2...s_s\}$ and a set of items $I = \{i_1, i_2, ... i_m\}$ occurring in these sequences. An *item* is a symbolic value. An *itemset I* $= \{i_1, i_2, ..., i_m\}$ is an unordered set of distinct items. For example, the itemset $\{a, b, c\}$ represents the sets of items a, b and c. A *sequence* is an ordered list of itemsets $s = \langle I_1, I_2, ... I_n \rangle$ such that $I_k \subseteq I$ ($1 \leq k \leq n$). For example, consider the sequence database depicted in Figure 1. It contains four sequences having respectively the *sequences ids* (SIDs) 1, 2, 3 and 4. In this example, each single letter represents an item. Items between curly brackets represent an itemset. For instance, the sequence $\langle\{a, b\},\{c\},\{f\},\{g\},\{e\}\rangle$ indicates that items a and b occurred at the same time, were followed successively by c, f, g and lastly e. A sequence $s_a = \langle A_1, A_2, ..., A_n \rangle$ is *contained in* another sequence $s_b = \langle B_1, B_2,..., B_m \rangle$ iff there exists integers $1 \leq i_1 < i_2 < ... < i_n \leq m$ such that $A_1 \subseteq B_{i1}$, $A_2 \subseteq B_{i2}$, ..., $A_n \subseteq B_{in}$ (denoted as $s_a \sqsubseteq s_b$). The *support of a subsequence* s_a in a sequence database *SDB* is defined as the number of sequences $s \in S$ such that $s_a \sqsubseteq s$ and is denoted by $sup(s_a)$.

Definition 1. The *problem of mining sequential patterns* in a sequence database *SDB* is to find all frequent sequential patterns, i.e. each subsequence s_a such that $sup(s_a) \geq$ *minsup* for a threshold *minsup* set by the user. For example, Figure 2 shows the 29 sequential patterns found in the database of Figure 1 for *minsup* = 2.

Several algorithms have been proposed to mine sequential pattern such as SPAM [4], PrefixSpan [5], and SPADE [6]. To reduce the number of sequential patterns found and find representative patterns, it was proposed to mine *closed* and *maximal sequential patterns*.

Definition 2. A *closed sequential pattern is* a frequent sequential pattern that is not strictly included in another sequential pattern having the same support [8, 9].

Definition 3. A *maximal sequential pattern* is a frequent sequential pattern that is not strictly included in another frequent sequential pattern. An equivalent definition is that a sequential pattern is maximal if it is a closed sequential pattern that is not strictly included in another closed sequential pattern.

Property 1. It can be easily seen that maximal patterns are a subset of the set of closed patterns and that closed patterns are a subset of the set of frequent sequential patterns. **Rationale.** This follows directly from the above definitions. **Example.** Consider the database of Figure 1 and *minsup* = 2. There are 29 sequential patterns (shown in Figure 2), such that 15 are closed (identified by the letter 'C') and only 10 are maximal (identified by the letter 'M').

Property 2. Maximal patterns are a lossless representation of all frequent sequential patterns (they allow recovering all frequent sequential patterns). **Proof.** By definition, a maximal sequential pattern has no proper super-sequence that is a frequent sequential pattern. Thus, if a pattern is frequent, it is either a proper subsequence of a maximal pattern or a maximal pattern. Figure 3 presents a simple algorithm for recovering all sequential patterns from the set of maximal sequential patterns. It generates all the subsequences of all the maximal patterns. Furthermore, a check is performed to detect if a sequential patterns has already been output (line 3) because a sequential pattern may be a subsequence of more than one maximal pattern.

3   The MaxSP Algorithm

The MaxSP algorithm is a pattern-growth algorithm inspired by PrefixSpan [5]. We therefore first briefly introduce the PrefixSpan algorithm. Then, we present the MaxSP algorithm and discuss optimizations.

3.1   Discovering Sequential Patterns by Pattern-Growth

PrefixSpan [5] is one of the most efficient sequential pattern mining algorithm. It offers several interesting properties such as being a pattern-growth approach, i.e. it discovers patterns directly without generating candidates. The pseudocode of the main steps of PrefixSpan is shown in Figure 4. PrefixSpan takes three parameters as input. The first parameter is a sequence database *SDB*. The second parameter is the user-defined minimum support threshold *minsup*. The third parameter is a prefix sequence *P*, which is initially set to the empty sequence $\langle\rangle$. The output of PrefixSpan is the set of frequent sequential patterns. PrefixSpan operates as follows. It first scans *SDB* once to calculate the support of each single item (line 1). Then, for each frequent item i, the algorithm outputs the sequential pattern $\langle\{i\}\rangle$ (line 2-4). For each frequent item i, a projection of the database SDB by i is performed to obtain a projected database SDB_i (line 5). Then, a recursive call is performed with parameters SDB_i, minsup and the concatenation of the prefix P with $\{i\}$ (line 6). The recursive call will then consider extending the pattern $\langle\{i\}\rangle$ with single items to form larger patterns. By the means of the recursive calls, the PrefixSpan algorithm recursively appends items one item at a time to discover larger patterns. The database projection operation is performed as follows.

SID	Sequences
1	$\langle\{a, b\},\{c\},\{f, g\},\{g\},\{e\}\rangle$
2	$\langle\{a, d\},\{c\},\{b\},\{a, b, e, f\}\rangle$
3	$\langle\{a\},\{b\},\{f\},\{e\}\rangle$
4	$\langle\{b\},\{f, g\}\rangle$

Fig. 1. A sequence database

Pattern	Sup.		Pattern	Sup.	
$\langle\{a\}\rangle$	3	C	$\langle\{b\},\{g\},\{e\}\rangle$	2	CM
$\langle\{a\},\{g\}\rangle$	2		$\langle\{b\},\{f\}\rangle$	4	C
$\langle\{a\},\{g\},\{e\}\rangle$	2	CM	$\langle\{b\},\{f, g\}\rangle$	2	CM
$\langle\{a\},\{f\}\rangle$	3	C	$\langle\{b\},\{f\},\{e\}\rangle$	2	CM
$\langle\{a\},\{f\},\{e\}\rangle$	2	CM	$\langle\{b\},\{e\}\rangle$	3	C
$\langle\{a\},\{c\}\rangle$	2		$\langle\{c\}\rangle$	2	
$\langle\{a\},\{c\},\{f\}\rangle$	2	CM	$\langle\{c\},\{f\}\rangle$	2	
$\langle\{a\},\{c\},\{e\}\rangle$	2	CM	$\langle\{c\},\{e\}\rangle$	2	
$\langle\{a\},\{b\}\rangle$	2		$\langle\{e\}\rangle$	3	
$\langle\{a\},\{b\},\{f\}\rangle$	2	CM	$\langle\{f\}\rangle$	4	
$\langle\{a\},\{b\},\{e\}\rangle$	2	CM	$\langle\{f, g\}\rangle$	2	
$\langle\{a\},\{e\}\rangle$	3	C	$\langle\{f\},\{e\}\rangle$	3	
$\langle\{a, b\}\rangle$	2	CM	$\langle\{g\}\rangle$	2	
$\langle\{b\}\rangle$	4		$\langle\{g\},\{e\}\rangle$	2	
$\langle\{b\},\{g\}\rangle$	3	C			

C = Closed   M = Maximal

Fig. 2. Sequential patterns found for minsup = 2 (right)

RECOVERY (*a set of maximal patterns M*)
1. **FOR** each sequential pattern $j \in M$
2.   **FOR** each subsequence j of M
3.   **IF** j has not been output
4.   **THEN** output j.

Fig. 3. Algorithm to recover all frequent sequential patterns from maximal patterns

---

Definition 4. The *projection of a sequence database SDB by a prefix P* is the projection of each sequence from SDB containing P by the prefix P.

Definition 5. The *projection of a sequence S by a prefix P* is the part of the sequence occurring immediately after the first occurrence of the prefix P in the sequence S. For instance, the projection of $\langle\{a\},\{c\},\{a\},\{e\}\rangle$ by the item a is the sequence $\langle\{c\},\{a\},\{e\}\rangle$ and the projection of $\langle\{a\},\{c\},\{b\},\{e\}\rangle$ by the prefix $\langle\{c\},\{b\}\rangle$ is $\langle\{e\}\rangle$.

Note that performing a database projection does not require to make a physical copy of the database. For memory efficiency, a projected database is rather represented by a set of pointers on the original database (this optimization is called *pseudo-projection*) [5]. Also, note that the pseudo-code presented in Figure 4 is simplified. The actual PrefixSpan algorithm needs to consider that an item can be appended to the current prefix P by *i-extension* or *s-extension* when counting the support of single items. An *i-extension* is to append an item to the last itemset of prefix P. An *s*-extension is to append an item as a new itemset after the last itemset of prefix P. The interested reader is referred to [5] for more details. The PrefixSpan algorithm is correct and complete. It enumerates all frequent sequential patterns thanks to the anti-monotonicity property, which states that the support of a proper supersequence X of a sequential pattern S can only be lower or equal to the support of S [2]. PrefixSpan is said to discover sequential patterns without candidate generation because only frequent items are concatenated with the current prefix P to generate patterns at each recursive call of the algorithm.

PrefixSpan (a sequence database *SDB*, a threshold *minsup*, a prefix P initially set to $\langle\rangle$)
1.   Scan *SDB* once to count the support of each item.
2.   **FOR** each item i with a support $\geq$ *minsup*
3. P' := **Concatenate**(P, i).
4.   **Output** the pattern P'.
5.   SDBi := **DatabaseProjection**(*SDB*, i).
6.   **PrefixSpan**(SDB$_i$, *minsup*, P').

Fig. 4. The main steps of the PrefixSpan algorithm

3.2   The MaxSP Algorithm

The main challenge to design a maximal sequential pattern mining algorithm based on PrefixSpan is how to determine if a given frequent sequential pattern generated by the PrefixSpan is maximal. A naïve approach would be to keep all frequent sequential patterns found until now into memory. Then, every time that a new frequent sequential pattern would be found, the algorithm would compare the pattern with previously found patterns to determine if (1) the new pattern is included in a previously found pattern or (2) if some previously found pattern(s) are included in the new pattern. The first case would indicate that the new pattern is not maximal. The second case would indicate that some previously found pattern(s) are not maximal. This approach is used for example in CloSpan [8] for closed sequential pattern mining. The drawback of this approach is that it can consume a large amount of memory if the number of patterns is large, and it is becomes very time consuming if a very large number of patterns is found, because a very large number of comparisons would have to be performed [7]. In this paper, we present a new checking mechanism that can determine if a pattern is maximal without having to compare a new pattern with previously found patterns. The mechanism is inspired by the mechanism used in the BIDE algorithm for checking if a pattern is closed [7]. In this subsection, we first introduce important definitions and then we present our solution. Note that in the following, we use sequences where itemsets contain single items (strings) for the sake of simplicity. Nevertheless, the definitions that we present can be easily extended for the general case of a sequence of itemsets containing multiple items (our implementation handle the general case of itemsets).

Definition 4. Let be a prefix P and a sequence S containing P. The *first instance of the prefix P in S* is the subsequence of S starting from the first item in S until the end of the first instance of P in S. For example, the first instance of $\langle\{a\},\{b\}\rangle$ in $\langle\{a\},\{a\},\{b\},\{e\}\rangle$ is $\langle\{a\},\{a\},\{b\}\rangle$.

Definition 5. Let be a prefix P and a sequence S containing P. The *last instance of the prefix P in S* is the subsequence of S starting from the first item in S until the end of the first instance of P in S. For example, the last instance of $\langle\{a\},\{b\}\rangle$ in $\langle\{a\},\{b\},\{b\},\{e\}\rangle$ is $\langle\{a\},\{b\},\{b\}\rangle$.

Definition 6. Let be a prefix P containing n items and a sequence S containing P. The *i-th last-in-last appearance* of P in S is denoted as LL_i and defined as follows. If $i = n$, it is the last appearance of the i-th item of P in the last instance of P in S. If $1 < i < n$, it is the last appearance of the i-th item of P in the last instance of P in S such that LL_i must appear before LL_{i+1}. For example, the first last-in-last appearance of $\langle\{a\},\{c\}\rangle$ in the sequence $\langle\{a\},\{a\},\{c\},\{e\},\{c\}\rangle$ is the second a, while the second last-in-last appearance in the sequence $\langle\{a\},\{a\},\{c\},\{e\},\{c\}\rangle$ is the second c.

Definition 7. Let be a prefix P containing n items and a sequence S containing P. Let P_{n-1} denotes the first n items of P. The *i-th maximum period of P in S* is defined as follows. If $1 < i < n$, it is the piece of sequence between the end of the first instance of P_{n-1} in S and the i-th last-in-last appearance of P in S. If $i = 1$, it is the piece of sequence in S before the first last-in-last appearance of P in S. For example, consider $S = \langle\{a\},\{b\},\{c\},\{d\}\rangle$ and $P = \langle\{a\},\{b\}\rangle$. The first maximum period of P in S is the empty sequence, while the second maximum period is $\langle\{b\},\{c\}\rangle$.

Based upon the above definitions, we propose a mechanism to determine if a frequent pattern P is maximal without maintaining previously found patterns in memory. The mechanism consists of verifying if P can be extended by appending items to form a larger frequent sequential pattern. If yes, it indicates that the pattern P is not maximal (by definition 3, a maximal pattern has no proper supersequence that is frequent). Otherwise, P is maximal and can be immediately output. The mechanism is implemented by two separate checks, which we respectively name *maximal-backward-extension check* and *maximal-forward-extension check*, and are defined as follows.

Definition 8. Let be a frequent sequential pattern P containing n items and a sequence database SDB. The pattern P has a *maximal-backward-extension* in SDB if and only if for an integer k ($1 \leq k \leq n$), there exists an item i having a support higher or equal to *minsup* in the k-th maximum periods of P in SDB_P.

Definition 9. Let be a frequent sequential pattern P containing n items and a sequence database SDB_P. The pattern P has a *maximal-forward-extension* in SDB if and only if there exist an item i having a support higher or equal to *minsup* in the projected database SDB_P.

We now demonstrate that the maximal-forward-extension check and maximal-back-extension-check is sufficient for determining if a pattern is maximal.

Property 3. A frequent sequential pattern P is a maximal sequential pattern for a sequence database SDB if and only if it has no maximal-forward-extension in SDB_P and no maximal-backward-extension in the projected database SDB_P. **Rationale.** By definition, a pattern is maximal if it has no proper supersequence that is frequent. Consider a sequential pattern $P=\langle\{a_1\},\{a_2\}, …,\{a_n\}\rangle$ containing n items. If there exists a proper supersequence of P that is frequent, it means that an item i can be added to P so that the resulting pattern P' would be frequent. It can be easily seen that the support of P' is the minimum of the support of i and the support of P. To count the support of P', we can consider SDB_P instead of SDB, since P' can only appear in a subset of the sequences where P appear. To detect if such an item i can be appended to P, we need to consider three cases: (1) an item i can be appended to P before a_1, (2) an item i can be appended to P between any a_x and a_y such that $y = x+1$ and $1 \le x \le n$, or (3) an item i can be appended to P after a_n. The first two cases are verified by the *maximum backward extension check*. Verifying if an item can be appended before a_1 consists of verifying if an item is frequent in the first maximum period of P in SDB_P. Verifying if an item can be appended between any a_x and a_y such that $y = x+1$ and $1 \le x \le n$ is done by verifying if an item appear frequently in the y-th maximum period of P in SDB_P. Finally, the third case can be verified by simply counting the support of items in SDB_P to see if an item is frequent (because SDB_P is projected by P and thus items in SDB_P are items that can be appended to P after a_n).

Having shown that maximal-forward-extension and maximal-back-extension can be used to determine if a pattern is maximal, we next explain how they can be incorporated in the search procedure. We show the modified procedure in Figure 5, which we name MaxSP. It takes the same parameters as PrefixSpan: a current database SDB, a threshold *minsup* and a current prefix P. The algorithm first initializes a variable *largestSupport* to 0, which will be used to store the support of the largest pattern that can be obtained by extending the current prefix (line 1). Then, the algorithm counts the support of each item in the current database (line 2). Then for each frequent item i, the algorithm appends i at the end of prefix P, to obtain a new prefix P' and the projected database SDB_i is created by projecting SDB_P by i (line 3-5). Next, the check for maximal-back-extension is performed (line 6). If there is no maximal-back-extension, the procedure MaxSP is recursively called to explore patterns starting by P' (line 7). This recursive call returns the largest support of patterns that can be created by appending items to P'. If the largest support is smaller than *minsup*, it means that there is no maximal-forward-extension (line 8). In this case, the pattern P' is output (line 9). Next, the largest support for the prefix P is updated (line 10). Finally, after all frequent items have been processed, the largest support of extensions of P is returned.

3.3   Optimizations

We performed three optimizations to improve the performance of MaxSP. First, we use pseudo-projections instead of projections to avoid the cost of making physical database copies (as suggested in PrefixSpan [5]). A second optimization is to remove infrequent items from the database immediately after the first database scan because they will not appear in any maximal sequential pattern. A third optimization concerns the process of searching for the maximal-backward-extensions of a prefix by scanning maximum periods. During the scan, item supports are accumulated. The scan can be stopped as soon as there is an item known to appear in *minsup* maximum periods, because it means that the prefix is not maximal. For large databases containing many sequences, we found that this optimization increase performance by about a factor of two. Note that, this optimization has similarity to the ScanSkip optimization proposed in the BIDE algorithm, that stop scanning sequences as soon as a pattern is determined to be non closed [7].

MaxSP (a sequence database *SDB,* a threshold *minsup,* a prefix *P* initially set to ⟨⟩)
1.   largestSupport := 0.
2.   Scan *SDB* once to count the support of each item.
3.   **FOR** each item *i* with a support ≥ *minsup*
4.   *P'* := **Concatenate**(*P, i).*
5.   SDB*i* := **DatabaseProjection**(*SDB, i*).
6.   **IF the pattern *P'* has no maximal-backward-extension in** SDB*i* **THEN**
7.   maximumSupport := **MaxSP** (SDB$_i$,   *minsup, P').*
8.   **IF** maximumSupport < minsup **THEN OUTPUT** the pattern   *P'.*
9.   **IF** support(*P'*) > *largestSupport* **THEN** *largestSupport := support(P')*
10. **RETURN** *largestSupport.*

Fig. 5. The pseudocode of the MaxSP algorithm

4   Experimental Evaluation

To evaluate the performance of the proposed algorithm, we performed a set of experiments on a computer with a third generation Core i5 processor running Windows 7 and 1 GB of free RAM. We compared the performance of MaxSP with BIDE [7], a state-of-the-art algorithm for closed sequential pattern mining, which also does not store intermediate candidates during the mining process. We do not compare the performance of MaxSP with PrefixSpan because BIDE was previously shown to be more than an order of magnitude faster than PrefixSpan [6, 8]. All algorithms are implemented in Java and memory measurements were done by using the Java API. Five real-life datasets *BMS, Snake, Sign, Leviathan* and *FIFA* with varied characteristics are used in the experiments. Table 1 summarizes their characteristics and data types. The experiments consisted of running MaxSP and BIDE on each dataset while decreasing the minsup threshold until either an algorithm became too long to execute or a clear winner was found. For each dataset, we recorded the execution time, memory usage and the number of patterns found for each algorithm. Figure 6 and 7 respectively show the number of patterns found and the execution times for each dataset. Figure 8 presents the maximum memory usage for *Sign, Snake* and *Leviathan.* The source code of compared algorithms and datasets can be downloaded from http://goo.gl/hDtdt.

Table 1. Datasets' Characteristics

dataset	sequence count	distinct item count	avg. seq. length (items)	type of data
BMS	59,601	497	2.51 (std = 4.85)	web click stream
Snake	163	20	60 (std = 0.59)	protein sequences
Sign	730	267	51.99 (std = 12.3)	language utterances
Leviathan	5,834	9,025	33.81 (std= 18.6)	book
FIFA	20,450	2,990	34.74 (std = 24.08)	web click stream

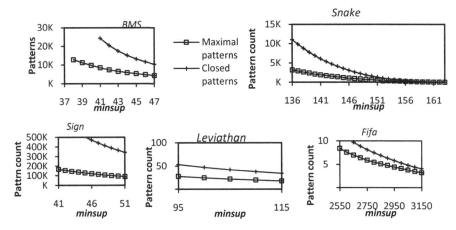

Fig. 6. Maximal pattern and closed pattern count

As it can be seen from the results, the number of maximal patterns is always considerably smaller than the number of closed patterns, and the gap increases quickly as *minsup* decreases. For example, for the *Sign* dataset, only 25 % of the closed sequential patterns are maximal for *minsup* = 47. Another example is *Snake*, where only 28 % of the closed patterns are maximal for *minsup* = 136. This confirms that mining maximal sequential patterns is more efficient in terms of storage space.

With respect to execution time, we can see that MaxSP is always faster than BIDE. The difference is large for sparse datasets such as *Sign* and *FIFA*. For example, for *Sign*, MaxSP was up to five times faster than BIDE. There are two reasons why MaxSP is faster. The first reason is how the maximal-backward-extension checking is performed. For each pattern found, MaxSP looks for items that could extend it with a support no less than minsup, while BIDE looks for items with a support equal to the support of the prefix. As soon as MaxSP or BIDE find an item meeting their respective conditions, they stop searching for backward extensions (the third optimization in MaxSP and the ScanSkip optimization in BIDE). Because the condition verified by BIDE is more specific and thus harder to meet, BIDE needs to analyze more sequences on average for backward extension checking, and this makes BIDE slower. The second reason is that BIDE needs to perform more write operations to disk for storing patterns because the set of closed sequential patterns is larger.

For the memory usage (cf. Figure 8), similar conclusions can be drawn. MaxSP generally uses less memory than BIDE. This is due to the fact that less sequences need to be scanned (as explained previously) and less patterns need to be created by MaxSP. Note that we did not show the memory usage for FIFA and BMS due to space limitation but results are similar as those of Leviathan, Snake and Sign.

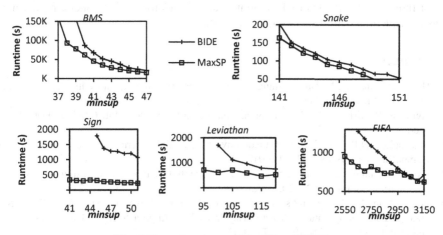

Fig. 7. Execution times of MaxSP and BIDE

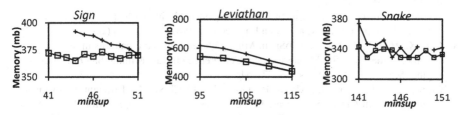

Fig. 8. Maximum memory usage for Sign, Leviathan and Snake

5   Conclusion

Maximal sequential pattern mining is an important data mining task that is essential for a wide range of applications. The set of maximal sequential patterns is a compact representation of sequential patterns. Several algorithms have been proposed for maximal sequential pattern mining. However, they are not memory efficient since they may produce too many redundant candidates and need to maintain a large amount of intermediate candidates in main memory during the mining process. To address these problems, we proposed a memory efficient algorithm to mine maximal sequential patterns named MaxSP (*M*aximal *S*equential *P*attern miner). It incorporates a novel checking mechanism consisting of verifying maximal-backward-extensions and maximal-forward-extensions, which allows discovering all maximal sequential patterns without storing intermediate candidates in memory nor producing redundant candidates. An experimental study on five real datasets shows that MaxSP is more memory efficient and up to five time faster than BIDE, a state-of-art algorithm for closed sequential pattern mining, and that the number of maximal patterns is generally much smaller than the number of closed sequential patterns. The source code of MaxSP and BIDE be downloaded from http://goo.gl/hDtdt as part of the SPMF opensource data mining software.

For future work, we plan to develop new algorithms for mining concise representations of sequential patterns and sequential rules [15, 16].

References

1. Han, J., Kamber, M.: Data Mining: Concepts and Techniques, 2nd edn. Morgan Kaufmann, San Francisco (2006)
2. Agrawal, R., Srikant, R.: Mining Sequential Patterns. In: Proc. Int. Conf. on Data Engineering, pp. 3–14 (1995)
3. Mabroukeh, N.R., Ezeife, C.I.: A taxonomy of sequential pattern mining algorithms. ACM Computing Surveys 43(1), 1–41 (2010)
4. Ayres, J., Flannick, J., Gehrke, J., Yiu, T.: Sequential PAttern mining using a bitmap representation. In: Proc. KDD 2002, Edmonton, Alberta, pp. 429–435 (2002)
5. Pei, J., Han, J., Mortazavi-Asl, B., Pinto, H., Chen, Q., Dayal, U., Hsu, M.: Mining Sequential Patterns by Pattern-Growth: The PrefixSpan Approach. IEEE Trans. Knowledge and Data Engineering 16(10), 1–17 (2001)
6. Zaki, M.J.: SPADE: An efficient algorithm for mining frequent sequences. Machine learning 42(1-2), 31–60 (2001)
7. Wang, J., Han, J., Li, C.: Frequent Closed Sequence Mining without Candidate Maintenance. IEEE Trans. on Know. and Data Engineering 19(8), 1042–1056 (2007)
8. Yan, X., Han, J., Afshar, R.: CloSpan: Mining closed sequential patterns in large datasets. In: Proc. of the Third SIAM International Conference on Data Mining, San Francisco, California, May 1–3 (2003) ISBN 0-89871-545-8
9. Gomariz, A., Campos, M., Marin, R., Goethals, B.: ClaSP: An Efficient Algorithm for Mining Frequent Closed Sequences. In: Pei, J., Tseng, V.S., Cao, L., Motoda, H., Xu, G. (eds.) PAKDD 2013, Part I. LNCS, vol. 7818, pp. 50–61. Springer, Heidelberg (2013)
10. García-Hernández, R.A., Martínez-Trinidad, J.F., Carrasco-Ochoa, J.A.: A new algorithm for fast discovery of maximal sequential patterns in a document collection. In: Gelbukh, A. (ed.) CICLing 2006. LNCS, vol. 3878, pp. 514–523. Springer, Heidelberg (2006)
11. Lin, N.P., Hao, W.-H., Chen, H.-J., Chueh, H.-E., Chang, C.-I.: Fast Mining Maximal Sequential Patterns. In: Proc. of the 7th International Conference on Simulation, Modeling and Optimization, Beijing, China, September 15-17, pp. 405–408 (2007)
12. Luo, C., Chung, S.: Efficient mining of maximal sequential patterns using multiple samples. In: Proc.5th SIAM int'l Conf. on Data Mining, Newport Beach, California (2005)
13. Lu, S., Li, C.: AprioriAdjust: An Efficient Algorithm for Discovering the Maximum Sequential Patterns. In: Proc. 2nd Int'l Workshop Knowl. Grid and Grid Intell. (2004)
14. Guan, E.-Z., Chang, X.-Y., Wang, Z., Zhou, C.-G.: Mining Maximal Sequential Patterns. In: Proc of the Second Int'l Conf. Neural Networks and Brain, pp. 525–528 (2005)
15. Fournier-Viger, P., Nkambou, R., Tseng, V.S.: RuleGrowth: Mining Sequential Rules Common to Several Sequences by Pattern-Growth. In: Proc. of the 26th Symposium on Applied Computing, Tainan, Taiwan, pp. 954–959. ACM Press (2011)
16. Fournier-Viger, P., Faghihi, U., Nkambou, R., MephuNguifo, E.: CMRules: Mining Sequential Rules Common to Several Sequences. Knowledge-based Systems 25(1), 63–76 (2012)

Improved Slope One Collaborative Filtering Predictor Using Fuzzy Clustering

Tianyi Liang[1], Jiancong Fan[1], Jianli Zhao[1,*],
Yongquan Liang[1], and Yujun Li[2]

[1] School of Information Science and Engineering,
Shandong University of Science and Technology, Qingdao 266510, China
`liangtee@126.com,zhaojianli@gmail.com,{fanjiancong,lyq}@sdust.edu.cn`
[2] Hisense State Key Laboratory of Digital Multi-media Technology
`liyujun@hisense.com`

Abstract. Slope One predictor, an item-based collaborative filtering algorithm, is widely deployed in real-world recommender systems because of its conciseness, high-efficiency and reasonable accuracy. However, Slope One predictor still suffers two fundamental problems of collaborative filtering : sparsity and scalability, and its accuracy is not very competitive. In this paper, to alleviate the sparsity problem for Slope One predictor, and boost its scalability and accuracy, an improved algorithm is proposed. Through fuzzy clustering technique, the proposed algorithm captures the latent information of users thereby improves its accuracy, and the clustering mechanism makes it more scalable. Additionally, a high-accuracy filling algorithm is developed as preprocessing tool to tackle the sparsity problem. Finally empirical studies on MovieLens and Baidu dataset support our theory.

Keywords: Slope One, fuzzy clustering, collaborative filtering, sparsity, scalability.

1 Introduction

Slope One predictor[1] is a kind of item-based collaborative filtering (CF)[2–4] algorithm proposed by Daniel Lemire and Anna Maclachlan in 2005. It is designed as a concise and understandable form to make it easy to implement and maintain. Many empirical studies prove Slope One predictor is high-efficient and its prediction accuracy can be comparable with some much more complex algorithms. Because of its simplicity and efficiency, it has been applied in many recommender systems, such as hitflip, Value Investing News and AllTheBests.

However, the extensive application also reveals several shortcomings of Slope One predictor:

- **Sensitive to data sparsity:** like most CF algorithms, the performance of Slope One predictor will decrease badly when the data is sparse.

---

* Corresponding author.

H. Motoda et al. (Eds.): ADMA 2013, Part I, LNAI 8346, pp. 181–192, 2013.

- **High algorithm complexity(Scalability):** suppose there are n users and m items, Slope One predictor requires $O(nm^2)$ time steps and $O(nm(m-1)/2)$ storage units to make recommendations. That makes it not very suitable for the large scale recommender system which needs to deal with millions of users and items.
- **Unremarkable accuracy:** the accuracy of Slope One predictor is not very outstanding, therefore it is often used as preprocessing or smoothing technique in practice.

As a promising branch of CF algorithm, Slope One predictor is worth being improved. In this paper, we adopt Fuzzy Clustering[5] technique to boost Slope One predictor. Our improved algorithm is :

- **Less sensitive to data sparsity**
- **High scalability**
- **High accuracy**
- **Still simplicity, easy to implement and maintain**

In the rest of this paper, we first provide a brief review of CF, and detailed descriptions of Slope One predictor and Fuzzy Clustering technique. Then in Section 3 we propose our algorithm, and comprehensively evaluate it by experiments in Section 4.

2 Background

2.1 State of the Art in Collaborative Filtering

To improve the scalability and alleviate the sparsity problem in CF, many approaches have been proposed. Sarwar et al. [6] proposed an Item-based CF that generates recommendations through comparing the similarity between items rather than users. The advantage of Item-based is the item similarity is relatively static, thus the computation of item similarity can be performed offline, which makes Item-based CF more scalable than User-based CF[2, 3]. Besides, Item-based and Use-based CF are also called Memory-based CF.

Model-based CF[2–4] is a family of algorithms which apply machine learning and data mining technique in CF to get better performance. Typical Model-based CF includes Regression-based[2, 3], Clustering-based[7], Classification-based[2] and MDP-based[2] etc. SVD[2–4, 8] and its variations (e.g. SVD++)[9, 10] which use Matrix Factorization technique to learn latent information from the original user-rating matrix are really popular in recent year because of its excellent performance in Netflix contest and KDD-Cup. Usually Model-based CF has more powerful performance and scalability than Memory-based CF, but the model-training process is expensive, and deploying it needs more domain knowledge.

Hybrid recommender[2, 3, 11] is widely employed in practice. A hybrid recommender usually blend several CF models, and some systems, such as Fab[12], combine Content-based model[13] with CF to get better performance. The research on multi recommender models ensemble has become a hot area.

2.2   Principle of Slope One Predictor

Slope One predictor works on the intuitive principle of a "popular differential" between items[1]. Concretely speaking, the "popular differential" reflects that how much better one item is liked than another, it can be measured through subtracting the mean ratings of two items. Formally the predictor is based on a simplified regression model : $f(x) = x + b$ where b is defined as the mean deviation of the item to be predicted and other items. The algorithm performs the computation on users who have rated these items. Given an user u , to predict his rating on item j , the mean deviation is calculate by equation 1:

$$dev_{j,i} = \frac{1}{|U_j \cap U_i|} \sum_{n \in U_j \cap U_i} (r_{n,j} - r_{n,i}) \tag{1}$$

And the final prediction is :

$$p_{u,j} = \frac{1}{|R_u|} \sum_{i \in R_u} (r_{u,i} + dev_{j,i}) \tag{2}$$

where U_i and U_j are respectively the sets of users who rated item i and item j . R_u is the set of ratings of user u. To improve the accuracy, Weighted Slope One[1] revises equation 2 by taking the number of ratings into consideration:

$$p_{u,j} = \frac{1}{\sum_{i \in R_u} Num_{j,i}} \sum_{i \in R_u} (r_{u,j} + dev_{j,i}) Num_{j,i} \tag{3}$$

Besides, Bi-Polar Slope One[1] improves accuracy by dividing items into user rated positively and negatively.

2.3   Fuzzy Clustering and Its Advantages

Clustering is a process of dividing data into different clusters and putting similar data elements into same cluster. Xue et al. [7] applied clustering technique to improve User-based CF from the following aspects:

- Increasing scalability: the scope of similarity calculation narrows to cluster rather than whole dataset.
- Increasing data density: the missing ratings in cluster can be smoothed by cluster mean rating.

For hard clustering technique, such as K-Means[5], each user must belong to exactly one clustering while in Fuzzy Clustering (FC) (or called soft clustering) each user can belong to more than one clusters. Given an user u , FC uses membership degree $w_{u,j}$ to represent the association strength between user u and cluster j. Suppose there are n users and k clusters, the results of FC satisfy the following three conditions simultaneously:

(1) For each user u and cluster j, $0 \leq w_{u,j} \leq 1$

(2) For each user u, $\sum_{j=1}^{k} w_{u,j} = 1$

(3) For each cluster j, $0 < \sum_{i}^{n} w_{u,j} < n$

Compared with hard clustering, fuzzy clustering is more suitable for the real-world recommender systems. For example, in a movie recommender system, a filmnik may not only likes action movies but also enjoys comedies. Putting him into only one cluster, "Action Fans Cluster" or "Comedy Fans Cluster", may be too rigorous, the better solution is letting him belonging to both of the two clusters.

3   Fuzzy Clustering-Based Slope One Predictor

3.1   Philosophy of Proposed Approach

As is described in Section 2.2, Slope One predictor is based on the "item popularity differential" principle which is measured by the mean deviation among ratings. The original deviation computing method (equation 1) is quite concise but not accurate enough, because it does not take the association between users into consideration, causing much valuable latent information of users are ignored. For instance, in a movie recommender system, users can be divided into groups according to their favorite movie genres, such as "action movie" group, "love movie" group and "commedy movie" group. For a movie, the popular differentials (the deviation values) about it in diffenent user groups reflect how much it is liked by these group of users. When to make predictions for a given user, if the system knows how much importance of each user group for him and translates the "importance" into numeric values, the popular differential can be calculated by weighted summation of each deviation from different user groups, and that will generate more accurate predictions. In order to implement this idea, our approach employs Fuzzy Clustering to divide users into different groups and uses the membership degrees as weight coefficients to adjust the deviations from different clusters, finally obtaining weighted mean deviations. Thus, formally, the equation 1 is updated to equation 4 :

$$dev_{j,i} = \sum_{k=1}^{K} (w_{u,k} \times sub\_dev_{j,i,k}) \tag{4}$$

where $w_{u,k}$ denotes the membership degree between user u and cluster k , $sub\_dev_{j,i,k}$ is the rating deviation between item j and i in cluster k. Compared with equation refl , the calculating method of sub_dev reduces calculation greatly. Details and complexity analysis of it are presented in Section 3.3.

Additionally, a quick and accurate filling algorithm is proposed to improve the performance of fuzzy clustering on sparse data, its details are described in Section 3.2. Finally, the framework of our algorithm (FC-SLP) is :

- **1. Data preprocess :** fill the sparse dataset using filling algorithm.

- **2. Cluster users :** perform fuzzy clustering on users.

- **3. Prediction:** generate predictions based on the works of step 1&2.

The symbols used in this paper are shown in Table 1:

Table 1. Symbols Used throughout This Paper

Symbol	Description
U_i	the set of users who rated item i
R_u	the rating set of user u
D	the dataset
$w_{u,j}$	the membership degree of user u to cluster j
c_j	the feature vector of centroid j
$wr_{m,j}$	the weighted mean rating of item m in cluster j, it is a feature value of c_j
RC_j	the rating set of cluster j
$r_{u,m}$	the rating on item m of user u
$\bar{r}_m$	the mean rating of item m
$sub\_dev_{j,i,k}$	the weight mean deviation of item j and item i in cluster k
$U_{bestC=j}$	for any user u in set $U_{bestC=j}$, $w_{u,j}$ is his **maximum** membership degree value.

3.2   Data Preprocess: Filling Algorithm

Filling the unknown ratings is the most direct way to densify the sparse data thereby boost the performance of fuzzy clustering. The effectiveness of this strategy mainly depends on the accuracy of the filling algorithm, thus, it is crucial for this strategy to select a appropriate filling algorithm. In our scheme, a high-accuracy filling algorithm (HAF) is proposed: formally, the items are divided into two sets : $D_{positive} = \{m \in D \mid \bar{r}_m \geq \bar{r}\}$ and $D_{negative} = \{m \in D \mid \bar{r}_m < \bar{r}\}$. Let $diff_{u,positive}$ and $diff_{u,negative}$ be the rating differentials of user u on "good" and "bad" items, they are computed respectively by equation 5 and 6 :

$$diff_{u,positive} = \frac{\sum_{m \in R_u \cap D_{positive}} (r_{u,m} - \bar{r}_m)}{|R_u \cap D_{positive}|} \tag{5}$$

$$diff_{u,negative} = \frac{\sum_{m \in R_u \cap D_{negative}} (r_{u,m} - \bar{r}_m)}{|R_u \cap D_{negative}|} \tag{6}$$

Finally the prediction is given by equation 7 :

$$p_{u,m} = \begin{cases} \bar{r}_m + diff_{u,positive} \ , & m \in D_{positive} \\ \bar{r}_m + diff_{u,negative} \ , & m \in D_{negative} \end{cases} \tag{7}$$

Evidently, through the filling algorithm, popular items (items with large quantity of ratings) will be filled with more accurate predictions than unpopular ones because popular items have more ratings. Thus, to improve the filling effectiveness, and keep the filling diversity, the wait-to-be-filled items are sampled according to the probability model :

$$ps_i = \frac{|U_i|}{\sum\limits_{n \in D} |U_n|} \tag{8}$$

where ps_i is the probability of item i to be sampled.

With the probability model, popular items have higher probability to be sampled, and unpopular items also have chance to be filled, which balances the accuracy and diversity.

3.3   Cluster Users through Fuzzy Clustering Technique

Define $E(C)$ is the object function :

$$E(C) = \sum_{i=1}^{n} \sum_{j=1}^{k} w_{u,j}^2 Similarity(u, c_j)^2 \tag{9}$$

where $Similarity(u, c_j)$ is the similarity calculation of user u and centroid j. The goal of our scheme is maximizing the function $E(C)$.

Cluster Users through Fuzzy Clustering

Input :
dataset D , cluster number K .
Output :
$C = \{c_1, c_2, ..., c_K\}$: a list of centroids.
$W_{N \times K} = \{w_{u,k}\}$: a membership degree matrix.
Procedure :
1. random select K users as initial centroids.
2. repeat:
3.   re-compute the centroid of each cluster by equation 10 .
4.   update matrix $W_{N \times K}$ by equation 11 .
5. until **E(C)** convergences.

For each feature value $wr_{m,j}$ of centroid c_j :

$$wr_{m,j} = \sum_{u \in U_m \cap n \in U_{bestC=j}} \frac{w_{u,j}}{\sum\limits_{n \in U_m \cap n \in U_{bestC=j}} w_{n,j}} \times r_{u,m} \tag{10}$$

For $w_{u,j}$ in matrix $W_{N \times K}$, it is updated by the following equation 11 :

$$w_{u,j} = \frac{Similarity(u, c_j)}{\sum\limits_{j=1}^{K} Similarity(u, c_j)} \tag{11}$$

where $Similarity(u, c_j)$ is calculated by Weighted-Pearson[3]:

$$Similarity(u, c_j) =$$
$$\begin{cases} S_{pearson}(u, c_j)\frac{|R_u \cap RC_j|}{50} & |R_u \cap RC_j| < 50 \\ S_{pearson}(u, c_j) & otherwise \end{cases} \tag{12}$$

After filling processing, the ratings of users consist of real ratings and filling ratings. Xue et al. [7] differentiated them by setting different weights, but the weights need to be tuned manually. In our scheme, the weights are determined by the user itself rather than empirical rules. Define $\lambda_{u,m}$ is the weight of rating m for user u :

$$\lambda_{u,m} = \begin{cases} 1 & m \in R_u \\ (\frac{|\hat{R}_u|}{|R_u|+|\hat{R}_u|})^2 & m \in \hat{R}_u \end{cases} \tag{13}$$

where $\hat{R}_u$ is the set of filling ratings of user u . Equation 14 is a modified Pearson Correlation which takes the λ into consideration:

$$S_{pearson}(u, c_j) =$$
$$\frac{\sum\limits_{m \in R_u \cap RC_j} \lambda_{u,m} \cdot (r_{u,m} - \bar{r}_i)(wr_{j,m} - \overline{wr}_j)}{\sqrt{\sum\limits_{m \in R_u \cap RC_j} \lambda_{u,m}^2 \cdot (r_{u,m} - \bar{r}_i)^2} \sqrt{\sum\limits_{m \in R_u \cap RC_j} (wr_{j,m} - \overline{wr}_j)^2}} \tag{14}$$

3.4 Prediction

The improved deviation calculation method has been represented by equation 4 where $sub\_dev_{j,i,k}$ is defined as equation 15:

$$sub\_dev_{j,i,k} = wr_{j,k} - wr_{i,k} \tag{15}$$

Finally, the prediction is given by equation 16:

$$p_{u,j} = \frac{1}{|R_u|} \sum\limits_{i \in R_u} (r_{u,i} + \sum\limits_{c=1}^{K} w_{u,c} \times (wr_{j,k} - wr_{i,k})) \tag{16}$$

Consider a database consists of n users and m items, FC-SLP algorithm consumes $O(km^2)$ time steps and $O(km(m-1)/2)$ storage units (k is the number of clusters) to make predictions comparing original Slope One's $O(nm^2)$ and $O(nm(m-1)/2)$. Due to k is far less than n , therefore the cost of computation and storage of FC-SLP algorithm is largely reduced, which makes it more scalable.

4 Experiments

4.1 Dataset, Evaluation Metric and Algorithms

To evaluate our algorithm more comprehensively, the experiments are conducted on two popular datasets: MovieLens and Baidu contest[16]. The dataset provided by Baidu is used to support the movie recommendation algorithm contest organized by Baidu in 2013. It contains real 1262741 ratings of 9722 users on 7889 movies. We randomly select 30% of users from the whole Baidu dataset to be experimental data. Details of them are shown in Table 2 :

Table 2. Details of MovieLens and Baidu datasets

	MovieLens	Baidu
No. of users	943	2917
No. of items	1682	7800
No. of ratings	100000	365639
Rating scale	1-5	1-5
Sparsity	6.3%	1.6%
Domain	Movie rating	Movie rating

We take RMSE (Root Mean Square Error) as evaluation metric. The formula definition of RMSE is :

$$RMSE = \sqrt{\frac{\sum_{i=1}^{n}(p_i - r_i)^2}{n}} \qquad (17)$$

A set of representative CF algorithms are chosen as comparisons, they are :

- User-based CF (UB)[3]
- Item-based CF (IB)[6]
- Cluster-based smoothing CF (CBS)[7]
- Weighted-Slope One (SLP)[1]
- SVD++[10]

4.2 Methodology

According to the principle of cross-validation, the dataset is divided into ten subsets (the ratings of each user are divided equally into ten parts), and each experiment is iterated for ten times. In each iteration, we randomly select N (0 < N < 10) subsets as test set and merge remaining ones as training set. For the experiments which need to continually change the percentage of training set, the value of N is from 1 to 9. The final results are the mean of ten times iterations.

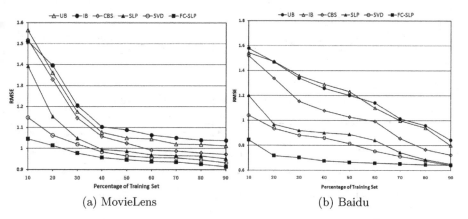

(a) MovieLens   (b) Baidu

Fig. 1. Evolution of RMSE according to percentage of training set

4.3   Results and Discussions

Experiment 1: Accuracy and Sparsity. In this experiment the percentage of training set is continually changing to simulate the different sparsity conditions. The results presented in Fig. 1 indicate that when the data is relatively dense (percentage>50%), the accuracy differentials among the Model-based CF(CBS, SLP, SVD++, FC-SLP) are not very obvious. However, as the data turns to be sparse, the differentials become large. At sparse conditions (percentage < 50%), only FC-SLP algorithm maintains reasonable accuracy, its performance markedly exceeds other algorithms. The excellent results largely because, through the fuzzy clustering technique, FC-SLP algorithm has the capacity to capturing the latent information of users, and the preprocessing of filling significantly improves the performance of fuzzy clustering on sparse data, thereby boost the accuracy of FC-SLP algorithm at sparsity condition. The CBS algorithm, which is based on k-means clustering, shows mediocre results. That is because, as is discussed in Section 2.2, the hard clustering technique is not very appropriate for CF systems. Besides, there is no data preprocessing mechanism in CBS algorithm, which causes the effectiveness of k-means clustering declines badly on sparse data, thereby drag the performance of CBS.

Experiment 2: Accuracy and Cluster Number. A serious of experiments are conducted to explore the correlation between the cluster number (K) and the performance of FC-SLP algorithm. Results shown in Fig. 2 demonstrate that the number has certain effect on the performance, but this is no linear relationship between cluster numbers and accuracy, the optimal K value is an empirical value. The reasons behind this phenomenon is : the small cluster number makes the latent information extracted by fuzzy clustering too general, inversely, the large clustering number makes the latent information too discrete.

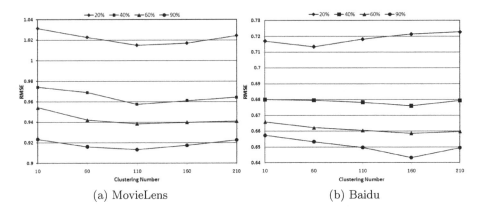

(a) MovieLens   (b) Baidu

Fig. 2. Evolution of RMSE according to cluster number

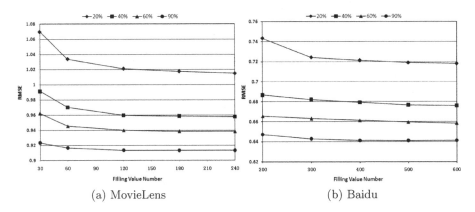

(a) MovieLens   (b) Baidu

Fig. 3. Evolution of RMSE according to filling value number

Experiment 3: Accuracy and Filling Value Number. Results depicted in Fig. 3 show that the number of filling values does affect the accuracy of FC-SLP algorithm. The accuracy keeps improving as the number of filling values increases, but the improvement will be more and more tinier. Considering the large quantity of filling values will extend the time of fuzzy clustering processing, thus, the number of filling values should not be too large.

Experiment 4: Comparison of Filling Algorithms. This experiment compares the accuracy of our proposed filling algorithm (HAF) and two typical CF filling algorithm : item average (IA) and item-user average (IUA). Results shown in Table 3 demonstrate that the proposed filling algorithm markedly outperforms other algorithms.

Table 3. RMSE of different filling algorithms

	MovieLens			Baidu		
	IA	IUA	HAF	IA	IUA	HAF
20%	1.0941	1.0622	0.9978	0.9794	0.9265	0.9187
40%	1.0433	0.9974	0.9852	0.9497	0.8962	0.8921
60%	1.0377	0.9825	0.9743	0.9018	0.8809	0.8723
90%	1.0261	0.9669	0.9573	0.7062	0.6730	0.6651

5 Conclusion

In this paper, we propose an improved Slope One predictor, which uses fuzzy clustering technique to alleviate the sparsity problem and boost scalability. A high-accuracy filling algorithm is developed as preprocessing tool to improve the effectiveness of fuzzy clustering on sparse data. Experiments on MovieLens and Baidu datasets demonstrate that our algorithm has outstanding prediction accuracy and scalability, what is more, it maintains high performance on sparse data.

On this basis, we aim to develop an automatic mechanism for setting optimal cluster number, replacing the pure empirical method in current scheme. The automatic mechanism is able to select the optimal cluster number through analyzing the dataset, without any manual intervention.

Acknowledgements. This work is supported by the national 973 plan project of China(No.2012CB724106), National Nature Science Foundation of China (No.61203305), Natural Science Foundation of Shandong Province (No.ZR2010FQ021, ZR2012FM003), Key science and technology project of Qingdao economic and Technological Development Zone(No.2013-1-25).

References

1. Lemire, D., Maclachlan, A.: Slope one predictors for online rating-based collaborative filtering. J. Society for Industrial Mathematics 5, 471–480 (2005)
2. Su, X., Khoshgoftaar, T.M.: A survey of collaborative filtering techniques. J. Advances in Artificial Intelligence, 4 (2009)
3. Cacheda, F., Carneiro, V., Fernndez, D., et al.: Comparison of collaborative filtering algorithms: Limitations of current techniques and proposals for scalable, high-performance recommender systems. J. ACM Transactions on the Web (TWEB) 5(1), 2 (2011)
4. Koren, Y., Bell, R.: Advances in collaborative filtering: Recommender Systems Handbook, pp. 145–186. M. Springer US (2011)
5. Han, J., Kamber, M., Pei, J.: Data mining: concepts and techniques. M. Morgan kaufmann (2012)

6. Sarwar, B., Karypis, G., Konstan, J., et al.: Item-based collaborative filtering recommendation algorithms. In: Proceedings of the 10th International Conference on World Wide Web, pp. 285–295. ACM (2001)

7. Xue, G.R., Lin, C., Yang, Q., et al.: Scalable collaborative filtering using cluster-based smoothing. In: Proceedings of the 28th Annual International ACM SIGIR Conference on Research and Development in Information Retrieval, pp. 114–121. ACM (2005)

8. Sarwar, B., Karypis, G., Konstan, J., et al.: Application of dimensionality reduction in recommender system-a case study. Minnesota Univ. Minneapolis Dept. of Computer Science (2000)

9. Ma, C.-C.: A Guide to Singular Value Decomposition for Collaborative Filtering (2008)

10. Koren, Y.: Factorization meets the neighborhood: a multifaceted collaborative filtering model. In: Proceedings of the 14th ACM SIGKDD International Conference on Knowledge Discovery and Data Mining, pp. 426–434. ACM (2008)

11. Wu, K.W., Ferng, C.S., Ho, C.H., et al.: A two-stage ensemble of diverse models for advertisement ranking. In: KDD Cup 2012 ACM SIGKDD KDD-Cup WorkShop (2012)

12. Balabanovi, M., Shoham, Y.: Fab: content-based, collaborative recommendation. J. Communications of the ACM 40(3), 66–72 (1997)

13. Pazzani, M.J., Billsus, D.: Content-based recommendation systems. In: Brusilovsky, P., Kobsa, A., Nejdl, W. (eds.) Adaptive Web 2007. LNCS, vol. 4321, pp. 325–341. Springer, Heidelberg (2007)

14. Gao, M., Wu, Z.: Personalized context-aware collaborative filtering based on neural network and slope one. In: Luo, Y. (ed.) CDVE 2009. LNCS, vol. 5738, pp. 109–116. Springer, Heidelberg (2009)

15. Wu, J., Li, T.: A modified fuzzy C-means algorithm for collaborative filtering. In: Proceedings of the 2nd KDD Workshop on Large-Scale Recommender Systems and the Netflix Prize Competition. ACM (2012)

16. The download link of Baidu contest dataset, http://pan.baidu.com/share/link?shareid=340221&uk=2000006609

Towards Building Virtual Vocabularies in the Semantic Web

Yunqing Wen[1], Xiang Zhang[2], Kai Shen[1], and Peng Wang[2]

[1] College of Software Engineering, Southeast University, Nanjing, China
{yunqing-wen,knshen}@seu.edu.cn
[2] School of Computer Science and Engineering, Southeast University, Nanjing, China
{x.zhang,pwang}@seu.edu.cn

Abstract. The development of ontologies in current Semantic Web is in a distributed and loosely-coupled way. Knowledge workers build their vocabularies in accessible web ontologies in their own manner. Two extreme cases are: many highly related concepts are defined separately in a set of tiny ontology fragments; while some massive ontologies defines a large set of concepts, which semantically belong to different areas. These cases bring a barrier to ontology reuse. In this paper, we propose an approach to semantically reorganizing concepts defined in various ontologies. We transform the reorganization problem to a graph clustering problem, and the result of reorganization is a set of virtual vocabularies for reuse. Experiments on a massive ontology repository show that our approach is feasible and efficient.

Keywords: ontology, virtual vocabulary, semantic web.

1 Introduction

Ontology formally represents a vocabulary consisting of a set of concepts within a domain, as well as relationships between them. Domain-dependent knowledge workers construct their ontologies in a distributed way, with different backgrounds and intentions. This leads to an observable phenomenon of ontological chaos: semantically-related concepts may be defined by different authors, and may be defined in different web-accessible ontology files with different URI prefix. In another word, vocabularies in current Semantic Web are not well-organized semantically.

This chaos brings a barrier to the reuse of ontologies in an efficient way. When a user needs to utilize previously defined vocabularies to describe data, he usually has to make a search in ontology search engine, and then visit a set of related ontology files developed by different organizations or individuals. In describing his data, some ontologies will be imported to construct a completed conceptual scheme of his data. This process of searching, parsing, understanding and integrating ontologies is quite time-consuming for the users.

A promising approach to organizing vocabularies is to break the boundary of ontology and reorganize concepts according to the strength of their relationships. Tightly-related classes or properties can be grouped together. For example,

H. Motoda et al. (Eds.): ADMA 2013, Part I, LNAI 8346, pp. 193–204, 2013.

a subClassOf or EquivalentClass relationship indicates that two classes are semantically highly related, and can be reorganized into a virtual group, although they may be defined in different ontologies. Besides, two semantically irrelevant classes can be reorganized into two separated virtual groups, although they may be defined in the same ontology. The research on ontology modularization[1,2] is a step towards reorganizing a single large ontology. Now we can step further towards a reorganization of all concepts in the Semantic Web to semantically cohesive virtual groups, which are called virtual vocabularies in this paper.

The contribution of this paper lies in two aspects: First, the notion of virtual vocabularies and its relation with ontology and vocabulary in the Semantic Web is clearly defined. Second, a Concept Relationship Graph is proposed to characterize the connection strength between concepts defined in various ontologies, and two clustering strategies are used to build virtual vocabularies. Virtual vocabularies will bring convenience for ontology reuse.

This paper is organized as follows: Section 2 gives a clear definition on the model of virtual vocabularies, and introduces the architecture of our approach. The definition and construction of Concept Relationship Graph is defined in Section 3. Two different clustering strategies on the graph model are discussed in Section 4. In Section 5, the rationality of reorganization of concepts is evaluated using a real dataset from Falcons search engine. Our work is compared with other related works in Section 6. Conclusion and future work are discussed in the last section.

2 Overview

In this section, the model of virtual vocabulary and the system architecture are given in section 2.1 and section 2.2 respectively to show what virtual vocabulary is and how to build virtual vocabulary.

2.1 Model of Virtual Vocabulary

The traditional hierarchy of ontology can be described with three levels (the first three in Fig.1). The bottom level is Term Level, in which each node

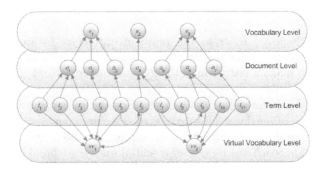

Fig. 1. Model of Virtual Vocabulary

represents a concrete class or a concrete property. Above the Term Level is Document Level, in which each node represents a physical file. The upmost level is Vocabulary Level, in which each node represents a specific namespace, in which all concepts URI having a fixed prefix.

Our work adds a Virtual Vocabulary Level underneath the Term Level, in which each node represents a virtual document that related to a topic. The virtual document not only includes the relevant class but also relevant property.

2.2 System Architecture

A three-level architecture is proposed to implement the Model of Virtual Vocabulary. *The RDF Parser* analyzes the original ontology files and extract all the classes properties and RDF Triples. *The CRG Builder* is used to construct Concept Relationship Graph(CRG) based on the classes and RDF Triples. *The Vocabulary Builder* consists of clustering methods to form virtual vocabularies.

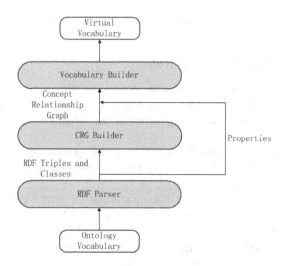

Fig. 2. System Architecture

3 Building Concept Relationship Graph(CRG)

In semantic web, there is a set of standard ontology language to clearly define the meaning of concepts and the relationships between the concepts. The most basic is *subClassOf* which describes the hierarchical structure of two concepts. It is important to know the relationships among different ontologies because of the large number of free defined ontologies. So the model of Concept Relationship Graph is given.

3.1   Definition

Based on the mined RDF Triples and classes from different ontologies, after applying some specific rules(described in 3.2) to these triples, it is possible to build a undirected labeled graph like Fig.3. Each node in Concept Relationship Graph is a class identified by the only URI, each edge represents the relationship between two classes.

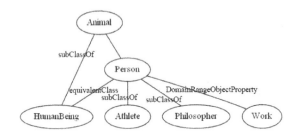

Fig. 3. An example of Concept Relationship Graph

3.2   Relationship Detection Rules

It is an obvious fact that some classes defined in different ontology have some implicit semantic relations, which often appear as patterns. Through observation, Eleven rules have been discovered to reveal the relationships between different classes.

Rule 1(equivalentClass Rule): Given the set of RDF Triples T, if ∃ $<a$, $owl:equivalentClass,b>∈ T$, build a relationship labeled by equivalentClass from class a to class b in CRG.

Rule 2(subClassOf Rule): Given the set of RDF Triples T, if ∃ $<a$, $owl:subClassOf, b>∈ T$, build a relationship labeled by subClassOf from class a to class b in CRG.

Rule 3(disjointWith Rule): Given the set of RDF Triples T, if ∃ $<a$, $owl:disjointWith, b>∈ T$, build a relationship labeled by disjointwith from class a to class b in CRG.

Rule 4(complementOf Rule): Given the set of RDF Triples T, if ∃ $<a$, $owl:complementOf, b>∈ T$, build a relationship labeled by complementOf from class a to class b in CRG.

Rule 5(allValuesFrom Rule): Given the set of RDF Triples T, if ∃ $< a$, $owl:complementOf,\_:1>(\_:1$ denotes the blank node), $<\_:1, rdf:type,$ $owl:Restriction>$, $<\_:1, owl:onProperty, member>$ and $<\_:1$ $owl:allValuesFrom,b>$, build a relationship labeled by allValuesFrom from class a to class b in CRG.

Rule 6(someValuesFrom Rule): Given the set of RDF Triples T, if ∃ $<a$, $rdfs:subClassOf, \_:1>$, $<\_ :1, rdf:type,owl:Restriction>$, $<\_ :1 owl:onProperty,$ $member>$, and $<\_ :1 owl:someValuesFrom, b>$, build a relationship labeled by someValuesFrom from class a to class b in CRG.

Rule 7(domainRangeObjectProperty Rule): Given the set of RDF Triples T, if $\exists$ $<p, rdfs:domain, a>$, $<p, rdfs:range, b>$, and the label of p is *owl:ObjectProperty*, build a relationship labeled by domainRangeObjectProperty from class a to class b in CRG.

Rule 8(domainRangeFunctionalProperty Rule): Given the set of RDF Triples T, if $\exists$ $<p, rdfs:domain\ a>$, $<p, rdfs:range, b>$, and the label of p is *owl:FunctionalProperty*, build a relationship labeled by domainRangeFunctionalProperty from class a to class b in CRG.

Rule 9(domainRangeInverseFunctionalProperty Rule): Given the set of RDF Triples T, if $\exists$ $<p, rdfs:domain, a>$, $<p, rdfs:range, b>$, and the label of p is *owl:InverseFunctionalProperty*, build a relationship labeled by domainRangeInverseFunctionalProperty from class a to class b in CRG.

Rule 10(domainRangeSymmetricProperty Rule): Given the set of RDF Triples T, if $\exists$ $<p, rdfs:domain, a>$, $<p, rdfs:range, b>$, and the label of p is *owl:SymmetricProperty*, build a relationship labeled by domainRangeSymmetricProperty from class a to class b in CRG.

Rule 11(domainRangeTranstiveProperty Rule): Given the set of RDF Triples T, if $\exists$ $<p, rdfs:domain, a>$, $<p, rdfs:range, b>$, and the label of p is *owl:TranstiveProperty*, build a relationship labeled by domainRangeTranstiveProperty from class a to class b in CRG.

3.3   Weighting Concept Relationship Graph

To apply some clustering methods to the Concept Relationship Graph, quantifying the relationships between two classes is a must. Each kind of relationships is assigned a weight ranges from 0 to 1, representing the closeness between two nodes. The 11 relationships can be divided into 4 categories.

The first is the *equivalentClass*, which means the two classes involved are identical. Empirically, it can be assigned the biggest value of 0.5.

The second describes the overlap relationship: *subClassOf* which denotes a hierarchical relation, *disjointWith* which denotes two notions have no intersection, *complementOf* which denotes two notions are complementary. Each of them assigning a value of 0.4.

The third describes the domain range relationship: *domainRangeObjectProperty domainRangeFunctionalProperty*, *domainRangeTransitiveProperty*, *domainRangeInverseFunctionalProperty*, *domainRangeSymmetricProperty*. Each of them gets 0.3.

The last are *someValuesFrom* and *allValuesFrom*, which denote the indirect relationship between classes, so they both get 0.2.

After categorizing these relationships, we give weights for 11 relationships in Table 1.

Table 1. Weighting CRG

Relationship	Weight
equivalentClass	0.5
complementOf	0.4
disjointWith	0.4
subClassOf	0.4
domainRangeObjectProperty	0.3
domainRangeFunctionalProperty	0.3
domainRangeTransitiveProperty	0.3
domainRangeInverseFunctionalProperty	0.3
domainRangeSymmetricProperty	0.3
allValueFrom	0.2
someValuesFrom	0.2

4 Building Virtual Vocabularies

When CRG is constructed, the next is to apply clustering methods to detect classes of similar notions to form virtual vocabularies. This paper uses two different cluster strategies: one is clustering by partitioning, and another is by hierarchical clustering.

4.1 CRG Clustering by Partitioning

In this paper, partition on the CRG is gained by using the K-Means algorithm. K-Means Cluster is first given by Macqueen[3,4]. The main idea of Partitioning Clustering Methods is to give the number of clusters in advance; it uses an iterative approach to improve the clusters by making the elements in the same cluster as similar as possible. Different K values may affect the result, if K is too small, the concepts within the same virtual document may be loosely related, if K is too big, it may bring a difficulty to reusing ontologies, so this paper aims to detect those most related concepts by seeking a proper K.

To apply the classic K-Means Clustering to CRG, we create an n-dimensional vector for each class in the graph by using N-step method.(Suppose the scale of the graph is n).

N-step means the set of neighbour within N steps. It extracts the closest classes around the observation. For example, the 1-step neighbour of Animal in Fig.3 are Person and Human Being, the 2-step neighbour of Animal contain Athlete, Philosopher and Work besides the 1-step neighbour.

Assume that the vector for the class $v_i = <x_1^i, x_2^i, \ldots, x_n^i>$, i=1,2,...,n, $x_j^i = \sum_{P_k(i,j)} = d_{i,e_1} \cdot d_{e_1,e_2} \cdots d_{e_x,j}$, j=1,2,...,n, where $P_k(i,j)$ is the set of the paths from node i to node j by definitely k steps, and $(i, e_1.e_2, \ldots, j)$ is a possible path from i to j, d_{e_i,e_j} is the weight of edge (i,j).

In Fig.3, After computing the vector for the node Animal, the vector <Animal,Person, HumanBeing, Athlete, Philosopher, Work> for Animal after considering 1-step neighbor is $< 0, 0.4, 0.4, 0, 0, 0 >$. The vector for Animal after considering 2-step neighbor is $< 0, 0.4, 0.4, 0.16, 0.16, 0.12 >$.

When working out vectors of all the classes, the virtual distance of two nodes in CRG can be defined as the following equation:

$$D_{i,j} = \sqrt{\sum_{k=1}^{n}(x_k^i - x_k^j)^2} \tag{1}$$

Considering that the scale of CRG n may be very large, it is necessary to reduce the dimension of vector to reduce time and space consumption.

4.2   CRG Clustering by Hierarchical Clustering

Hierarchical clustering is a method of cluster analysis. It seeks to build a hierarchy of clusters. Generally there are two types of strategies for hierarchical clustering: Agglomerative ("bottom up" approach) and Divisive ("top down" approach).

Considering the large size of our input, this paper uses a agglomerative algorithm proposed by M. E. J. Newman[5], which is much faster.

The algorithm is based on the idea of optimizing modularity. Given a graph $G(V, E)$, to represent the quality of division of clusters,we can define a quality function or modularity Q[6]:

$$Q = \sum_{i}(e_{ii} - a_i^2) \tag{2}$$

Where $e_{i,j}$ is the fraction of edges in the network that connect vertices in group i to those in group j, and $a_i = \sum_j e_{ij}$, for all possible j.

Beginning with a state in which each vertex is the sole member of one of n communities, we repeatedly choose a pair of communities and join them together, at each step making the join result in the greatest increase (or smallest decrease) in Q, until all vertexes is joined or not existing any edge between groups.

To apply fast Newman algorithm to the Concept Relationship Graph, we should define the similarity between ontologies. Here we let the similarity be the weight that defined in section 3.2.

5   Experiments

5.1   Dataset

The dataset we used is a real dataset from Falcons search engine[7]. We filtered 131618 classes and 4823 properties from a total number of 212 physical ontology files through Jena. By using the methods discussed above, we constructed a CRG containing 17208 edges.

5.2   Quality of Virtual Vocabularies

To evaluate whether the clustering is reasonable enough, and also to compare the Partitioning and hierarchical Clustering method, this paper proposed 2 criteria, one measures the correlation among clusters, another measures the balance degree.

The first criterion is connectedness[8]. It compares the number of shared edges between two clusters and the number of the total edges in CRG.

$$connectedness = \frac{|\{e(i,j)|i \in p_m, j \in p_n, m \neq n\}|}{|\{e(i,j)\}|} \tag{3}$$

Where $|\{e(i,j)\}|$ is the number of edges in CRG, $|\{e(i,j)|i \in p_m, j \in p_n, m \neq n\}|$ is the number of edges across two clusters. A low connectedness indicates a tightly cohesive and loosely coupled partition. However, connectedness cannot reflect whether a clustering is well-balanced. For example, a clustering consisting of a singleton and a large cluster with all the other vertices in it, its connectedness is very low, but it is not well-balanced.

The second criterion is coefficient of variation (CV):

$$CV = \frac{\sigma}{\mu} \tag{4}$$

$$\mu(p_i) = \frac{\sum_i |p_i.\text{sents}|}{|\{p_i\}|} \tag{5}$$

$$\sigma(p_i) = \sqrt{\frac{1}{|p_i|} \cdot \sum_i (|p_i.\text{sents}| - \mu(p_i))^2} \tag{6}$$

$\mu(p_i)$ is the average size for all partitions. $\sigma(p_i)$ measures the discrepancy of $\mu(p_i)$ partition and current partition. $p_i.\text{sents}$ represent the size of ith partition. CV measures whether there existing big differences among clusters size.

5.3   Evaluation on Partitioning

We experiment on different size of dataset with Partitioning Clustering. We extract a class set and an edge set from original dataset with different size. The number of clusters has been defined in advance. We use some method to compress the number of attributes into 1000 to save the time.

In Table 2, with number of classes growing, time used for clustering increases but decreases sharply when number of classes is 330. The indicator CV fluctuates around 20 when the number of classes varies, which shows the partitions result is quite unbalanced. The connectedness seems high for some cases, while others not. It suggests this method sometimes can behave well, and sometimes not.

Table 2. Result for partitioning cluster

# of classes	# of clusters	time(s)	CV	connectedness
10000	1872	762	24.803	0.0517
20000	1910	2124	24.949	0.119
30000	1585	2659	22.586	0.1808
50000	739	2063	17.057	0.2462
60000	477	2349	15.231	0.2247
80000	550	3140	16.740	0.2449
100000	330	1789	17.607	0.0019
131618	589	3400	23.636	0.0027

Table 3. Result for hierachical clustering

# of classes	# of clusters	time(s)	CV	connectedness
10000	1872	5	1.975	0.0083
20000	1910	22	2.518	0.0078
30000	1585	61	2.563	0.0074
50000	739	272	1.707	0.0069
60000	477	335	1.183	0.0067
80000	550	925	1.562	0.0064
100000	330	3076	1.864	0.0134
131618	589	5620	2.751	0.0084

5.4 Evaluation on Hierarchical Clustering

We experiment on different size of dataset with Hierarchical Clustering, in the same way. The result is shown in Table 3.

Fast Newman algorithm time approximation is $O(n^2)$[5], so the time used is reasonable with the input larger and larger. Also, we find the all connectedness are less than 0.02, showing the clustering result is high cohesion inside the group and low coupling among groups. Besides, the CV indicators all are less than 3.0, illustrating the size of clusters we clustered are relatively balanced.

5.5 Comparison

Fig. 4 illustrates the fact that the clusters got from hierarchical clustering are more balanced than that from partition.

Fig. 5 shows from prospects of indicator of connectedness, the partition from hierarchical clustering is rather low for most cases, while Partitioning are not. Although with number of classes greater than 100000, connectedness of partitioning clustering method is quite low and better than that of hierarchical clustering, meaning clustersgot from partitioning clustering are tightly cohesive and loosely coupled, it is not well-balanced since some clusters have a few classes, while others have a lot.

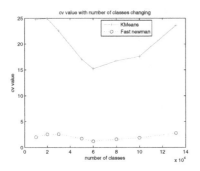

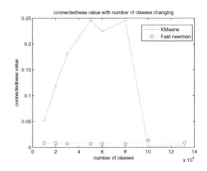

Fig. 4. Comparison of CV **Fig. 5.** Comparison of Connectedness

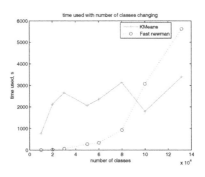

Fig. 6. Comparison of Time Consumption

Fig. 6 shows that the actual time cost of Hierarchical Clustering method are consistent with the theoretical value $O(n^2)$,while the time cost of Partitioning Clustering are more stable because of the fixed size of feature vector (the time for dimension reduction excluded). The growth of time for Partitioning Clustering is slower at the expense of a worse CV and connectedness.

In addition, partitioning method requires to be set the number of clusters in advance, while hierarchical clustering not.

6   Related Works

In the last few years, lots of work has been done for improving the reusability of ontology. The idea of reusing ontology knowledge is referred to by many different names by different authors. In [9], J. Seidenberg and A. Rector proposed some basic segmentation algorithm such as upwards traversal of the hierarchy, downwards traversal of the hierarchy or inclusion of the sibling classes in the hierarchy, and a property filtering method is proposed for the extract concepts above to constrain the segment size. In [1], Paul Doran introduced a definition of ontology modules from the reuse perspective and an approach to module extraction based on such a definition. Bernardo proposed the notions of conservative extension, safety and module for a very general class of logic-based ontology languages in [2]. In [10],Elena emphasized the need for a context and task sensitive treatment of ontologies and argue for the need for ontology-reuse methodologies which optimally exploit human and computational intelligence to effectively operationalize reuse processes. In [11], the author distinguishes among ontology reuse, ontology module reuse, ontology statement reuse, and ontology design pattern reuse. He also provides many guidelines for reusing ontological resources.

7 Conclusion and Future Works

This paper proposes an efficient way to reorganize the ontology in order to improve the reusability of ontology. By constructing a Concept Relationship Graph, the implicit semantic relations among classes and properties among different ontology vocabularies are detected. By applying some clustering methods to CRG, virtual vocabularies are formed from ontology. Based on a real dataset, this paper illustrate the whole process and give the performance of clustering methods by experiments, which proves that our approach appears to be feasible.

In this paper, the weight for CRGs relationship is given with empirical experience. In future, we will learn out the good and reliable weights for CRGs relationship by applying Machine Learning methods to it. To evaluate the partitioning algorithm better, we'll take cohesiveness into consideration in future. In addition, the virtual vocabulary can be embedded into the ontology search engine.

Acknowledgement. The work is supported by the NSFC under Grant 61003055, 61003156, 61300200, and by NSF of Jiangsu Province under Grant BK2011335. We would like to thank Xing Li for his valuable suggestions and work on related experiments.

References

1. Doran, P., Tamma, V., Iannone, L.: Ontology module extraction for ontology reuse: an ontology engineering perspective. In: Proceedings of the Sixteenth ACM Conference on Conference on Information and Knowledge Management, pp. 61–70. ACM (2007)
2. Grau, B.C., Horrocks, I., Kazakov, Y., et al.: Modular reuse of ontologies: Theory and practice. Journal of Artificial Intelligence Research 31(1), 273–318 (2008)
3. MacQueen, J.: Some methods for classification and analysis of multivariate observations. In: Proceedings of the Fifth Berkeley Symposium on Mathematical Statistics and Probability, vol. 1, pp. 281–297 (1967)
4. Zhang, H.: Extensions to the K-means algorithm for clustering large data sets with categorical Values. Data Mining and Knowledge Discovery (2), 283–304 (1998)

5. Newman, M.E.J.: Fast algorithm for detecting community structure in networks. Physical Review E 69(6), 66133 (2004)
6. Newman, M.E.J., Girvan, M.: Finding and evaluating community structure in networks. Physical Review E 69(2), 026113 (2004)
7. Cheng, G., Ge, W., Qu, Y.: Falcons: searching and browsing entities on the semantic web. In: Proceedings of the 17th International Conference on World Wide Web, pp. 1101–1102. ACM (2008)
8. Schlicht, A., Stuckenschmidt, H.: Towards Structural Criteria for Ontology Modularization. In: The 1st International Workshop on Modular Ontologies (2006)
9. Seidenberg, J., Rector, A.: Web ontology segmentation: Analysis, classification and use. In: Proc. 15th Int. Conf. on World Wide Web (WWW 2006), Edinburgh, Scotland, May 23-26 (2006)
10. Simperl, E.: Reusing ontologies on the Semantic Web: A feasibility study. Data & Knowledge Engineering 68(10), 905–925 (2009)
11. Surez-Figueroa, M.C.: NeOn Methodology for building ontology networks: specification, scheduling and reuse. Informatica (2010)

Web Mining Accelerated with In-Memory and Column Store Technology

Patrick Hennig, Philipp Berger, and Christoph Meinel

Hasso-Plattner-Institut
University of Potsdam, Germany
{patrick.hennig,philipp.berger,office-meinel}@hpi.uni-potsdam.de

Abstract. Current web mining approaches use massive amounts of commodity hardware and processing time to leverage analytics for today's web. For a seamless application interaction, those approaches have to use pre-aggregated results and indexes to circumvent the slow processing on their data stores e.g. relational databases or document stores. The upcoming trend of in-memory, column-oriented databases is widely used to accelerate business analytics like financial reports, but the application on large text corpora remains unaffected. We argue that although in-memory, column-oriented stores are tailor-made for traditional data schemes, they are also applicable for web mining applications that mainly consists of raw text informations enriched with limited semantic meta data. Thus, we implement a web mining application that stores every information in a pure main memory data store. We experience an acceleration of current web mining queries and identify new opportunities for web mining applications. To evaluate the performance impact, we compare the run-time of general web mining tasks on a traditional row-oriented, disc-based database and a column-oriented, in-memory database using the example of BlogIntelligence, which serves exemplary for web mining applications.

Keywords: blog analysis, web mining, data mining, in-memory, column-layout.

1 Introduction

Since the information overload problem occurs in the mid of the 90s [8], the amount of available information in the Word Wide Web grows exponentially. Hence, the research area of web mining develops and gains increasing importance in today's businesses and life. Web mining essentially is the task of deriving knowledge out of the available information on the internet and delivering it to the user [7]. As the amount of data in the web continues to growth new challenges of data processing occur that have to be handled by today's web mining applications.

In general, web mining is the appliance of data mining techniques to web documents [4]. It tries to derive new knowledge of the vast pool of available data

H. Motoda et al. (Eds.): ADMA 2013, Part I, LNAI 8346, pp. 205–216, 2013.
© Springer-Verlag Berlin Heidelberg 2013

by automatically extracting and discovering information. According to Kosala et al. [7], web mining comprises the following sub tasks:

- **Resource finding** includes the retrieval of web documents based on the user's need like a Google search.
- **Information selection and pre-processing** consists of tasks of natural language processing and information abstraction like extracting receipts from web documents.
- **Generalization** is the automatic discovery of new patterns across a set of web pages like clustering or trend detection.
- **Analysis** reasons the recognized patterns and gives interpretations.

We introduce a tailor-made web mining application focused on weblogs and social media, called BlogIntelligence. During the development of this application we tested different kind of web-scale data stores. Based on our experience from various analytical algorithms, we identified the need for a relational data schema. Although it is hard to apply recursive link analysis algorithms, it supports delivers a high performance for aggregates.

Based on BlogIntelligence, we identify the necessity to compare the benefits of different data stores by looking at typical tasks of web applications. BlogIntelligence helps us to discover the variety of tasks reaching from selecting the next urls to clustering of web pages in communities.

We argue that based on our experience, document stores offer too limited analysis capabilities and traditional disc-based relational databases cannot handle the massive amount of data produced by the social web.

Finally, we conclude that an in-memory, column-oriented data store as proposed by Plattner [12] offers better analytical performance and opens new ways of analytics. Therefore it is necessary that all analytical data can be stored in main-memory. Due to the massive amount of cheap main memory that is meanwhile available this is not a problem anymore.

This paper is structured as follows. In the next section, we describe related data store techniques with their pros and cons. In Section 3, we give an introduction to BlogIntelligence which is an exemplary web mining application. Section 4 describes the different application areas, which the in-memory technology extremely accelerates the computing time and where it has possible shortcomings. Furthermore, we propose new analysis techniques that extend the feature set of today's web mining applications. We test our assumptions in Section 5 by comparing the execution time of essential queries of a row-oriented and column-oriented database. We give a short outlook and propose adaption for the tested data store in Section 6. Finally, Section 7 summarizes our work and conclude our results.

2 Related Work

The related approaches to speed up the store and analytical components of web mining application roughly consists of three areas.

First, the traditional row-oriented disc-based database approaches like Post-gres [9] and MySql [14]. These are based on a B-Tree structure and optimized for accessing patterns of traditional hard-drives. Further, they offer index mechanisms that are beneficial for data access and query processing (eg. cube constructs). Nevertheless, those structures are pre-aggregated and need a high amount of processing time to be created. In addition, traditional databases fail to scale up to the massive data load of the web.

The second area consists of distributed data stores that use a large number of commodity hardware like Google BigTable [3] or Apache Cassandra [5]. These are specially adapted to handle massive amounts of data and to provide fast search access. Hence, each inserted data gets preprocessed and categorized into the tailor-made meta structure. Although the access performance of those data stores is remarkable, the analytical algorithms applicable are limited or need to process the whole data set at once like MapReduce PageRank [1].

Thirdly, the usage of complex index building applications that aggregate beforehand data for predefined questions like Apache Lucene[1]. The idea behind those applications is to incorporate with traditional databases and deliver fast query results for various kinds of analytical queries. Furthermore, they offer the possibility to integrat natural language processing stages into their index building process. Nevertheless, complex analysis algorithms like topical clustering, ranking, or equally complex aggregations need to be facilitated by external components.

In contrast to the related work, we argue that with the availability of massive, inexpensive main memory, web mining can be leveraged by in-memory database with a tremendous performance gain. The advantage of those databases is that the data access is equally fast than the data aggregation. We implement and test web mining operations for an in-memory database and equally for a traditional database (see Section 5).

3   Blog Intelligence

Blog Intelligence is a web mining application tailor-made for blog mining with the objective to map, and ultimately reveal, content-oriented network-related structures of the *blogosphere* by employing an intelligent *blog crawler*. As described in [10], BlogIntelligence is able to harvest the pool of millions of interconnected blogs, called *blogosphere*.

We want to gain a better user experience and discover new analyses by using the benefits of in-memory computing column-oriented store for the unusual application area of web mining. The identified benefits are discussed and finally evaluated in this paper. We look at some critical parts in the overall system to evaluate the different techniques.

---

[1] http://lucene.apache.org/

4   Application Areas

In general, a web mining application consists of three components that are responsible for the three main tasks: crawling data, storing data, and analyzing data. The crawling of data is the actual process of downloading web pages from the internet. Next, storing the data is the task of managing the massive amount of data and prepare it for analytical queries. Finally, analyzing data consists of running diverse text mining and natural language processing algorithms including community discovery, influencer identification, and topic extraction.

The performance of the crawling and analyzing component depend dramatically on the performance of the intermediate data store. We try to understand the relation between the data store and the other two components, and the possible performance gain through a in-memory, column-oriented data store. Hence, we introduce the characteristics of these components using the example of *BlogIntelligence*. In addition, we give an overview of new analyses ways enabled by the usage of in-memory technology.

4.1   Crawling

The major prerequisite of web mining tools, like BlogIntelligence, is harvesting web pages. These crawling activities deliver the data that is necessary for the underlying analytics. During the development of our focused web mining application, *BlogIntelligence*, we experienced major benefits of the in-memory technology in both areas.

The crawling of web pages consists essentially of downloading pages and selecting new urls to crawl. The BlogIntelligence framework uses an intelligent and scalable tailor-made blog-crawler [10] to harvest blog pages.

Especially inserting data is a common task for harvesting weblogs. By design, the insertion costs of column-oriented database are comparable high. This is caused by the distribution of the column values in the main memory, which results in high insert costs. In contrast, a row-oriented layout enables the database to write one line sequentially into the main memory without caring about any specific place for the column values. Anyway, since each website is only inserted once, this disadvantage is almost negligible. An evaluation of the execution time of insert webpages is given in Section 5.

Beside inserting pages, selecting new pages is another frequently executed task during harvesting to identify the best pages for crawling. The complexity of the selection of new urls varies among different use cases. A general crawler simply selects the next not visited url from the database, whereas a specialized crawler has to ensure more complex selection constraints.

A constraint can dictate a specific priority for each web page depending on the rank of the source page, the last visiting date of the page itself, the expected content of the page, the type of the page or even restrictions of the content provider. We describe and evaluate an example query of our focused web crawler, which is part of BlogIntelligence, in Section 5.

4.2   Analytics

The analytics of BlogIntelligence are at a magnitude more complex than the crawling. In the following, the major algorithm types that get applied in BlogIntelligence are introduced.

One typical application area is a link analysis algorithm. The most prominent algorithm is PageRank [11]. It calculates a rank for a web page based on the rank of all incoming web pages. For Blog Mining this gets even more important since it is very important to select the most important and interesting blog. Therefore such a ranking function can be improved to take the structued data of the blogosphere into account as described in the previous work [6]. Caused by the inherent complexity of recursive ranking algorithms in databases, we simplified the example ranking query to an subcomponent of the rank calculation. Therefore we want to demonstrate this simply by calculating a count of incoming links of blogs. We expect this select and count aggregation to perform fast on a column-oriented layout.

Sentiment analysis handles the problem of sentiment extraction from sentences in the web page's text. Thereby, an algorithm applies a set of predefined language specific rules that identify whether a word has a positive or negative meaning. These extraction as well as an entity extraction is done asynchronously by inserting data into the database. Therefore these aspects can be retrieved afterwards pretty fast via SQL and can be used in very simple way for other calculations. Nevertheless, this is a database-specific extension that can also be executed by external application.

In addition by inserting new data into the database some additional structures can be filled and therefore kept up-to-date all the time. For example a compressed Document-Term matrix helps to calculate similar terms or similar documents based on the well-known TF-IDF [13] measure. In the evaluation part some execution runs are shown for this analysis.

Compared to link analysis algorithms, blog rank algorithms incorporate a set of additional factors. Another metric BlogIntelligence provides is to rank blogs according to the consistency of the content they are writing about. For example if an author is writing about the same topic all the time, he should have a big knowledge in this topic. This is accomplished by looking at the usage of the tf-idf measure.

Last but not least, another important part is to identify top emerging trends within the blogosphere. Since the metric is basically specified the user can get trends for his own topic space and change the metric according to his special interests.

4.3   New Application Area

In former times the analytics were carefully designed beforehand. Afterwards the execution of these analytics produced the results, which got visualized in a meaningful way. The execution of the complete analytics often took several days, even with a limited data set.

As a result two major problems occurred. The first major problem of this life cycle is that it contradicts the ever changing nature of the World Wide Web. Therefore, the results are out-dated by the time they are produced.

The second major problem is that the analytics have to be set up-front. This can be hidden from the end-user by offering smart filtering options for the analytical results. Nevertheless, the user is not able to change the underlying metrics and get immediate feedback of his change via a freshly calculated result.

In addition, it is very important to give the user the possibility to adapt metrics according to his own interest. If a user wants to get the best and most interesting blogs, he often does not want to use a general metric. Since no results are pre-aggregated, the user can change a metric until it fits to his needs. The user can even limit the analyses to a specified topic space.

Furthermore, it was not feasible to ask the database questions like the following:

– Which are the blogs with the most incoming links in a specific topic?
– Who are the authors with the most posts writing about politics?
– What are the most recent posts which the user is interested in?
– How is a topic discussed in its community or within a certain time frame or within the most popular blogs?

By using in-memory technology together with a column store we are able to answer these questions without knowing the question beforehand. As well, we can immediately provide results for his latest analytical questions at design time.

In order to get a better impression, we want to take a deeper look at the basic extraction, crawling and analyses tasks in the next section.

5   Evaluation

In this Section we evaluate the performance of the PostgreSQL database[2] and an in-memory database from SAP on the presented web mining operations.

5.1   Data Set

The data set for this evaluation consists of 73 221 376 webpages. 2 473 898 of these webpages are parsed. Parsed webpages are already downloaded and processed by the web mining applications that results in additional informations like raw text, feed items, authors, categories and other semantic data.

After the classification BlogIntelligence identified 15 327 blogs with 818 865 posts. The link graph of the web is also represented as a table that consists of 200 000 000 entries. The space used for the two main tables, webpage and link, is 200 gigabyte. Hereby, the webpage table uses 151 gigabytes and the link table uses 49 gigabytes.

All tables are stored as column-oriented tables with a primary index. Some columns are compressed by the database which leads to the small amount of

---

[2] http://www.postgresql.org/

main memory needed for the dataset. As a big advantage of the column store the table containing all blog pages, the webpage table, is able to handle 79 columns including the original HTML page, the extracted text, authors, the publishing date as well as other information.

This data set is the result of a 3-week-run from August 2012. We use this data set because we like to provide real results to the users of our prototype and measure the performance more accurate by using a data set that is not cleaned or adapted in any way.

5.2 Setup

Both databases get evaluated on the same server hardware. The database server has 32 cores one terabyte of main memory. The full file system size is 20 terabyte. The file system is located on a EMC storage, which uses SSDs for storing. During the execution of the experiments the server is exclusively used by the databases and only the minimal SUSE Linux basic daemons are running beside the database.

Since the in-memory database is operating exclusively on main-memory the time to load data into main memory is negligible. Therefore the tables has to be re-constructed from the disk only if the database server was shutdown completely. Fore more details about the concepts of an in-memory database the Book of Prof. Hasso Plattner [12] provides deeper insights.

For this setting we want to look at different aspects that specially depend on the performance of the data stores for web mining applications.

5.3 Crawling

Heavy Inserts. As already mentioned a column-oriented layout is not the best choice for processing a lot of different insert statements. Nevertheless it provides such an improvement benefit in the analytical part that this can be neglect. However, the in-memory database from SAP used in the experiment is optimized to absorb these disadvantages.

For the experiment 200 000 insert commands were sent to the database, filled with generated content. The web page table has columns from almost every type, from binary columns over integers to big text columns.

For the execution of the 200 000 insert commands a parallelized Java application running with 50 threads is used together with a JDBC driver for both databases. The execution results are shown in the following Table 1.

As we can see by looking at Table 1 the database is able to absorb the low insert performance of a traditional column-oriented data store. In average the in-memory database is still 20 seconds faster than the traditional row-oriented layout.

Selection of Next Job. The selection of the next urls to crawl, called job, is one of the most frequently executed tasks during crawling. As described in

Table 1. Execution of 200 000 insert queries

Run	SAP Hana	PostgreSQL
1	1m 29s 155ms	3m 8s 466ms
2	2m 32s 622ms	3m 8s 466ms
3	2m 32s 622ms	3m 50s 583ms
4	3m 53s 147ms	3m 22s 176ms
5	3m 58s 291ms	4m 29s 566ms

Section 4.1, a focused crawler for blogs has to incorporate complex constraints for the selection.

In former times the complex selection process as described previously was a pretty time consuming task. With the change to an in-memory database this task can be accomplished within a few seconds. The actual crawler uses a query with more than 10 constraints. We simplified the query to one constraint that uses the pre-calculated fetchtime for a URL 1.1. The fetchtime is calculated while the URL was inserted based on our own heuristic, when it is most promising to download this URL.

Listing 1.1. Selection of next job

```
SELECT ID FROM WEBPAGE
WHERE FETCHTIME < ( currentTimeInMillis )
ORDER BY SCORE DESC LIMIT 10000
```

The results of this experiment is shown in Table 2 after several executions on each database.

Table 2. Selection of next urls

Run	SAP Hana	PostgreSQL
1	0m 21s 919ms	3m 46s 666ms
2	0m 21s 554ms	3m 46s 007ms
3	0m 21s 832ms	3m 41s 213ms
4	0m 21s 745ms	3m 56s 505ms
5	0m 21s 811ms	3m 56s 198ms

By looking at Table 2 we can see that the selection of the next urls is up to 8 times faster with the in-memory database. Furthermore in the real scenario this difference would even more important.

5.4 Analytics

Blogs with the Most Incoming Links. As already mentioned the ranking got very simplified for this experiment, by counting the incoming links and order them accordingly by the top blog with the most incoming links.

Listing 1.2. SQL Query most incoming links

```
SELECT COUNT(*) as incomLinks, toHost
FROM link
GROUP BY toHost ORDER BY COUNT(*) DESC
```

The execution time for each database of this experiment is shown in Table 1.3.

Table 3. Execution of the link query

Run	SAP Hana	PostgreSQL
1	0m 11s 800ms	1m 46s 140ms
2	0m 11s 955ms	1m 43s 245ms
3	0m 11s 735ms	1m 44s 058ms
4	0m 11s 780ms	1m 42s 443ms
5	0m 11s 940ms	1m 45s 194ms

Table 1.3 shows that the execution time of this query could in average decreased by factor 10.

Big Blogs with Additional Information. At the first look it sounds pretty easy to retrieve information from the biggest blogs. But in order to get additional information of each blog, a join between the table containing blogs and the table containing links is necessary. This makes it more complicated for the database to handle.

Listing 1.3. Blogs with additional information

```
SELECT POSTTITLE, POSTAUTHOR
FROM WEBPAGE INNER JOIN (

SELECT toUrl, COUNT(*) as links
FROM link
GROUP BY toUrl

) as link
ON link.toUrl = webpage.ID
WHERE type='POST'
ORDER BY links desc
```

Table 4. Execution of the addition info query

Run	SAP Hana	PostgreSQL
1	1m 12s 544ms	208m 02s 737ms
2	0m 1s 722ms	199m 51s 321ms
3	0m 1s 868ms	233m 42s 876ms
4	0m 2s 243ms	201m 33s 128ms
5	0m 1s 925ms	214m 55s 057ms

The execution time of this experiment is shown in Table 4. The measurement reveals a significant difference in the execution time. The in-memory database is faster by a few orders of magnitude. This difference is caused by the layout of the database. Since we are just looking at columns, it is not a problem if a table has lot of columns. For a row-oriented database this is a big problem. Therefore the PostgreSQL database has to touch 150 gigabyte of data which of course takes some time. Another interesting point is, that the in-memory database obviously is performing some kind of caching. Thus, the execution gets really fast if the database stays untouched.

Calculate Similar Terms / Similar Documents. In addition we take a short excursion at text mining area that is essential for web mining applications. As discussed, the tested database provides a compressed Document-Term matrix and the functionality to calculate similar terms or similar documents with the well-known tf-idf measure.

Performing a search for similar terms can be accomplished on this data set within short time. The following Table 5 shows the execution time for finding similar terms for given terms. In this experiment the text-mining index contains 400 million entries with 7 million unique terms.

Table 5. Execution of similar term search

terms	time
Apple	1025 ms
Apple or Microsoft	1057 ms
Obama	630 ms
Merkel or Obama	950 ms
Politik or Wirtschaft or Bildung	1620 ms

The experiments showed that the in-memory technology together with a column-oriented layout performs faster on simplified web mining tasks than a traditional row-oriented disc-based store. We expect that this performance gain applies also for the compound analytical tasks of web mining applications. Further, insert commands, which are expected to be cost-intensive tasks in a column-oriented layout, also perform very fast due to database optimizations specific to SAP HANA.

6   Future Work

In this paper we evaluated some basic data store task for crawling and analysis. Although we show performance improvements for these tasks, we need to dive deeper into complex analysis algorithms and their behavior on our storage. Thus, we investigating current analysis approaches and try to transfer them to our system. The following two directions support this objective and promise insight about complex tasks.

In order to benefit from this technique in the future, some additional functionality should be improved inside the database. As mentioned above, the internal Document-Term matrix helps to calculate text analysis in almost real time based on the latest data. Thus, we suggest to use similar structures to store a up-to-date clustering for blogs and related terms. A hierarchical clustering fit these requirements, because the amount of clusters is not known at creation time. This clustering provides the possibility to access a certain hierarchy level or the subtree of a specific topic space. We want to build a prototype using such a structure and evaluate this concerning the real-time performance.

Furthermore, we currently investigate a personalized rank editor. Ranking mechanisms for blog posts are a very interesting research topic as stated by Bross et al. [2]. Nevertheless, it is not possible to define one universal rank for every user of a search engine. Thus, we try to integrate the user into the ranking process and let the user decide what is important. The user gets the ability to define his own ranking formula based on intuitive factors. A first prototype of this ranking editor is already available at BlogIntelligence[3].

Since the user expects immediately feedback when defining a new ranking equation, the computational effort is tremendous. For example, if a user wants to search for blog posts ranked by a quotient between incoming and outgoing links, the whole link table has to be scanned every time. The system is still under construction, but we were able to get the first prototype running really fast. With a traditional database this would be unthinkable. We want to improve that functionality in the future and provide as many as possible ranking criteria for the editor.

In addition, we are working on a personalized trend detection interface. Hereby, a user can search for trends in his topic of interest and can select a certain time window to detect trends. This helps to understand discussions and trends in more detail and gives the user the possibility to predict trends in the future.

7   Conclusion

As shown in this paper we were able to apply the in-memory technology in conjunction with a column-oriented layout and gained an for the tested web mining application including crawling, analysis and visualization of big amount of web pages in this case blogs. We emphasize that the whole crawled web data including extracted meta data is stored in main memory without helping structures on disc. In addition, the optimization of the database used in the experiment is able to absorb the low performance when inserting thousands of tuples. This helps to make use of the tremendous performance improvements of the in-memory database when executing analytics.

Furthermore, we demonstrate that advanced analysis like similar term extraction can also be performed very fast by integrating the specific data structure like a document term matrix directly into the database.

---

[3] http://www.blog-intelligence.com

Finally, the change from an traditional row-based database to an in-memory column-oriented database provides BlogIntelligence complete new kind of analysis like the described ranking editor where users are able define their own ranking equation and immediately get the results for their ranking equation. BlogIntelligence is now able to provide personalized real-time analyses for each user separately according to the user's special topic of interest.

References

1. Bahmani, B., Chakrabarti, K., Xin, D.: Fast personalized pagerank on mapreduce. In: Proceedings of the 37th SIGMOD International Conference on Management of Data, pp. 973–984 (2011)
2. Bross, J., Richly, K., Kohnen, M., Meinel, C.: Identifying the top-dogs of the blogosphere. Social Netw. Analys. Mining 2(1), 53–67 (2012)
3. Chang, F., Dean, J., Ghemawat, S., Hsieh, W.C., Wallach, D.A., Burrows, M., Chandra, T., Fikes, A., Gruber, R.E.: Bigtable: A distributed storage system for structured data. ACM Transactions on Computer Systems (TOCS) 26(2), 4 (2008)
4. Etzioni, O.: The world-wide web: quagmire or gold mine? Communications of the ACM 39(11), 65–68 (1996)
5. Hewitt, E.: Cassandra: the definitive guide. O'Reilly Media, Incorporated (2010)
6. Bross, J., Kohnen, M., Richly, K., Kohnen, M., Meinel, C.: Identifying the top dogs of the blogosphere. Social Network Analysis and Mining. Springer LNSN (2011)
7. Kosala, R., Blockeel, H.: Web mining research: A survey. ACM Sigkdd Explorations Newsletter 2(1), 1–15 (2000)
8. Maes, P., et al.: Agents that reduce work and information overload. Communications of the ACM 37(7), 30–40 (1994)
9. Momjian, B.: PostgreSQL: introduction and concepts, vol. 192. Addison-Wesley (2001)
10. Hennig, P., Berger, P., J.B.C.M.: Mapping the blogosphere - towards a universal and scalable blog-crawler. In: Proceedings of the Third IEEE International Conference on Social Computing (Social Com2011), pp. 672–677. IEEE CS, MIT, Boston, USA (2011)
11. Page, L., Brin, S., Motwani, R., Winograd, T.: The pagerank citation ranking: bringing order to the web (1999)
12. Plattner, H.: A course in In-Memory Data Management. Springer, Berlin (2013)
13. Sparck Jones, K.: A statistical interpretation of term specificity and its application in retrieval, pp. 132–142 (December 1988)
14. Widenius, M., Axmark, D., MySQL, A.: MySQL reference manual: documentation from the source. O'Reilly Media, Incorporated (2002)

Constructing a Novel Pos-neg Manifold for Global-Based Image Classification[*]

Rong Zhu[1], Jianhua Yang[2], Yonggang Li[1], and Jie Xu[1]

[1] School of Information Engineering, Jiaxing University, Jiaxing, China
[2] School of Computer Science and Technology, Zhejiang University, Hangzhou, China
{zr,jhyang}@zju.edu.cn

Abstract. For the task of global-based image classification, we construct an image manifold, i.e., a pos-neg manifold, based on the solving strategies of two-class classification problem, which includes a positive sub-manifold and a negative one. We also present an improved globular neighborhood based locally linear embedding (an improved GNLLE) algorithm, fully taking account of the big differences between the positive and negative category images, thus the data distance calculation defined in the high-dimensional space can be translated into the one on the image manifold with lower dimensionality. Moreover, to simplify the distance measure between two nonlinear sub-manifolds, we put forward a clustering-based method to determine a manifold center for each sub-manifold. Experimental results on the real-world Web images show that the proposed method can improve the classification performance significantly.

Keywords: Global-based image classification, image manifold, two-class classification problem, distance measure, manifold center.

1 Introduction

The progresses from image acquisition and storage technology have generated large-scale multimedia data. Especially in this 'big data' era, countless digital devices and various sensors which act as major data sources have been producing enormous multimedia data, and most of which is the image data. On the Internet, online photo management and sharing applications (i.e., Flickr) report thousands of uploaded images per minute, and provide users more facilities for accessing and exchanging images. Hence users have pursed more and more powerful abilities to utilize Web images. On the other hand, researchers also have been keen on the study of developing advanced mining tools for deeper information processing. Therefore, how to effectively analyze complex Web images and reveal their high-level semantic information have become an important issue in many research fields, including data mining, computer vision, and multimedia information processing, etc.[1-2].

[*] Project supported by the Zhejiang Provincial Nonprofit Technology and Application Research Program of China (No. 2012C21020).

H. Motoda et al. (Eds.): ADMA 2013, Part I, LNAI 8346, pp. 217–228, 2013.
© Springer-Verlag Berlin Heidelberg 2013

Classification is a critical research topic in data mining, and the effectiveness of classification results will affect the performance of subsequent processing. But most of the traditional classification approaches are failure in classifying Web images since they only consider low-level visual features but not high-level semantic information. Recently, a development trend for image classification is region- or object-based image classification[3-5]. Compared to the global-based image classification, the region- or object-based image classification has stronger representation capability for semantic information, but its effectiveness depends on the fact that an image should have one clear object(s) and the object(s) should be well segmented. However, it is difficult to distinguish the object(s) from the background by most existing segmentation methods. And what's more, users usually are not professional photographers, so the uploaded images sometimes are not clear. Through extensive analysis, we found that for those kinds of Web images, the region- or object-based image classification will not be available. (1) The subject of an image is not a specific object, but an abstract concept, such as landscape images. (2) An image contains too many objects, or the object(s) has blurred boundary. (3) An image has no explicit subject, such as some abstract images.

On the other hand, the results of psychological research[6] show that human can understand images without knowing any information about the object(s). Thus more global features will be depended on, rather than the visual features only related to the object(s). Datta et al.[7] mentioned that the extraction of the global features was a shortcut for obtaining semantic information and thereby it makes practical significance for image retrieval research. It is very plausible, in Web image retrieval, that the global semantic information is preliminary estimated before the optimizations participate in. Theoretically, such a strategy, i.e., 'forest first tree second', is efficient in many cases. Therefore, the research on the global-based image classification not only has a large practical potential but also faces new technology challenges.

2   Related Works

According to the former research results, to obtain better classification performance for the region- or object-based image classification problems, certain hierarchical models were introduced into the classification process and the feature extraction was conducted on the object and scene levels, respectively. But for the global-based image classification, such classification strategies are not suitable since no effective method exists to recognize the object(s) or the methods are quite expensive to segment the object(s) from images. Therefore, many global features should be extracted from an image for the global-based image classification and thus it generates the high-dimensional image features. It is difficult indeed, to mine, analyze, and deal with the image data in the high-dimensional space.

Fortunately, in 2000, Seung and Lee[8] pointed out that the data having common characteristics can be considered as lying on or closing to a low-dimensional nonlinear manifold embedded in the high-dimensional space. Taking account of the hypothesis that the low-dimensional nonlinear manifold is human cognitive-based and without semantic gap, the mappings of the high-dimensional data on the manifold can be regarded as high-level semantic information, i.e., an abstract representation for the original image data. In 2002, Wang[9] stated that the data in one category is harmonious with the continuity rule in the low-dimensional space. Specifically, there exists a gradual change of the data belonged to one category, while any data satisfies this gradual change can be grouped into the same category. This makes the assumption more reasonable that any two data belonged to one category are usually nearby in the low-dimensional space. Therefore, we suppose that the unified semantic information of the image data belonged to one category can be revealed from a low-dimensional nonlinear manifold with specific topological structure.

Recently, more and more attentions have been paid to the research of manifold learning in image classification. For example, Huo et al.[10] reviewed the ideas and numerical performance of the existing manifold learning methods, as well as some potential applications in image data. Kim et al.[11] presented a framework for integrating spatial and spectral information, and applied a hierarchical spatial-spectral recognition method to construct the nonlinear manifold for hyperspectral data. Huh et al.[12] proposed a new method named as discriminative topic model (DTM) for classification which brings the neighboring data closer together and separates the non-neighboring data from each other, to preserve the whole manifold structure as well as the local consistency. To deal with time and memory limitations faced in the large-scale problems, Farajtaba et al.[13] provided a novel approximate Newton's method for maintaining the local and global consistency of the manifold structure. However, up to now, there is little research about developing manifold learning-based methods for the global-based image classification.

In this paper, we propose a novel Web image classification method based on manifold learning. At first, use the improving GNLLE algorithm to reduce the high dimensionality of the global features extracted from the image data in the training image set. Then, construct an image manifold in the low-dimensional space for the image data included in the positive and negative categories, based on the solutions of the two-class classification problem. Finally, construct a classifier according to the distance measure between each manifold center and the image data to be classified.

3   A Pos-neg Manifold Construction

3.1   Motivation

In data classification research, two-class classification problem has received many attentions from researchers in recent years, because of its extendibility to the multi-class classification problem[14-16]. It is a feasible way to introduce the solutions of the two-class classification problem into image classification research. In fact, most of the image classification problems can be converted into the two-class classification ones. For example, for recognizing face images, Timotius et al.[17] proposed a new two-class classification method based on kernel principal component analysis (KPCA) and support vector machine (SVM). Wen et al.[18] built a two-class emotion classification system to classify four emotion states (i.e., joy, anger, grief, and fear), using the Tabu search algorithm and the Fisher classifier. Pan et al.[19] presented a novel two-class classification method based on dominant color descriptor, to classify traditional Chinese paintings and calligraphy images. But it is also a new study to construct an image manifold based on the solutions of the two-class classification problem, and furthermore apply it to the global-based image classification. One relative research first appeared in the works proposed by Xu et al.[20]. They proposed a classification method that uses locally linear embedding (LLE) first for dimensionality reduction and then uses SVM for classifier construction, to apply for the indoor and outdoor images from the Internet. However, this method has one defect: since the topological structure of a nonlinear manifold merely reflects the regularities and correlations of the image data, the LLE algorithm can get the embedding manifold in the low-dimensional space correctly only if an image set is evenly distributed. But in practical applications, the overall distribution of an image set is usually uneven, which can be seen as a composition of many discrete subsets.

3.2   Image Manifold Definition

The underlying assumption is: the image data belonged to the positive and negative categories in the training image set lie on the respective nonlinear manifolds (i.e., a positive manifold is built for the image data in the positive category, while a negative manifold is built for the image data in the negative category). When a testing image data arrives, if it belongs to the positive category, it must lie on the positive manifold based on the continuity rule, and its distribution is accordance with the topological structure of the positive manifold; otherwise, this image data belongs to the negative category and lies on the negative manifold, and its distribution is accordance with the topological structure of the negative manifold. Hence, we can classify Web images via calculating the distances between the image data from the two low-dimensional nonlinear manifolds, and the testing image data will be easily classified.

Definition 1. A pos-neg manifold $M_{p\text{-}n}$ is constructed for an image set (it consists of the image data belonged to the positive and negative categories) in the high-dimensional space $\mathbf{R}^D$, defined as a combination of two low-dimensional nonlinear sub-manifolds, i.e., $M_{p\text{-}n} = \{M_{pos}, M_{neg}\}$. Where M_{pos} is a positive sub-manifold, corresponded to the image data in the positive category; M_{neg} is a negative sub-manifold, corresponded to the image data in the negative category, and these two sub-manifolds are disjoint, i.e., $M_{pos} \cap M_{neg} = \Phi$.

Different from many existing manifold learning-based classification methods[21,22] that only construct a low-dimensional nonlinear manifold in one global coordinate system, for the image data belonged to various semantic categories, here a pos-neg manifold is constructed for the image data included in each semantic category. Obviously, if we directly construct a low-dimensional nonlinear manifold for all image data, a poor classification result will be obtained, and the more the number of semantic categories, the more overlaps among manifolds, the worse the classification results would be. Therefore, constructing a pos-neg manifold for the image data belonged to one semantic category, fully considering the differences between the image data in the positive and negative categories, will not only achieve high classification accuracy for the global-based image classification, but also have powerful manifold expansibility for the image data with more semantic categories.

3.3 An Improved GNLLE Algorithm

As mentioned above, the aim of the pos-neg manifold construction is to make the data distance the smaller the better within the positive or negative sub-manifold, and meanwhile make the distance the farther the better between the positive and negative sub-manifolds. However, the major difficulty of Web image classification is that in the network environment, natural images usually have large intra-category changes and small inter-category changes[5]. Zhu et al.[23] proposed a method of globular neighborhood based locally linear embedding (GNLLE) to optimize the neighborhood construction step in LLE. But this method would still be affected by the singular points and the noise. We present an improved GNLLE algorithm to extract the discrimination enhanced features and nonlinearly reduce the dimensionality of the image data. Moreover, a credibility detection method proposed by Chang et al.[24] is applied to calculate the possibility of the outliers for the neighbors, and then both the credibility and category information are added into the distance calculation formula, to update the neighbors in original neighborhood area.

Definition 2. Let $X=\{x_1, x_2, ..., x_N\}$ be an image set in the high-dimensional space $\mathbf{R}^D$. N image data are assumed to lie on a low-dimensional pos-neg manifold $M_{\text{p-n}}$ with intrinsic dimensionality d $(d < <D)$. Suppose that x_i and x_j $(i, j=1, 2, ..., N; i \neq j)$ are two image data in X, and then the distance between x_i and x_j is defined as follows:

$$d(x_i, x_j) = \|x_i - x_j\| / \sqrt{p(x_i)p(x_j)} \tag{1}$$

where $p(x_i)$ is the average distance between x_i and other image data in X; $p(x_j)$ is the average distance between x_j and other image data in X.

Definition 3. Let $X=\{x_1, x_2, ..., x_N\}$ be an image set in the high-dimensional space $\mathbf{R}^D$. N image data are assumed to lie on a low-dimensional pos-neg manifold $M_{\text{p-n}}$ with intrinsic dimensionality d $(d < <D)$. Assume that these image data are divided into C semantic categories, and furthermore ascribed to the positive category or the negative one according to the two-class classification problem. Suppose that x_i $(i=1, 2, ..., N)$ is an image data in X and x_{ik} $(i, k=1, 2, ..., N)$ is its neighbor, and then the distance between x_i and x_{ik} is defined as follows:

$$d'(x_i, x_{ik}) = \|x_i - x_{ik}\| / \sqrt{p(x_i)p(x_{ik})} + \alpha m(x_i)(1 - \sqrt{s(x_i)s(x_{ik})}) \tag{2}$$

where $p(x_i)$ and $p(x_{ik})$ are denoted as above; a is an adjustment coefficient; $m(x_i)$ denotes the maximum distance between x_i and other image data within the neighborhood area of x_i; $s(x_i)$ and $s(x_{ik})$ are the credibility values of x_i and x_{ik}, respectively, where $0 \leq s(x_i), s(x_{ik}) \leq 1$ (they are equal to 1 when x_i and x_{ik} are both in the positive or negative category, and the smaller the values of $s(x_i)$ and $s(x_{ik})$, the larger possibility of the outliers they would be, and vice versa).

An Improved GNLLE Algorithm
Let $X=\{x_1, x_2, ..., x_N\}$ be an image set including N image data in the high-dimensional space $\mathbf{R}^D$. These image data are assumed to lie on a low-dimensional pos-neg manifold $M_{\text{p-n}}$ with intrinsic dimensionality d $(d < <D)$. Suppose that the mappings of N image data in the low-dimensional space $\mathbf{R}^d$ is denoted as $Y=\{y_1, y_2, ..., y_N\}$.

Step 1: For each image data x_i (i=1, 2, ..., N) in X, construct its globular neighborhood using the globular radius r, and calculate the distance between x_i and x_j (j=1, 2, ..., N; $i{\neq}j$) in X (Eq.(1)). If there exist p_i image data within the globular neighborhood area, these image data are selected as the neighbors of x_i.

Step 2: Reselect the neighbors of x_i by calculating the distance between x_i and x_{ik} (i, k=1, 2, ..., N) within the globular neighborhood area of x_i (Eq.(2)), and the number of the updated neighbors are denoted as q_i.

Step 3: Calculate the reconstruction weight w_{ik} using the q_i neighbors of x_i, and minimize the local reconstruction error for x_i as follows:

$$\varepsilon_i = \left\| x_i - \sum_{k=1}^{q_i} w_{ik} x_{ik} \right\|^2 \tag{3}$$

where w_{ik} is subject to the constraints: $\sum_{k=1}^{q_i} w_{ik} = 1$ and w_{ik}=0 (if x_{ik} is not in the globular neighborhood area of x_i). After conducting the above steps for all N image data in X, a weight matrix W (=$[w_{ij}]_{N \times N}$) is formed by the reconstruction weights.

Step 4: Calculate the best low-dimensional embedding Y based on the weight matrix W, and minimize the cost function as follows:

$$\varepsilon(Y) = \sum_{i=1}^{N} \left\| y_i - \sum_{k=1}^{q_i} w_{ik} y_{ik} \right\|^2 \tag{4}$$

where Y is subject to the constraints: $\sum_{i=1}^{N} y_i = 0$ and $\sum_{i=1}^{N} y_i y_i^{\mathrm{T}} / N = I$. Based on the weight matrix W, a sparse, symmetric, and positive semi-definite cost matrix M (=$[m_{ij}]_{N \times N}$) is given by $M = (I-W)^{\mathrm{T}}(I-W)$. Then Eq.(4) can be rewritten as follows:

$$\varepsilon(Y) = \left\| Y(I-W) \right\|^2 = tr(Y^{\mathrm{T}}(I-W)^{\mathrm{T}}(I-W)Y) = tr(Y^{\mathrm{T}} M Y) \tag{5}$$

where $tr()$ denotes the trace function of matrix. At last, by the Rayleigh-Ritz theorem, Y can be obtained by finding the eigenvectors with the smallest (nonzero) eigenvalues of the cost matrix M.

3.4   Classifier Design

To simplify the classification process on the pos-neg manifold, we absorb the idea of the clustering methods, i.e., use a manifold center (it is obtained by the calculation principle of aggregation center) to represent one sub-manifold, and then the distance measure between two nonlinear manifolds can be defined based on the data distance between two manifold centers.

Definition 4. Suppose that Y_{pos}=$\{y_{p,1}, y_{p,2}, ..., y_{p,N_p}\}$ is the mappings of N_p image data belonged to the positive category; Y_{neg}=$\{y_{n,1}, y_{n,2}, ..., y_{n,N_n}\}$ is the mappings of N_n image data belonged to the negative category, and both of them are assumed to lie on the pos-neg manifold M_{p-n} in the low-dimensional space $\mathbf{R}^d$. Let the positive sub-manifold M_{pos}

be the embedded manifold of Y_{pos}; the negative sub-manifold M_{neg} be the embedded manifold of Y_{neg}, and then the manifold centers of the two nonlinear sub-manifolds are denoted as S_{pos} and S_{neg}, respectively.

To calculate the manifold center for each nonlinear sub-manifold, we should minimize the following cost functions:

$$\varepsilon(Y_{pos}) = \sum_{i=1}^{N_p} \left\| y_{p,i} - S_{pos} \right\|^2 \tag{6}$$

$$\varepsilon(Y_{neg}) = \sum_{i=1}^{N_n} \left\| y_{n,i} - S_{neg} \right\|^2 \tag{7}$$

Compute Eq.(6) based on the matrix norm properties ($\varepsilon(Y_{pos})$ for instance):

$$\varepsilon(Y_{pos}) = tr \sum_{i=1}^{N_p} (y_{p,i} - S_{pos})(y_{p,i} - S_{pos})^T \tag{8}$$

where $(y_{p,i} - S_{pos})(y_{p,i} - S_{pos})^T$ is decomposed to $y_{p,i} y_{p,i}^T - y_{p,i} S_{pos}^T - S_{pos} y_{p,i}^T + S_{pos} S_{pos}^T$, thus, $y_{p,i} S_{pos}^T + S_{pos} y_{p,i}^T$ should be the larger the better. Due to $(S_{pos} y_{p,i}^T)^T = y_{p,i} S_{pos}^T$, Y_{pos} is desired to satisfy the constraints: $\arg\max \sum_{i=1}^{N_p} (y_{p,i} S_{pos}^T)$ and $S_{pos} S_{pos}^T = 1$, and then Eq.(8) can be rewritten by maximizing the following formula:

$$\sum_{i=1}^{N_p} (y_{p,i} S_{pos}^T)^T (y_{p,i} S_{pos}^T) = S_{pos} (\sum_{i=1}^{N_p} y_{p,i}^T y_{p,i}) S_{pos}^T \tag{9}$$

As a result, the value of the manifold center S_{pos} of the positive sub-manifold M_{pos} will be obtained by the eigenvector corresponded to the maximum eigenvalue of the matrix $\sum_{i=1}^{N_p} y_{p,i}^T y_{p,i}$. Similarly, the value of the manifold center S_{neg} of the negative sub-manifold M_{neg} will be obtained by the eigenvector corresponded to the maximum eigenvalue of the matrix $\sum_{i=1}^{N_n} y_{n,i}^T y_{n,i}$.

Just as the above pos-neg manifold M_{p-n}, a classifier is designed to distinguish the image data in one semantic category from the image data belonged to other semantic categories, here a testing image data can be classified into one of the semantic categories based on the distance measure-based classification strategy. The distances between the image data to be classified and two sub-manifolds (i.e., M_{pos} and M_{neg}) on the M_{p-n} are calculated as follows:

$$d_{pos}(y', M_{pos}) = \left\| y' - S_{pos} \right\| \tag{10}$$

$$d_{neg}(y', M_{neg}) = \left\| y' - S_{neg} \right\| \tag{11}$$

where y' is the mapping of the testing image data x' in the low-dimensional space $\mathbf{R}^d$; S_{pos} and S_{neg} are the manifold centers of the positive and negative sub-manifolds, respectively. Thus, the testing image data x' will be classified into the nearest the positive category or the other.

Definition 5. Let x' be a testing image data in the high-dimensional space $\mathbf{R}^D$, and y' be its mapping on a low-dimensional nonlinear manifold with intrinsic dimensionality d ($d \ll D$). Suppose that there exist C semantic categories in the training image set (i.e., $\omega_1, \omega_2, ..., \omega_N$). The classifier for the image data belonged to the ith (i=1, 2, ..., C) semantic category can be defined by a discriminated function $f_i(x')$ as follows:

$$f_i(x') = d_{\text{pos}}(y', M_{\text{pos},i}) - d_{\text{neg}}(y', M_{\text{neg},i}) = \left\| y' - S_{\text{pos},i} \right\| - \left\| y' - S_{\text{neg},i} \right\| \quad (12)$$

where $M_{\text{pos},i}$ and $M_{\text{neg},i}$ are the positive and negative sub-manifolds within the posneg manifold $M_{\text{p-n}}$ (it is constructed for the image data included in the ith semantic category), respectively; $S_{\text{pos},i}$ and $S_{\text{neg},i}$ are their manifold centers. When $f_i(x') \geq 0$, $x' \in \omega_i$ (it is classified into the positive category); otherwise $x' \notin \omega_i$ (it is classified into the negative category).

4   Experiments

4.1   Experimental Preparation

The following experiments were used to verify our image classification method. Hardware platform was: CPU (T2350 1.86GHZ), memory (2G); software tools were: VC ++6.0 and MATLAB7.0 mixed programming. Web images used in the experiments were all downloaded from the Internet.

The image set contained five semantic categories, 3000 images in a total, where the 'Beach', 'Grass', 'Sunrise/Sunset', 'Building', and 'Garden' images are 600, respectively. The image size after preprocessed is 126×189 or 189×126. Through the feature analysis of Web images, 128 feature dimensions were extracted, including color histogram in HSV color space (64 dimensions), color moments in RGB color space (9 dimensions), color moments in LUV color space (9 dimensions), Tamura texture features (6 dimensions), Gabor texture features (24 dimensions), wavelet-based texture features (3 dimensions), Hu invariant moments (7 dimensions), and edge direction histogram (6 dimensions).

We randomly selected 2/3 images from each semantic category as the training image set, and the rest 1/3 as the testing image set. As mentioned above, here the classification of multiple semantic categories was regarded as a two-class classification problem, thus we designed a classifier for each semantic category. For instance, the 'Beach' images can be distinguished from other images conveniently, when the 'Beach' classifier was constructed, where the 'Beach' images were seen as the positive category, while the 'Grass', 'Sunrise/Sunset', 'Building' and 'Garden' images were viewed as the negative category.

The experiments were performed five times, and the average correct classification rate (CCR) and the average error classification rate (ECR) were set to be the evaluation criterion, which could be defined as follows:

$$\text{CCR}_i = N_{i\text{-c}} / N_i \quad (13)$$

$$\text{ECR}_i = N_{i\text{-e}} / N_i \quad (14)$$

where N_i denotes the total number of the images included in the ith semantic category (the positive category); $N_{\bar{i}}$ denotes the total number of the images not included in the ith semantic category (the negative category); $N_{i\text{-}c}$ denotes the number of the images correctly classified into the ith semantic category (they are in the positive category); $N_{\bar{i}\text{-}e}$ denotes the number of the images wrongly classified into the ith semantic category (they are actually not in the positive category).

4.2   Parameter Setting

In the improved GNLLE algorithm-based manifold learning method, the dimensionality of the low-dimensional nonlinear manifold d is an important parameter. Since image classification accuracy does not increase any more when the dimensionality is about 12 (whenever the globular radius r takes different values), thus the intrinsic dimensionalities of the positive and negative sub-manifolds were both set to 12. In the definition of distance measure, an adjustment coefficient a is used to control the combined degree of category information, and here it was set to 0.5. In addition, the data sampling rate in the data resampling step is set to 80 images per second. The data resampling step can also be omitted when system processing time is limited, but obviously, which will affect the performance of the image classification based on manifold learning.

4.3   Results and Analysis

A performance comparison was taken among our method (the improved GNLLE + manifold center), Yao et al. [25] method (LLE + mean), and the other method (LLE + aggregation center). The above manifold learning-based image classification methods are all based on the two-class classification problem solving strategies.

Table 1. Comparison of the average correct classification rate (CCR) among three image classification methods

Semantic categories	Average correct classification rate (CCR)		
	LLE1	LLE2	GNLLE
Beach	0.687	0.783	0.854
Grass	0.736	0.823	0.889
Sunrise/Sunset	0.764	0.873	0.919
Building	0.801	0.884	0.960
Garden	0.748	0.816	0.892

LLE1:LLE+mean;LLE2:LLE+aggregation center; GNLLE: the improved GNLLE+manifold center

Table 2. Comparison of the average error classification rate (ECR) among three image classification methods

Semantic categories	Average error classification rate (ECR)		
	LLE1	LLE2	GNLLE
Beach	0.407	0.261	0.138
Grass	0.452	0.330	0.165
Sunrise/Sunset	0.533	0.380	0.184
Building	0.315	0.165	0.062
Garden	0.426	0.230	0.109

LLE1: LLE+mean; LLE2: LLE+aggregation center; GNLLE: the improved GNLLE+manifold center

Table 1 and Table 2 show that, among three image classification methods, ours achieve the highest average CCR and the lowest average ECR. Compared with the other two methods (LLE + aggregation center, the improved GNLLE + manifold center), the method (LLE + mean) that applies mean values to calculate manifold distances is easily affected by the exceptional values, thus it obtains the poorest classification result. Moreover, because the improved GNLLE algorithm is not sensitive to uneven data distribution and has better robustness to the outliers, it is superior to the other method (LLE + aggregation center) when classifying the images with multiple semantic categories.

From another point of view, among five semantic categories, since the images included in 'Building' usually have dark colors and abundant lines, they are easily distinguished with the images in other semantic categories. When constructing the 'Building' classifier, because there exists a large difference between the positive and negative categories, and the positive and negative sub-manifolds are relatively independent in the low-dimensional space, all of the three methods obtain better classification results. However, for the images in 'Beach', some images have similarity with the ones in 'Sunrise/Sunset' and 'Garden', thus they have the lowest average CCRs. Moreover, when classifying the images in 'Sunrise/Sunset', due to the fact that a part of images in other semantic categories are easily misclassified into this category, the images in 'Sunrise/Sunset' get the highest average ECRs.

5   Conclusions

In this paper, we propose a novel global-based image classification method based on manifold learning, taking account of the strategies in two-class classification problem solving. An improved GNLLE algorithm is first applied to establish an image manifold (it includes a positive sub-manifold and a negative sub-manifold) in the low-dimensional space; then a classifier is designed according to the distance measure between two nonlinear manifolds, and new image can be classified into the corresponding category in the light of the minimum distance priority principle. The experimental results show that, our method not only has better data separability, but also obtains higher classification performance.

However, this method still needs to be improved. For instance, when the improved GNLLE algorithm conducts dimensionality reduction, although its parameter is a weak one and its value has little effect on the classification performance, this parameter is still an empirical value. Furthermore, the overall execution speed would be slower for the construction of the image manifold and some additional optimizing operations, although here the computational complexity and the wrong results from the redundant data can be reduced. Hence, developing parallel algorithms for the manifold learning-based method would be one of our future study subjects. In addition, more distinguishable information needed to be added into manifold center determination and data distance calculation.

References

1. Kamde, P.M., Algur, S.P.: A Survey on Web Multimedia Mining. Int. J. Multimedia & Its Applications 3(3), 72–84 (2011)
2. Bhatt, C.A., Kankanhulli, M.S.: Multimedia Data Mining: State of the Art and Challenges. Multimedia Tools and Applications 51(1), 35–76 (2011)
3. Zeng, Z.Y., Yao, Z.Q., Liu, S.G.: An Efficient and Effective Image Representation for Region-Based Image Retrieval. In: 2nd International Conference on Interaction Sciences: Information Technology, Culture and Human, pp. 429–434 (2009)
4. Devasena, C.L., Sumathi, T., Hemulatha, M.: An Experiential Survey on Image Mining Tools, Techniques and Applications. Int. J. Computer Science and Engineering 3(3), 1155–1167 (2011)
5. Zhu, R., Yao, M., Ye, L.H., Xuan, J.Y.: Learning a Hierarchical Image Manifold for Web Image Classification. J. Zhejiang Univ.-Sci. C 13(10), 719–735 (2012)
6. Biedeman, I.: Aspects and Extensions of a Theory of Human Image Understanding. In: Proc. of Computational Process in Human Vision: An Interdisciplinary Perspective, pp. 370–428 (1998)
7. Datta, R., Joshi, D., Li, J., Wang, J.Z.: Image Retrieval: Ideas, Influences, and Treads of the New Age. ACM Computing Surveys 40(2), 1–60 (2008)
8. Seung, H.S., Lee, D.: The Manifold Ways of Perception. Science 290(5500), 2268–2269 (2000)
9. Wang, S.J.: Bionic (Topological) Pattern Recognition- a New Model of Pattern Recognition Theory and Its Applications. Acta Electronica Sinica 30(10), 1417–1420 (2002)
10. Huo, X.M., Ni, X.L., Smith, A.K.: A Survey of Manifold-Based Learning Methods: Emerging Nonparametric Methodology (2007), http://citeseerx.ist.psu.edu/viewdoc/summary?doi=10.1.1.95.903
11. Kim, W., Chen, Y.C., Crawford, M.M., Tilton, J., Ghosh, J.: Multiresolution Manifold Learning for Classification of Hyperspectral Data (2009), http://citeseerx.ist.psu.edu/viewdoc/summary?doi=10.1.1.205.5783
12. Huh, S., Fienberg, S.: Discriminative Topic Modeling Based on Manifold Learning. In: ACM SIGKDD Conference on Knowledge Discover and Data Ming
13. Farajtaba, M., Rabiee, H.R., Shaban, A., Soltani-Farani, A.: Efficient Iterative Semi-Supervised Classification on Manifold. In: IEEE International Conference on Data Mining, pp. 228–235 (2011)

14. Tax, D.M.J., Duin, R.P.W.: Using Two-Class Classifiers for Multiclass Classification. In: 16th International Conference on Pattern Recognition, vol. 2, pp. 124–127 (2002)
15. Aly, M.: Survey on Multiclass Classification Methods (2005), http://citeseerx.ist.psu.edu/viewdoc/summary?doi=10.1.1.175.107
16. Chen, W., Metz, C.E., Giger, M.L., Drukker, K.: A Novel Hybrid Linear/Nonlinear Classifier for Two-Class Classification: Theory, Algorithm, and Applications. IEEE Trans. on Medical Imaging 29(2), 428–441 (2010)
17. Timotius, I.K., Setyawan, I., Febrianto, A.A.: Two-Class Classification with Various Characteristic Based on Kernel Principal Component Analysis and Support Vector Machines. Maraka J. Technology Series 15(1), 96–100 (2011)
18. Wen, W.H., Liu, G.Y., Xiong, X.: Feature Selection Model and Generalization Performance of Two-Class Emotion Recognition Systems Based on Physiological Signal. Computer Science 38(5), 220–223 (2011) (in Chinese)
19. Pan, W.G., Bao, H., He, N.: Novel Binary Classification Method for Traditional Chinese Paintings and Caligraphy Images. Computer Science 39(3), 257–260 (2012) (in Chinese)
20. Xu, Z.J., Yang, J., Wang, M.: A New Nonlinear Dimensionality Reduction for Color Image. J. Shanghai Jiaotong University 38(12), 2063–2072 (2004) (in Chinese)
21. Lu, D., Weng, Q.: A Survey of Image Classification Methods and Techniques for Improving Classification Performance. International J. Remote Sensor 28(5), 823–870 (2007)
22. Jun, G., Ghosh, J.: Nearest-Manifold Classification with Gaussian Processes. In: 20th International Conference on Pattern Recognition, pp. 914–917 (2010)
23. Zhu, R., Yao, M.: Image Feature Optimization Based on Nonlinear Dimensionality Reduction. J. Zhejiang University Science A 10(12), 1720–1737 (2009)
24. Chang, H., Yeung, D.Y.: Robust Locally Linear Embedding. Pattern Recognition 39(6), 1053–1065 (2006)
25. Yao, L.Q., Tao, Q.: One Kind of Manifold Learning Method for Classification. PR & AI 18(5), 542–545 (2005) (in Chinese)

3-D MRI Brain Scan Feature Classification Using an Oct-Tree Representation

Akadej Udomchaiporn[1], Frans Coenen[1], Marta García-Fiñana[2],
and Vanessa Sluming[3]

[1] Department of Computer Science, University of Liverpool, Liverpool, UK
{akadej,coenen}@liv.ac.uk
[2] Department of Biostatistics, University of Liverpool, Liverpool, UK
m.garciafinana@liv.ac.uk
[3] School of Health Science, University of Liverpool, Liverpool, UK
vanessa.sluming@liv.ac.uk

Abstract. This paper presents a procedure for the classification of specific 3-D features in Magnetic Resonance Imaging (MRI) brain scan volumes. The main contributions of the paper are: (i) a proposed Bounding Box segmentation technique to extract the 3-D features of interest from MRI volumes, (ii) an oct-tree technique to represent the extracted sub-volumes and (iii) a frequent sub-graph mining based feature space mechanism to support classification. The proposed process was evaluated using 210 3-D MRI brain scans of which 105 were from "healthy" people and 105 from epilepsy patients. The features of interest were the left and right ventricles. Both the process and the evaluation are fully described. The results indicate that the proposed process can be effectively used to classify 3-D MRI brain scan features.

Keywords: Image mining, 3-D Magnetic Resonance Imaging (MRI), Image segmentation, Oct-tree representation, Image classification.

1 Introduction

Image mining involves a number of challenges of which the most significant relates to the representation of the image data in a format that allows the effective application of data mining techniques. The nature of the image representation will affect both the efficiency and effectiveness of the data mining. We can divide the domain of image mining into whole image mining and Region Of Interest (ROI) mining where the distinction is that the second is directed as some specific sub-image present across an image collection. In this paper we consider ROI image mining, more specifically we consider 3-D ROI image mining, thus Volume Of Interest (VOI) image mining. We propose a Bounding Box technique to identify specific VOIs within am image collection and an oct-tree formalism with which to represent the identified VOIs. The oct-tree representation in turn can then be processed using a frequent sub-graph mining process which can then be used to generate a feature space that is compatible with the application of data mining (classification) techniques.

H. Motoda et al. (Eds.): ADMA 2013, Part I, LNAI 8346, pp. 229–240, 2013.
© Springer-Verlag Berlin Heidelberg 2013

To act as a focus for the work we consider 3-D Magnetic Resonance Imaging (MRI) data of the human brain. Note that a 3-D MRI scan comprises a sequence of 2-D "slices". The VOIs in this case are the lateral (left and right) ventricles. The ventricles are fluid-filled open spaces at the centre of the brain; there are four ventricles in a human brain, but in this paper we only consider the lateral ventricles. An example of a 3-D MRI brain scan is shown in Figure 1 where the lateral ventricles are the dark areas at the centre of the brain. The dataset used to evaluate the proposed process was composed of 210 MRI brain scans of which 105 were from "healthy brains" and the remaining 105 were from epilepsy patients. The application goal of the research is to classify MRI brain scans as either epilepsy or non-epilepsy according to the nature of the ventricles (the nature of the lateral ventricles are considered to be indicators of the presence of conditions such as epilepsy). The proposed process is as follow. First we apply our Bounding Box technique to isolate the ventricles. We then represent each ventricle using an oct-tree formalism. A graph mining technique is then used to identify frequently occurring sub-oct-trees. The identified sub-oct-trees are then used to define a feature space from which a feature vector representation can be produced (one vector per image) to which established data mining techniques can be applied.

The rest of the paper is organised as follows. Section 2 introduces some previous work. In Section 3 the classification process is described in detail. The experimental set-up whereby the process was evaluated, and the results obtained, are presented in Sections 4 and 5 respectively. Finally, the paper is concluded in Section 6 with a summary of the main findings.

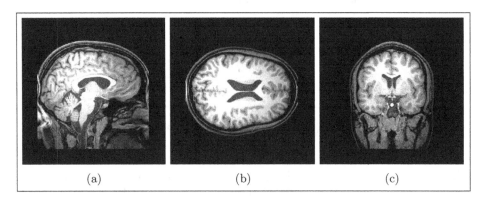

Fig. 1. Example of a 3-D brain MRI scan: (a) Sagittal (SAG) plane; (b) Transverse (TRA) plane; (c) Coronal (COR) plane

2   Previous Works

Some of the published work on the segmentation of brain images has common aspects with the work described in this paper. For example in [6,7] a Modified Spectral Segmentation algorithm, founded on a multiscale graph decomposition, was proposed to segment the corpus callosum (another feature present in MRI brain scans). Experimentation was conducted using 76 MRI data sets; the results indicated that their algorithm could detected the corpus callosum more accurately than when using existing segmentation techniques. However, the work described in [7] was directed at 2-D data (specifically, the *midsagittal* slice of a MRI volume). The work described in [14] is of particular interest with respect to the work described in this paper because they also used an oct-tree conceptualisation from which a feature vector representation was extracted using frequent sib-graph mining. The work described in [14] was also directed at brain ventricles however in the context of Alzheimer's disease. In the context of data capture a technique was presented in [14] to segment both the lateral ventricles and the third ventricles using an oct-tree decomposition, different to that presented in this paper, coupled with a dynamic thresholding technique. This segmentation technique automatically found the most suitable threshold value for each brain image. It was argued that this tended to produce a more accurate result. However, the work described in [14] was focussed on classification accuracy (no evaluation was conducted concerning the segmentation). It is also worth nothing that the process of dynamic thresholding is time consuming. Some other reported work on MRI brain scan segmentation can be found in [17] where a "hand-segmentation" approach was proposed to extract brain ventricles from 3-D MRIs. The approach was implemented using active shape models [3] and level set methods [15]. However, as in the case of [14], no evaluation of the segmentation was included in [15].

Some of the published work on image classification using graph based representations also has common aspects with the work presented in this paper. For instance, in [6] a quad-tree technique was used to represent the corpus callosum. The work also used a frequent sub-graph mining technique to discover frequently occurring sub-graphs in the quad-trees using the well-known Gspan algorithm [18] and the Average Total Weighing (ATW) scheme proposed in [13]. Experiments were conducted to classify MRI brain scans as epilepsy or non-epilepsy (the same application domain as considered in this paper), musicians or non-musicians, and left-handedness or right handedness. With respect to the epilepsy data set a best classification accuracy of 86.32% was obtained; however, as noted above, the technique was only applied in the 2-D context. In the case of [14] (see above) their oct-tree representation technique was used to represent ventricles. The work also used a frequent sub-graph mining algorithm to identify frequently occurring sub-octrees which were then translated into a vector space representation (in a similar manner to that described in this paper). Experiment were conducted to classify individual brain scans with respect to Alzheimer's disease (a dataset of 166 images was used) and with respect to level of education (dataset of 178 images). A best accuracy of 74.2% was obtained for the former and a best accuracy of 77.2% for the latter. To the best knowledge of the authors, there has been a very little work on the application of image mining techniques in the context of ventricles other than that reported in [14] although the latter was directed at the classification of Alzheimer's disease.

3   The Classification Process

Our proposed classification process is illustrated in Figure 2. With reference to the figure a segmentation process is first applied so as to extract the VOI (the lateral ventricles). Secondly an image decomposition process is used to generate oct-trees (one per ventricle). Next frequently occurring sub-graphs are identified and used to define a feature space from which feature vectors are then generated (one per ventricle) using a feature selection mechanism. Any one of a number of classifier generators can then be applied to the feature vector represented data. Each sub-process (indicated by a rectangular box in Figure 2) is described in more detail on the following sub-sections.

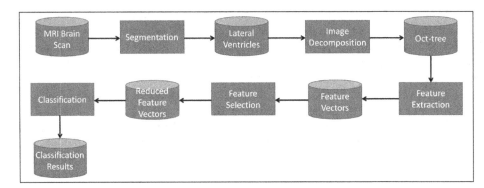

Fig. 2. The Proposed Classification Process

3.1   Segmentation

The identification of VOIs in image data is an important step in image analysis of all kinds. The accuracy with which the nature of a VOI is captured directly affects the effectiveness of any subsequent analysis. The key point of image representation, in the context of data mining, is to remove those elements which will not contribute to the effectiveness of any subsequent analysis, while retaining those that will. In the context of the 3-D MRIs of the human brain used as a focus for the work described in this paper some image preprocessing was first conducted, namely: (i) slice capture and registration, (ii) contrast enhancement.

As indicated in Figure 1, 3-D brain scan MRIs are recorded in three planes: (i) Sagittal (left to right), (ii) Coronal (front to back) and (iii) Transverse (top to bottom). There are a number of software tools which can be used to view and extract slices from 3-D MRI data files. For the work described in this paper the MRICro[1] software system was used. After capturing a collection of image slices, using MRICro, a registration process was applied so that all slices conformed to the same reference framework. Note that the process only requires one set of slices (to evaluate the overall process slices from the Sagittal plane were used).

---

[1] http://www.mccauslandcenter.sc.edu/mricro/

For contrast enhancement a thresholding technique [2] was used. This technique is more effective when an object's colours are obviously different to their background colours. However, the technique can still be used in the case where the object and background colours are not noticeably clear. For example, in Figure 3 it can be seen that the ventricles are represented by the dark areas towards the middle of the image, surrounded by brain tissue which appears as grey (or white) matter. Generally, the contrast between the ventricle and other parts of the brain is easily noticeable, but in some slices (such as SAG slice number 160 shown in Figure 3(c)) it is difficult to identify the boundary of the ventricle because there are grey shades within the ventricle area. In this case, the thresholding technique will enhance the contrast so as to aid the identification of the ventricles. During thresholding, each pixel's brightness is compared to a predefined threshold. If the pixel is considered to be part of the VOI the pixel colour is set to some predefined distinguishing colour, otherwise it will be identified as background and set to an alternative predefined colour. The key success of the thresholding process is the selection of the threshold value. In the work described here the image processing suite of functions available in Matlab[2] was used. Using the Matlab suite the threshold value can be automatically assigned by the software or manually set by a human user. With respect to the work described here the selected threshold value was manually set to 0.30. This was selected after the effect of a range of threshold values ($\{0.28, 0.29, 0.31, 0.32, \ldots\}$) had been manually observed with respect to a reasonable number of different cases by a domain expert. As a result, if a pixel was darker than 0.30 (of interest) it was set to black, otherwise (not of interest) it was set to white. The brain MRI slices shown in Figure 3 are again shown in Figure 4 after the thresholding technique has been applied.

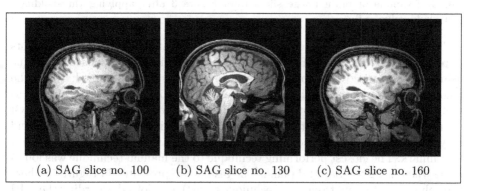

| (a) SAG slice no. 100 | (b) SAG slice no. 130 | (c) SAG slice no. 160 |

Fig. 3. Example of brain image slices in the SAG plane

Once the contrast enhancement was complete the VOIs (the verticals) could be segmented. To this end the Bounding Box technique was developed by the authors. This comprised three steps: (i) define a bounding box that is expected to encompass

---

[2] http://www.mathworks.co.uk/products/matlab/

the ventricles of interest with respect to all relevant slices in the given MRI volume, (ii) for each slice collect the black pixels (voxels) and (iii) apply appropriate noise removal. The required bounding box is rectangular in shape and defined by the coordinates of its corners. To ensure that the bounding box is likely to encompasses all the ventricle voxels of interest it needs to be defined in such a manner that it is considerably larger than the expected ventricle area (so that nothing is missed). All black pixels are collected from each slice that is located within the bounding box. Because the bounding box is defined so that a considerably larger area than the expected ventricle area is covered some black pixels located outside the ventricle area (noise pixels) will also be collected. These are therefore removed in the final step using a simple noise reduction technique whereby the black pixels that are not connected to the largest group of connected pixels are simply removed. In other words the largest group of pixels is assumed to be the ventricles.

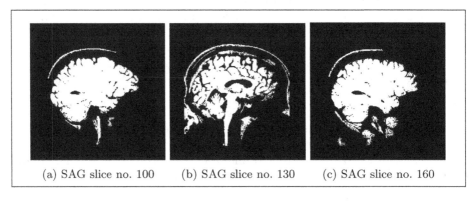

| (a) SAG slice no. 100 | (b) SAG slice no. 130 | (c) SAG slice no. 160 |

Fig. 4. Example of brain image slices from Figure 3 after applying thresholding (threshold value = 0.30)

To evaluate the proposed Bounding Box segmentation technique experiments were conducted using 85 MRI brain scans. By applying the proposed technique using two sampling directions, two sets of volumes were obtained: (i) in the Sagittal plane and (ii) in the Transverse plane. From the experiments it was found that the volumes obtained were close to those obtained manually. Of course the manually estimated volumes do not provide a "gold standard", and may themselves be flawed due to human error. However they did provide a benchmark. The closest performing technique to the manual technique was found to be when using the Sagittal plane. The results are presented in Figure 5, where the volumes estimated from each technique (mm^3) are plotted according the MRI scan identification numbers sorted according to their associated ventricle size.

The difference in volume (mm^3) between the manual and the proposed techniques against the average is plotted in Figure 6. The mean difference (bias estimate) and the 95% range of agreement (calculated as the mean difference between +2SD and -2SD) is represented by the continuous horizontal lines. From the figure it can be seen that the mean difference between the manually estimated volumes and the volumes collected by the Bounding Box technique when used in the Sagittal plane is the smallest (1.10 mm^3) with a standard deviation of 1.53 mm^3. Note that the difference in volume between the manual and the Bounding Box technique increases as the overall volume of the ventricle increases. Although the proposed techniques produced good results there were some limitations. Firstly, the bounding box had to be initially manually defined thus requiring some resource in order to ensure that the bounding box was not too small. Secondly the technique requires application of a noise removal process which had the effect of increasing the overall runtime of the algorithm. On the positive side the idea behind the technique was simple, easy to implement, and effective. Note that the dataset used in this experiments (comprising 85 MRI brain scans) was a subset of the dataset used in the experiments to evaluate the entire classification process. This was because manual identification of VOIs is a time consuming process hence the authors were only able to manually process 85 MRI brain volumes.

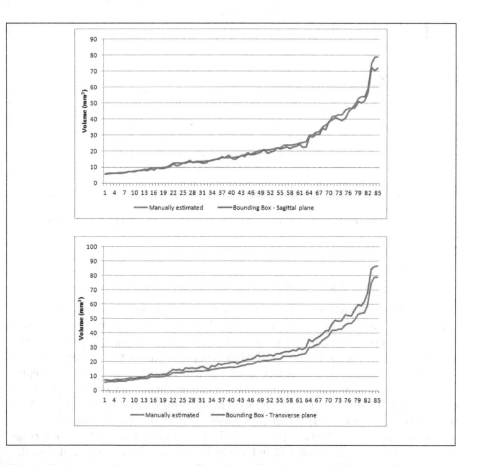

Fig. 5. Comparisons between manually estimated volumes and volumes collected using the proposed automated techniques

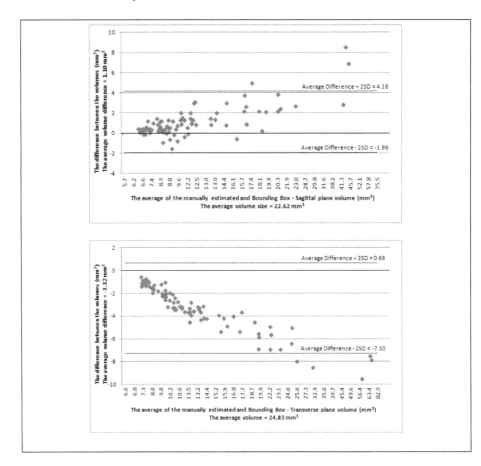

Fig. 6. Levels of agreement in volume estimation between the manual and automated techniques

3.2   Image Decomposition

The objective of image decomposition is to represent an image in some hierarchical format. There are many types of image decomposition, common mechanisms use data structures such as oct-trees, quad-trees and scale space representations [5]. With respect to the work described in this paper, the oct-tree representation was adopted. An oct-tree is a tree data structure which can be used to represent 3-D images that have been recursively subdividing it into eight equal sized octants [12]. Each node in the tree holds image data related to its octant. With respect to the work described in this paper a binary encoding was used. If an octant was part of a ventricle (coloured black) the node was set to 1, otherwise it was set to 0. If an octant did not comprise a homogenous colour it was decomposed further. The process continued until a user defined "maximum depth" was reached. Note that the lateral ventricles consist of two ventricles (left and right), thus two oct-trees were generated with respect to each image which (for convenience) are joined at the root nodes.

3.3   Feature Extraction, Selection, and Classification

In the final stage of the process the oct-tree represented images are processed further to form a feature vector representation to which standard classification techniques can be applied. This is an idea first proposed in [6,7] in the context of 2-D MRI brain scans and subsequently used by a number of other researchers such as [9] in the context of 2-D retina images and [14] with respect to 3-D MRI scans. The first element in this part of the process was thus to apply Frequent sub-graph Mining (FSM) to the oct-tree data. A number of FSM algorithms have been proposed, such as: (i) Gspan [18], (ii) AGM [11], and (iii) FFSM [10]. The Gspan algorithm was used with respect to the work described here. The output of FSM is a set frequently occurring sub-graphs together with their occurrence counts. Typically a large number of sub-graphs are generated many of which are redundant (do not serve to discriminate between classes). Feature selection techniques are typically used to reduce the overall number of identified frequent sub-graphs. With respect to the work presented in this paper the feature selection mechanisms available within the Waikato Environment Knowledge Analysis (WEKA) data mining workbench [8] were used.

4   Experimentation

This section described the experimental set up used to evaluate the proposed process described above (the results obtained are presented in Section 5 below). The experimentation was conducted using three classification methods: (i) Naive Bayes [1], (ii) Support Vector Machine (SVM) [4], and (iii) Decision Trees (C4.5) [16]. All of them are provided within WEKA [8]. With respect to the Gspan algorithm four different minimum support thresholds were used to define frequent sub-graphs ($\{20\%, 30\%, 40\%, 50\%\}$). As a result, four sets of feature vectors were generated and used as inputs for each classifier. Results were produced using Ten-fold Cross Validation (TCV). Recall from the introduction to this paper that the image set used for evaluation purposes comprised 210 MRIs obtained from the Magnetic Resonance and Image Analysis Research Centre at the University of Liverpool. Each scan consisted of 256 two dimensional (2-D) parallel image slices in each plane. The "resolution" of each image slice was 256 x 256 pixels with colour defined using an 8-bit gray scale (thus 256 colours). This was the same data set as used by El Sayed et al. as reported in [7], however El Sayed et al. investigated the potential of the corpus callosum as an indicator of epilepsy (as opposed of the ventricles) and only considered 2-D representations.

5   The Classification Results

The classification results obtained are shown in Table 1. The metrics used to evaluate performance of the classification methods are accuracy (Accu.), sensitivity

(Sens.) and Specificity (Spec.). The T values in the first column are the Gspan minimum support thresholds. From Table 1 it can be seen that the best classification accuracy was from the SVM classifier coupled with $T = 30\%$, while the worst was obtained using Naive Bayes coupled with $T = 50\%$. The relation between classification accuracy and support threshold for each classifier is shown in the graph presented in Figure 7. From the graph it is obvious that the best classifier for this dataset is SVM, and the best support threshold is $T = 30\%$. The classification accuracy with respect to all the classifiers considered tended to decrease as the support threshold increased. It was conjectured that this was because significant sub-graphs were not discovered using Gspan when high support thresholds were adopted.

Table 1. Classification results obtained using Naive Bayes, Support Vector Machine (SVM), and Decision Trees

T(%)	Naive Bayes			SVM			Decision Trees		
	Accu.	Sens.	Spec.	Accu.	Sens.	Spec.	Accu.	Sens.	Spec.
20	64.34	67.25	66.80	68.53	70.40	68.23	67.83	70.58	68.67
30	68.53	70.40	69.23	72.34	75.67	70.45	70.45	74.28	73.20
40	65.13	68.57	70.34	70.45	72.34	69.23	65.87	70.47	65.40
50	61.56	65.40	64.15	62.28	66.96	60.67	62.80	67.05	64.15

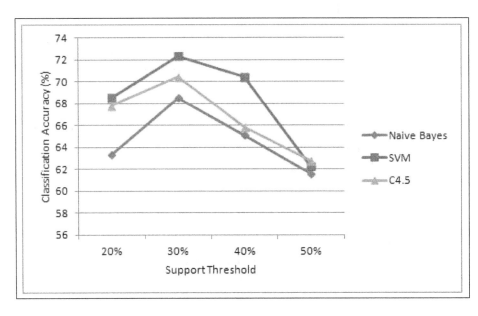

Fig. 7. Relation between classification accuracy and support threshold for each classifier

6   Conclusions

In this paper an approach to 3-D MRI brain scan feature classification has been proposed. The main contributions were the Bounding Box segmentation technique, the oct-tree representation technique and the use of FSM to identify features. In the reported experimental study the Bounding Box technique when applied in the Sagittal plane produced the best segmentation outcomes. In the context of the overall proposed classification process (using the oct-tree representation and FSM) promising outcomes were produced. The results reported in [7] were better than those reported here, however the work in [7] was directed at a 2-D representation of the corpus callosum which may be a better indicator of epilepsy. Likewise, despite using a similar technique, the results reported in [14] were slightly better than those reported in this paper. However, the work in [14] considered not only the lateral but also the "third" ventricle. Moreover, the datasets in [14] (Alzheimer's disease and level of education) were different to those used in this paper. The lateral and third ventricles may be better indicators of Alzheimer's disease and level of education than epilepsy.

The next stage of our research will focus on alternative methods of representing 3-D MRI brain scan features so that machine learning techniques can be applied. The intention is also to consider the use of dynamic thresholding techniques, as used in [14], to determine whether this helps improve the effectiveness of the segmentation process. Regarding the classification process some weighting technique (such as that proposed in [13]) could be applied to the FSM process in order to improve its operation.

References

1. Bramer, M.: Principles of Data Mining. Springer (2007)
2. Burger, W., Burge, M.J.: Digital Image Processing: An algorithmic Introduction Using Java. Springer (2008)
3. Cootes, T., Taylor, C., Cooper, D., Graham, J.: Active shape models-their training and application. Computer Vision and Image Understanding 61(1), 38–59 (1995)
4. Cortes, C., Vapnik, V.: Support-vector Networks. Machine Learning 20(3), 273–297 (1995)
5. Da, L., Costa, F., Cesar Jr., R.M.: Shape Analysis and Classification: Theory and Practice. CRC Press (2001)
6. Elsayed, A., Coenen, F., Jiang, C., García-Fiñana, M., Sluming, V.: Corpus Callosum MR Image Classification. In: Proceedings AI 2009, pp. 333–348. Springer (2009)
7. Elsayed, A., Coenen, F., Jiang, C., García-Fiñana, M., Sluming, V.: Corpus Callosum MR Image Classification. Knowledge-Based Systems 23(4), 330–336 (2010)
8. Hall, M., Frank, E., Holmes, G.: The WEKA Data Mining Software: An Update. ACM SIGKDD Explorations Newsletter 11(1), 10–18 (2009)
9. Ahmad Hijazi, M.H., Jiang, C., Coenen, F., Zheng, Y.: Image Classification for Age-related Macular Degeneration Screening Using Hierarchical Image Decompositions and Graph Mining. In: Gunopulos, D., Hofmann, T., Malerba, D., Vazirgiannis, M. (eds.) ECML PKDD 2011, Part II. LNCS, vol. 6912, pp. 65–80. Springer, Heidelberg (2011)

10. Huan, J., Wang, W., Prins, J.: Efficient Mining of Frequent Subgraphs in the Presence of Isomorphism. In: Proceedings of The Third IEEE International Conference on Data Mining, pp. 549–552. IEEE Comput. Soc. (2003)

11. Inokuchi, A., Washio, T., Motoda, H.: An Apriori-Based Algorithm for Mining Frequent Substructures from Graph Data. In: Zighed, D.A., Komorowski, J., Żytkow, J.M. (eds.) PKDD 2000. LNCS (LNAI), vol. 1910, pp. 13–23. Springer, Heidelberg (2000)

12. Jackins, C.L., Tanimoto, S.L.: Oct-trees and Their Use in Representing Three-dimensional Objects. Computer Graphics and Image Processing 14(3), 249–270 (1980)

13. Jiang, C., Coenen, F.: Graph-based Image Classification by Weighting Scheme. In: Applications and Innovations in Intelligent System XVI, pp. 63–76 (2009)

14. Long, S., Holder, L.B.: Graph-based Shape Shape Analysis for MRI Classification. International Journal of Knowledge Discovery in Bioinformatics 2(2), 19–33 (2011)

15. Osher, S., Sethian, J.A.: Fronts Propagating with Curvature-Dependent Speed: Algorithms Based on Hamilton-Jacobi Formulations. Journal of Computational Physics 79(1), 12–49 (1988)

16. Quinlan, J.R.: C4.5: Programs for Machine Learning. Morgan Kaufmann Publishers (1993)

17. Rousson, M., Paragios, N., Deriche, R.: Implicit Active Shape Models for 3D Segmentation in MR Imaging. In: Barillot, C., Haynor, D.R., Hellier, P. (eds.) MICCAI 2004. LNCS, vol. 3216, pp. 209–216. Springer, Heidelberg (2004)

18. Yan, X.: gSpan: Graph-based Substructure Pattern Mining. In: Proceeding of The IEEE International Conference on Data Mining, pp. 721–724. IEEE Comput. Soc. (2002)

Biometric Template Protection Based on Biometric Certificate and Fuzzy Fingerprint Vault

Weihong Wang, Youbing Lu, and Zhaolin Fang

College of Computer Science, Zhejiang University of Technology, Hangzhou, China
{wwh,fzl}@zjut.edu.cn, pez1420@163.com

Abstract. Biometric Certificate (BC) is a kind of data structure that binds user identity and biometric template, which is able to be applied to access control and identity authentication for various applications like electronic transactions in network environment. A critical issue in biometric system is that may be suffered from biometric template attack, such as "cross-matching attack", "hill climbing attacks" etc. Hence, it is extremely important to provide high security and privacy for biometric template in BC. This paper implemented a biometric template protection scheme in BC using fuzzy fingerprint vault and fingerprint-based pseudo random number generator (FBPRNG) technique. First, the fingerprint keys are derived from fingerprint template through FBPRG and fingerprint template is encrypted by fingerprint keys to keep it secret. Second, the fingerprint keys are hidden using the fingerprint-based fuzzy vault scheme, and then storing encrypted fingerprint template and fuzzy vault into BC. Finally, this scheme is implemented on open source CA software called EJBCA. The result of experiment shows that the scheme can not only generate the self-certified fingerprint keys, but also effectively secure fingerprint template.

Keywords: fingerprint template protection, biometric certificate, fuzzy vault, EJBCA, PKI.

1 Introduction

Biometric cryptography technology is being widely used in the place like governments, businesses and organizations of access control, authentication on insecure public network. Biometric system is essentially a technical system that to recognize a person by using his/her any uniquely biometric features information. In general, biometric associates a permanent relationship with that person and can never be changed. Several important problems in biometric system should be taken into account. One of the most critical issues is that biometric template may be suffered from a variety of attack, e.g. "cross-matching attack" [1] (the biometric system which uses the same biometric feature), "hill climbing attacks" [2], "the correlation attack" [3]. Since biometric information (fingerprint, iris, voice, ect.) are unique to individuals, if biometric template was leaked, it would not able to be revoked. Therefore, it is a critically confronted issue for biometric systems that how to effectively secure the biometric template.

H. Motoda et al. (Eds.): ADMA 2013, Part I, LNAI 8346, pp. 241–252, 2013.
© Springer-Verlag Berlin Heidelberg 2013

A biometric authentication system which typically operates in two stages, namely, registration and verification procedure. In registration procedure, fingerprint feature template, which can be divided into centralized storage and distributed storage in terms of the storage location. For current biometric applications like based-fingerprint information, fingerprint template generally stored in the database or smart card [4-7]. In terms of centralized storage, it is of great convenience to collect fingerprint feature template of user and store it in a centralized database, and the advantage is that system maintenance is relatively simple. The drawback is that the efficiency in fingerprint template matching and system security is difficult to be guaranteed, e.g. vulnerable to Trojan attacks on backend server. Distributed storage mainly stored the template in an intelligent card (smart card). The smart card also stores extra information such as digital certificates and its private keys. In verification procedure, system is capable of comparing on-site collection of biometric templates with user's feature template in registration procedure. If matching has been successful, then user is granted privileges to access some restricted service, otherwise it means user identification failed.

This paper on the basis of the concept of distributed storage bind fingerprint feature template, digital certificate and fingerprint keys that is derived from fingerprint-based pseudo random generator (FBPRG) because key generated by general random number generator are not able to guarantee key self-certified, all of which are stored in a data structure called biometric certificate for ensuring that only the biometric data of user can generate fingerprint key. Biometric Certificate is a kind of data structure that binds user identity and biometric template, which allows users to authenticate electronic transactions or verify a person's identity in network environment by using biometric data and whose format is also referenced and based on the X.509 and ANSI X.9 standards [9-11].

This work implemented a biometric template protection scheme in BC based on fuzzy fingerprint vault and FBPRNG. First, the fingerprint key is derived from fingerprint feature template through FBPRG and fingerprint feature template is encrypted by fingerprint key to keep it secret. Second, the fingerprint key is hidden using the fuzzy vault scheme, and then storing encrypted fingerprint template and fuzzy fingerprint vault into BC. At last, we implement the fingerprint protection scheme on an open source CA software called EJBCA.

2 Biometric Template Protection Based on BC and Fuzzy Fingerprint Vault

2.1 Biometric Certificate

PKI (public key infrastructure) is a promising technique to satisfy security demand in the electric payment and identity authentication. In cryptography, a PKI is an arrangement combines public keys with user identities by means of certificate authority (CA). The key aim of CA is to manage and distribute X.509 digital certificate, which contains mainly certificate version, a serial number, expiration dates, public-key, extensions and the digital signature of CA.

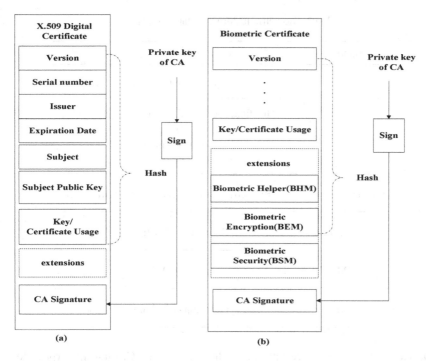

Fig. 1. The structure of general X.509 (a) and biometric certificate (b)

EJBCA is an open source PKI certificate authority built on EJB technology. It is a robust, high performance, platform independent, flexible, and component based CA to be used standalone or integrated in other J2EE applications [12]. EJBCA is divided into several functional components, which are distributed with different host machines according to different layers where they are located in [14]. These layers include client layer, WEB tier, business tier and data tier. Business tie contains two major components RA (registered authority) and CA, in which RA provides an interface between user and CA that obtain and authenticate the user's identity in order to request a certificate from CA. In this paper, RA component is redeveloped to issue user's biometric certificate. Second, RA collects fingerprint template by using acquisition equipment and then add biometric feature and its protection information into the extension attribute of X.509 digital certificate when user need to request a certificate. The above information demand to be digitally signed by the EJBCA that generated data structure called a biometric certificate. Fig. 1 shows the structure of general X.509 certificate (a) and biometric certificate (b).

Biometric feature template extension in BC consists of three modules, which are biometric helper module (BHM), biometric encryption module (BEM), biometric security module (BSM) respectively. BHM includes mainly the version of biometric feature, security option, and the type of biometric feature, creation date, expiration date and other helper information, which are generated by EJBCA according to the real application environment. BEM store the biometric feature template data that is encrypted by biometric keys (like fingerprint key is derived from FBPRG). BSM are used for binding the biometric keys and encoded biometric template to form a fuzzy vault that can hide biometric keys.

Table 1. Biometric feature template protection

Field	Notes
Version	Current version of biometric certificate
Security option	This field specifies the security applied to the BHM. 0:BHM is plaintext 1:BHM is encrypted
Biometric Type	Biometric Type (fingerprint, face, iris, etc.)
Validation Date	Biometric validation date
BSM	Fuzzy Vault
BEM	Encrypted biometric feature template

In previous section, we have introduced the concept of BC. First, fingerprint feature information is extracted from user to generate fingerprint feature template by using acquisition equipment, and then encode fingerprint feature template. Second, the fingerprint key is generated from fingerprint template through FBPRG, and encoded fingerprint template is encrypted by fingerprint key to generate BEM module of BC. Fingerprint key are organically bound with encoded fingerprint template, which can form a fuzzy vault called BSM modules in BC. Fig. 2 shows the block diagram of the proposed fingerprint template protection scheme.

2.2   Fingerprint-Based Pseudo Random Number Generator

Random number generator is a method or device generating seems no correlation sequence using some algorithms, physical signals and environmental noise. Almost all pseudo-random numbers on computer are not a true random number but simply repeating the cycle of the relatively large number of columns, it is generated according to a certain algorithm and seed value.

Fingerprint template in BC need to be encrypted, by which generating random and high security key is an interesting and critical issue. General Pseudo random number generator [15] is not able to reveal the source of key; however, the biometric key derived from human biometric template not only represents a random value, but also indicates that the key where it comes from and implements it's self-certified.

The minutiae point of fingerprint template can be translated, rotated, scaled, and then generating fixed-length keys through a certain hash algorithms.

1) Translation

Point P is any a minutiae point in the set of fingerprint template. P' is a direction vector of translation. T is a point after a translation. Translation is an operating that translates every minutiae point a constant distance in a specified direction. Let $P = (x, y)$, $P' = (x', y')$, $T = (\Delta x, \Delta y)$

$$P' = (x', y') = P + T = (x + \Delta x, y + \Delta y) \tag{1}$$

2) Rotation

Let point P is a minutiae point after a translation. R is a matrix of rotation.

Let $P' = P \times R$, $R = \begin{bmatrix} \cos\theta & -\sin\theta \\ \sin\theta & \cos\theta \end{bmatrix}$,

A rotation is a circular movement of a minutiae point $P(x, y)$ around a center point $O(O_x, O_y)$ of rotation by an angle θ. It will get

$$x' = (x - O_x)\cos\theta - (y - O_y)\sin\theta + O_x \tag{2}$$

$$y' = (y - O_y)\sin\theta + (x - O_x)\cos\theta + O_y \tag{3}$$

Let $O(O_x, O_y) = O(0,0)$, $P' = (x', y')$, then

$$x' = x\cos\theta - y\sin\theta \tag{4}$$

$$y' = x\sin\theta + y\cos\theta \tag{5}$$

3) Scaling

Scaling is a liner transformation that enlarges or diminishes minutiae point $P(x, y)$ that has been translated and rotated.

Z is a scaling matrix. Let $Z = \begin{bmatrix} z_x & 0 \\ 0 & z_y \end{bmatrix}$, $P' = Z \times P$

$$x' = z_x * x \tag{6}$$

$$y' = z_y * y \tag{7}$$

At last, stitching all minutiae points after being translated, rotated and scaled to generate a fingerprint-based pseudo random number, FPBRNG encodes it into a fixed-length fingerprint keys by using hash algorithm (like MD5, SHA .e.g.). Hash algorithm always produces the same output for given the same input. Therefore, although fingerprint feature template has been transformed, the topology of Fingerprint feature minutiae remains the same and key is able to be self-certified. Another advantage of FPBRNG is that a fingerprint feature template can generate different fingerprint keys, once the key is compromised, it can cancel the key to re-generate a new fingerprint digital key.

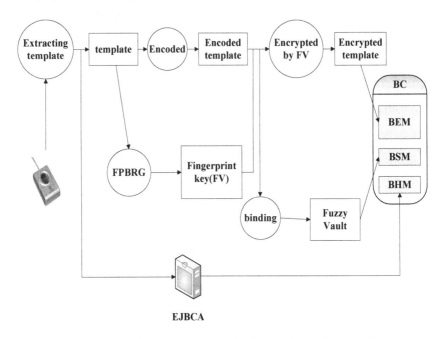

Fig. 2. Fingerprint template protection based on biometric certificate and fuzzy vault

2.3   Fuzzy Vault Coding

In this section, we present fuzzy fingerprint vault scheme, which is a biometric cryptosystem that secures both the secret key and the fingerprint template represented as an unordered set by binding them within a cryptographic framework. A player *Alice* may place a secret key k in a vault and "lock" it using an unordered set A. If *Bob* plans to "unlock" the vault using an unordered set B of the similar length, he may get k if A is overlapped with B substantially [14]. The security of fuzzy vault is relied mainly on infeasibility of the polynomial reconstruction problem. Considering the precision and operation speed of the computer, we use the finite field $GF(p)$ for constructing the vault and p must be a prime number and offer large universe to ensure vault security.

1) The template fingerprint image I is acquired by fingerprint scanner, and we will extract fingerprint minutiae points from I after processing. The template minutiae set $(x_i, y_i, \theta_i), i \in \{1, 2, ... n\}$, in which n is the number of minutiae in I. Then, fingerprint digital key K used to encrypt biometric template is derived from fingerprint-based pseudo random number generator.

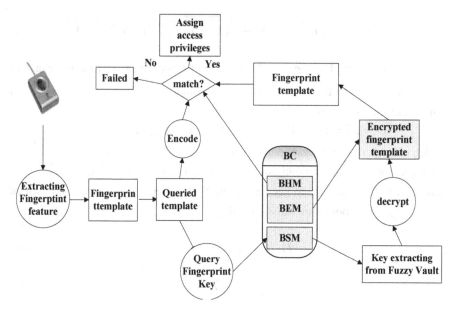

Fig. 3. The process of retrieving fingerprint template in BC

2) Cyclic Redundancy Check (CRC) is a method of the simple parity bit checking. It is commonly used in communication network to check for errors in raw data. In our scheme, hence, it needs to use CRC technique on secret key K to generate a 16-bit error correcting codes (e) and Stitching K with e to form a new key K'. It [22] pointed out that determining a person's identity that needs at least 12 fingerprint minutiae, so we choose the greatest power of polynomial value as 11. This is similar with the fact that shamir key distribution (n, k) threshold schemes proposed in literature [6]. The constructed polynomial is given by:

$$P(x) = r_0 + r_1 x + r_2 x^2 + r_3 x^3 + r_4 x^4 + \cdots + r_n x^{11} = \sum_{i=0}^{11} r_i x^i \qquad (8)$$

We will take 128 bits key as an example. So, K' can be divided into 9 segments $\kappa'_0 \kappa'_1, \kappa'_2 ... \kappa'_8$ in the field GF (P). κ'_0 is a CRC error-correcting code and each κ'_i takes 2-byte that corresponds to a specific coefficient r_i in polynomial. r_9 r_{10} and r_{11} need to be generated randomly.

3) Each point in the set of minutiaes (x_i, y_i) , $i \in \{1,2,....n\}$ should be compressed to the range between [0,255]. In order to keep an appropriate distance between genuine points with chaff points, we choose k minutiae according to the following rules. $D(a,b)$ denote the euclidean distance between a and b . δ is a preset threshold value.

$$D((x_i, y_i), (x_j, y_j)) = \sqrt{(x_i - x_j)^2 + (y_i, y_j)^2} \geq \delta \qquad (9)$$

We will concatenate horizontal and vertical coordinates (x, y) of minutiae, namely $g_i = x_i \mid y_i$. g_i will be projected on the polynomials. The final set of projection of genuine points are given by

$$G = \{(g_1, P(g_1)), (g_2, P(g_2)), ..., (g_k, P(g_k))\}$$

which consists of a significant part of fuzzy vault.

4) Then, we generate randomly a distributed uniform set $(c_i, d_i), i \in \{1, 2, ...m\}$ called chaff points that need to keep a certain distance with genuine points, in which $c_i = c_{ix} \mid c_{iy}$. Chaff points are satisfied with given rules.

$$D(g_i, c_i) \geq T_1 \qquad (10)$$

$$D(c_i, c_j) \geq T_1 \qquad (11)$$

These chaff set need to satisfy the requirement for not lie on polynomial ($P(c_i) \neq d_i, i \in \{1, 2, ...m\}$) and number of genuine points need to be far more than the actual number of chaff points. The order of genuine points $\{(g_i, P(g_i))\}, i \in \{1...k\}$. Union of genuine and chaff set are composed of fuzzy fingerprint vault to be stored BSM module in biometric certificate.

At last, Biometric certificate can be sent to verifier side when user needs to achieve some privilege for various applications in open network.

2.4   Vault Decoding

When the server (verifier) receives client's biometric certificate, fingerprint template in BC need to be retrieved for verifying client's identity. The procedure of retrieving the fingerprint template composes of as follows. Fig 3 depicts the block diagram of retrieving fingerprint template in BC.

1) First, client tries to acheive query fingerprint feature information to generate fingerprint feature template by using fingerprint acquisition equipment. Fingerprint template is transformed to generate query template by using the same encoding method with encryption. Next, we can calculate the distance between each point (q_x, q_y) in query template and mixed points (genuine points and chaff points) $v_i = (v_{ix}, v_{iy})$ in fuzzy vault of biometric certificate

$$D((v_{ix}, v_{iy}), (q_x, q_y)) \leq T_2 \qquad (12)$$

If D less than preset threshold value T_2, we can put it into candidates set $C = \{(c_i, f_i)\}$. If the number of points in C is less than 12, then key recovery failed. We will select randomly a subset $C' = \{(c_i', f_i')\}$ consisting of 12 points from set C to reconstruct coefficients of polynomial. In this paper, we use a simply recursive way to quickly solve the coefficients of the interpolation polynomial:

$$P'(x) = \frac{(x - c_2')(x - c_3')\ldots(x - c_n')}{(c_1' - c_2')(c_1' - c_3')\ldots(c_1' - c_n')} f_1' + \cdots$$
$$+ \frac{(x - c_1')(x - c_2')\ldots(x - c_n')}{(c_n' - c_1')(c_n' - c_2')\ldots(c_n' - c_{n-1}')} f_n' \tag{13}$$

The result polynomial is given by (14):

$$P'(x) = r_0' + r_1'x + r_2'x^2 + \cdots + r_n'x^{11} \tag{14}$$

2) Each coefficient r_i' of polynomial $P'(x)$ is a 16-bit binary string, stitched together to form a 128-bit secret key K''. If the result of computing K'' of CRC-16 check codes is r_0', then it denotes that decoding vault is succeeded. We stitched r_1'、$r_2' \ldots r_8'$ to form secret key K'. K' is used for decrypting the biometric template in BC that compare with encoded Query template. If the match is successful, user is assigned appropriate access privileges of biometric system, otherwise access is denied.

3   Experiment and Result Analysis

X.509 digital certificate is described and encoded based on ASN (abstract syntax notation). The extension attribute in version 2 or 3 of X.509 certificate has following structure:

Extension := SEQUENCE
{
    ExtnID   OBJECT IDENTIFIER,
    Critical   BOOLEAN DEFAULT FALSE,
    ExtnValue   OCTET STRING
}

ExtnID denotes an extension OID (Object Identifier)
Critical denotes that extension is critical or not
extnValue denotes the value of extension attribute

Subject Directory Attributes extension (subject directory attribute) field in certificate consists mainly certificate owners basic properties, like birthday, address and other information, whose OID is "2.5.29.9". This paper carried out in this extension, defining the OID as "1.3.6.1.5.5.7.9.6" and adding biometric template field information. Biometric certificate issued by EJBCA is showed in Fig 4.

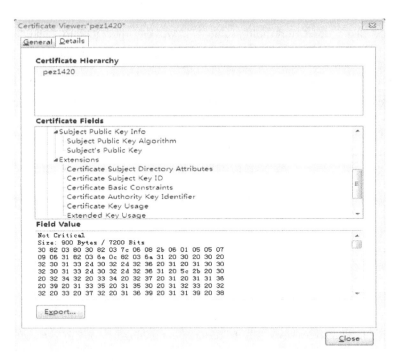

Fig. 4. Biometric certificate issued by EJBCA

In order to evaluate the performance of proposed scheme, 50 different user's fingerprint database are captured with   U.are.U 4000 micro fingerprint sensors in different environment. Each user's fingers were randomly selected 4 fingers, and repeatedly collected 4 images per finger (Total 800 images and approximately of size 328×364). For simplicity, the minutiae coordinates (x, y) are linearly mapped to 8-bit range $(\lfloor x/1.4 \rfloor, \lfloor y/1.4 \rfloor)$ for horizontal and vertical dimensions. The number of minutiae points $k = 20$, chaff points $m = 200$ respectively. The fingerprint key derived from FBPGNG, the length of which is 128-bit and use AES encryption algorithm. Generator polynomial of CRC is denoted as $G(x) = x^{16} + x^5 + x + 1$. Prime number P in finite field $GF(P)$ is defined $P = 2^{16} + 1 = 65537$.

We choose images with the most minutiae as a template for 4 images per finger to generate a fuzzy vault (BSM). False Accepted Rate (FAR) and False Rejection Rate (FRR) are using evaluating the performance of proposed method in different matched minutiae points N .The result is listed in Table 2.

The false rejection was generated mainly as follows: one is that fingerprint image collected by sensors, whose quality is so poor that is not able to extract enough genuine minutiaes; another is that the number of matched minutiae points in query is less than fingerprint template. As listed in Table 2, it can see that with the increasing of different matched minutiae points, it is obvious that FRR is increasing at the same time.

Table 2. Performance evaluation

Different matched minutiae points	7	9	10	12
FRR	7.5%	10.2%	14.72%	16.83%
FAR	0.2%	0%	0%	0%

4   Conclusion

In this paper, we implement a fingerprint template protection scheme based on biometric certificate (BC) and fuzzy vault. Biometric Certificate is a kind of data structure that binds user's identity information and biometric feature template, in which fingerprint keys used to encrypt fingerprint template is derived from fingerprint-based pseudo random number generator (FBPRNB). BC combines the advantages of PKI and biometric authentication, which enhances the association between user and fingerprint keys. The smaller capacity of three modules BHM, BSM and BEM in BC, the more efficient transmission in network environment. Research work in the next step will be to study how to effectively compress the capacity of biometric template. Meanwhile, BC can be applied to e-commerce platform to identify the user's identity information because BC can ensure the unique with user's identity.

Acknowledgment. This research was supported by the National Natural Science Foundation of China (60873033), the Natural Science Foundation of Zhejiang Province (R1090569 and LY12F02039) and the State Key Laboratory of Software Development Environment Open Fund (SKLSDE-2012KF-05).

References

1. Jain, A.K., Nandakumar, K., Nagar, A.: Template Security. EURASIP Journal on Advances in Signal Processing 2008, 17 (2008)
2. Jain, A.K., Ross, A., Pankanti, S.: Biometrics: A Tool for Information Security. IEEE Transactions on Information Forensics and Security 1(2), 125–143 (2006)
3. Scheirer, W., Boult, T.: Cracking fuzzy vaults and biometric encryption. In: Biometric Symposium, Baltimore, MD, USA, pp. 1–6 (September 2007)
4. Clancy, T., Kiyavash, N., Lin, D.: Secure smartcard-based fingerprint authentication. In: Proc. ACM SIGMM Workshop on Biometrics Methods and Applications, California, pp. 45–52 (2003)
5. Sanchez-Reillo, R., Mengibar-Pozo, L.: Microprocessor Smart Cards with Fingerprint User Authentication. IEEE AESS Systems Magazine 18(3), 22–24 (2003)
6. Sun, H.W., Lam, K.Y., Gu, M., Sun, J.G.: Improved fingerprint-based remote user authentication. In: Communications, Circuits and Systems, ICCCAS 2007, pp. 472–475 (2007)
7. Jun, E.-A., Kim, J.G., Jung, S.W., Lee, D.H.: Extended Fingerprint-based User Authentication Scheme Using Smart Cards in Education IPTV. In: Information Science and Applications, ICISA, pp. 1–8 (2011)

8. Moon, D., Chae, S.H., Kim, J.N.: A Secure fingerprint template generation algorithm for smart card. In: 2011 IEEE International Conference on Consumer Electronics (ICCE), pp. 719–720 (2011)
9. Chung, Y., Moon, K.: Biometric Certificate based Biometric Digital Key Generation with Protection Mechanism. In: Frontiers in the Convergence of Bioscience and Information Technologies, pp. 709–714 (2007)
10. Jo, J.-G., Seo, J.-W., Lee, H.-W.: Biometric Digital Signature Key Generation and Cryptography Communication Based on Fingerprint. In: Preparata, F.P., Fang, Q. (eds.) FAW 2007. LNCS, vol. 4613, pp. 38–49. Springer, Heidelberg (2007)
11. Li, C., Xing, Y., Niu, X.X., Yang, Y.X.: Identity Authentication Scheme Based on Biometric Certificate. Computer Engineering 33(20), 159–161 (2007)
12. Ejbca-design (2013), http://sourceforge.net/projects/ejbca/
13. Ejbca (2013), http://www.ejbca.com
14. Zhang, L.Y., Liu, Q.H., Liu, M.: Research and application of EJBCA based on J2EE. In: Wang, W., Li, Y., Duan, Z., Yan, L., Li, H., Yang, X. (eds.) Integration and Innovation Orient to E-Society. IFIP, vol. 251, pp. 337–345. Springer, Boston (2007)
15. Xu, D.C., Li, B.L.: A Pseudo-random Sequence Fingerprint Key Algorithm Based on Fuzzy Vault. In: Proceedings of the 2009 IEEE International Conference on Mechatronics and Automation, China, Changchun, pp. 2421–2425 (2009)
16. Shamir, A.: How to share a secret. Communications of the ACM 22(11), 612–613 (1979)
17. Pankanti, S., Prabhakar, S., Jain, A.K.: On the individuality of fingerprint. IEEE Transactions on Pattern Analysis and Machine Intelligence 24(8), 1010–1025 (2002)
18. Jules, A., Sudan, M.: A Fuzzy vault scheme. In: Proc. IEEE Int. Symp. on Information Theory, Lausanne, Switzerland, pp. 408–408 (2002)
19. Uludag, U., Pankanti, S., Jain, A.K.: Fuzzy vault for fingerprints. In: Kanade, T., Jain, A., Ratha, N.K. (eds.) AVBPA 2005. LNCS, vol. 3546, pp. 310–319. Springer, Heidelberg (2005)
20. Uludag, U., Jain, A.: Securing fingerprint template: Fuzzy Vault with Helper Data. In: Proceedings of CVPR Workshop on Private Research in Vision, USA, pp. 163–169 (2006)
21. Tan, T.Z., Zhang, H.Y.: Improved Fuzzy Vault fingerprint encryption scheme. Application Research of Computers 29(6), 2208–2210 (2012)
22. Moon, K.Y., Moon, D., Yoo, J.H., Cho, H.S.: Biometrics Information Protection using Fuzzy Vault Scheme. In: 2012 Eighth International Conference on Signal Image Technology and Internet Based Systems, pp. 124–128 (2012)

A Comparative Study of Three Image Representations for Population Estimation Mining Using Remote Sensing Imagery

Kwankamon Dittakan[1], Frans Coenen[1], Rob Christley[2], and Maya Wardeh[2]

[1] Department of Computer Science,
University of Liverpool, Liverpool, L69 3BX, United Kingdom
[2] Institute of Infection and Global Health,
University of Liverpool, Leahurst Campus,
Chester High Road, CH64 7TE Neston, Cheshire, United Kingdom
{dittakan,coenen,robc,maya.wardeh}@liverpool.ac.uk

Abstract. Census information regarding populations is important with respect to many governmental activities such as commercial, health, communication, social and infrastructure planning and development. However, the traditional census collection methods (ground surveys) are resource intensive in terms of both cost and time. The resources required for census collection activities are particularly high in rural areas which feature poor communication and transport networks. In this paper the interpretation of high-resolution satellite imagery is proposed as a low cost (but less accurate) alternative to obtaining census data. The fundamental idea is to build a classifier that can label households according to "Family size" which can then be used to generate a census estimation. The challenge is how best to translate the raw satellite data into a form that captures key information in such a way that an appropriate classifier can still be built. Three different representations are considered: (i) Colour Histogram, (ii) Local Binary Pattern and (iii) Graph-based. The representations were evaluated by generating census information using test data collected from a rural area to the northwest of Addis Ababa in Ethiopia.

Keywords: Satellite Image Analysis and Mining, Data Mining Applications, Population Estimation Mining.

1 Introduction

Census data collection is an important governmental activity with respect to a wide range of applications such as the provision of public services (education, health care, transportation and water and electricity supply). The traditional approaches to census collection are typically undertaken using either questionnaires or door-to-door surveys. In the case of questionnaires each householder within a given region receives a questionnaire either by post or electronically, members of the household complete the questionnaire and return it (again either by post or electronically). In the case of door-to-door surveys field staff visits every household so as to "interview" the household. Once the data has been collected it needs to be processed, another resource intensive activity.

H. Motoda et al. (Eds.): ADMA 2013, Part I, LNAI 8346, pp. 253–264, 2013.

Whatever the case, traditional census collection methods are expensive in terms of both time and cost. The resource required is exacerbated with respect to rural area where communication and transport networks tend to not be as sophisticated as in urban areas. In this paper an alternative approach to census collection is proposed whereby population counts are obtained remotely using a classifier applied to satellite imagery. The approach has particular application in the context of sparsely populated rural areas. The fundamental idea is to train a classifier, using labelled satellite imagery of households, which can then be applied to obtain large area population estimates (census data). The proposed approach offers a number of advantages: (i) the resource required, in terms of time and cost, is significantly less than in the case of the traditional methods; (ii) it is efficient (traditionally census data is typically collected following a five or ten year cycle, using the proposed approach census data can be collected as required); (iii) it is non-intrusive in the sense that populations do not know it is happening (populations tend to be suspicious of census collection activities); and (iv) the collected census data is immediately in a format that allows for further processing. Although our proposed approach offers many advantages the drawback is that it is not as accurate as the traditional ("on ground") census collection methods. However, as will be demonstrated later in this paper the error margin is not significant.

The main challenge for satellite image classification (and it can be argued for image mining in general) is how best to represent satellite image data so that machine learning techniques can be applied. The amount of data available in a satellite image collection is typically too large for it to be used in its entirety thus a representation is required that serves to capture the salient information but in a reduced form. At the same time the representation needs to be compatible with the classification process which is usually founded on a feature vector representation. Three satellite image representations are therefore considered in this paper: (i) Colour Histogram, (ii) Local Binary Pattern (LBP) and (iii) Graph-based.

The rest of this paper is organised as follows. In Section 2 some related work is briefly presented. Section 3 then provides a description of the proposed census mining framework. A brief overview of the proposed image segmentation process is presented in Section 4. In Section 5 details of each of the proposed image representations are presented. The performance of the proposed census mining framework, using test data collected from a rural area to the northwest of Addis Ababa in Ethiopia, is then considered in Section 6. Finally Section 7 provides a summary and some conclusions.

2   Previous Work

Population estimation has been the subject of significant research work. Population estimation using satellite images offers the benefit over alternative population estimation methods in that the information is: (i) up-to-date, (ii) (at least to an extent) reliable and (iii) obtained at low cost [1]. Satellite imagery has been used with respect to population estimation in many ways. For example the relationship between light frequency (from night-time satellite imagery) and the distribution of populations has been investigated in [18]. Correlation analyse was used in [2] to analyse the relationship between light intensity from night-light satellite images and population density. Correlation analysis was also applied in [14] to study the connection between remote sensing variables (such as spectral signatures and vegetation indices) and population density with respect to block regions in the city of Indianapolis in The USA. In [10] satellite mage texture features were extracted to provide for a population density estimation model. In [19] Multi-Spectral satellite images data was used to obtain population estimates with respect to the city of Belo Horizonte in Brazil. However, the concept of applying machine learning to satellite imagery for population estimation, to the best knowledge of the authors, has not been previously considered.

A necessary precursor for satellite image classification, and image classification in general, is that the key properties or characteristics of the image set are first extracted. This is typically achieved using some feature extraction method. The idea is to transform the input image into a set of features to produce a feature vector representation. Image features can be categorise into two groups: (i) general feature and (ii) domain-specific feature. General features are application independent and included features such as colour, texture and structure. Domain-specific features are application dependent, for example elements of the human face as used in face recognition. The work presented in this paper is founded on three basic general features: (i) Colour, (ii) Texture and (iii) Structure. Colour features are typically used in the context of colour image representation methods [8], for example using colour histogram. Colour histograms are obtained by discretising the image colours and counting the number of pixels in each of the *quantisation bins* [22]. Colour histograms offer advantages of: (i) rotation and translation invariance, (ii) computational simplicity and (iii) low storage requirements [11]. Texture features give information about the spatial arrangement of colours and/or intensities in a given image. There are many texture extraction methods that may be adopted such as the use of Gray Level Co-occurence Metrics (GLCMs), texture spectrums and wavelet transforms [9,24]. The texture extraction method of interest with respect to the work described in this paper is the use of Local Binary Patterns (LBPs), as first proposed in [16], where each pixel is define using the relative grayscale of its neighbourhood pixels [15]. Texture features tend to be robust with respect to image rotation, illumination change and occlusion [21]. Structure features are commonly used to describe the "geometry" of the relative position of the elements containing within a given image. Image structure is often represented using a graph representation [20,7]. Examples of graph-based structural image feature representations include Attributed Graphs (AGs), Function Describe Graphs (FDGs) and Quadtrees. The advantages of graph-based representations are their general applicability [3] and their invariance to rotation and translation [12].

3   Census Mining Framework

An overview of the proposed process for population estimation mining is presented in this section. A schematic of the framework is shown in Figure 1. The framework comprised three phases (as represented by the rectangular boxes): (i) image segmentation, (ii) image representation and (ii) classification. The input to the first phase is a large scale satellite image of a given area covering (say) a number of villages. This image is first coarsely segmented and then finely segmented in order to identify individual households. This segmentation process was described in detail in [5]; however, for

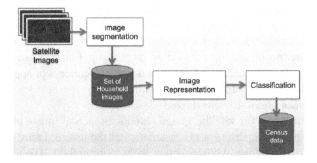

Fig. 1. Proposed population estimation mining from satellite imagery framework

completeness, a brief overview of the process is presented in Section 4 below. During image representation the identified household pixel data is translated into a representation that allows for the application of a classifier. Three alternative representations are considered and compared in this paper, these are described in Section 5. After the households have been segmented and appropriately represented the classification phase may be commenced. This is relatively straight forward given that established techniques can be adopted, a back propagation neural network learning method was uitiised with respect to the evaluation presented in Section 6.

4   Image Segmentation

An overview of the adopted satellite image segmentation process is presented in this section (further detail can be found in [5]). The aim of the segmentation is to identify individual households from within satellite imagery. Figure 2 illustrates the image segmentation process. This process consists of three individual steps: (i) coarse segmentation, (ii) image enhancement and (iii) fine segmentation. During the first step (the left rectangle in Figure 2) the input satellite images of a given area are coarsely divided into a series of sub-images each covering a small group of households (between 1 and 4). Each sub-image is then enhanced (the middle rectangle in Figure 2) so as to facilitate the fine segmentation. During the fine segmentation process (the right rectangle in Figure 2) the enhanced coarse segmented sub-images are further segmented so as to

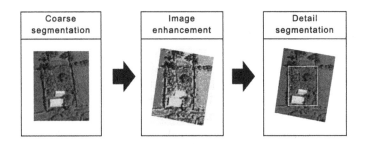

Fig. 2. Schematic illustration the segmentation process (Phase 1 of the overall process)

identify individual households. The end result is a set of individual household images. (Note that the image segmentation process was implement using the MATLAB, matrix laboratory, workbench[1].)

5 Image Representation

This section considers the three different household image representations which form the central theme of this paper: (i) Colour Histogram, (ii) Local Binary Pattern (LBP) and (iii) Graph-Based. Each is discussed in the flowing three sub-sections, Sub-Sections 5.1 to 5.3 respectively.

5.1 Colour Histogram

As noted in Section 2 colour histograms offer advantages of computational simplicity and tolerance against small changes such as rotation, translation and scaling. Colour histograms are wildly used in the context of image analysis and understanding, and hence were adopted for consideration with respect to the work described in this paper. More specifically the colour histogram concept was used to represent individual households in terms of colour distribution. The X-axis of the histogram comprised a number of "bins" such that each bin represented a "colour range". The Y-axis of the histogram then represented the number of pixels falling into each bin. Seven different type of histogram were considered: (i) red channel histogram, (ii) green channel, (iii) blue channel, (iv) hue channel, (v) saturation channel, (vi) value channel and (vii) grayscale. Each histogram was comprised 32 bins, giving a total of 224 ($7x32$) features. In addition 13 colour statistical distribution metrics were extracted, these are listed in Table 1. Thus a feature vector of length 237 ($224 + 13 = 237$) could be generated for each household image.

Table 1. Additional colour based statistical features

#	RGB color channel descripton	#	HSV color channel description	#	Grayscale channel description
1	Average red	6	Average hue	11	Mean of grayscale
2	Average green	7	Average saturation	12	Standard deviation of
3	Average blue	8	Average value		grayscale
4	Mean of RGB	9	Mean of HSV	13	Average of grayscale
5	Standard deviation of RGB	10	Standard deviation of HSV		histogram

5.2 Local Binary Pattern

As already noted in Section 2, LBPs have also been wildly adopted since they are easy to calculate, robust to illumination changes and are an effective texture description mechanism. Consequently the use of LBPs was also considered with respect to the work described in this paper. In the context of the individual household images each image was first transform into grayscale. A 3×3 pixel window was then considered. For each

---

[1] http://www.mathworks.com

window the grayscale value of the centre pixel, the threshold value, was compared with surrounding eight neighbour values. A 1 was reccorded if the grayscale value was greater than the threshold and a 0 otherwise. The eight binary digits recorded in this way then formed an eight bit number, the LBP for the window. Using LBPs in this way $2^8 = 256$ different patterns can be formed. If the number of sampling points (neighbours) is P and the pixel distance from the centre point is R. The notation $LBP(P,R)$ is used to describe variations of the LBP approach. Three different variations were considered: (i) 8 sampling points with a radius of 1 ($LBP(8,1)$), (ii) 8 sampling points with a radius of 2 ($LBP(8,2)$) and (iii) 8 sampling points with a radius of 3 ($LBP(8,3)$). Again additional statistical features were used to augment the LBP representation. Three categories of texture statistic were identified: (i) entropy features (E), (ii) grey-level occurrence matrix features (M) and (iii) wavelet transform features (W). Table 2 lists the statistical features generated (the letter in parenthesis in each case indicates the category of the feature). Thus the LBP based representation could result in feature vectors of length 266 ($2^R + 10 = 2^8 + 10 = 256 + 10 = 266$).

Table 2. Additional texture based statistical features

#	Description	#	Description	#	Description
1	Entropy (E)	7	Average approximation	10	Average diagonal
2	Average Local Entropy (E)		coefficient matrix, cA (W)		coefficient matrix, cD (W)
3	Contrast (M)	8	Average horizontal		
4	Correlation (M)		coefficient matrix, cH (W)		
5	Energy (M)	9	Average vertical		
6	Homogeneity (M))		coefficient matrix, cV (W)		

5.3 Graph-Based Structure

Graph based techniques allow for the structural representation of image objects. In the context of the census estimation application a quad-tree representation was adopted on the grounds that this has been frequently used with respect to many other image mining applications. The desired graph-based representation, for each household image, was generated using a four step process: (i) quadtree decomposition, (ii) tree construction, (iii) frequent subgraph mining and (iv) feature vector transformation. Each household image was first "cropped" to a size of 128x128 pixel (best appropriate size) surrounding the main building of the household by detecting the roof of the building (the largest contiguous homogenous region within the image). Then the cropped household images were recursively subdividing into four quadrate until a minimum 8×8 pixel decomposition was arrived at. During the second step the decomposition was cast into a quad tree data structure. Each node was labelled with one of 32 ranged potential mean grayscale values for the associated quadrant. A set of identifiers, $\{1,2,3,4\}$, representing the north west (NW), north east (NE), south west (SW) and south east (SE) quadrants of a decomposition respectively were used as edge labels.

Although graph-based representations provide for a powerful and flexible structural image model they do not readily lend themselves to input with resect to classification algorithms [17]. However, we can use frequent sub-graph mining techniques to identify frequently occurring sub-graphs across the collection of household image quad trees and consider these frequent sub-graphs to be features within a standard feature space representation. There were a number of different frequent subgraph mining approaches than could have been adopted. From the literature the most frequently referenced frequent sub-graph mining algorithm is the gSpan algorithm [23]. This was therefore adopted with respect to the work described in this paper. Gspan uses the concept of a support threshold σ to differentiate a frequently occurring sub-graph from an infrequently occurring sub-graph. If the occurrence count (the *support*) for a subgraph is greater than σ then the subgraph is identified as a frequently occurring subgraph.

6 Evaluation

In this section, the evaluation and comparison of the proposed population estimation mining approaches is presented. Extensive evaluation has been conducted so as to compare the operation of the three different representations and their variations, however this paper only reports the most significant results obtained (there is insufficient space to allow for the presentation of all the results obtained). For evaluation purposes, so as to compare the operation of the three different representations and their variations, a back propagation neural network learning method (multilayer perceptron) was used for the classification coupled with Ten-fold Cross Validation (TCV). In each case the effectiveness of the classification was recorded in terms of: (i) accuracy (AC), (ii) area under the ROC curve (AUC), (iii) precision (PR), (iv) sensitivity (SN) and (v) specificity (SP). Note also that χ-squared feature selection was applied to reduce the overall size of the feature space in each case so that the top k features could be selected (the most appropriate value for k was found to be 35). The remainder of this section is organised as follows. Sub-section 6.1 gives an overview of the test data used. Sub-section 6.2 then itemises the results obtained.

6.1 Test Data

To act as a focus for the research a case study was considered directed at a rural area within the Ethiopian hinterland, more specifically two data sets were collected with respect to two villages (Site A and Site B) located within the Horro district in the Oramia Region of Ethiopia (approximately 300 km north-west of Addis Abba) as shown in Figure 3[2]. Site A was bounded by the parallels of latitude 9.312650N and 9.36313N, and the meridians of longitude 37.123850E and 37.63914E and Site B was by the parallels of latitude 9.405530N and 9.450000N, and the meridians of longitude 36.590480E and 37.113550E. Using the know bounding latitudes and longitudes of our two test sites appropriate satellite imagery was extracted from Google Earth. The images were originally obtained using the GeoEye satellite with a 50 centimetre ground resolution. The satellite images for Site A were obtained during the rainy season (lot of green colouring) and were released by Google Earth on 22 August 2009 and those for Site B were obtained during the dry season (lot of brown colouring) and released on 11 February 2012. On-ground household data (including family size and household latitude and longitude) was collected by University of Liverpool field staff in May 2011 and July 2012. The minimum and maximum family size were 2 and 12 respectively, the mean was 6.31,

---

[2] https://maps.google.co.uk/

the medium were 6 and standard deviation was 2.56. These two data sets then provided the training and test data required to evaluate our proposed census collection system. The benchmark household family size was separated into three classes: (i) Small family size (less than or equal to 5 people), (ii) Medium family size (between 6 and 8 people in the family) and (iii) Large family size (more than 8 people). Some statistics concerning the class distributions for the Site A and B data sets are presented in Table 3.

Fig. 3. The test site location: Horro district in Ethiopia

For the colour histogram representation technique, three data sets were produced: colour histogram (CH), colour statistics (CS) and combine (CH+CS). For the LBP representation seven data sets were produced: LBP(8,1), LBP(8,2), LBP(8,3), texture statistics (TS), combine (LBP(8,1) + TS), combine (LBP(8,2) + TS) and combine (LBP(8,3) + TS). For the graph-based representation a range of σ values, $\{10, 20, 30, 40, 50\}$, were used. The number of sub-graphs generated in each case is presented in Table 4. From the table it can be seen that, as would be expected, the number of identified subgraphs decreases as the value for σ increases (and vice-versa).

Table 3. Class label distribution for data set Site A and data set Site B

Sites	Small family	Medium family	Large family	Total
Site A	28	32	10	70
Site B	19	21	10	50
Total	47	53	20	120

Table 4. Number of identified sub-graphs produced using a range of σ values with respect to the Site A and B data

Support threshold	Site A	Site B
$\sigma = 10$	757	420
$\sigma = 20$	149	119
$\sigma = 30$	49	60
$\sigma = 40$	24	39
$\sigma = 50$	12	19

6.2   Results

The obtained results are presentation in Table 5. Based on these results the following observations can be made:

- With respect to the colour histogram based representation best results were obtained using CH (AUC = 0.754) for Site A and CH + CS (AUC = 0.765) for Site B.

- With respect to the LBP representation best results tended to be produced using LBP(8,1) for both Site A (AUC = 0.88) and Site B (AUC = 0.792).
- With respect to the graph-based structure representation, $\sigma = 10$ produced the best results with respect to both Sites A and B (AUC = 0.730 and 0.744 respectively).

Table 5. Comparison of different variations of the proposed histogram, LPB and Graph-based representations in terms of classification performance

Types	Techniques	Site A					Site B				
		AC	AUC	PR	SN	SP	AC	AUC	PR	SN	SP
Histogram	CH	0.614	0.754	0.615	0.614	0.761	0.560	0.753	0.568	0.560	0.723
	CS	0.400	0.550	0.404	0.400	0.629	0.480	0.645	0.466	0.480	0.693
	CH+CS	0.557	0.741	0.551	0.557	0.713	0.580	0.755	0.602	0.580	0.728
LBP	LBP(8,1)	0.718	0.880	0.758	0.771	0.856	0.600	0.792	0.599	0.600	0.764
	LBP(8,2)	0.614	0.765	0.618	0.614	0.725	0.600	0.705	0.600	0.600	0.762
	LBP(8,3)	0.586	0.718	0.583	0.586	0.710	0.660	0.672	0.679	0.660	0.774
	TS	0.414	0.502	0.379	0.414	0.695	0.500	0.650	0.602	0.500	0.754
	LBP(8,1)+TS	0.757	0.848	0.754	0.757	0.847	0.600	0.768	0.601	0.600	0.757
	LBP(8,2)+TS	0.643	0.770	0.645	0.643	0.746	0.580	0.736	0.580	0.580	0.742
	LBP(8,3)+TS	0.557	0.706	0.553	0.557	0.686	0.620	0.638	0.639	0.620	0.748
Graph-based	$\sigma = 10$	0.586	0.730	0.582	0.586	0.739	0.580	0.744	0.584	0.580	0.756
	$\sigma = 20$	0.500	0.588	0.499	0.500	0.669	0.480	0.606	0.483	0.480	0.698
	$\sigma = 30$	0.400	0.495	0.391	0.400	0.607	0.480	0.648	0.479	0.480	0.726
	$\sigma = 40$	0.357	0.397	0.344	0.357	0.542	0.320	0.521	0.344	0.320	0.656
	$\sigma = 50$	0.314	0.506	0.330	0.314	0.588	0.280	0.459	0.309	0.280	0.643

To conduct a further performance comparison, with respect to the recorded AUC values, the Friedman's test was applied [6]. The AUCs for all 15 classification techniques (Colour Histogram (3), LBP (7) and Graph-based (5)) for both the Site A and the Site B data are listed in Table 6 (columns two and three); the number in parentheses in each case indicates the overall "ranking" of each individual result with respect to the two data sets. The average rank (AR) is given in the fourth column, this is the mean value of the rankings for each classification technique. The Friedman test statistic is based on the AR values, and is calculated using equation 1, where N is the number of data sets (2), K is the total number of classification technique considered (15) and r_i^j is the rank of classification technique j on data set i.

$$\chi_F^2 = \frac{12N}{K(K+1)} \left[\sum_{j=1}^{K} AR_j^2 - \frac{K(K+1)^2}{4} \right] \qquad (1)$$

The Friedman test statistic and corresponding p value are presented in the first row of Table 6 . The Friedman test statistic (25.70) and the significance threshold ($p < 0.005$) indicate that the null hypothesis, that there is no difference between the techniques, can be rejected. The highest recorded AUC value for each data set is indicated in bold font in Table 6 indicating that the LBP(8,1) representation produced the overall best performance (AR = 1.0), while the graph-based representation with $\sigma = 40$ produced the worst overall result (AR = 14.5).

Table 6. Area Under the Curve (AUC) results

Friedman test statistic = 25.70 ($p < 0.005$)			
Techniques	Site A	Site B	AR
CH	0.754 (5)	0.753 (4)	4.5
CS	0.550 (11)	0.645 (11)	11
CH+CS	0.741(6)	0.755 (3)	4.5
LBP(8,1)	**0.880** (1)	**0.792** (1)	1
LBP(8,2)	0.765 (4)	0.705 (7)	5.5
LBP(8,3)	0.718 (8)	0.672 (8)	8
TS	0.502 (13)	0.650 (9)	11
LBP(8,1)+TS	0.848 (2)	0.768 (2)	2
LBP(8,2)+TS	0.770 (3)	0.736 (6)	4.5
LBP(8,3)+TS	0.706 (9)	0.638 (12)	10.5
$\sigma = 10$	0.730 (7)	0.744 (5)	6
$\sigma = 20$	0.588 (10)	0.606 (13)	11.5
$\sigma = 30$	0.495 (14)	0.648 (10)	12
$\sigma = 40$	0.397 (15)	0.521 (14)	14.5
$\sigma = 50$	0.506 (12)	0.459 (15)	13.5

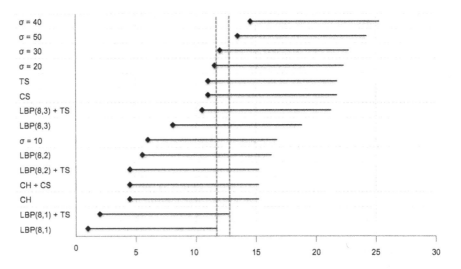

Fig. 4. AR comparison between the proposed image classification approach using Nemenyi's post hoc test with $\alpha = 0.05$

A post hoc Nemanyi test [4,13] was applied to each class distribution to evaluate the performance the approaches. The performance between individual approaches are significantly different if their ARs are different by more than a Critical Difference (CD) value that can be calculated using equation 2 where the critical difference level α ($\alpha = 0.05$) and the value $q_{\alpha,\infty,K}$ is based on the Studentised range statistic.

$$CD = q_{\alpha,\infty,K} \sqrt{\frac{K(K+1)}{12N}} \qquad (2)$$

The significant diagram in Figure 4 illustrates the AUC performance rank of the proposed image classification technique with Nemenyi's critical difference (CD) tail (the calculated CD for the diagram is 10.72). The diagram shows the classification techniques listed in ascending order of ranked performance on the y-axis, and the image classification techniques's average rank across all two data sets displayed on the x-axis. The LBP(8,1) approach produced the best results with $AR = 1$, which was significantly better than graph-based representation ($\sigma = 30, 50, 40$) with $AR = 12$, 13.5 and 14.5, respectively.

7   Conclusion

A framework for population estimation mining using satellite imagery has ben proposed. The framework can operate with a number of different representations, three different categories of representation were considered: (i) Colour Histogram, (ii) Local Binary Pattern (LBP) and (iii) Graph-based. These were used to encode individual segmented household images so that a classifier generation process could be applied. Experiments were conducted using the test data collected from two villages located in a rural part of the Ethiopian hinterland. The main findings with respect to the evaluation (using a neural network learning method and a χ-squared top k feature selection algorithm with $k = 35$) were: (i) with respect to the Colour Histogram based representations the best result was obtained using CH for the Site A (rainy season) data and CH + CS for the Site B (dry season) data, (ii) with respect to the LBP representation the best result was produced using LBP(8,1) for both Site A and Site B, and (iii) with respect to the Graph-based structure representation, $\sigma = 10$ produced the best results with respect to both the Site A and Site B data. However, the LBP(8,1) representation produced the best overall result. For future work the research team intend to conduct a large scale population estimation exercise using the proposed framework.

References

1. Alsalman, A.S., Ali, A.E.: Population Estimation From High Resolution Satellite Imagery: A Case Study From Khartoum. Emirates Journal for Engineering Research 16(1), 63–69 (2011)
2. Cheng, L., Zhou, Y., Wang, L., Wang, S., Du, C.: An estimate of the city population in china using dmsp night-time satellite imagery. In: IEEE International Geoscience and Remote Sensing Symposium, IGARSS 2007, pp. 691–694 (2007)
3. Cook, D.J., Holder, L.B.: Substructure discovery using minimum description length and background knowledge. Journal of Artificial Intelligence Research 1(1), 231–255 (1994)
4. Demšar, J.: Statistical comparisons of classifiers over multiple data sets. J. Mach. Learn. Res. 7, 1–30 (2006)
5. Dittakan, K., Coenen, F., Christley, R.: Towards the collection of census data from satellite imagery using data mining: A study with respect to the ethiopian hinterland. In: Bramer, M., Petridis, M. (eds.) Proc. Research and Development in Intelligent Systems XXIX, pp. 405–418. Springer, London (2012)
6. Friedman, M.: A Comparison of Alternative Tests of Significance for the Problem of m Rankings. The Annals of Mathematical Statistics 11(1), 86–92 (1940)

7. Ganea, E., Brezovan, M.: Graph object oriented database for semantic image retrieval. In: Catania, B., Ivanović, M., Thalheim, B. (eds.) ADBIS 2010. LNCS, vol. 6295, pp. 563–566. Springer, Heidelberg (2010)

8. Han, J., Ma, K.: Fuzzy color histogram and its use in color image retrieval. IEEE Transactions on Image Processing 11(8), 944–952 (2002)

9. Hung, C.C., Pham, M., Arasteh, S., Kuo, B.C., Coleman, T.: Image texture classification using texture spectrum and local binary pattern. In: IEEE International Conference on Geoscience and Remote Sensing Symposium, IGARSS 2006, July 31-August 4, pp. 2750–2753 (2006)

10. Javed, Y., Khan, M.M., Chanussot, J.: Population density estimation using textons. In: 2012 IEEE International Geoscience and Remote Sensing Symposium, IGARSS, pp. 2206–2209 (2012)

11. Jeong, S., Won, C.S., Gray, R.M.: Image retrieval using color histograms generated by gauss mixture vector quantization. Computer Vision and Image Understanding 94(3), 44–66 (2004)

12. Jouili, S., Tabbone, S.: Towards performance evaluation of graph-based representation. In: Jiang, X., Ferrer, M., Torsello, A. (eds.) GbRPR 2011. LNCS, vol. 6658, pp. 72–81. Springer, Heidelberg (2011)

13. Lessmann, S., Baesens, B., Mues, C., Pietsch, S.: Benchmarking classification models for software defect prediction: A proposed framework and novel findings. IEEE Trans. Software Engineering 34(4), 485–496 (2008)

14. Li, G., Weng, Q.: Using Landsat ETM+ imagery to measure population density in Indianapolis, Indiana, USA. Photogrammetric Engineering and Remote Sensing 71(8), 63–69 (2005)

15. Liang, P., Li, S.F., Qin, J.W.: Mulit-resolution local binary patterns for image classification. In: 2010 20th International Conference on Wevelet Analysis and Pattern Recognition, pp. 164–169 (2010)

16. Ojala, T., Pietikainen, M., Maenpaa, T.: Multiresolution gray-scale and rotation invariant texture classification with local binary patterns. IEEE Transactions on Pattern Analysis and Machine Intelligence 24(7), 971–987 (2002)

17. Ozdemir, B., Aksoy, S.: Image classification using subgraph histogram representation. In: 2010 20th International Conference on Pattern Recognition (ICPR), pp. 1112–1115 (2010)

18. Pozzi, F., Small, C., Yetman, G.: Modeling the distribution of hunman population with nighttime satellite imagery and gridded population of the world. In: Conference on Pecora 15/Land Satellite Information IV/ISPRS Comission I/FIEOS (2002)

19. Reis, I.A., Silva, V.L., Reis, E.A.: Adjusting population estimates using satellite imagery and regression models. In: Anais XV Simpsio Brasileiro de Sensoriamento Remoto, SBSR, pp. 830–837 (April 5, 2011)

20. Sanfeliu, A., Alquézarb, R., Andrade, J., Climent, J., Serratosa, F., Vergésa, J.: Graph-based representations and techniques for image processing and image analysis. Pattern Recognition 35(3), 639–650 (2002)

21. Song, C., Li, P., Yang, F.: Multivariate texture measured by local binary pattern for multispectral image classification. In: Proc. IEEE International Conference on Geoscience and Remote Sensing Symposium, IGARSS 2006, pp. 2145–2148 (2006)

22. Swain, M.J., Ballard, D.H.: Color indexing. Int. J. Comput. Vision 7(1), 11–32 (1991)

23. Yan, X., Han, J.: gspan: graph-based substructure pattern mining. In: Proceedings of the 2002 IEEE International Conference on Data Mining, ICDM 2003, pp. 721–724 (2002)

24. Zhang, Y., He, R., Jian, M.: Comparison of two methods for texture image classification. In: Second International Workshop on Computer Science and Engineering, WCSE 2009, pp. 65–68 (2009)

Mixed-Norm Regression for Visual Classification

Xiaofeng Zhu[1], Jilian Zhang[1], and Shichao Zhang[1,2,*]

[1] College of CS & IT, Guangxi Normal University, Guilin, 541004, China
[2] The Centre for QCIS, Faculty of Engineering and Information Technology,
University of Technology Sydney, Australia
{zhuxf,zhangjl,zhangsc}@mailbox.gxnu.edu.cn

Abstract. This paper addresses the problem of multi-class image classification by proposing a novel multi-view multi-sparsity kernel reconstruction (MMKR for short) model. Given images (including test images and training images) representing with multiple visual features, the MMKR first maps them into a high-dimensional space, e.g., a reproducing kernel Hilbert space (RKHS), where test images are then linearly reconstructed by some representative training images, rather than all of them. Furthermore a classification rule is proposed to classify test images. Experimental results on real datasets show the effectiveness of the proposed MMKR while comparing to state-of-the-art algorithms.

Keywords: image classification, multi-view classification, sparse coding, Structure sparsity, Reproducing kernel Hilbert space.

1 Introduction

In image classification, an image is often represented by its visual feature, such as HSV (Hue, Saturation, Value) color histogram, LBP (Local Binary Pattern), SIFT (Scale invariant feature transform), CENTRIST (CENsus TRansform hISTgram), and so on. Usually, different representations describe different characteristics of images. For example, CENTRIST [19] is a suitable representation for place and scene recognition.

Recent studies (e.g., [19]) have shown that although an optimal representation (such as SIFT) is better for some given tasks, it might no longer be optimal for the others. Moreover, a single visual feature is not always robust to all types of scenarios. Give an example illustrated in Figure 2, we can easily classify the two figures (e.g., Figure 2.(a) and 2.(b)) into the category IRIS according to the extracted local feature. However, we maybe not easily make the same decision while giving their global feature, such as HSV. Actually, in this case, we may category two figures (e.g., Figure 2.(a) and 2.(c)) into IRIS. According to our observation, we cannot category Figure 2.(c) into IRIS since the captions in Figure 2.(c) makes the classification difficult.

In contrast, literatures (e.g., [23,22]) have shown that representing image data with multiple features really reflects the specific information of image data. Moreover, this case is complementary each other and helpful for disambiguation. For example, the local feature HSV is less robust to the changes in frame rate, video length, captions.

---

* Corresponding author.

H. Motoda et al. (Eds.): ADMA 2013, Part I, LNAI 8346, pp. 265–276, 2013.

(a)   (b)   (c)

Fig. 1. A illustration on image IRIS with different representations

SIFT is sensitive to changes in contrast, brightness, scale, rotation, camera viewpoint, and so on [7,24].

Aforementioned observation motivates us to combine several visual features (rather than a single type of visual feature) to perform image visual classification for discriminating each class best from all other classes. In the machine learning domain, learning with multiple representations is well known as multi-view learning (MVL) or multi-modality learning [1].

Using multi-view learning brings clear advantages over traditional single-view learning: First, multi-view learning is more effective than generating a single model via considering all attributes at once, especially when the weaknesses of one view complement the strengths of the others [4]. In many application areas, such as bioinformatics and video summarization, literatures have shown that multimedia classification can achieve greatly benefit from multi-view learning [14,15]. Second, different information about the same example in multi-view learning can help solve other issues, such as transfer learning and semi-supervised learning [19]. Therefore, multi-view learning is becoming popular in real applications [6,10,20], such as web analysis, object recognition, image classification, and so on.

However, previous studies on multi-view learning contain at least two following drawbacks. First, multi-view learning employ all the views for each data point without considering the individual characteristics of each data point. For example, sometimes a data point can be described well with several representations and can not be added any other. In this case, we really expect to select the best suitable views according to its characteristics. Second, in real application, image datasets are often corrupted by noise, but existing multi-view learning approaches have difficulty for dealing with noisy observations [3]. Therefore, we expect to remove the noise or redundancy from the training data for selecting appropriated views for each image.

In this paper we focus on the issue of multi-class image classification by combining multiple visual features. That is, given a test image representing with multiple views (or visual features), the learnt classifier should be designed to decide which class the test image belongs to. Specifically, in the proposed multi-view multi-sparsity kernel reconstruction (MMKR) model, multi-view learning is employed to perform image classification, aiming at minimizing reconstruction error across all the views. The MMKR performs kernel reconstruction in a RKHS, in which each test image is linearly reconstructed by training images coming from a few object categories, for capturing nonlinear relationship between each test image and training images. Moreover, a new multi-sparsity regularizer, which concatenates an ℓ_1-norm with a Frobenius norm (F-norm for short), is embedded into the proposed objective function of the MMKR to achieve following advantages, such as selecting training images from a few object categories to reconstruct the test image via the F-norm regularizer, and removing noise in visual features via the ℓ_1-norm regularizer. Meanwhile, we devise a novel iterative algorithm to efficiently solve the derived objective function of the MMKR, and then theoretically prove that such a objective function converges to its global optimum via the proposed algorithm. Finally, experimental results on challenging real datasets show the effectiveness of the proposed MMKR by comparing to state-of-the-art algorithms.

The remainder of the paper is organized as below: Preliminary is described in Section 2, followed by the proposed MMKR approach in Section 3 and its optimization in Section 4. The experimental results are reported and analyzed in Section 5 while Section 6 concludes the paper.

2   Preliminary

For clarity, we summarize the notations used in this paper in Table 1.

Table 1. Notations used in this paper

Notation	Description
feature space	uppercase italic
scalar	lowercase italic
column vector	lowercase bold
matrix	uppercase bold
transpose	superscript T
inverse of matrix	superscript -1

Besides, the ℓ_p-norm of a vector $\mathbf{v} \in \mathbf{R}^n$ is defined as $\|\mathbf{v}\|_p = \left(\sum_{i=1}^{n} |v_i|^p \right)^{\frac{1}{p}}$, where v_i is the ith element of $\mathbf{v}$. If $p = 0$ then we get the "pseudo norm" (a.k.a., ℓ_0-norm) which is defined as the number of non-zero elements in $\mathbf{v}$. The $\ell_{r,p}$-norm over a matrix $\mathbf{M} \in \mathbf{R}^{n \times m}$ is defined as $\|\mathbf{M}\|_{r,p} = \left(\sum_{i=1}^{n} \left(\sum_{j=1}^{m} \|m_{ij}\|^r \right)^{\frac{p}{r}} \right)^{\frac{1}{p}}$, where m_{ij} is the element of the ith row and j^{th} column.

3   Approach

Given a set of training images $\mathbf{X}$, each image is represented with V visual features (or views) and described by one of C object categories appeared in $\mathbf{X}$. We denote $\mathbf{x}_c^v$ ($\mathbf{x}_c^v \in$

$R^{m_v}, c = 1, ..., C, v = 1, ..., V$) as the v-th view of an image in c-th object category. $\mathbf{X}_c^v$ ($\mathbf{X}_c^v \in R^{m_v \times n_c}$) is a set of training images associated with the c-th object category and represented by the v-th view. We also denote $\sum\limits_{v=1}^{V} m_v = M$, and $\sum\limits_{c=1}^{C} n_c = N$, where N is training size and m_v is the dimensionality of the v-th view.

3.1 Objective Function

Sparse learning distinguishes important elements from unimportant ones by assigning the codes of unimportant elements as zero and the important ones as non-zero. This enables that sparse learning reduces the impact of noises and increase the efficiency of learning models [11]. Thus it has been embedded into various learning models, such as sparse principal component analysis (sparse PCA [26]), sparse non-negative matrix factorization (sparse NMF [9]), and sparse support vector machine (sparse SVM [17]), in many real applications [2,7], including signal classification, face recognition and image analysis [18]. In this paper, we cast multi-view image classification as multi-view sparse learning in the RKHS.

Given the v-th visual feature of a test image $\mathbf{y}^v$ ($\mathbf{y}^v \in R^{m_v}$), we first search for a linear relationship between $\mathbf{y}^v$ and the v-th visual feature of training images. For this, we consider to build a reconstruction process as: $f(\mathbf{y}^v) = \sum\limits_{c=1}^{C} \mathbf{y}_c^v \mathbf{w}_c^v$, where $\mathbf{w}_c^v \in R^{n_c}$ is the v-th view reconstruction coefficient.

To perform the reconstruction process in multi-view learning, we expect to minimize reconstruction error across all the views. To avoid the issue of over-fitting as well as to obtain sparse effect, we propose a regularizer leading to multiple sparsity into the framework of sparse learning, i.e., the proposed multi-sparsity regularizer includes an ℓ_1-norm and an F-norm for achieving the element sparsity and the block sparsity respectively. The objective function is defined as:

$$\min_{\mathbf{w}_c^1, ..., \mathbf{w}_c^V} \sum_{v=1}^{V} \left\| \mathbf{y}^v - \sum_{c=1}^{C} \mathbf{X}_c^v \mathbf{w}_c^v \right\|_2^2 \tag{1}$$

$$+\lambda_1 \sum_{v=1}^{V} \sum_{c=1}^{C} |w_c^v| + \lambda_2 \sum_{c=1}^{C} \sqrt{\sum_{v=1}^{V} (w_c^v)^2}$$

where λ_1 and λ_2 are trade-off parameters. The first term in Eq.1 is to minimize the reconstruction error through all views. The last two terms are introduced to avoid the issue of over-fitting and to pursue multi-sparsity.

For convenience, we denote: $\tilde{\mathbf{x}}_c^v = [0, ..., 0, (\mathbf{x}_c^v)^T, 0, ..., 0]^T$, $\tilde{\mathbf{y}}^v = [0, ..., 0, (\mathbf{y}^v)^T, 0, ..., 0]^T$, $\tilde{\mathbf{w}}_c^v = [0, ..., 0, (\mathbf{w}_c^v)^T, 0, ..., 0]^T$, and $\tilde{\mathbf{W}} = [(\tilde{\mathbf{w}}_1)^T, ..., (\tilde{\mathbf{w}}_C)^T]^T \in R^{M \times V}$, where $\tilde{\mathbf{w}}_c \in R^{n_c \times V}$, where both $\tilde{\mathbf{x}}_c^v (\in R^M)$ and $\tilde{\mathbf{y}}^v (\in R^M)$ are a one-dimensional column vector with the $(\sum\limits_{i=1}^{v-1} m_i + 1)$-th to the $\sum\limits_{i=1}^{v} m_i$-th elements being nonzero. Therefore, Eq.1 can be converted as:

$$\min_{\tilde{\mathbf{W}}} \| \tilde{\mathbf{Y}} - \tilde{\mathbf{X}} \tilde{\mathbf{W}} \|_F^2 + \lambda_1 \| \tilde{\mathbf{W}} \|_1 + \lambda_2 \sum_{c=1}^{C} \| \tilde{\mathbf{W}}_c \|_F \tag{2}$$

where $\|.\|_F$ denotes F norm, $\tilde{\mathbf{X}}_c \in R^{M \times n_c}$, $\tilde{\mathbf{Y}} \in R^{M \times V}$ and $\tilde{\mathbf{Y}} \in R^{n_c \times V}$.

However, Eq.2 is developed for image classification in original space. Motivated by the fact that kernel trick can capture nonlinear similarity, which has been demonstrated to reduce feature quantization error and boost learning performance, we use a nonlinear function ϕ^v in each view v to map training images and test images from original space to a high-dimensional space, e.g., the RKHS, via defining $k(x_i, x_j)^v = \phi(x_i^v)^T \phi(x_j^v)$ for some given kernel functions k^v, where $v = 1, ..., V$. That is, given a feature mapping function $\phi : R^M \rightarrow R^K$, $(M < K)$, both training images and test images in feature space R^M are mapped into a RKHS R^K via ϕ, i.e., $\tilde{\mathbf{X}} = [\tilde{\mathbf{x}}_1, ..., \tilde{\mathbf{x}}_M] \rightarrow \phi(\tilde{\mathbf{X}}) = [\phi(\tilde{\mathbf{x}}_1), ..., \phi(\tilde{\mathbf{x}}_M)]$. By denoting $\mathbf{A} = \phi(\tilde{\mathbf{X}})^T \phi(\tilde{\mathbf{Y}})$ and $\mathbf{B} = \phi(\tilde{\mathbf{X}})^T \phi(\tilde{\mathbf{X}})$, we convert the objective function defined in the original space (see eq.2) to the objective function of the proposed MMKR as:

$$\min_{\tilde{\mathbf{W}}} \|\mathbf{A} - \mathbf{B}\tilde{\mathbf{W}}\|_F^2 + \lambda_1 \|\tilde{\mathbf{W}}\|_1 + \lambda_2 \sum_{c=1}^{C} \|\tilde{\mathbf{W}}_c\|_F \tag{3}$$

where $A \in R^{K \times V}$ and $B \in R^{K \times N}$.

According to the literatures, e.g.,[8], the λ_1-norm regularizer generates the element sparsity, whose sparsity is in single element of $\tilde{\mathbf{W}}$, and benefits for removing noise by assigning its codes as sparse, i.e., 0. The F-norm regularizer generates the block sparsity, whose sparsity is through the whole block, i.e., zero through the whole object category in this paper. Thus the F-norm regularizer enables the object categories with the block sparsity (i.e., sparsity in each code through the whole objective category) not to be involved into the reconstruction process. By inducing the multi-sparsity regularizer, only a few training images from representative object categories are used to reconstruct each test image. Meanwhile, removing noise is also considered.

3.2 Classification Rule

By solving the objective function in Eq.3, we obtain the optimal $\tilde{\mathbf{W}}$. According to the literature in [21], for each view v, if we use only the optimal coefficients $\mathbf{W}_c^v$ associated with the c-th class, we can approximate the v-th view $\mathbf{y}^v$ of the test image as $\phi(\mathbf{y}^v) = \phi(\mathbf{X}_c^v)\mathbf{W}_c^v$. Then the classification rule is defined as in favor of the class with the lowest total reconstruction error through all the V views:

$$c^* = \arg\min_c \sum_{v=1}^{V} \theta_v \|\phi(\tilde{\mathbf{y}}^v) - \phi(\tilde{\mathbf{X}}_c^v)\mathbf{w}_c^v\|_2^2 \tag{4}$$

where θ_v, $(c = 1, ..., V$ and $\sum_{v=1}^{V} \theta_v = 1)$ is the weight measuring the confidence of the v-th view in the final decision. We only simply set $\theta_v = \frac{1}{V}$ in this paper.

4   Optimization

Eq.3 is convex, so it admits the global optimum. However, its optimization is very challengeable because both the $\|\tilde{\mathbf{W}}\|_F$-norm and the $\|\tilde{\mathbf{W}}\|_1$-norm in Eq.3 are convex but non-smooth. In this section we propose a simple algorithm to optimize Eq.3.

By setting the derivative of Eq.3 with respect to $\tilde{\mathbf{w}}_i$ ($1 \leq i \leq V$) as zero, we obtain:

$$(\mathbf{B}^T\mathbf{B} + \lambda_1\mathbf{E}_i + \lambda_2\mathbf{D})\tilde{\mathbf{w}}_i = \mathbf{B}^T\mathbf{a}_i \tag{5}$$

where $\mathbf{E}_i$ is a diagonal matrix with the k-th diagonal element as $\frac{1}{2|\tilde{w}_k^i|}$ and $\mathbf{A} = \{\mathbf{a}_1, ..., \mathbf{a}_V\}$. $\mathbf{D} = diag(\mathbf{D}_1, ..., \mathbf{D}_C)$, the '*diag*' is the diagonal operator and each $\mathbf{D}_c$ ($c = 1, ..., C$) is also a diagonal matrix with the j-th diagonal element as $D_{j,j} = \frac{1}{2\|\tilde{\mathbf{w}}_c\|_F}$, $j = 1, ..., n_c$.

By observing Eq.5, we find that both $\mathbf{E}_i$ and $\mathbf{D}$ depend on the value of $\tilde{\mathbf{W}}$. In this paper, following the literatures [12,24], we design a novel iterative algorithm (i.e., Algorithm 1) to optimize Eq.3 and then prove its convergence. Here we introduce Theorem 1 to guarantee that Eq.3 monotonically decreases in each iteration of Algorithm 1.

We first give a lemma as follows:

Lemma 1. *For any positive values α_i and β_i, $i = 1, ..., m$, the following holds:*

$$\sum_{i=1}^{m} \frac{\beta_i^2}{\alpha_i} \leq \sum_{i=1}^{m} \frac{\alpha_i^2}{\alpha_i} \Longleftrightarrow \sum_{i=1}^{m} \frac{(\beta_i+\alpha_i)(\beta_i-\alpha_i)}{\alpha_i} \leq 0$$

$$\Longleftrightarrow \sum_{i=1}^{m}(\beta_i - \alpha_i) \leq 0 \Longleftrightarrow \sum_{i=1}^{m}\beta_i \leq \sum_{i=1}^{m}\alpha_i \tag{6}$$

Theorem 1. *In each iteration, Algorithm 1 monotonically decreases the objective function value in Eq.3.*

Proof. According to the sixth step of Algorithm 1, we denote $\mathbf{W}^{[t+1]}$ as the results of the $(t + 1)$-th iteration of Algorithm 1, then we have:

$$\tilde{\mathbf{W}}^{[t+1]} = \min_{\tilde{\mathbf{W}}} \frac{1}{2}\|\mathbf{A} - \mathbf{B}\tilde{\mathbf{W}}\|_F^2 + \lambda_1 \sum_{i=1}^{V} \tilde{\mathbf{W}}_i^T E_i^{[t]} \tilde{\mathbf{W}}_i$$

$$+ \lambda_2 \sum_{c=1}^{C} tr((\tilde{\mathbf{W}}_c)^T(\mathbf{D}_c)^{[t]}\tilde{\mathbf{W}}_c) \tag{7}$$

then we can obtain:

$$\frac{1}{2}\|\mathbf{A} - \mathbf{B}(\tilde{\mathbf{W}}^{[t+1]})^T\|_F^2 + \lambda_1 \sum_{i=1}^{V} \tilde{\mathbf{W}}_i^T E_i^{[t]} \tilde{\mathbf{W}}_i$$

$$+ \lambda_2 \sum_{c=1}^{C} tr((\tilde{\mathbf{W}}_c)^T(\mathbf{D}_c)^{[t]}\tilde{\mathbf{W}}_c)$$

$$\leq \frac{1}{2}\|\mathbf{A} - \mathbf{B}(\tilde{\mathbf{W}}^{[t]})^T\|_F^2 + \lambda_1 \sum_{i=1}^{V} \tilde{\mathbf{W}}_i^T E_i^{[t]} \tilde{\mathbf{W}}_i \tag{8}$$

$$+ \lambda_2 \sum_{c=1}^{C} tr((\tilde{\mathbf{W}}_c)^T(\mathbf{D}_c)^{[t]}\tilde{\mathbf{W}}_c)$$

which indicates that:

$$\frac{1}{2}\|\mathbf{A} - \mathbf{B}(\tilde{\mathbf{W}}^{[t+1]})^T\|_F^2 + \lambda_1 \sum_{i=1}^{M} \sum_{j=1}^{N} \frac{((\tilde{w}_i^j)^{[t+1]})^2}{2\|(\tilde{w}_i^j)^{[t]}\|_2}$$

$$+ \lambda_2 \sum_{c=1}^{C} \frac{\|(\tilde{\mathbf{W}}_c)^{[t+1]}\|_F^2}{2\|(\tilde{\mathbf{W}}_c)^{[t]}\|_F}$$

$$\leq \frac{1}{2}\|\mathbf{A} - \mathbf{B}(\tilde{\mathbf{W}}^{[t]})^T\|_F^2 + \lambda_1 \sum_{i=1}^{M} \sum_{j=1}^{N} \frac{((\tilde{w}_i^j)^{[t]})^2}{2\|(\tilde{w}_i^j)^{[t]}\|_2} \qquad (9)$$

$$+ \lambda_2 \sum_{c=1}^{C} \frac{\|(\tilde{\mathbf{W}}_c)^{[t]}\|_F^2}{2\|(\tilde{\mathbf{W}}_c)^{[t]}\|_F}$$

Substituting β_i and α_i with $((\tilde{w}_i^j)^{[t+1]})^2$ (or $\|(\tilde{\mathbf{W}}_c)^{[t+1]}\|_F$) and $((\tilde{w}_i^j)^{[t]})^2$ (or $\|(\tilde{\mathbf{W}}_c)^{[t]}\|_F$) in Lemma 1, we have:

$$\frac{1}{2}\|\mathbf{A} - \mathbf{B}(\tilde{\mathbf{W}}^{[t+1]})^T\|_F^2 + \lambda_1\|\tilde{\mathbf{W}}^{[t+1]}\|_1 + \lambda_2 \sum_{c=1}^{C} \|(\tilde{\mathbf{W}}_c)^{[t+1]}\|_F$$

$$\leq \frac{1}{2}\|\mathbf{A} - \mathbf{B}(\tilde{\mathbf{W}}^{[t]})^T\|_F^2 + \lambda_1\|\tilde{\mathbf{W}}^{[t]}\|_1 + \lambda_2 \sum_{c=1}^{C} \|(\tilde{\mathbf{W}}_c)^{[t]}\|_F^2 \qquad (10)$$

This indicates that Eq.3 monotonically decreases in each iteration of Algorithm 1. Therefore, due to the convexity of Eq.3, Algorithm 1 can enable Eq.3 to converge to its global optimum.

Algorithm 1. The proposed method for solving Eq.3.

Input: $\mathbf{A}, \mathbf{B}, \lambda_1$ and λ_2;

Output: $\tilde{\mathbf{W}} \in R^{N \times V}$;

1 Initialize $t = 1; \tilde{\mathbf{W}}^{[1]}$;

2 **repeat**

3   Update the k-th element in the diagonal matric $\mathbf{E}_i^{[t+1]}$ via $\frac{1}{2|(\tilde{w}_k^i)^{[t]}|}$;

4   Update the c-th diagonal matrix in the diagonal matrix $\mathbf{D}^{[t+1]}$ via
    $(D_{j,j})^{[t]} = \frac{1}{2\|(\tilde{\mathbf{W}}_c)^{[t]}\|_F}$;

5   for each i,$1 \leq i \leq C$,

6         $\tilde{\mathbf{W}}_i^{[t+1]} = (\mathbf{B}^T\mathbf{B} + \lambda_1\mathbf{E}_i^{[t]} + \lambda_2\mathbf{D}^{[t]})^{-1}\mathbf{B}^T\mathbf{a}_i$;

7     $t = t+1$;

8 **until** *No change on the objective function value in Eq.3*;

---

5 Experiments

To evaluate the effectiveness of the proposed MMKR, we apply it and several state-of-the-art methods to multi-class object categorization on real datasets [13], such as 17 category and Caltech101 respectively.

17 category dataset (Flower for short) consists of 17 species of flowers with 80 images per class, and totally contains 1360 images. Each image is extracted with seven types of visual features, i.e., seven views. In our experiments, we randomly generate ten samples, in which each sample includes 1124 training images and 136 test images. Caltech101 dataset (Caltech for short) consists of 8189 images divided into 102 flower classes. Each class consists of 40-250 images. Each image is extracted with seven views. In our experiments, we randomly generate ten samples, in which each sample consists of 15 images per classes as training images, 15 images per classes as test images.

The comparison algorithms include algorithm KMTJSRC [21] only considering the block sparsity in RKHS, algorithm KSR [5] only considering the element sparsity in RKHS, the representatives of multiple kernel learning (MKL) methods, e.g., [16].

In our experiments, we obtain kernel matrices by computing $exp(-\chi^2(x, x')/\mu$, where μ is set to be the mean value of the pairwise χ^2 distance on training set.

In the following parts, first, we test parameters' sensitivity of the proposed MMKR according to the variation on parameters λ_1 and λ_2 in Eq.3, aiming at achieving its best performance. Second, we compare the MMKR with comparison algorithms in terms of average accuracy, i.e., classification accuracy averaged over all classes.

5.1 Parameters' Sensitivity

In this subsection we test different settings on parameters (i.e., λ_1 and λ_2 in Eq.3) in our proposed model, and set the value of them varying as $\{0.01, 0.1, 1, 10, 100\}$. The performance on average accuracy of the MMKR is illustrated in Fig.2.

It is clear that the proposed MMKR is sensitive to parameters' setting, which is the same as the conclusion on sparse learning models in [25]. As can be seen from the experimental results, the maximal difference between the best average accuracy and the worst one is about 30.66% and 70.52% for Flower and Caltech respectively. However, we also find the best performance is always obtained in cases with moderate value on both the λ_1 and the λ_2. For example, while the value of parameters' pair (λ_1, λ_2) is (1, 1) for both dataset Flower and dataset Caltech, our MMKR achieves the best average accuracy. Actually, according to our experiments, these cases lead to both the element sparsity (via the λ_1) and the block sparsity (via the λ_2). This illustrates it is feasible to select some training images from a few object categories to perform multi-class image classification.

5.2 Comparison

In this subsection, we set the values of parameters for the compared algorithms by following the instructions in their papers. For all the algorithms, we repeat each sample ten runs. We record the best performance on each combination of their parameters'

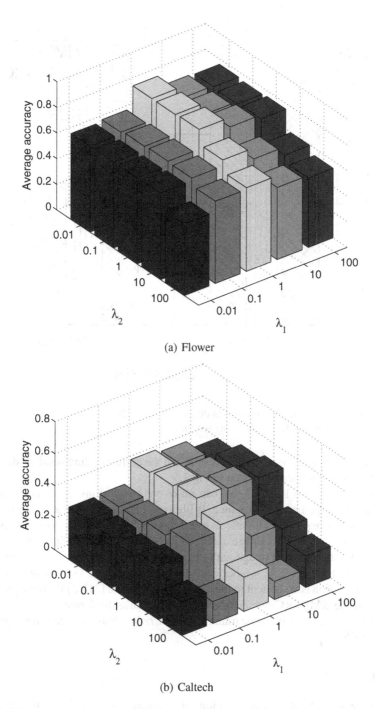

(a) Flower

(b) Caltech

Fig. 2. Average accuracy on various parameters' setting at different datasets

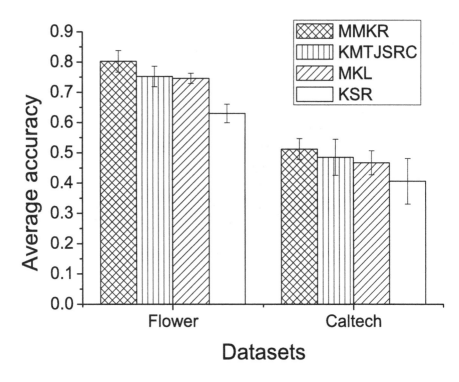

Fig. 3. Average accuracy on all algorithms at different datasets. Note that, the error bars shown in the bars represent standard deviation among ten runs.

setting in each run, and report average results and the corresponding standard deviation in ten runs. The results are illustrated in Fig.3.

As can be seen from Figure 3, we can make our conclusions as: 1) The proposed MMKR achieves the best performance. It illustrates that our MMKR is the most effective for multi-class image classification in our experiments. This occurs because the MMKR performs multi-class image classification via deleting noise in training data as well as representing the test image with only some training images from a few object categories. 2) The KMTJSRC outperforms traditional multiple kernel learning methods. This conclusion is consistent to the ones in the literature [21]. 3) Both the proposed MMKR and the KMIJSRC outperform the KSR because the former two methods reconstruct the test image with some training images, rather than using all training images used in the KSR.

6   Conclusion

In this paper we address the issue of multi-class image classification by first mapping the images (including training images and test images) into a RKHS. In the RKHS, each test image is linearly reconstructed with training images from a few object categories.

Meanwhile, removing noise is also considered. Then a classification rule is proposed by considering the derived reconstruction coefficient. Finally, experimental results show that the proposed method outperforms state-of-the-art algorithms. In the future, we will extend the proposed method into the scenario of multi-label image classification.

Acknowledgements. This research has been supported by the Australian Research Council (ARC) under large grant DP0985456; the Nature Science Foundation (NSF) of China under grants 61170131, 61263035, and 61363009; the China 863 Program under grant 2012AA011005; the China 973 Program under grant 2013CB329404; the Guangxi Natural Science Foundation under grant 2012GXNSFGA060004; and the Guangxi "Bagui" Teams for Innovation and Research.

References

1. Blum, A., Mitchell, T.: Combining labeled and unlabeled data with co-training. In: Annual Conference on Computational Learning Theory, pp. 92–100 (1998)
2. Boureau, Y.-L., Roux, N.L., Bach, F., Ponce, J., LeCun, Y.: Ask the locals: multi-way local pooling for image recognition. In: International Conference on Computer Vision, pp. 2651–2658 (2011)
3. Chen, N., Zhu, J., Xing, E.: Predictive subspace learning for multi-view data: A large margin approach. In: Advances in Neural Information Processing Systems (NIPS), vol. 23 (2010)
4. Dhillon, P.S., Foster, D., Ungar, L.: Multi-view learning of word embeddings via cca. In: Neural Information Processing Systems, pp. 9–16 (2011)
5. Gao, S., Tsang, I.W.-H., Chia, L.-T.: Kernel sparse representation for image classification and face recognition. In: Daniilidis, K., Maragos, P., Paragios, N. (eds.) ECCV 2010, Part IV. LNCS, vol. 6314, pp. 1–14. Springer, Heidelberg (2010)
6. Geng, B., Tao, D., Xu, C.: Daml: Domain adaptation metric learning. IEEE Transactions on Image Processing (99), 1 (2010)
7. Hou, C., Nie, F., Yi, D., Wu, Y.: Feature selection via joint embedding learning and sparse regression. In: IJCAI, pp. 1324–1329 (2011)
8. Jenatton, R., Audibert, J.-Y., Bach, F.: Structured variable selection with sparsity-inducing norms. Journal of Machine Learning Research 12, 2777 (2011)
9. Kim, J., Monteiro, R., Park, H.: Group sparsity in nonnegative matrix factorization. In: SIAM International Conference on Data Mining (2012)
10. Kumar, A., DauméIII, H.: A co-training approach for multi-view spectral clustering. In: International Conference on Machine Learning, pp. 393–400 (2011)
11. Mairal, J., Bach, F., Ponce, J., Sapiro, G.: Online learning for matrix factorization and sparse coding. Journal of Machine Learning Research 11, 19–60 (2010)
12. Nie, F., Huang, H., Cai, X., Ding, C.: Efficient and robust feature selection via joint l2,1-norms minimization. In: Neural Information Processing Systems, pp. 1813–1821 (2010)
13. Nilsback, M.-E., Zisserman, A.: A visual vocabulary for flower classification. In: IEEE Conference on Computer Vision and Pattern Recognition, pp. 1447–1454 (2006)
14. Owens, T., Saenko, K., Chakrabarti, A., Xiong, Y., Zickler, T., Darrell, T.: Learning object color models from multi-view constraints. In: IEEE Conference on Computer Vision and Pattern Recognition, pp. 169–176 (2011)
15. Quadrianto, N., Lampert, C.H.: Learning multi-view neighborhood preserving projections. In: International Conference on Machine Learning, pp. 425–432 (2011)
16. Rakotomamonjy, A., Bach, F.R., Canu, S., Grandvalet, Y.: Simplemkl. Journal of Machine Learning Research 9, 2491–2521 (2008)

17. Tan, M., Wang, L., Tsang, I.W.: Learning sparse svm for feature selection on very high dimensional datasets. In: International Conference on Machine Learning, pp. 1047–1054 (2010)
18. Wang, H., Nie, F., Huang, H., Ding, C.: Feature selection via joint embedding learning and sparse regression. In: Proceedings of IEEE Conference on Computer Vision and Pattern Recognition, pp. 3097–3012 (2013)
19. Wu, J., Rehg, J.M.: Centrist: A visual descriptor for scene categorization. IEEE Transactions on Pattern Analysis and Machine Intelligence 33(8), 1489–1501 (2011)
20. Xia, T., Tao, D., Mei, T., Zhang, Y.: Multiview spectral embedding. IEEE Transactions on Systems, Man, and Cybernetics, Part B: Cybernetics 40(6), 1438–1446 (2010)
21. Yuan, X., Yan, S.: Visual classification with multi-task joint sparse representation. In: IEEE Conference on Computer Vision and Pattern Recognition, pp. 3493–3500 (2010)
22. Zhu, X., Huang, Z., Cui, J., Shen, H.T.: Video-to-shot tag propagation by graph sparse group lasso. IEEE Transactions on Multimedia 15(3), 633–646 (2013)
23. Zhu, X., Huang, Z., Wu, X.: Multi-view visual classification via a mixed-norm regularizer. In: Pei, J., Tseng, V.S., Cao, L., Motoda, H., Xu, G. (eds.) PAKDD 2013, Part I. LNCS, vol. 7818, pp. 520–531. Springer, Heidelberg (2013)
24. Zhu, X., Huang, Z., Yang, Y., Shen, H.T., Xu, C., Luo, J.: Self-taught dimensionality reduction on the high-dimensional small-sized data. Pattern Recognition 46(1), 215–229 (2013)
25. Zhu, X., Shen, H.T., Huang, Z.: Video-to-shot tag allocation by weighted sparse group lasso. In: ACM Multimedia, pp. 1501–1504 (2011)
26. Zou, H., Hastie, T., Tibshirani, R.: Sparse principal component analysis. Journal of Computational and Graphical Statistics 15(2), 265–286 (2006)

Research on Map Matching
Based on Hidden Markov Model

Jinhui Nie[1], Hongqi Su[1,*], and Xiaohua Zhou[2]

[1] School of Mechatronical Engineering, China University of Mining and Technology, Beijing 100081, China
[2] School of Computer Science and Engineering, Beihang University, Beijing 100191, China

Abstract. Map matching is the procedure for determining the sequence of road links a vehicle has traveled on using the GPS data collected by sensors. Low sampling frequency and high offset noises are the main problems that the map matching algorithm needs to solve. In this study, the authors proposed a map matching algorithm based on the Hidden Markov Model (HMM). Naively matching the GPS sampling points with noise to the nearest road will result in some unreasonable map matching results, while this algorithm takes into account the location information suggested by GPS point and the road link transition probability. Also no more traffic information is needed in the procedure, which has a high accuracy and generalization ability. The algorithm was test with the real-word GPS data on a complex road network. The performance of the algorithm was found to be sufficiently accurate and efficient for the actual projects.

Keywords: Map matching, Hidden Markov Model, traveling track, road network.

1 Introduction

Real-time and accurate map matching algorithm has been focused greatly in the area of intelligent transportation in recent years. In the vehicle navigation systems, for example, the real-time map matching algorithm can determine the vehicle's position at any time. In such circumstance, the efficiency and accuracy are the key factors for the navigation services. Probe vehicle technology is an emerging method for the collection of traffic data. It is widely used in road states computing, analysis of driving law and other related areas. Since the process of data collection is insensitive to environment, time, weather, and other factors, a probe vehicle can collect even one hundred million records in one day. Before using the data obtained from the process of collection, we should match them to the road link accurately and efficiently.

In this paper, we put forward a map-matching algorithm based on HMM [1]. This algorithm takes many GPS points into consideration in the matching process. It can match several points at the same time and a lot of adjacency information of multiple road links is joined. This algorithm can not only reduce the computing complexity, but also improve the reasonableness, accuracy and efficiency.

The rest of this paper is organized as below. Section 1 describes the related researches on map matching algorithms. Section 2 describes the problems existing in

H. Motoda et al. (Eds.): ADMA 2013, Part I, LNAI 8346, pp. 277–287, 2013.

current map matching algorithms and the solutions to solve these problems based on HMM. Section 3 illustrates the definitions such as probability of output, transition probability and path selection, evaluating the algorithm's results by quantification. The conclusion and future research focus are described in section 4.

2   Related Work

The input of map-matching algorithm should be the location information that GPS points marked, adjacency information of the road link as well as other vehicle driving information. And the key issue is how to infer the vehicle travel route by synthesizing all aspects information quickly and accurately.

White et al (2000) elaborated on the common problems in map matching, and then proposed a map-matching algorithm based on the vehicle traveling direction [2]. However, the algorithm assumes that vehicle is traveling along the road links has at least one sampling point, which is difficult to guarantee in areas with high density road network. Greenfeld (2002) proposed a map matching algorithm which does not require the vehicle traveling direction and speed information, the algorithm infers vehicle location mainly via cables between GPS sampling points, the angle and distance between the motor vehicle routes. This algorithm improved the matching accuracy significantly, but needs the support of high-frequency GPS sampling data and may produce error match results affected by GPS points with large deviation. Lou et al (2009) proposed a map matching algorithm especially for low-frequency GPS data. This algorithm determines the sequence of candidate road links for each GPS point firstly, and then connected the candidate road links which can make up the shortest path. Then these connects are the match result [3]. The algorithm runs better for the low-frequency GPS data, but did not deal with circumstance that the distance between two adjacent GPS points is too long. Moreover, efficiency of this algorithm is not idealistic enough.

Cao Wen et al (2011) proposed a map-matching algorithm based on Zernike moments shape referring the Curve similarity theory. This algorithm introduces a Zernike moments to describe the shape of road curve, and then Infer the vehicle trajectory via calculating the similarity of shape of road curve and curve between GPS points [4]. The algorithm makes full use of the historical trajectory, and can achieve high accuracy in the case of that the sampling frequency is lower and the algorithm has more ambiguous solutions. But the algorithm is relatively complex; it is difficult to be applied to the actual road network. Tomio et al (2012) put forward the concept of road links selection [5] based on the matching algorithm of Kojima and Hato et al (2004), they modified the loopholes in the original algorithm, Improved the matching accuracy greatly. However, it is difficult to quantify road links' costs accurately in large-scale road network.

3   Map Matching Problems

Map matching is applied to determine the sequence of road links a vehicle has traveled on by eliminating GPS point errors and determining the road links that the current GPS point belongs to [6].Figure 1 shows the map matching process.

In Figure 1, the motor vehicle travels along *Link₁*, *Link₂*. P_1, P_2, P_3 are three consecutive sampling GPS points. While the nearest road segment away from P_1 is *Link₃* (5.9m), and the nearest road away from P_1, P_2 is *Link₂* with distance of 9.5m and 7.1m. Based on map-matching algorithm, we may determine that *Link₂* and *Link₃* should be the road segments. But we can see from figure 1 that there is no collection between *Link₂* and *Link₃* and even no the shortest reachable route, so the map-matching result is wrong.

With analysis above, we may conclude that map-matching algorithms based on shortest distance are usually unreliable. In order to achieve better matching result, topological information of the road network should be combined with the position information in the map-matching process [7].

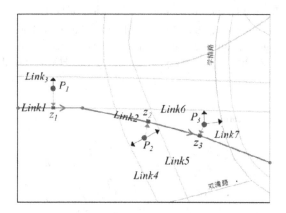

Fig. 1. Schematic diagram of map-matching

In this paper, we presents a HMM-based map matching algorithm to recover the whole sequence of roads that matching to a given sequence of GPS points, which combines topological information of the road network with the position context. The HMM-based method can be described by a quintuple available as formula (1),

$$\lambda = (R, P, A, B, \pi) \tag{1}$$

Set $T = \{t_1, t_2, t_3 \cdots t_n\}$ is a time sequence. Given $P = \{p_{t_1}, p_{t_2}, p_{t_3} \cdots p_{t_n}\}$, or the finite set of possible observations of HMM, where p_{t_i} is a sampling GPS points sequence collected at time t_i. Then $R_i = \{r_1, r_2, r_3 \cdots r_k\}$ is the sequence of road links corresponding to GPS point p_{t_i} at time t_i, it is the finite set of possible states of HMM.

$A = \{a(t_i, t_{i+1}), t_i \in T\}$ is the distribution matrix of transition probability at time t_i, or the transition probability of HMM [8], where

$a(t_i, t_{i+1}) = p(r_{i+1}|r_i)$ is the migration probability [9] from route r_i to route

r_{i+1} .

Then $B = \{b(t_i, r_k), t_i \in T \ \ r_k \in R_i\}$ is distribution matrix of output

probability at time t_i, or the observation probability of HMM [10], where

$b(t_i, r_k) = p(p_{t_i}|r_k)$ is the probability of GPS sampling point p_{t_i} corresponding

to road link r_k .

Set $\pi = \{\pi_k\}$ is the probability distribution of p_{t_1} matching to a sequence of

road links at time t_1, that is, the initial state distribution of HMM.

Map-matching problem is a decoding problem of the three basic HMM problems [11], which aims to get a sequence that guarantees the observed value to be a maximum probability. Figure 2 shows a running example of the HMM-based map matching method.

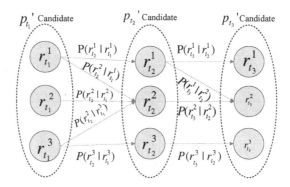

Fig. 2. An example of HMM-based map matching method

A HMM-based map matching method is illustrated in Figure 1, the input and output of this algorithm are described as follow,

Input: Road network R, a sequence of T GPS observation points. $T : p_{t_1} \rightarrow p_{t_2} \rightarrow \ldots \rightarrow p_{t_n}$.

Output: The matching sequence of road links $P : r_{t_1}^{j_1} \rightarrow r_{t_2}^{j_2} \rightarrow \cdots \rightarrow r_{t_n}^{j_n}$.

Each GPS point in T has a corresponding sequence of candidate road links, which is the finite set [12] of possible states of HMM. We use output probability to denote the likelihood of GPS point corresponding to a certain sequence of road links [13]. The transition probability of HMM [14] is the migration probability from current route to another route in candidate sequence.

4   HMM for Map Matching

4.1   Observation Probability

Observation probability or output probability of HMM, gives a probability that a GPS point matches a candidate point computed based on the distance between the two points [15]. In figure 1, for example, the output probability of P_l is p (p $|L$ $i n$ k) .

The shortest distance between P_l and $Link_l$ is $||P_l-Z_l||_{\text{Euclid Distance}}$, Given that Z_l is the projection point of P_l on $Link_l$. This distance is the offset error caused by the GPS point deviation. In this research, we randomly selected 1000 time-continuous GPS points, then manually determined their matching route sequence by way of MapInfo [12] and calculated the offset distance of each point. Assumed that the link direction is the positive direction, we provide that GPS points that fall on the right of the link has a positive error value, and GPS points that fall on the left of the link has a positive error value. By Practical Analysis, we found that error value of GPS points is generally between 0-50m, which is related to the travel speed. Large error value when the travel speed is high and smooth, relatively small when the speed is slow and volatile. In order to facilitate the analysis, we divided the error range of -35m to 45m into 7 sections with interval of 10m, and count the sample frequency falls on the sections. The result is shown in Figure 3.

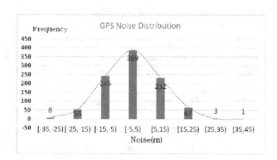

Fig. 3. GPS noise distribution

According the experimental result in Figure 4, GPS offset errors generally obey the normal distribution [16] with mean 0. Based on the research from VanDiggelen et al [17], probability of matching between GPS point and a certain link (or the output probability) can be calculated with formula (2).

$$p(p_{t_i}|r_k) = \frac{1}{\sqrt{2\pi}\sigma_z} e^{-0.5\left(\frac{||p_{t_i}-z_{t_i}||_{EU}}{\sigma_z}\right)^2} \qquad (2)$$

where σ_z is the standard deviation of the GPS measurement, estimation formula is as follow,

$$\sigma_z = \sqrt{\frac{1}{n-1}\left(x_k - \overline{x}\right)^2} \qquad (3)$$

And X_k is the offset distance between the GPS point and the candidate road chain, $\overline{X}$ is the mean value of offset distance, n is the number of the GPS points. In this paper, the estimated value of σ_z is 3.92.

Considering that GPS points offset error is generally less than 50m [18], we draw a circle with a radius of 100m and the current point as the center. We provide that all the output probability of road chain that outside the circle is 0. We can also use equation (2) to calculate the initial probability, which is $\pi_k = p(p_{t_1}|r_k)$.

4.2   State Transition Probabilities

Each GPS point has a sequence of candidate road links, transition probability gives a probability of the vehicle driving from a crosspoint to another one on the road network [19]. For example, in Figure 1, P_1's candidate sequence of road links is $R_1=\{Link_1,Link_3\}$, P_2's candidate sequence of road links is $R_2=\{Link_2,Link_4,Link_5\}$. One possible route selection is $Link_1 \rightarrow Link_2$ with the probability of $p(Link_2|Link_1)$, which is the transition probability [20]. We may discover from figure 1 that transition probability from a link to another, such as $Link_3 \rightarrow Link_2$, is 0. This is because vehicles can not drive from $Link_3$ to $Link_1$ during the period from t_1 to t_2 for normal driving conditions. So in the choice of links, this paper prefers links that do not so distanced with adjacent GPS points.

Given that Z_t is the projection point of P_t on a road link, Z_{t+1} is the projection point of P_{t+1} on road link r_{k+1}. $\|P_{t+1}-P_t\|_{Euclid\ Distance}$ is the Euclid distance between P_t and P_{t+1}, $\|Z_{t+1}-Z_t\|_{route}$ is the driving distance between Z_t and Z_{t+1} along road link r_k and r_{k+1}. Let d_t be the difference of the two distance, the formula [21] is as follow.

$$d_t = \left| \left\| p_{t+1} - p_t \right\|_{EU} - \left\| Z_{t+1} - Z_t \right\|_{route} \right| \tag{4}$$

By formula (4), we calculate the difference between distance of the 1000 points we selected in this paper and route distance. Analyzing the calculation results by the frequency analysis method, we get the Frequency chart as follow in Figure 4. This difference value roughly obeys exponential distribution [22], so the fitting formula should be formula (5).

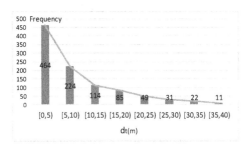

Fig. 4. Frequency chart of difference

$$p(d_t) = \beta \cdot e^{-\beta \cdot d_t} \tag{5}$$

By calculation, the fitted value of β is 1.8.

As sampling intervals between some GPS points is large, candidate road links of adjacent GPS point may not be adjacent to each other which is shown in figure 5.

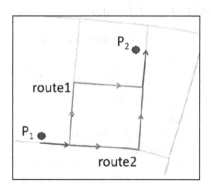

Fig. 5. Path Selection

In the figure 6, the vehicle possible driving route can be $P_1\rightarrow route_1\rightarrow P_2$ and $P_1\rightarrow route_2\rightarrow P_2$ when the vehicle driving from P_1 to P_2. In the absence of other additional information, we can not determine which route should be selected [23]. Therefore in this paper, we select the road link sequence with the shortest distance between the two points as the match result. In calculating the transition probability, middle road link will merge with the reachable candidate rode link of GPS points.

In areas with a highly dense road network, there are many actually unreachable route combinations in sequence of candidate route of adjacent GPS points [24]. In Figure 1, for example, vehicles can not drive along the route from $Link_3$ to $Link_2$ beyond 1 minute, they need to drive around another 12 km. So, in this paper, when we calculate the transition probability of a road link, we set the transition probability of unreachable route combinations within a certain time to 0, instead of calculate route distances of all the route combinations.

4.3 Infer of Map Matching Route

Map-matching problem is a decoding problem of the three basic HMM problems. It aims to determine the sequence of road links [16] (The state sequence [25, 26] of HMM) based on the collected GPS sampling data, which assures the probability of GPS points' occurrence to be maximum according to the gathered GPS points. The Viterbi algorithm with basic idea of dynamic programming [17] is used to solve this problem. The steps of Viterbi can be described as follow,

Input: GPS survey time series $T = \{t_1, t_2, t_3 \cdots t_n\}$, GPS sampling points sequence $P = \{p_{t_1}, p_{t_2}, p_{t_3} \cdots p_{t_n}\}$.

Step 1: Calculate the sequence of candidate road links Set_{t_1} of point p_{t_1}.

Step 2: For $\forall r \in Set_{t_1}$, calculate via formula (2): $\delta_{t_1}(r) = p(p_{t_1}|r)$.

Step 3: Set $i = 2, end = null$;

Step 4: Calculate $\delta_{t_i}(r) = \max_{r_{i-1}}\left\{\delta_{t_{i-1}}(r_{t_{i-1}}) \cdot p\left(p_{t_i}|r\right) \cdot p\left(r|r_{i-1}\right)\right\}$ by formula (2) and (5), and take notes on the 2-tuple $\left(r_{pre}, r\right)$, and r_{pre} is the prior road link when $\delta_{t_i}(r)$ gets the maximum value. Then, set $end = \arg\ \max_r\left\{\delta_{t_i}(r)\right\}$.

Step 5: Set $i = i + 1$. if $i \leq n$, go to step 4, otherwise go to step 6.
Step 6: Traversing the tuple record by step 4 in order of inverted time series, find and record the prior road link r_{pre}, then the inverse sequence of road links is the final result, that is, the map-matching result.
In order to improve efficiency of the algorithm furthermore, we divide our map into small grid [27], each gird has one-to-many map relationship with road links [28]. Additionally, we abstract our map into a directed graph with the road links as the nodes, and the intersection as the side. This will accelerate the search of next route chain, quickly and easily calculate transition probabilities and complete the screening of unreachable chains [29].

4.4   Map-Matching Result and Algorithm Evaluation

The experimental data used in this study is part of traveling information in Beijing in April 2012. Data acquisition frequency is 1 Hz. Ten millions of data records are collected approximately per day. Most of the data distributed on the trunk road within the fifth ring, small part of the data were collected on the suburban highway. Each data record contains the timestamp, and basic information such as latitude and longitude.

To improve the versatility of algorithm, sampling GPS data with a 5-s to 30-s interval is adopted as experimental data in this study. Figure 6 shows the distribution of our gathered data.

Fig. 6. GPS Data Distribution on the road network of Beijing

In this paper, since the ultimate goal of this paper is to determine the trajectory of the vehicle and the road link that a GPS point belongs to, the matching result does not contain the actual location on the route chain. The data format is shown in Table 1,

Table 1. Matching result

	time	longitude	latitude	Link
P_1	t_1	116.349416	40.023143	$Link_1$
P_2	t_2	116.350160	40.022875	$Link_2$
P_3	t_3	116.350714	40.022942	$Link_2$

This algorithm costs 4 hours and 15 minutes processing in a hadoop cluster with13 nodes, basically meet the efficiency requirements of massive data processing.

We use *ARR = length of correctly matched route/ total length of correct route* [22] to express map matching accuracy, as shown in figure 7.

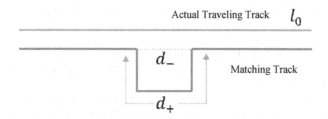

Fig. 7. Matching precision

d_- is the ignored miles because of matching error, similarly d_+ is the added mils in the error. The **precision** rate formula is shown as below [30],

$$accuracy = \frac{d_+ + d_-}{l_0} \times 100\% \qquad (6)$$

Precision rate of our algorithm is 89.3% verified by the gathered probe vehicle data of April. The experiment results demonstrate that our map-matching algorithm based on HMM performs very well in terms of matching accuracy and executing efficiency.

5   Conclusions and Future Research

In this paper, we summarize the existing map matching algorithms and propose a new method based on Hidden Markov Model. This algorithm takes the matching probability from the current GPS point to a candidate route chain as the probability of output. While the probability from one route to another one in the candidate data set is taken as the transition probability. The optimal matching route will be obtained based on the Viterbi algorithm. The algorithm can meet the application demand on efficiency and accuracy according to the experimental result. But some parameters such as σ and β in this algorithm are obtained only by one sample. It should be optimized in the future study.

References

1. Miwa, T., Kiuchi, D., Yamamoto, T., et al.: Development of Map Matching Algorithm for Low Frequency Probe Data. Transportation Research Part C: Emerging Technologies 22, 132–145 (2012)
2. Duda, R.O., Hart, P.E., Stork, D.G.: Pattern Classification. Wiley-Interscience (2012)
3. Sun, Y.R., Huang, B., Wang, L.N., et al.: Vector map matching navigation method with anti-scale transformation. Zhongguo Guanxing Jishu Xuebao 21(1) (2013)
4. Hunter, T., Abbeel, P., Bayen, A.M.: The path inference filter: Model-based low-latency map matching of probe vehicle data. In: Frazzoli, E., Lozano-Perez, T., Roy, N., Rus, D. (eds.) Algorithmic Foundations of Robotics X. STAR, vol. 86, pp. 591–607. Springer, Heidelberg (2013)
5. Bierlaire, M., Chen, J., Newman, J.: A probabilistic map matching method for smartphone GPS data. Transportation Research Part C: Emerging Technologies 26, 78–98 (2013)
6. Li, Y., Liu, C., Liu, K., Xu, J., He, F., Ding, Z.: On Efficient Map-Matching According to Intersections You Pass By. In: Decker, H., Lhotská, L., Link, S., Basl, J., Tjoa, A.M., et al. (eds.) DEXA 2013, Part II. LNCS, vol. 8056, pp. 42–56. Springer, Heidelberg (2013)
7. Ren, M., Karimi, H.A.: 9 Multisensor Map Matching for Pedestrian and Wheelchair Navigation. Advanced Location-Based Technologies and Services 209 (2013)
8. Zhu, L., Bao, Y., Wang, S.-G., Zhou, Q., Bao, X.: Map-Matching Compatible with Junction Adjusting in Vehicle Navigation System. In: Qian, Z., Cao, L., Su, W., Wang, T., Yang, H., et al. (eds.) Recent Advances in CSIE 2011. LNEE, vol. 129, pp. 451–458. Springer, Heidelberg (2012)
9. Levin, R., Kravi, E., Kanza, Y.: Concurrent and robust topological map matching. In: Proceedings of the 20th International Conference on Advances in Geographic Information Systems, pp. 617–620. ACM (2012)
10. Wang, D., Geng, X., Zhu, S.: Research on a Map-matching Method for a GNSS-based ETC System. Journal of Highway and Transportation Research and Development (English Edition) 6(4), 96–102 (2012)
11. Lou, Y., Zhang, C., Zheng, Y., et al.: Map-matching for Low-sampling-rate GPS Trajectories. In: Proceedings of the 17th ACM SIGSPATIAL International Conference on Advances in Geographic Information Systems, pp. 352–361. ACM (2009)
12. Wang, M., Bao, X., Zhu, L., et al.: A Map-Matching Method Using Intersection-Based Parallelogram Criterion. Advanced Materials Research 403, 2746–2750 (2012)
13. Zheng, K., Zheng, Y., Xie, X., et al.: Reducing uncertainty of low-sampling-rate trajectories. In: 2012 IEEE 28th International Conference on Data Engineering (ICDE), pp. 1144–1155. IEEE (2012)
14. Yang, L., Wang, R., Yan, H.: Urban road network matching method based on stable stroke hierarchical structure (2013)
15. Chen, J., Bierlaire, M.: Probabilistic multimodal map-matching with rich smartphone data. Journal of Intelligent Transportation Systems (just-accepted, 2013)
16. Yu, X., Yang, Q., Rong, X.: An Efficient Stereo Matching Algorithm Based on Intensity Weighted Correlation. In: Zhong, Z. (ed.) Proceedings of the International Conference on Information Engineering and Applications (IEA 2012). LNEE, vol. 217, pp. 399–406. Springer, Heidelberg (2013)
17. Wei, H., Wang, Y., Forman, G., et al.: Fast Viterbi map matching with tunable weight functions. In: Proceedings of the 20th International Conference on Advances in Geographic Information Systems, pp. 613–616. ACM (2012)

18. Zhao, Y., Qin, Q., Li, J., et al.: Highway map matching algorithm based on floating car data. In: 2012 IEEE International Geoscience and Remote Sensing Symposium (IGARSS), pp. 5982–5985. IEEE (2012)
19. Zhang, T., Yang, D., Yang, Y., et al.: Vehicle Behavior Pattern Recognition at Merging and Diverging Sections for Expressways in Map-matching. China Journal of Highway and Transport 6, 017 (2010)
20. Yang, S., Wu, C., Dong, G.: Improvement of the Map-Matching Algorithm Based on WSN Localization Algorithm and Bearing Information. Journal of Yunnan University of Nationalities (Natural Sciences Edition) 1, 14 (2011)
21. Lu, W., Zhang, Y., Chen, L., et al.: Map-matching algorithm research based on computational geometry. Machinery Design & Manufacture 1, 18 (2012)
22. Liu, Y., Yao, E., Li, X., et al.: Map-matching Accuracy of Probe Vehicle System. Journal of Highway and Transportation Research and Development, S1 (2011)
23. Li, R., Cao, W.: The Application of Improved Genetic Algorithm in Map Matching. Hydrographic Surveying and Charting 6, 021 (2011)
24. Yuan, Y.M., Guan, W., Qiu, W.: Map matching algorithm for inner suburban freeway based on handover location technique. Journal of Jilin University (Engineering and Technology Edition) 41(5), 1240–1245 (2011)
25. Wang, J., Yu, X.: A Map-matching Arithmetic of Approximate Reasoning. Computer Simulation 8, 074 (2010)
26. Zhu, J., Du, S., Ma, L., et al.: Merging Grid Maps Via Point Set Registration. International Journal of Robotics and Automation 28(2) (2013)
27. Ng, M.K.: Markov chains: models, algorithms and applications. Springer (2013)
28. Bonneville, R., Jin, V.X.: A hidden Markov model to identify combinatorial epigenetic regulation patterns for estrogen receptor α target genes. Bioinformatics 29(1), 22–28 (2013)
29. Hsu, D., Kakade, S.M., Zhang, T.: A spectral algorithm for learning hidden Markov models. Journal of Computer and System Sciences 78(5), 1460–1480 (2012)

A Rule-Based Named-Entity Recognition
for Malay Articles

Rayner Alfred[1], Leow Ching Leong[1], Chin Kim On[1], Patricia Anthony[2],
Tan Soo Fun[1], Mohd Norhisham Bin Razali[1], and Mohd Hanafi Ahmad Hijazi[1]

[1] Center of Excellence in Semantic Agents,
School of Engineering and Information Technology, Universiti Malaysia Sabah,
Jalan UMS, 88400, Kota Kinabalu, Sabah, Malaysia
[2] Department of Applied Computing, Faculty of Environment, Society and Design,
Lincoln University, Christchurch, New Zealand
ralfred@ums.edu.my, dragon_july14@hotmail.com,
patricia.anthony@lincoln.ac.nz,
{kimonchin,soofun,hishamrz, hanafi}@ums.edu.my

Abstract. A Named-Entity Recognition (NER) is part of the process in Text
Mining used for information extraction. This NER tool can be used to assist user in identifying and detecting entities such as person, location or organization.
Different languages may have different morphologies and thus require different
NER processes. For instance, an English NER process cannot be applied in
processing Malay articles due to the different morphology used in different languages. This paper proposes a Rule-Based Named-Entity Recognition algorithm
for Malay articles. The proposed Malay NER is designed based on a Malay
part-of-speech (POS) tagging features and contextual features that had been implemented to handle Malay articles. Based on the POS results, proper names
will be identified or detected as the possible candidates for annotation. Besides
that, there are some symbols and conjunctions that will also be considered in
the process of identifying named-entity for Malay articles. Several manually
constructed dictionaries will be used to handle three named-entities; Person,
Location and Organizations. The experimental results show a reasonable output
of 89.47% for the F-Measure value. The proposed Malay NER algorithm can be
further improved by having more complete dictionaries and refined rules to be
used in order to identify the correct Malay entities system.

Keywords: Named Entity Recognition, Malay Named Entity Recognition,
Rule-based, Information Extraction.

1   Introduction

Natural Language Processing (NLP) is one of the important fields in Computer
Science. Basically, it analyzes text that is based on both a set of theories and a set of
technologies [1]. NLP initially started at the late 1940s when machine translation was
first used to decrypt enemy codes during World War II. However, not many researches in NLP were conducted until the 1980s. There are a lot of fields that apply

H. Motoda et al. (Eds.): ADMA 2013, Part I, LNAI 8346, pp. 288–299, 2013.
© Springer-Verlag Berlin Heidelberg 2013

the NLP technologies such as Cross Language Information Retrieval, Information Extraction and Question-Answering [1, 24]. Most recent studies focus on Information Extraction (IE).

IE is a process that extracts information from unstructured text to provide more useful information. One of the sub-tasks of IE is known as a Named Entity Recognition (NER) process. A Named Entity Recognition process was a popular discussion at the Sixth Message Understanding Conference (MUC-6) [5, 6]. The NER process helps users to produce a more meaningful corpus by identifying proper names in the corpus and classifying them into groups such as person, organization, locations and etc. For example, the query "Steve Job" should not check only on the word "Steve" or "Job". The word "Job" may lead to the process of searching for similar word such as "occupation". Hence it will lead to other meaning instead of "Steve Job", the co-founder of Apple Inc.

The implementation of the NER algorithm for NLP is normally influenced by the domain of the studies. A domain-specific NER application may not be applicable for recognizing named-entities on other specific domains such as restaurant guides. For instance, AbGene [2], Abner [3] and BioNer [4] will not perform well in processing military articles as they are designed for different domains. In addition to that, different languages may require different techniques in recognizing the named entity. For instance, detecting the types of named entity for articles written in English language could easily be done by detecting the proper nouns. Proper nouns usually start with a capital letter. It is used to represent a unique named entity such as people, location, organization and etc. However, such methods may not be applicable to be applied for articles written in Arabic language as it does not contain such unique symbols that can be used to detect the named entity [11]. This is because most languages differ morphologically from other languages. In short, the implementation of the Named Entity Recognition depends upon the domain of studies and also the type of languages used.

There are a few NER systems that exist for various types of languages such as English, Indonesia, Arabic, Hindu and etc. However there is no existing system that is design to detect types of named entity in Malay language. Hence, in this paper, a rule-based Malay NER framework will be proposed that is designed to assist users in identifying types of named entity in order to improve the process of retrieving articles written in Malay language more effectively and efficiently.

This paper is organized as followed. Section 2 describes some of the works related to named entity recognition methods. Section 3 describes the general overview of the proposed rule-based named entity recognition for Malay language. Section 4 outlines the experimental setup and discusses the results obtained and finally Section 5 concludes this paper.

2   Types of NER

Algorithms for named-entity recognition (NER) systems can be classified into three categories; rule-based, machine learning and hybrid [8]. A Rule-Based NER algorithm detects the named entity by using a set of rules and a list of dictionaries that are manually pre-defined by human. The rule-based NER algorithm applies a set of rules in order to extract pattern and these rules are based on pattern base for location names, pattern base for organization name and etc. The patterns are mostly made up from grammatical, syntactic and orthographic features [8]. In addition to that, a list of dictionaries is used to speed up the recognition process. However, the types of dictionaries affect the performance of the NER systems and these dictionaries normally include the list of countries, major cities, companies, common first names and titles [7].

Next, a machine-learning NER algorithm normally involves the usage of machine learning (ML) techniques and a list of dictionaries. There are two types of ML model for the NER algorithms; supervised and unsupervised machine learning model. Unsupervised NER does not require any training data [9, 10]. The objective of such method is to create the possible annotation from the data. This learning method is not popular among the ML methods as this unsupervised learning method does not produce good results without any supervised methods. Unlike unsupervised NER methods, supervised NER methods require a large amount of annotated data to produce a good NER system. Some of the ML methods that had been used for NER algorithm includes artificial neural network (ANN) [11], Hidden Markov Model (HMM) [12], Maximum Entropy Model (MaxEnt) [13], Decision Tree [14], Support Vector Machine [15] and etc. ML methods are applicable for different domain-specific NER systems but it requires a large collection of annotated data. Hence, this might require high time-complexity to preprocess the annotate data.

Finally, a hybrid named entity recognition algorithm implements both the rule-based and machine learning methods [16]. Such method will produce a better result. However, the weaknesses of the rule-based are still unavoidable in this hybrid system. A domain-specific NER algorithm may need to customize the set of rules used to recognize different types of named entity when the domain of studies is changed.

3   A Rule-Based Named-Entity Recognition Algorithm for Malay Language

In this paper, a rule-based NER for Malay language will be proposed. In this work, a rule-based is applied instead of the machine learning technique due to the lack of annotated corpus resources for Malay language that can be used as a training data. Creating a large annotated dataset for Malay language is also time-consuming. The proposed rule-based NER for Malay language consists of three major steps. The first step is the tokenization. The purpose of the tokenization process is to split the sentences into tokens. For instance, the sentence "Pengerusi KMR telah sampai di Kuala Lumpur hari ini." will be converted into several tokens as shown in Table 1. The sentence is split into words, punctuation and numbers.

The second step involves the part-of-speech tagging (POS) process. In order to retrieve the part-of speech tagging, a Rule-Based Part of Speech (RPOS) tagger has been implemented. RPOS tagger is a simple rule-based POS tagger for Malay languages that applies a POS tag dictionary and affixing rules in order to identify the word definition [17]. The flow of the RPOS tagger is shown in Fig. 1.

The rule-based NER for Malay language is basically implemented based on the rule-based POS tagging process for Malay language and contextual features rules. The contextual features rules are studied and proposed for Iban and Indonesia Languages [19][20] which are almost similar to Malay language. For instance, when the part-of-speech tag for the current word shows that the current word is proper noun, then a specific rule will be applied to this current word in order to determine whether it is an entity or not. In other words, the rules are built based on the POS-tagging contexts. In this work, these rules are designed to detect three major types of named entities that include a person, an organization and a location. For instance, the tokenized words will be initially evaluated using the POS tag dictionary. There are roughly more than 8,700 words in the tag dictionary which are retrieved from the Thesaurus Bahasa Melayu and stored manually in the POS tag dictionary. If the POS tag dictionary returns more than one tag results, the best tag will be chosen according to the predefined rules [17]. The rules that are used to determine the tag are shown in Table 2. Table 3 shows the list of POS tag for Malay language words. However, if there is no word in the dictionary that can match the current word, then the affixing rules will be applied to determine whether the word is a noun, an adjective or a verb type of word as shown in Tables 3, 4 and 5. For instance, the result of this POS tagging process for the sentence "Pengerusi KMR telah sampai di Kuala Lumpur hari ini." is shown in Table 1.

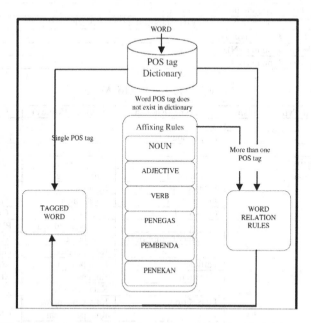

Fig. 1. The flow of the Rule-Based Part of Speech Tagging for Malay Language

Table 1. Part-of-Speech tagging result

Word	POS-tagging
Pengerusi	\<POS>NNP\</POS>
KMR	\<POS>NNP\</POS>
telah	\<POS>AUX\</POS>
sampai	\<POS>VB\</POS>
di	\<POS>IN\</POS>
Kuala	\<POS>NNP\</POS>
Lumpur	\<POS>NNP\</POS>
hari	\<POS>RB\</POS>
ini	\<POS>NN\</POS>
.	\<POS>PNC\</POS>

Table 2. Word Type Relations

Word Type	Valid Sequences of Word Types
Noun (NN)	adjective (JJ), adverb (RB), verb (VB), noun (NN), preposition (IN)
Verb (VB)	auxiliary (AUX), adverb (RB), noun (NN), penekan (PEN), pembenda (BND)
Adjective (JJ)	penguat (GUT), preposition (IN)
Adverb (RB)	verb (VB), preposition (IN), adjective (JJ), noun (AUX)
Direction (DR)	noun (NN), preposition (IN)
Preposition (IN)	noun (NN), verb (VB), adjective (JJ)
Auxiliary (AUX)	adjective (JJ), verb (VB), preposition (IN)
Cardinal number	noun (NN)
Penekan (PEN)	adverb (RB), noun (NN), conjunction (CC)
Pembenda (BND)	conjunction (CC), noun (NN)
Conjunction (CC)	noun (NN), verb (VB), preposition (IN), adjective (JJ)
Penguat (GUT)	adjective (JJ)
Interrogative (WP)	noun (NN), verb (VB)
Pangkal ayat (PNG)	noun (NN)

Table 3. POS Tag List for Malay

Word Type (English)	Subtype (English)	Subtype (Malay)	Tag
Noun			NN
	Proper noun		NNP
Verb			VB
Adjective			JJ
Function	Conjunction	Kata hubung	CC
	Interjection	Kata seru	UH
	Interrogative	Kata Tanya	WP
	Command	Kata perintah	CO
		Kata pangkal ayat	PNG
	Auxiliary (Amplifier)	Kata bantu	AUX
		Kata penguat	GUT
	Particles	Kata penegas	RP
	Negation	Kata naïf	NEG
		Kata pemeri	MER
	Preposition	Kata sendi name	IN
		Kata pembenar	BNR
	Direction	Kata arah	DR
	Cardinal number	Kata bilangan	CD
		Kata penekan	PEN
		Kata pembenda	BND
	Adverb	Adverb	RB

Table 4. Noun Affixing Identification Rules

Rules	Prefix	Next Character	Sequences of character	Suffix May end with	Suffix
1a	Pe	ny, ng, r, l and w	a-z	an	-
1b	Pem	b and p	a-z	an	-
1c	Pen	d, c, j, sy and z	a-z	an	-
1d	Peng	g, kh, h, k and vowel	a-z	an	-
1e	Penge	-	a-z (3 to 4 character)	an	-
1f	pel or ke	-	a-z	an	-
1g	Juru, maha, tata, pra, swa, tuna, eka, dwi, tri, panca, pasca, pro, anti, poli, auto sub, supra	-	a-z	-	-
1h	not started with me, meng, mem, menge, ber, be, di, diper	-	a-z	-	an, at, in, wan, wati, isme, isasi, logi, tas, man, nita, ik, is, al

Table 5. Adjective Affixing Idenfitication Rules

Rules	Prefix	Next Character	Sequences of character	Suffix May end with	Suffix
2a	ter, se, bi	-	a-z	-	-
2b	ke	-	a-z	an	-
2c	not starting with di and men	-	a-z	-	In, at, ah, iah, sequences of vowels then with and consequences of consonants ending with i

Table 6. Verb Affixing Identification Rules

Rules	Prefix	Next Character	Sequences of character	Suffix May end with	Suffix
3a	me	ny, ng, r, l, w, y, p, t, k, s	a-z	-	-
3b	mem	b, f, p and v	a-z	kan and i	-
3c	men	d,c, j, sy, z, t and s	a-z	kan and i	-
3d	meng	g, gh, kh, h, k and vowel	a-z	-	-
3e	menge	-	a-z (3 to 4 character)	an	-
3f	memper or diper	-	a-z	kan or i	-
3g	ber	not r	a-z	kan or an	-
3h	bel	-	a-z	-	-
3i	ter	not r	a-z	-	-
3j	ke	-	a-z	-	an
3k	-	-	a-z	-	i or kan
3l	di or diper	-	a-z	kan or i	-

3.1   Rules for Identifying a Person-Entity

In this work, the person-entity is recognized based on the person's titles and these person's titles are identified based on the standard titles used in Malay and English language. If the word is a person title, then the rest of the proper noun word is known as a person's name. These titles include "Yang Teramat Mulia", "Yang Amat Berhormat" and "Dr". For instance, in this sentence, "Dr. Tan Boon Keong telah tiba di Sabah hari ini", the word "Dr." is a title of a person. Hence "Tan Boon Keong" will be recognized as a person. Other than that, there are other pattern recognition methods that can be used to detect a person entity such as "A. Monhagen". If the word starts with a single character followed by proper nouns then it is recognized as a person name. The example of such pattern is "M. Night Shyamalan".

When the person's title is not included in the person's name, then a different set of rules will be applied to identify the patterns of name. Since the name of a person is highly dependent on the ethnic group or the nationality of a person, a different set of rules will need to be defined to handle a person entity for the Malaysian people. For instance, Malaysian Chinese people usually start their names with their surname, Malaysian Malay and Indian people start with their first names and followed by their father's name. Besides that, the gender of a person can also be determined based on the person's name. For instance, for Malay people, they use "Bin" for male (e.g., Ali Bin Ahmad) and "Binti" for female (e.g., Helmi Binti Yunus) in their names. Indian people use" A/L" which stands for "Anak Lelaki" or "Son Of" in English and "A/P" which stands for "Anak Perempuan" or "Daughter of" in English for their names. In some cases, for Sarawak native, they use the word "Anak" in their naming convention. Hence, by detecting such patterns will assist the process of recognizing the person entity.

Besides that, a person entity can also be recognized by using the preposition features. These preposition features include "oleh" (e.g., means "by"). For instance, given a sentence, "Hadiah ini disedia oleh En. Lim" (e.g., This present prepared by Mr. Lim), the person entity can be identified by looking at the word "oleh".

3.2   Location Rule

The location entity can be identified by looking at the location's prefixes and the usage of the "Di" preposition in the sentence. The words "Jalan", "Lorong", "Taman" and "Persiaran", are commonly used for the location's prefixes. For instance, "Lorong Kinabalu", "Persiaran Damai" and "di Ranau" are some of locations named entity that can be identified as "Kinabalu", "Damai" and "Ranau".

3.3   Organization Rule

Prefixes and the suffixes of organizations' names may be used to identify the named entity for an organization. The prefix and suffix used for organization are shown in table 6. For instance, "Syarikat Buku" is known as an organization because of the word "Syarikat" is identified as an organization's prefix. "Hong Leong Bank" is also known as an organization due to the word "Bank". Besides that, there is another pattern that can be used to recognize an organization entity which is "Persatuan Pelayaran Malaysia (MYA)". In this case, "MYA" will be recognized as an organization entity as it is an abbreviation of "Persatuan Pelayaran Malaysia". Hence, it can be concluded that when a string of words, that appears before the substring that consists of "(" and ")" symbols, is identified as an organization entity, then the abbreviation that appears within the parentheses (e.g., MYA) will be considered as an organization entity too.

Table 7. List of contextual features rules

Feature	Example
Location Prefix	Jalan, Bukit, Kampung
Preposition that usually followed by location	Di, ke
Organization prefix	Syarikat, Kelab, Persatuan
Organization suffix	Sdn. Bhd.,
Person prefix	Tan, Lim
Person middle	Bin, binti, a/p, a/l, anak
Person title	Dato Paduka, Tun
Preposition that usually followed by person	Oleh

Table 7 shows the list of rules that had been applied for the three categories. There are some exceptions that need to be handled in recognizing all these types of named entity. This is because there are some named entities that do not fulfill any rules that are predefined. For example, the proposed rules do not handle the entity recognition process for the phrase "Jabatan Keselamatan dan Kesihatan Pekerjaan" in which it consists of the word "dan" (e.g., sometimes "&" is used) and the word "dan" is spelled in a small letter. In order to overcome such weaknesses, a list of dictionaries shall be used to handle these types of named entity that are difficult to be detected or recognized by using the proposed rules.

4   Experimental Setup

In order to evaluate the effectiveness of the proposed rule-based named-entity recognition algorithm for Malay language, four different categories articles had been retrieved from two local Malay websites (http://www.bernama.com/bernama/v7/bm/ and http://www.mstar.com.my/). There are a total of 155 articles retrieved in General category, 143 articles retrieved in Economic category, 35 articles retrieved in Politic category and lastly 30 articles retrieved from Sport category. The main purpose of this experiment is to identify the types of patterns that are failed to be identified by the proposed NER algorithm. The NER for these three types of named entity (e.g., person, location and organization) will be evaluated based on three measures which are Recall, precision and F-measure as proposed in MUC [19].

$$Recall = \frac{Correct + 0.5 * Partial}{Possible} \quad (1)$$

$$Precision = \frac{Correct + 0.5 * Partial}{Actual} \quad (2)$$

$$F - Measure = \frac{Recall * Precision}{0.5 * (Recall + Precision)} \quad (3)$$

In this work, in Equation (1), the term Correct represents the number of correct annotations produced by the proposed NER algorithm for Malay language and the Partial term shows the number of partially correct annotations. For instance, given an entity in two words "Barack Obama", the proposed NER algorithm should be able to identify these two words as a Person entity. However, if the proposed NER algorithm is only able to annotate either "Barack" or "Obama" as a Person entity, then it is called a partially correct annotation. A manually tagged annotation used for training purposes is called as "Possible" term. The term Actual shows the actual number of annotated entity that should be produced by the proposed NER algorithm for Malay language. In short, the produced annotated entity can be categorized into Correct, Partially Correct or Incorrect entity.

5   Results and Discussions

The results of the proposed NER for Malay language are comparable to other NER algorithm for other language [20-22] in which the obtained F-measure is 89.47 %. Nevertheless, the performance of the proposed NER can further be improved by reformulating the rules used in these experiments. Table 8 shown below indicates some of the incorrect/missing annotations and partially correct annotations obtained from the experiment.

Based on the errors produced by the proposed NER, it can be concluded that the proposed set of rules produced by the Malay named-entity recognition algorithm is not complete. It is also due to the fact that the lists of words stored in the dictionaries are not complete. For instance, an organization entity "U-Mobile" is successfully annotated as an Organization entity because the company name "U-Mobile" does not have any organization prefixes or organization suffixes. Besides that, this word does not exist in any of the dictionaries. Not all of the actual named entities (NEs) are written starts with a capital letter (e.g., "i-City"). Hence, this makes the process of NER more complicated. Other than that, some of the NEs are ambiguous. For example, the word "Medan" is mostly used to refer as a location in Indonesia. However, in this experiment, the term "Medan anak Nunying" is actually identified as a person entity. Other than symbols or non-capital letters, the proposed NER should also be able to handle numbering symbols. However, there are some location entities that contain numbering symbols such as "Kampung Baru 30". The proposed NER only manages to detect "Kampung Baru" as a location entity instead of "Kampung Baru 30". In short, most of the annotations that are made partially correct can be solved by analyzing the present rules in more details.

Table 8. List of errors

Word	Error
<LOCATION>Kampung Baru</LOCATION> 30	Annotation is partially correct
Emiriyah <PERSON>Arab Bersatu</PERSON>	Wrong annotation
<PERSON> Datuk Seri Najib Tun Razak Selasa</PERSON>	Annotation is partially correct
Hishammudin	Missing
i-City	Missing
<LOCATION>AS</LOCATION> 350 B3	Wrong annotation
Ameika Syarikat	Missing
<PERSON>Jade Gallery</PERSON>	Wrong annotation
Di <LOCATIO>Papan Utama</LOCATION>	Wrong annotation
U-Mobile	Missing
S.Manikavasagam	Missing

6   Conclusion

This paper has proposed the first effort to generate a NER algorithm for Malay language. Based on the results obtained, the proposed Malay NER algorithm requires some adjustments for improvements in the predefined rules and also in the dictionaries used for the named-entity recognition process. The main challenge of implementing an acceptable Malay NER is to keep updating all libraries used up-to-date. There should be an effective way to ensure that the list of dictionaries used is always updated. Thus, updating the dictionaries manually is not a good option. The creation of ontology technology for semantic web usage might be helpful in producing a better list of dictionary for organization or location entities. The morphological features of Malay language are so rich and complex and this also contributes to the difficulties of implementing an effective Malay NER algorithm.

For future works, more additional rules should also be implemented and tested to handle more complex Malay sentence structure (e.g., "Jenny @ Jita Eyir"). A Malay NER algorithm should also be able to detect named entity based on existing online knowledge-based in order to produce a more robust Malay NER system. Other named entities such as time, date and percentage should also be considered in implementing a more complete and effective Malay NER system in the future.

Acknowledgments. This work has been partly supported by the LRGS and RAGS projects funded by the Ministry of Higher Education (MoHE), Malaysia under Grants No. LRGS/TD/2011/UiTM/ICT/04 and RAG0008-TK-2012.

References

1. Liddy, E.D.: Natural Language Processing. In: Encyclopedia of Library and Information Science, 2nd edn., Marcel Decker Inc., NY (2001)
2. ABGene, ftp://ftp.ncbi.nlm.nih.gov/pub/tanabe/AbGene/
3. Abner, http://pages.cs.wisc.edu/~bsettles/abner/

4. Song, Y., Eunji, Y., Eunju, K., Gary, G.L.: POSBIOTM-NER: A machine learning approach for bio-named entity recognition. In: Proceedings of the EMBO Workshop on Critical Assessment of Text Mining Methods in Molecular Biology (2004)
5. Ralph, G., Beth, S.: Message Understanding Conference-6: A Brief History. In: The Proceedings of the 16th International Conference on Computational Linguistics (COLING), pp. 466–471. Center for Sprogteknologi, Copenhagen (1996)
6. Ralph, G.: The NYU system for MUC-6 or Where's the Syntax. In: Inthe Proceedings of Sixth Message Understanding Conference (MUC-6), Fairfax, Virginia, pp. 167–195 (1995)
7. Wakao, T., Gaizaukas, R., Wilks, Y.: Evaluation of an algorithm for the Recognition and Classification of Proper Names. In: The Proceedings of COLING 1996 (1996)
8. Alireza, M., Liily, S.A., Ali, M.: Named Entity Recognition Approaches. Proceedings of IJCSNS International Journal of Computer Science and Network Security 8(2), 339–344 (2008)
9. Micheal, C., Yoram, S.: Unsupervised models for Named Entity Classification. In: The Proceedings of the Joing SIGDAT Conference on Empirical Methods in Natural Language Processing and Very Large Corpora (1999)
10. Kim, J., Kang, I., Choi, K.: Unsupervised Named Entity Classification Models and their Ensembles. In: Proceedings of the 19th International Conference on Computational Linguistics, pp. 1–7 (2002)
11. Naji, F.M., Nazlia, O.: Arabic Named Entity Recognition Using Artificial Neural Network. Journal of Computer Science 8(8), 1549–3636 (2012) ISBN 1549-3636
12. Daniel, M.B., Scoot, M., Richard, S., Ralph, W.: Nymble: a highperformance learning name-finder. In: The Proceedings of the Fifth Conference Applied Natural Language Processing, pp. 194–201. USA Morgan Kaufmann Publishers Inc., San Francisco (1997)
13. Benajiba, Y., Rosso, P., BenedíRuiz, J.M.: ANERsys: An Arabic Named Entity Recognition System Based on Maximum Entropy. In: Gelbukh, A. (ed.) CICLing 2007. LNCS, vol. 4394, pp. 143–153. Springer, Heidelberg (2007)
14. Bechet, F., Nasr, A., Genet, F.: Tagging Unknown Proper Names Using Decision Trees. In: Proceedings of the 38th Annual Meeting of the Association for Computational Linguistics (2002)
15. Wu, Y.-C., Fan, T.-K., Lee, Y.-S., Yen, S.-J.: Extracting Named Entities Using Support Vector Machines. In: Bremer, E.G., Hakenberg, J., Han, E.-H(S.), Berrar, D., Dubitzky, W. (eds.) KDLL 2006. LNCS (LNBI), vol. 3886, pp. 91–103. Springer, Heidelberg (2006)
16. Srihari, R., Niu, C., Li, W.: A Hybrid Approach for Named Entity and Sub-Type Tagging. In: Proceedings of the Conference on Applied Natural Language Processing (ANLP 2000), pp. 247–254 (2000)
17. Alfred, R., Mujat, A., Obit, J.H.: A Ruled-Based Part of Speech (RPOS) Tagger for Malay Text Articles. In: Selamat, A., Nguyen, N.T., Haron, H. (eds.) ACIIDS 2013, Part II. LNCS, vol. 7803, pp. 50–59. Springer, Heidelberg (2013)
18. Douthat, A.: The Message Understanding Conference Scoring Software User's Manual. In: Proceedings of the 7th Message Understanding Conference (MUC-7) (1998)
19. Budi, I., Bressan, S., Wahyudi, G., Hasibuan, Z.A., Nazief, B.A.A.: Named Entity Recognition for the Indonesian Language: Combining Contextual, Morphological and Part-of-Speech Features into a Knowledge Engineering Approach. In: Hoffmann, A., Motoda, H., Scheffer, T. (eds.) DS 2005. LNCS (LNAI), vol. 3735, pp. 57–69. Springer, Heidelberg (2005)

20. Yong, S.F., Bali, R.M., Alvin, Y.W.: NERSIL: The Named-Entity Recognition System for Iban Language. In: PACLIC, pp. 549–558 (2011)
21. Asharef, M., Omar, N., Albared, M.: Arabic Named Entity Recognition in Crime Documents. Journal of Theoretical and Applied Information Technology 44(1), 1–6 (2012)
22. Ferreira, E., Balsa, J., Branco, A.: Combining rule-based and statistical methods for named entity recognition in Portuguese. In: Actas da 5ª Workshop emTecnologias da Informação e da Linguagem Humana (2007)
23. Alfred, R., Kazakov, D., Bartlett, M., Paskaleva, E.: Hierarchical Agglomerative Clustering for Cross-Language Information Retrieval. International Journal of Translation 19(1), 139–162 (2007)

Small Is Powerful! Towards a Refinedly Enriched Ontology by Careful Pruning and Trimming

Shan Jiang*, Jiazhen Nian, Shi Zhao, and Yan Zhang**

Department of Machine Intelligence,
Peking University Key Laboratory on Machine Perception,
Ministry of Education Beijing 100871, P.R. China
{jsh,nian,z.s}@pku.edu.cn,
zhy@cis.pku.edu.cn

Abstract. In this paper, we study how to better merge a WordNet-like ontology with an online encyclopedia. We first eliminate the noises with some heuristic rules, and then adopt a domain-dependent strategy to trim the encyclopedia structure. Finally, we integrate entities from the trimmed structure into the original ontology, and construct a refinedly-enriched ontology. The experimental results show that this ontology can achieve better performance than the original version as well as a coarsely-enriched version constructed without pruning and trimming.

1 Introduction

Ontologies are widely used in text mining [27], information retrieval [17] and many applications of Natural Language Processing [16]. Nevertheless, as numerous new vocabularies are emerging and many of them are becoming popular in daily life, it gradually far exceeds an expert-edited ontology's ability to keep pace with the emergence of neologisms. On the other hand, when a new word becomes popular, the enthusiastic voluntary editors are likely to introduce the word into online encyclopedias instantaneously. Thus it is beneficial for an ontology to borrow knowledge from online encyclopedias.

This thing is even more important for Chinese ontologies. As one branch of Sino-Tibetan languages, Chinese has its own characteristics. For example, a significant difference between Chinese and many other languages is that the Chinese words are not separated from each other by the space symbol in writing. As a result, word segmentation is always the primary step when processing Chinese text and the quality of segmentation will inevitably affect the following procedures. If important words are absent in the ontology for the segmentation, the meaning of some sentences may even be misunderstood. Thus constructing a powerful ontology of keeping up with the growth of language is of significance.

However, it is not necessary to merge the entire encyclopedia into the expert-edited ontology. The quality of the information provided by an online encyclopedia is rather uneven due to the collaborative way of editing and the discrepancy

---

* Now at UIUC.
** Corresponding author.

H. Motoda et al. (Eds.): ADMA 2013, Part I, LNAI 8346, pp. 300–312, 2013.
© Springer-Verlag Berlin Heidelberg 2013

of editors' ability in writing and editing. If we take all the data without pruning and trimming, the low-quality data in the online encyclopedia will impair the quality of the combined lexical repository.

In this paper, we study how to better construct an enriched linguistic ontology which incorporates the merits of an expert-edited Chinese ontology, namely Chinese Concept Dictionary (CCD), and an online Chinese encyclopedia, namely Baidu Baike. As one of the most influential Chinese ontologies, CCD [30][11] is a WordNet-like semantic lexicon of contemporary Chinese. CCD is well structured from the viewpoint of computational lexicology and is widely used in Chinese NLP. Baidu Baike[1] is provided by a top Chinese search engine and attracts hundreds of millions web users, containing more than six million articles.

By trimming and pruning the information from Baidu Baike, we construct a refinedly-enriched ontology. At the same time, we construct a coarsely-enriched ontology by merging all the entities from Baidu Baike to CCD and employ both of them in real applications. The experimental results show that the refinedly-enriched ontology can achieve better performance than the coarsely-enriched one meanwhile it greatly reduces the time and space consumption.

The remainder of this paper is organized as follows. In Section 2 some notations and definitions are introduced. The details of the data pruning and trimming are discussed in Section 3. The ontology merging method is presented in Section 4. The experimental results are shown in Section 5 and the related works are reviewed in Section 6. Finally, we conclude the paper in Section 7.

2 Notation and Definition

Concept is the basic component of CCD, and represented as a synonym set:

Ontology Concept: *An ontology concept ς in CCD is denoted as a tetrad $\langle Syn, Def, Hyper, Hypo \rangle$, where Syn is the synonym set, Def is the definition of the concept, Hyper and Hypo denote the hypernym and the hyponym set of the concept respectively.*

In this paper, we employ Baidu Baike, which is one of the most influential online encyclopedias in Chinese and has a similar editing manner as English Wikipedia. In each article, editors interpret the entity from different aspects and link some words appearing in the article to other relevant articles in the encyclopedia. Editors also give some further semantic information of an entity such as categories, related entities, etc. There are also entries for the editing information, such as who have edited the article and how many times an editor has edited it.

Encyclopedia entity: *An encyclopedia entity ϵ is denoted as a sextuple $\langle Title, Des, Cat, Rel, Edt, Cite \rangle$, where Title is the title of the article representing the entity, Des is the description, Cat is the category set of the entity, Rel denotes the set of the related entities. Edt refers to the editing information and is represented by a set of two-tuples $\langle Edtr, Num \rangle$ which means editor Edtr has edited the article for Num times. Cite is the article set that this article cites.*

---

[1] http://baike.baidu.com/

3   Data Pruning and Trimming

To keep the balance between the vocabulary coverage and the practicability of the new ontology, we need to pick out the entities with more importance and higher quality. We first eliminate the noise entities with some statistical features and then decide a certain selection proportion for each domain. Finally, we evaluate the importance and the edit quality of each entity article, choose the more valuable entities and send them to the merging stage.

3.1   Noise Elimination

There are lots of low-quality articles in an online encyclopedia. For example, this article[2] is meaningless in terms of "words". If these entities are embodied in the new linguistic ontology, noises will be brought in and the quality of the ontology will be hurt.

Illegal symbols give explicit indication. If the title of an entity article contains illegal symbols (e.g. "&" and "%"), we can infer that the article is ill-edited. The length of the title is another useful sign. Figure 1 shows the distribution of the word length in CCD. It can be observed that most words in CCD contain less than 6 characters. Thus, only if the length of an article's title is less than a threshold (which is denoted by θ_{wl} in Algorithm ?? and set to 6 in this paper), it will be selected as a qualified word for the following process (Science domain is an exception due to long scientific terminologies). The edit-times is also a good indicator. If the edit-times of an article is less than a threshold θ_{et}, it will be discarded. In this paper, θ_{et} is empirically set to 5.

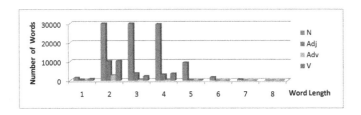

Fig. 1. Distribution of word length in CCD

3.2   Domain-Dependent Selection

It varies heavily between the numbers of words in different domains in Baidu Baike. The largest domain (Geography) is about 5 times larger than the smallest one (Economy) due to the voluntary editors' enthusiastic behavior. If we adopt a domain-independent method for entity selection, domains like Geography will dominate the new ontology. Therefore we adopt a domain-dependent strategy to avoid such a heavy bias.

---

[2] http://baike.baidu.com/view/3620572.htm?fr=iml

If a domain grows very fast, people will pay much attention to this domain in their daily life and entities in this domain are likely to be commonly used. If a domain grows very slow, why not keep all the new entities since they will not spend much space? Thus we introduce a hyperbola whose transverse axis is parallelized with the y-axis and takes the top branch as a parameter to calculate the selection proportions of the domains:

$$\frac{y^2}{(\frac{med}{max})^2} - \frac{(x-t)^2}{med^2 - 1} = 1. \tag{1}$$

Here x denotes the variable of the growth speed, and t is set to be the average of the second and the third largest values of the growth speed for all the domains. max and med are the maximum and median values of the growth speed respectively. Thus $y = \sqrt{1 + \frac{(x-t)^2}{med^2-1}} \cdot \frac{med}{max}$ is used as one component of the final formula for the selection proportion.

The CCD-coverage is another factor to be considered. If CCD has covered a domain very well, the structure of this domain should be kept and the entities should be chosen more rigorously to avoid producing latent damage. Thus, the selection proportion of a domain d $Pro(d)$ is set to be:

$$Prd(d) = \sqrt{1 + \frac{(G(d)-t)^2}{med^2 - 1}} \cdot \frac{med}{max} \cdot \sqrt{1 - C(d)}, \tag{2}$$

$G(d)$ is the growth speed of domain d. $C(d)$ is its CCD-coverage, which is calculated as $C(d) = \frac{|\{\epsilon|\epsilon \in d \wedge \epsilon \in CCD\}|}{|\{\epsilon|\epsilon \in d\}|}$, where $\epsilon \in CCD$ means that $\epsilon.Title$ is contained by a certain synonym set in CCD.

3.3 Article Importance

As is observed in [9], links in Wikipedia are typically based on words naturally occurring in one article and linked to another "relevant" article. Similarly to the Web, links in Wikipedia can be considered as voting for authority. These observations are also applicable to Baidu Baike. Furthermore, not every vote is equivalent. Baidu Baike has 12 domains and each domain will have its own representative entities, so the recommendation is domain-biased as well.

In this paper, a domain-biased Pagerank score is employed to evaluate the importance of an article, which is denoted as $P_d(\epsilon)$ and calculated by:

$$P_d(\epsilon) = (1-\alpha) \sum_{\rho \in \{\rho'|\epsilon \in \rho'.Cite\}} \frac{P_d(\rho)}{\sum_{\epsilon' \in \rho.Cite} R_{\rho\epsilon'}} \cdot R_{\rho\epsilon} + \alpha, \tag{3}$$

where α is the damping factor and

$$R_{\rho\epsilon} = \begin{cases} l(\rho, \epsilon) \cdot \tau, \ if \ \rho \ and \ \epsilon \ are \ in \ the \ same \ domain; \\ l(\rho, \epsilon), \quad else. \end{cases} \tag{4}$$

$l(\rho, \epsilon)$ is the number of links from ρ to ϵ and τ is a parameter used to enhance the influence of an internal link within a domain whose value should be greater than 1. Before each iteration, the sum of P_d values is normalized to 1.

3.4   Edit Quality

Edit-quality is different from article importance. The importance of an article is intrinsically determined by the connotation and extension of the concept rather than how the editors interpret it, while the edit-quality of an article depends on the editors' ability in writing and editing. We observe that articles' edit-quality and editors' editing-ability have a mutual reinforcement relationship. Therefore, we borrow the key idea of the HITS algorithm [4] to estimate articles' edit-quality and editors' editing-ability.

Let $a(E_i)$ denote the editing-ability of editor E_i and $q(\epsilon_j)$ denote the edit-quality of article P_j, then:

$$a(E_i) = \sum_{k=0}^{n} t_{ik} \cdot q(\epsilon_k), \tag{5}$$

$$q(\epsilon_j) = \sum_{r=0}^{m} t_{rj} \cdot a(E_r). \tag{6}$$

where t_{ij} denotes that editor E_i edits article P_j for t_{ij} times, namely $\langle E_i, t_{ij} \rangle \in \epsilon_j$ and ϵ_j is the entity represented by article P_j. n is the number of articles and m is the number of editors. It is necessary to normalize during each iteration. That is to say, let $\sum_{i=1}^{m} a(E_i) = 1$ and $\sum_{j=1}^{n} q(\epsilon_j) = 1$.

Finally, we get the edit-quality scores $\{q(\epsilon_1), q(\epsilon_2), ..., q(\epsilon_n)\}$. For each encyclopedia entity ϵ whose title is not included by CCD, if its importance $P_d(\epsilon)$ and edit-quality $q(\epsilon)$ are both in the top range of its domain, it will be sent to the next stage for ontology merging.

4   Ontology Merging

In the ontology merging stage, we map entities selected from the online encyclopedia to CCD and construct a new ontology.

4.1   Sources of Hypernym Relation Extraction

The key point of the merging process is to extract hyponymy relations from the online encyclopedia. In Baidu Baike, a section called "name card" is given in the first part of each entity article. This section includes the description and a brief introduction. The description usually offers important clues to figure out the higher level semantic concept. For instance, "A 是B的一种"("A is a kind of B") is a common pattern appearing in the description. Other semantic relations such as *LocatedIn*, *BornInYear* can be extracted as well. In this paper, we mainly focus on the hyponymy (*IsA*) relation [7]. To automatically extract hyponym-hypernym pairs from the sentences of description, we summarize five *IsA* patterns in Chinese which are given in Table 1.

To be honest, the pattern-based method is not valid all the time. Sometimes it is really intractable to automatically extract hyponym-hypernym pairs when the sentence structure is complicated. To tackle these cases, we explore the category

Table 1. Patterns for hyponymy extraction

Chinese patterns	English translation
A 是\为\指 + 一 + 量词 + B	A is a B
A 是\为\指 B 的一[种—类]	A is a kind of B
A 是\为\指 B 之一	A is one of B
A 被归类是\为 B [的一种]	A is categorized to B
A 属于 B [之一]	A belongs to a kind of B

information given by the editors. If an entity is subsumed by a category, it is plausible that the entity and the category have a hyponymy relation because "*A belongs to category B*" implies that *A* is a kind of *B*.

However, some categories are not limited to the hyponymy relation but more likely to be "tags". So it is necessary to discriminate those tag-style categories and prevent them from being included in the hypernym candidates. We observe that the semantic distance between a tag-style category and a hypernym-style category is usually farther than that between two hypernym-style categories. Therefore, it is probable that the most suitable candidate will appear in a branch of categories which has the closest semantic relation with ϵ.

4.2   Ontology Merging Algorithm

Based on the above observations, namely, the existence of *IsA* pattern in entity description and the relation between entity and its category, the procedure of the ontology merging operation is summarized in Algorithm 1. The merged ontology Ω is initially set to O, the concept set of CCD. The first-level domain names of Baidu Baike are all included in CCD. Therefore, we can add the second-level category names under these existing domains in CCD directly. After that, we map the entities into the refinedly-enriched ontology Ω.

Mapping Entities in a Semi-adaptive Order. The mapping order affects the accuracy of the mapping results of some entities. For example, given two entities ϵ_1 and ϵ_2, the title of ϵ_1 is in the category set of ϵ_2 and $\epsilon_1.Title$ happens to be the most suitable candidate for the hypernym of ϵ_2. If we map ϵ_2 before ϵ_1 is handled, ϵ_2 will be added to an ill-suited position. To surmount this problem, the ideal way is to implement topological ordering for the directed graph constructed by taking entities as vertex and the *IsCategoryOf* relations between entities as edges. Unfortunately, in real datasets, the graph turns out to be cyclic and the topological ordering method is infeasible. Thus we have to map the entities in a semi-adaptive order to make better use of the category information.

We divide the entity set E into two parts. The first part E_1 is composed of the entities whose titles are in the category set of others. And the remainders are put into the other part E_2. We map the entities in E_1 to Ω ahead of dealing with the entities in E_2. In this way, most of the category names of the entities will be included in Ω when dealing with E_2, so the probability that the most

Algorithm 1. Algorithm for ontology merging

---

input : Encyclopedia entity set E, CCD concept set O
output: The merged ontology Ω
$\Omega \leftarrow O$, $E_1 \leftarrow \emptyset$, $E_2 \leftarrow \emptyset$
add categories of the second level of the encyclopedia to Ω
foreach $\epsilon \in E$ **do**
    if $\exists \epsilon_0 \in E$ s.t. $\epsilon.Title \in \epsilon_0.Cat$ **then**
        $E_1 \leftarrow E_1 \cup \{\epsilon\}$
        $Sub(\epsilon) \leftarrow Sub(\epsilon) \cup \{\epsilon_0\}$
    else $E_2 \leftarrow E_2 \cup \{\epsilon\}$
foreach $\epsilon \in E_1$ **do**
    $f(\epsilon) \leftarrow |\{c|c \in \epsilon.Cat \wedge c$ is included in $\Omega\}|$
build a max-heap H of elements in E_1 according to their f value
while H is not empty **do**
    $\epsilon_t \leftarrow$ the top element of H
    $map(\epsilon_t, \Omega)$
    if $\epsilon_t \in \Omega$ **then**
        foreach $\epsilon'_t \in Sub(\epsilon_t)$ **do**
            $f(\epsilon'_t) \leftarrow f(\epsilon'_t) + 1$
    remove ϵ_t from H
    update H
foreach $\epsilon \in E_2$ **do**
    $map(\epsilon, \Omega)$

proc $map(\epsilon, \Omega)$
extract ϵ's hypernym h from $\epsilon.Des$ with the pattern-based method
if $h \neq \emptyset \wedge h \in \Omega$ **then**
    add ϵ into Ω as the subordinate node of h
else
    $C \leftarrow \{\varsigma|\varsigma \in \Omega \wedge \exists c(c \in \epsilon.Cat \wedge c \in \varsigma.Syn)\}$
    if $C \neq \emptyset$ **then**
        $lso \leftarrow$ the lowest super-ordinate of elements in C
        while $\exists \varsigma(\varsigma \in C \wedge \varsigma$ is a descendant of $lso)$ **do**
            foreach $\varsigma' \in \varsigma.Hypo$ **do**
                calculate $SP(\epsilon, \varsigma')$
            $lso \leftarrow \arg\max_{\varsigma' \in \varsigma.Hypo} SP(\epsilon, \varsigma')$
        add ϵ into Ω as the subordinate node of lso

---

suitable category is absent in Ω so that the number of entities added in ill-suited positions will be decreased.

For each entity ϵ in E_1, we predefine a function $f(\epsilon)$ to denote the number of ϵ's categories which are included in Ω. Then we build a max-heap H of the elements in E_1 according to their f values. While H is not empty, we recursively pop the top element ϵ_t of H, which represents the entity with the largest number of categories in Ω among the unhandled entities in E_1, and map it to Ω with procedure $map(\epsilon, \Omega)$ which will be discussed later. In each round, after ϵ_t is successfully mapped to Ω, the f values of the entities whose category sets contain $\epsilon.Title$ are increased by 1 and H should be updated.

After all the elements in E_1 are processed, we proceed to E_2. Because none of the elements of E_2 is in the category of any others, the mapping of an element will not affect others. Thus, in this step we just map the elements in turn.

Mapping Procedure. In the mapping procedure $map(\epsilon, \Omega)$, we first use the patterns to extract hypernym of ϵ. If the method works and the extracted hypernym h is in ontology Ω, $\epsilon.Title$ is subsumed as a hyponym of h. Otherwise, we utilize the category information to find the hypernym of ϵ. Concepts in ontology Ω whose synsets contain any of the categories of ϵ are put in set C. As is mentioned above, we attempt to find a branch of categories which has the closest semantic relation with ϵ. The information provided by the related entities of ϵ will be served as navigation and a formula is predefined to calculate the semantic proximity between a branch and the entity ϵ:

$$SP(\epsilon, r) = \begin{cases} \frac{\sum_{\varsigma \in \Phi} \sum_{w \in \epsilon.Rel} sim(w,\varsigma)}{|\Phi|}, & if(|\Phi| \neq 0); \\ 0, & else. \end{cases} \quad (7)$$

Here, $\Phi = \{\varsigma' | \varsigma' \in C \wedge \varsigma' \text{ is a descendant of } r\}$ and r is the root of the branch. $sim(w, \varsigma)$ [8] is the semantic similarity between two concepts:

$$sim(w, \varsigma) = \begin{cases} 1 & if(w = \varsigma) \\ \frac{log\frac{depth(w)+depth(\varsigma)}{len(w,\varsigma)}}{log(2(maxD+1))} & if((w \neq \varsigma) \wedge w \in \Omega \wedge \varsigma \in \Omega) \\ 0 & if((w \neq \varsigma) \wedge (w \notin \Omega \vee \varsigma \notin \Omega)) \end{cases} \quad (8)$$

in which $len(w, \varsigma)$ is the length of the shortest path between ϵ and ς in the ontology Ω. $depth(\varsigma) = len(\varsigma, \varsigma_r)$, where ς_r is the root of Ω. And $maxD$ is the maximum depth in Ω.

We find the lowest super-ordinate (or most specific common subsumer) of the elements in C and denote it as lso. The SP value between ϵ and every branch whose root is one of the child nodes of lso is calculated. Then the search scope is narrowed to the branch with the maximum SP score. We repeat the above procedure until the branch contains no node in C except the root.

5 Experiments

5.1 Data and Resource

In this paper, we use the version of CCD updated in June 2009, which is the latest version. It contains 103,736 nouns, 17,458 verbs, 17,825 adjectives and 3,894 adverbs. The Baidu Baike data is crawled from Oct. 25th 2011 to Oct. 27th 2011. The original dataset is composed of 3,905,910 entities, which mainly represent nouns. In our experiment, we only employ the nouns and construct a refinedly-enriched ontology of 176,978 concepts. For the coarse version, we integrate all the entities from Baidu Baike into CCD without trimming. The damping factor α in Equation 3 is set to 0.15 and τ in Equation 4 is set to 1.5.

5.2 Case Study

Table 2 gives some examples. In the second column is the CCD concept, and the encyclopedia entities added to it are shown in the third column. Case 1 is selected from the Science domain, in which 6 subjects on Artificial Intelligence are found. Case 2 is chosen from the Life domain. We can see that several kinds of wheaten foods which are famous in China are added. Generally speaking, a large proportion of the newly added words are popular and commonly used.

Table 2. Case study in the refinedly-enriched ontology

#	Ontology concept	Encyclopedia entity
1	AI 人工智能 (Artificial Intelligence)	语音识别系统(speech recognition system);模拟退火算法(simulated annealing algorithm);图灵测试(Turning Test);复杂系统(complex system); 概念图(concept map);智能机器人(intelligent robots)
2	面团 面糊 面食 (cooked wheaten food)	排骨面(noodles with pork ribs);花卷(steamed rolls);拉条(pulled noodles);燃面(spicy oiled noodle);涡阳干扣面(Woyang noodle); 花馍(coloured streamed bread)

5.3 Quality Evaluation of the New Ontologies

We conduct a manual evaluation of the two ontologies. If the encyclopedia entity is attached under the most suitable concept node, it will get 4 points. If the node is better suited to be an ancestor rather than parents, it will get 3 points. If the semantic relation between the entity and the concept is close but the semantic relation is not hypernym-hyponym type, it will get 2 points. Otherwise, 1 point.

500 cases are randomly sampled from the refinedly-enriched ontology and the coarsely-enriched version respectively, and a score is given to each case manually. The average score of the refined ontology and the coarse one are 3.274 and 2.882 respectively. The distribution of the scores is shown in Table 3.

Table 3. Distribution of scores in manual evaluation

Score	4	3	2	1
Refined	242	170	71	17
Coarse	173	151	120	56

5.4 Experimental Comparison in Text Mining

In the comparison of text mining, a feature selection method named "TCRL" [8] is utilized. TCRL model learns a semantic concept collection from a WordNet-like ontology according to the characteristics of a given document collection and represents the documents adaptively.

To find out which step in the data pruning and trimming process contributes more to the enhancement of the new ontology, we construct three "partially-refined" ontologies and employ them in the experiment as well. They are denoted as "NE", "PI" and "EQ" respectively. "NE" stands for the ontology which is constructed by merging the remaining entities from Baidu Baike after noise elimination to CCD. "PI" is made from CCD and a subset of Baidu Baike which is retained after noise elimination and domain-dependent trimming based on page importance. The construction of "EQ" is similar to "PI", while the data trimming is based on edit quality rather than page importance.

As far as we know, there is no standard Chinese dataset available for text mining task. So we make use of a collection crawled from one of the most popular news website "Sohu News" [3] in Dec. 2009. This collection contains 1,165,452 news documents and are manually categorized into 13 pre-defined classes. We use the classes of *Business, Culture, Health, IT* and *Military* and randomly select 1,000 documents from each class.

We perform two types of text mining tasks, namely, clustering and classification. For classification, two types of the F-measure scores [29] are calculated to evaluate the result, namely Macro-average F-measure score (F^{ma}) and Micro-average F-measure score (F^{mi}). For clustering, purity is employed as the evaluation standard.

Results and Analysis. The θ in the TCRL is set to 0.6. LibSVM [5] with a linear kernel is employed for the classification. For clustering, k-means is run for three times and the average *Purity* score is calculated.

The results of the experiment are shown in Table 4. "CLA" stands for classification and "CLU" means clustering. We can see that the refinedly-enriched ontology outperforms the coarse one, especially in clustering. It indicates that the feature space learned from the refined ontology has a stronger capability of discriminating the documents. The best result in classification is achieved by "NE". One possible reason is that the SVM classifier is robust enough to the redundant semantic information. Compared with the refined ontology, "PI" and "EQ" perform better in classification but worse in clustering. The decline of the performance in clustering indicates that redundant information may be brought in when filtering based on page importance or edit quality.

Table 4. Experimental results of text mining

		Coarse	NE	PI	EQ	Refined	CCD
CLA	F^{ma}	.716	**.722**	.719	.718	.718	.705
	F^{mi}	.860	**.867**	.864	.863	.862	.846
CLU	*Purity*	.663	.672	.670	.817	**.825**	.780

Time and Space Consumption. The time and space consumption are shown in Table 5. The row of "LOAD" is the time and space consumption of loading the ontology and building the hierarchical structure based on hyponymy relation. All the experiments are run on a PC with Intel Core 2 Duo processor at 2.53Hz

---

[3] http://news.sohu.com

Table 5. Time and space consumption

		Coarse	NE	PI	EQ	Refined	CCD
	LOAD	31.438	4.047	1.984	1.641	1.375	0.875
Time	CLA	25.516	21.078	18.281	17.609	16.859	14.938
(sec.)	CLU	4.453	3.844	3.469	3.250	3.109	2.891
	LOAD	217,512	172,632	122,088	98,700	80,220	50,892
Space	CLA	104,132	78,448	71,268	71,076	68,572	48,184
(KB)	CLU	48,740	44,824	43,396	42,468	41,716	39,22

and 3.24 GB RAM. Generally, the time and space consumption of classification and clustering have positive correlation with the dimensionality.

6 Related Work

Researchers have done extensive study on enhancement of lexical knowledge bases. Some tools [14] and methodologies [2,13] are proposed to align lexical resources with heterogeneous structures in different languages. Pociello et al. design and develop a multilingual lexical knowledge base, named Basque Word-Net [20]. Ramírez et al. automatically construct a Spanish-Japanese-English thesaurus based on Wikipedia and WordNet [22]. BabelNet is also a multilingual semantic network integrating knowledge from WordNet and Wikipedia [18]. Anderka et al. discuss the quality flaws in user-generated content when improving Wikipedia [1]. Melo and Weikum construct a large-scale multilingual lexical database by extending WordNet in over 200 languages [15]. Mandhani and Soderland use template patterns to extract relationships and enrich the ontology [12]. Subramaniam et al. propose algorithms for automatically merging a taxonomy into another in an asymmetry manner [24]. YAGO [25] is an extensible ontology automatically extracting knowledge from Wikipedia and WordNet.

As an important semantic lexicon, WordNet [6] has attracted attention from researchers. Bentivogli and Pianta [3] extend WordNet with syntagmatic information. Navigli et al. [19] extend WordNet with domain concepts by a system for word sense disambiguation named OntoLearn. Shi and Mihalcea [23] integrate WordNet with FrameNet and VebNet. Veale and Hao enrich WordNet with the cultural associations by mining explicit similes from the Internet [26]. Ponzetto and Navigli [21] present link WikiTaxonomy to WordNet and maximize the structural overlap between WordNet's taxonomy and Wikipedia's Category. Kozareva and Hovy [10] propose an algorithm to learn patterns from the Web hyponym-hypernym pairs and reconstruct parts of the WordNet taxonomy.

The above works mainly focus on the ontology enhancement and reconstruction, whereas our work pays more attention to the data pruning and trimming. Wu and Weld [28] refine the Wikipedia ontology and integrate its infobox-class schemata with WordNet. Their work casts the problem of ontology refinement as a machine learning problem while ours refines the encyclopedia data from different aspects and analyze the influence of each factor.

7   Conclusions

In this paper, we aim at how to better construct an enriched linguistic repository by merging an expert-edited ontology and an online encyclopedia. We find a refinedly-enriched version through data pruning and trimming which can achieve better performance than both the coarse version and the original CCD in text mining.

As a future work, we plan to make more efforts to identify hyponym-hypernym pairs more precisely and improve the data pruning and trimming. Another possible direction is to reconstruct some domain-biased ontologies according to the characteristics of different domains.

Acknowledgements. This work was supported by NSFC with Grant No. 61073081 and 61370054, and 973 Program with Grant No.2014CB340405.

References

1. Anderka, M., Stein, B., Lipka, N.: Predicting quality flaws in user-generated content: The case of wikipedia. In: Proc. of SIGIR 2012, pp. 981–990 (2012)
2. Artola, X., Soroa, A.: Elhisa: An architecture for the integration of heterogeneous lexical information. Natural Language Engineering 14(2) (April 2008)
3. Bentivogli, L., Pianta, E.: Extending wordnet with syntagmatic information. In: Proc. of GWC 2004, pp. 47–53 (2004)
4. Borodin, A., Roberts, G.O., Rosenthal, J.S., Tsaparas, P.: Finding authorities and hubs from link structures on the world wide web. In: Proc. of WWW 2001, pp. 415–429 (2001)
5. Chang, C.C., Lin, C.J.: LIBSVM: a library for support vector machines (2001)
6. Fellbaum, C.: WordNet: An Electronic Lexical Database. The MIT Press, Cambridge (1998)
7. Hearst, M.A.: Automatic acquisition of hyponyms from large text corpora. In: Proc. of COLING 1992, pp. 539–545 (1992)
8. Jiang, S., Bing, L., Sun, B., Zhang, Y., Lam, W.: Ontology enhancement and concept granularity learning: Keeping yourself current and adaptive. In: Proc. of KDD 2011, pp. 1244–1252 (2011)
9. Kamps, J., Koolen, M.: Is wikipedia link structure different? In: Proc. of WSDM 2009 (2009)
10. Kozareva, Z., Hovy, E.: A semi-supervised method to learn and construct taxonomies using the web. In: Proc. of EMNLP 2010, pp. 1110–1118 (2010)
11. Liu, Y., Yu, S., Yu, J.: Building a bilingual wordnet-like lexicon: the new approach and algorithms. In: Proc. of COLING 2002 (2002)
12. Mandhani, B., Soderland, S.: Exploiting hyponymy in extracting relations and enhancing ontologies. In: Proc. of WIIAT 2008, pp. 325–329 (2008)
13. McCrae, J., Spohr, D., Cimiano, P.: Linking lexical resources and ontologies on the semantic web with lemon. In: Antoniou, G., Grobelnik, M., Simperl, E., Parsia, B., Plexousakis, D., De Leenheer, P., Pan, J. (eds.) ESWC 2011, Part I. LNCS, vol. 6643, pp. 245–259. Springer, Heidelberg (2011)
14. McGuinness, D., Fikes, R., Rice, J., Wilder, S.: An environment for merging and testing large ontologies. In: Proc. of KR 2000 (2000)

15. Melo, G.D., Weikum, G.: Towards a universal wordnet by learning from combined evidence. In: Proc. of CIKM 2009, pp. 513–522 (2009)
16. Morato, J.Á., Marzal, M., Lloréns, J., Moreiro, J.: Wordnet applications. In: Proc. of GWC 2004 (2004)
17. Navigli, R., Crisafulli, G.: Inducing word senses to improve web search result clustering. In: Proc. of EMNLP 2010, pp. 116–126 (2010)
18. Navigli, R., Ponzetto, S.P.: Babelnet: Building a very large multilingual semantic network. In: Proc. of ACL 2010, pp. 216–225 (2010)
19. Navigli, R., Velardi, P., Cucchiarelli, A., Neri, F.: Extending and enriching wordnet with ontolearn. In: Proc. of GWC 2004, pp. 279–284 (2004)
20. Pociello, E., Agirre, E., Aldezabal, I.: Methodology and construction of the basque wordnet. Language Resources and Evaluation 45(2), 121–142 (2011)
21. Ponzetto, S.P., Navigli, R.: Large-scale taxonomy mapping for restructuring and integrating wikipedia. In: Proc. of IJCAI 2009, pp. 2083–2088 (2009)
22. Ramírez, J., Asahara, M., Matsumoto, Y.: Japanese-spanish thesaurus construction using english as a pivot. In: Proc. of IJCNLP 2008, pp. 473–480 (2008)
23. Shi, L., Mihalcea, R.: Putting pieces together: Combining framenet, verbnet and wordnet for robust semantic parsing. In: Gelbukh, A. (ed.) CICLing 2005. LNCS, vol. 3406, pp. 100–111. Springer, Heidelberg (2005)
24. Subramaniam, L.V., Nanavati, A.A., Mukherjea, S.: Enriching one taxonomy using another. IEEE TKDE 22(10) (October 2010)
25. Suchanek, F.M., Kasneci, G., Weikum, G.: Yago: a core of semantic knowledge. In: Proc. of WWW 2007, pp. 697–706 (2007)
26. Veale, T., Hao, Y.: Enriching wordnet with folk knowledge and stereotypes. In: Proc. of GWC 2008 (2008)
27. Wang, P., Hu, J., Zeng, H., Chen, Z.: Using wikipedia knowledge to improve text classification. Knowledge and Information Systems 19(3), 265–281 (2009)
28. Wu, F., Weld, D.S.: Automatically refining the wikipedia infobox ontology. In: Proc. of WWW 2008, pp. 635–644 (2008)
29. Yang, Y., Liu, X.: A re-examination of text categorization methods. In: Proc. of SIGIR 1999, pp. 42–49 (1999)
30. Yu, J., Yu, S., Liu, Y., Zhang, H.: Introduction to chineses concept dictionary. In: Proc. of ICCC 2001, pp. 361–366 (2001)

Refine the Corpora Based on Document Manifold

Chengwei Yao[1], Yilin Wang[2], and Gencai Chen[1]

[1] College of Computer Science and Technology, Zhejiang University, China
{yaochw,chengc}@zju.edu.cn
[2] NG7 2RD, Nottingham, University of Nottingham, UK
zeroe2009@gmail.com

Abstract. Nowadays, it is quite challenging to track and utilize overwhelming news information generated by internet. One approach is using topic models, such as pLSI, LDA, LPI, LapPLSI, LTM etc, to discover news topics automatically. However, in many real applications, the topics inferred by all these kinds of models are not much useful, because there are always a proportion of the documents actually belong to no topics. In this paper, we proposed a new technique to refine the document corpora before topic modeling. Inspired by manifold theory, we use Laplacian eigenmaps to discover the submanifold structure of the document space, and try to find those documents with loose relations to other documents, then exclude them from the corpora. Experiments show that topic models combined with our algorithm can improve the quality of the topics significantly.

Keywords: topic model, manifold, graph Laplacian, document clustering.

1   Introduction

With the booming of internet, huge amount of news information has generated by lots of kinds of applications based on web sites and mobile devices, such as news portals, social networks, instant communication systems and so on. How can we track, analyze, and utilize such overwhelming information effectively and efficiently?

One of the most popular and convincing ways is using topic modeling technologies to discover the topics by inferring from latent semantic space of the documents. Among those canonical topic models, Latent Semantic Indexing(pLSI)[10][11] and Latent Dirichlet Allocation(LDA)[4] adopt graphical models to infer probabilities of latent variables based on Euclidean space, and Locality Preserving Indexing(LPI)[5][6], Laplacian Probabilistic Latent Semantic Indexing(LapPLSI)[7], Locally consistent Topic Modeling(LTM) [8][9] go further to model the document space as a submanifold embedded in the ambient space.

Although all the topic models presented above have achieved great success in many cases, they have to assign every document to at least one topics. However, in many real applications of News Topic Discovering, many documents in the corpus do not have to be assigned to a topic. Someone might say, more or less, a news document belongs to a category, say politics, sports, science and technology, entertainment, and

H. Motoda et al. (Eds.): ADMA 2013, Part I, LNAI 8346, pp. 313–322, 2013.
© Springer-Verlag Berlin Heidelberg 2013

so on. But it is not sufficient for news tracking and analyzing. In our real practices, top news(or hot news) have strong keywords cooccurrence, while some other news belong to a larger category which have no relation to each other. For example 'community news', may have some degree of word cooccurrence, but not strong. That is why many topics are useless in our practices when we directly apply the topic models described above.

In this paper we propose a new technique to refine the document corpora based on document manifold before topic modeling. Our strategy is using Laplacian eigenmaps[13] to discover the submanifold structure of the document space, so that we can just focus on the documents that have strong keywords cooccurrence, and ignore other documents that have ambiguous probability distributions along the latent components. According to this strategy, intuitively, we discover the local neighborhood structure in the document space, which need to been constructed as a submanifold embedded in the ambient space[15]. We use graph Laplacian to approximate submanifold structure, and then solved it by computing eigenvectors. From the eigenvectors, we can get characteristic of every documents point on the submanifold and exclude those lower valuable points. Our work is based on the lapPLSI and the later works [18][22][23][24][25].

The rest of the paper is organized as follows: In section 2, short review of lapPLSI and its later works is given to depict basic ideas behind this kind of models. Section 3 introduces our motivation why we adopt such strategy. The detail technique is describe in section 4 and then give the experimental results in section 5. At the end of this paper, we bring out our conclusion and discussion in section 6.

2   Background

As one of the most elegant topic models, LDA[4] and it descendants[19][20][21] have achieved successfully in topic modeling by discover the hidden variables in the Euclidean space. However, recent studies suggest that the documents are usually sampled from a nonlinear low-dimensional manifold which is embedded in the high-dimensional ambient space. Based on this assumption, LapPLSI[7] and LTM[8][9] has been proposed.

In the two models, P_D is represent the distribution of $d \in D$, which lives in the very high dimensional space, according to the size of dictionary. And if $d_1, d_2 \in D$ are *close* in the intrinsic geometry of P_D, then $P(z|d_1)$ and $P(z|d_2)$ are similar to each other. In other words, the conditional probability distribution $P(z|d)$ varies smoothly along the geodesics in the intrinsic geometry of P_D. Based on this manifold assumption[1], various kinds of dimensionality reduction algorithms can be applied[2][3].

Define $f_k(d) = P(z_k|d)$ to be the conditional Probability Distribution Function, and $\mathcal{M} \subset \mathbb{R}^M$ to be compact submanifold which embedding in the original high dimensional space P_D. Directly solve $\|f_k\|_M^2$, the smoothness of f_k, is quite difficult. There is a convincing approach can be adopted, which using a nearest neighbor graph and then construct graph Laplacian to discretely approximate $\|f_k\|_M^2$.

Firstly, the algorithms build weigh matrix W to represent the relations of the points in the nearest neighbor graph, as follows:

$$W_{ij} = \begin{cases} \cos(d_i, d_j) \text{ or } 1, & \text{if } d_i \in N_p(d_j) \text{ or } d_j \in N_p(d_i) \\ 0, & \text{otherwise} \end{cases} \quad (1)$$

where $N_p(d_i)$ denotes the set of p nearest neighbors of d_i. Then there are two ways to discrete approximate the Beltrami operator Δ_M.

In LapPLSI: It construct graph Laplacian $L = D - W$, where $D_{ii} = \sum_j W_{ij}$, and discrete approximation of $\|f_k\|_M^2$ is computed as follows:

$$\mathcal{R}_k = \frac{1}{2} \sum_{i,j=1}^{N} \left(P(z_k|d_i) - P(z_k|d_j) \right)^2 W_{ij} = \mathbf{f}_k^T L \mathbf{f}_k \quad (2)$$

In LTM: By using Kullback-Leibler Divergence (KL-Divergence), which is considered as better measurement of distance, the approximation of smoothness along the geodesics in the intrinsic geometry of data is as follows:

$$\mathcal{R} = \frac{1}{2} \sum_{i,s=1}^{N} \left(D(P_i(z)) \| D(P_s(z)) + D(P_s(z)) \| D(P_i(z)) \right)^2 W_{is} \quad (3)$$

where $D(P_i(z)) \| D(P_s(z))$ is the KL-Divergence between two distributions $P_i(z)$ and $P_s(z)$, which is given below:

$$D(P_i(z)) \| D(P_s(z)) = \sum_z P_i(z) \log \frac{P_i(z)}{P_s(z)} \quad (4)$$

Then both models use these measurements of smoothness along the geodesics as the regularization of pLSI to overcome its overfitting problems[14]. Many experiments show that LapPLSI and LTM can achieve not only higher accuracy but also better performance than traditional topic models based on Bayesian network.

3   Motivation

Although LapPLSI, LTM and the later works have improved the quality of topics, they will meet the bottleneck in many real applications when they try to assign all data points to some classes, where a proportion of these points are regarded as noise. What if we try to cluster the points as the figure 1(left) shows.

In common sense, the points in the red circles should be clustered as figure 1(right) shows, and consider the points outside the red circle to be noises and do not assign them to any classes. So it is natural to find a way to exclude noisy points or some lower valuable points in advance before clustering algorithms will be deployed.

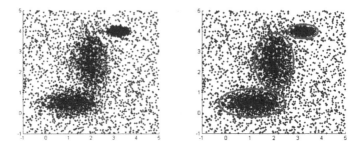

Fig. 1. Demonstration points to be clustered with random noise

How to refine the points? Our intuition also come from the LapPLSI and LTM. Based on manifold assumption, the relation between two points is preserved by local manifold structure, which can be approximated by Laplacian Eigenmaps. To encourage this thought, we visualize the eigenvectors of Laplacian matrix L from a sample dataset in our real application by contour plot, as figure 2 shows.

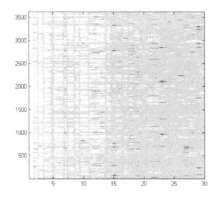

Fig. 2. Contour plot of eigenvectors of Laplacian Matrix from a real dataset

The x axis represents the latent components, and y axis is the points to be analyzed. As the Figure shows, there are some areas with red spot which means, in that areas, the points have much higher value corresponding to those latent components. In other words, these points have significant meaning to those latent components. On the contrary, the points with ambiguous distribution along the latent components in Laplacian Eigenmaps will have higher probability to be separated in submanifold, therefore they need not to be clustered. Our strategy is to find these lower valuable data points and exclude them from the dataset.

4   The Algorithm

LapPLSI and LTM use the submanifold structure of documents space as regularization[16] to the pLSI, and combine EM[12] algorithm to infer the latent topics. In our technique, we also apply the same structure, but use it as a baseline to evaluate the relations of documents, and exclude some irrelevant documents from corpus.

Suppose there are M documents in the corpus. Let P_D to be the distribution of $d \in D$, and $d_1, d_2 \in D$ are in the high dimensional space which is governed by the size of dictionary. We try to build manifold, where $P(z|d_1)$ and $P(z|d_2)$ are similar to each other if d_1, d_2 are *close* in the intrinsic geometry of P_D. z is the topics, the latent components, which govern the dimensionality of the manifold. Let $f_k(d) = P(z_k|d)$ to be the conditional Probability Distribution Function. and a reasonable criterion for choosing a *good* map from D to z is to minimize the following objective function[13]

$$\sum_{i,j=1}^{N} \left\| P(z|d_i) - P(z|d_j) \right\|^2 w_{ij} = tr(\mathbf{F}^T \mathbf{L} \mathbf{F}) \tag{5}$$

where $\mathbf{F}^{(l)} = [\mathbf{f}_1(d_i), \mathbf{f}_2(d_i), ..., \mathbf{f}_K(d_i)]^T$, K is the number of topics, and w_{ij} is entries of weighted similar matrix W of adjacency graph. $L = D - W$ called graph Laplacian, where $D_{ii} = \sum_j W_{ij}$. The solution is then provided by the eigenvectors of the corresponding smallest eigenvalues of the generalized eigenvalue problem $L\mathbf{f} = \lambda D\mathbf{f}$[17].

After we get eigenvectors matrix V, which is $M \times K$ matrix, the geometric features of the documents in manifold are revealed. Each row of V represent $P(z|d_i)$. Like figure 2 shows,   each row has different characteristic, which shows how strong the document is related to the latent components. That can help us select those unvalued documents d_i where each $P(z_k|d_i), k \in \{1,2,...,K\}$ are not so different. It is natural to compute its standard deviation the   as below

$$S_i = \left(\frac{1}{K} \sum_{k=1}^{K} \left(P(z_k|d_i) - \overline{P(z|d_i)} \right)^2 \right)^{\frac{1}{2}} \tag{6}$$

where $\overline{P(z|d_i)} = \frac{1}{K} \sum_{k=1}^{K} P(z_k|d_i)$

Then we exclude the $1 - l$ percent of   documents which has lowest S_i, because lower S_i means $P(z_k|d_i)$ are not too much various along the latent components. l can be chosen by experience according to the specific application. The detail procedure of the algorithm is as follows:

— We get document-term matrix from the corpora, and the entries of the matrix are just the counts of terms occurrence. Here we do not adopt the tfidf algorithm is that submanifold structure of documents space will characterize the features better.

- We construct the adjacency graph $G(V,E)$ by K nearest neighbors schema, where $Node_i$ and $Node_j$ is connected if $Node_i$ is among K nearest neighbors of $Node_j$ and vice versa. $Node_i$ represents the document d_i in the document-term matrix. Then we choose the weight by using cos metric like equation(1) based on adjacency graph G, and get similar matrix W.
- Get the normalized graph Laplacian $L = I - D^{1/2}WD^{-1/2}$, and compute generalized eigenvalue from $\mathbf{Lf} = \lambda \mathbf{Df}$ and select first K eigenvectors corresponding to K smallest eigenvalue, then we get eigenvector matrix V.
- Compute standard deviation of each rows of V, and sort them by descending order, select first l percent of points to be the collection of the new corpus. In our experiments, we select l to be Golden Ratio, 0.618. However, in different applications, l will be quite different according to the dataset at hand.
- At last, with new corpus, we can chose whatever topic models we like to infer the topics.

Pseudo-code of the algorithm is as follows:

Algorithm: Refine the Corpus Based on Document Manifold

Input: M documents with a vocabulary size N
   The percent of documents of the corpus to be selected: l
   The number of components K, when select eigenvectors
   The number of nearest neighbors p
Output: The new corpus with $M \times l$ documents

1: Compute the similar graph matrix W by using the cos metric as Eqn.(1);
2: $D_{ii} = \sum_j W_{ij}$;
3: $L = I - D^{1/2}WD^{-1/2}$;
4: Compute first K eigenvector $v_1, v_2, ... v_K$ of L, and let $V = [v_1, v_2, ... v_K]$;
5: Compute standard deviation of each rows and then get $S = [S_1, S_2, ... S_M]^T$ as Eqn.(6);
6: Sort S by descending order, get document indices of   the first l percent of S, and construct the new corpus with the new document indices.

---

5   Experiments

5.1   Datasets

We conduct our experiments using Reuters_21578[1] and TDT2[2] document datasets, which are widely used in the experiments of topic modeling and documents classification. The Reuters corpus contains 21578 documents which are grouped into 135 cluster,

---

[1]   Reuters-21578 corpus is at `http://www.daviddlewis.com/resources/testcollections/reuters21578/`
[2]   Nist Topic Detection and Tracking corpus at
   `http://www.nist.gov/speech/tests/tdt/`

and TDT2 corpus consists of data collected during the first half of 1998 and taken from 6 sources, including 2 newswires (APW, NYT), 2 radio programs (VOA, PRI) and 2 television programs (CNN, ABC).

In our experiments, we pick first 7800 documents from Reuters dataset and first 8741 documents from TDT2 dataset. From certain point of view, the corpus we selected have been well organized, as each document has its own label. However, relations between document to document are various, even if these documents come from the same cluster. We will demonstrate later, the quality of the clusters has much space to be improved.

5.2   Experiments Design and Evaluation Metric

We design the experiments into 2 procedures, in order to prove the corpus refinement algorithm is effective. The first procedure is we just randomly pick 61.8% of the documents from Reuters and documents from TDT2, then run pLSI and LTM separately. The second procedure is using our refinement algorithm to select 61.8% of the documents from Reuters and documents from TDT2, then run pLSI and LTM again. And we compare the results from two procedures, see what is the difference.

The key question is how to evaluate the difference. The most natural and popular metric is *perplexity*[4]. The perplexity, used by convention in language modeling, is monotonically decreasing in the likelihood of the test data, and is algebraically equivalent to the inverse of the geometric mean per-word likelihood. A lower perplexity means better generalization performance. The formula of perplexity is as bellow:

$$Perplexity(D) = \exp\left\{ -\frac{\sum_{d=1}^{M} \log p(\mathbf{w}_d)}{\sum_{d=1}^{M} N_d} \right\} \tag{7}$$

where $\mathbf{w}_d$ is words vector of document d.

5.3   Results

As Table 1, Table 2 and Figure 3 shows, the results are encouraging. By select the same proportion of documents from the same datasets, the perplexity achieved by our refinement algorithm is significant lower than the random selection. That means the algorithm can excluding some lower valuable documents, and make final topics to be higher valuable, even if the topic modeling algorithms are the same.

Another interesting result is that LTM is more effective than pLSI when using the TDT2, while using Reuters, LTM is better in most cases but not so significant. We think the reason is relate to the datasets, and more specific analyzing this point is not the objective of this paper.

Table 1. The   result of 'perplexity' by using Reuters_21578

# Topics	Randomly select *l* percent documents		Select *l* percent documents by the refinement algorithm	
	pLSI	LTM	pLSI	LTM
5	1251.39	1336.13	997.51	900.03
10	1068.05	1104.37	821.43	753.70
15	973.78	967.60	712.73	689.97
20	920.41	894.20	649.96	644.21
25	871.49	847.49	604.98	602.49
30	838.93	788.39	517.06	578.64
35	788.20	728.80	536.10	558.86

Table 2. The result of 'perplexity' by using TDT2

# Topics	Randomly select *l* percent documents		Select *l* percent documents by the refinement algorithm	
	pLSI	LTM	pLSI	LTM
5	4274.57	3599.99	3053.87	2632.58
10	3409.83	2820.62	2448.99	2063.07
15	2936.87	2533.23	2157.07	1851.48
20	2718.77	2358.09	1942.52	1670.84
25	2574.27	2210.96	1799.93	1571.32
30	2436.68	2105.32	1686.25	1504.33
35	2321.33	2006.47	1582.74	1511.99

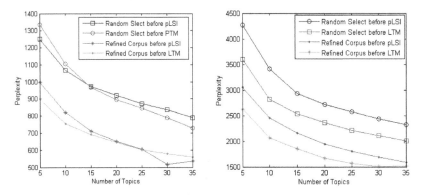

Fig. 3. Perplexity result from Reuters(left) and perplexity result from TDT2(right)

6   Conclusion and Discussion

There are so many researchers try hard to improve the performance of Topic model-ing algorithms. Nearly all these algorithms are applied to handle all the data points at hand. However, in many real applications such as News Topics Discovering, users just want to get some hot news topics and pass other irrelevant documents. To meet this requirement, we propose a new technique to refine the document corpora based on document manifold, and try to exclude such lower valuable documents from being analyzed. The experiments show that improvement to the quality of topics is significant by using our algorithm. Nevertheless, there is long way to shorten the gap about the conception of 'Topic' between meaning from statistics and meaning from the human sense. The algorithm we proposed in this paper provide an alternative way to meet this end.

Acknowledgments. This work is supported by 'Electronic Newspaper Analyzing Project'(Project No. H20121225) and 'Demonstration of Digital Medical Service and Technology in Destined Region'(Grant No. 2012AA022814).

References

1. Belkin, M.: Problems of Learning on Manifolds. PhD thesis, University of Chicago (2003)
2. Belkin, M., Niyogi, P.: Laplacian eigenmaps and spectral techniques for embedding and clustering. In: NIPS, vol. 14 (2001)
3. Belkin, M., Niyogi, P., Sindhwani, V.: Manifold regularization: A geometric framework for learning from examples. Journal of Machine Learning Research 7, 2399–2434 (2006)
4. Blei, D., Ng, A., Jordan, M.: Latent Dirichlet Allocation. Journal of machine Learning Research (2003)
5. He, X., Cai, D., Liu, H., Ma, W.-Y.: Locality preserving indexing for document representation. In: Proc. 2004 Int. Conf. on Research and Development in Information Retrieval (SIGIR 2004), Sheffield, UK, pp. 96–103 (July 2004)
6. Cai, D., He, X., Han, J.: Document clustering using locality preserving indexing. IEEE Transactions on Knowledge and Data Engineering 17(12), 1624–1637 (2005)
7. Cai, D., Mei, Q., Han, J., Zhai, C.: Modeling Hidden Topics on Document Manifold. In: Proc. 2008 ACM Conf. on Information and Knowledge Management (CIKM 2008), Napa Valley, CA (October 2008)
8. Cai, D., Wang, X., He, X.: Probabilistic dyadic data analysis with local and global consistency. In: Proceedings of the 26th Annual International Conference on Machine Learning (ICML 2009), pp. 105–112 (2009)
9. Cai, D., He, X., Han, J.: Locally Consistent Concept Factorization for Document Clustering. IEEE Transactions on Knowledge and Data Engineering 23(6), 902–913 (2011)
10. Hofmann, T.: Probabilistic latent semantic indexing. In: Proc.1999 Int. Conf. on Research and Development in Information Retrieval (SIGIR 1999) (1999)
11. Hofmann, T.: Unsupervised learning by probabilistic latent semantic analysis. Machine Learning 42(1-2), 177–196 (2001)
12. Neal, R., Hinton, G.: A view of the EM algorithm that justifies incremental, sparse, and other variants. In: Learning in Graphical Models. Kluwer (1998)
13. Lee, J.M.: Introduction to Smooth Manifolds. Springer, NewYork (2002)
14. Si, L., Jin, R.: Adjusting mixture weights of Gaussian mixture model via regularized probabilistic latent semantic analysis. In: Ho, T.-B., Cheung, D., Liu, H. (eds.) PAKDD 2005. LNCS (LNAI), vol. 3518, pp. 622–631. Springer, Heidelberg (2005)

15. Zhang, D., Chen, X., Lee, W.S.: Text classification with kernels on the multinomial manifold. In: Proceedings of the 28th Annual International ACM SIGIR Conference on Research and Development in Information retrieval (SIGIR 2005), pp. 266–273 (2005)

16. Zhu, X., Lafferty, J.: Harmonic mixtures: combining mixture models and graph-based methods for inductive and scalable semi-supervised learning. In: Proceedings of the 22nd International Conference on Machine Learning (ICML 2005), pp. 1052–1059 (2005)

17. Sha, F., Saul, L.: Analysis and extension of spectral methods for nonlinear dimensionality reduction. In: International Workshop on Machine Learning, vol. 22 (2005)

18. Cai, D., He, X.: Manifold Adaptive Experimental Design for Text Categorization. IEEE Transactions on Knowledge and Data Engineering 24(4), 707–719 (2012)

19. Blei, D., Lafferty, J.: Dynamic topic models. In: Proceedings of the 23rd International Conference on Machine Learning (2006)

20. Wang, C., Blei, D., Heckerman, D.: Continuous time dynamic topic models. In: Uncertainty in Artificial Intelligence (UAI 2008) (2008)

21. Wang, C., Paisley, J., Blei, D.: Online variational inference for the hierarchical Dirichlet process. Artificial Intelligence and Statistics (2011)

22. Zhang, L., Chen, C., Bu, J., Chen, Z., Cai, D., Han, J.: Locally Discriminative Coclustering. IEEE Transactions on Knowledge and Data Engineering 24(6), 1025–1035 (2012)

23. Bu, J., Xu, B., Wu, C., Chen, C., Zhu, J., Cai, D.: Unsupervised face-name association via commute distance. In: ACM Multimedia (ACM-MM 2012) (2012)

24. Zhu, J., Ma, H., Chen, C., Bu, J.: Social Recommendation Using Low-Rank Semi-definite Program. In: AAAI 2011 (2011)

25. Liu, X., Song, M., Zhao, Q., Tao, D., Chen, C., Bu, J.: Attribute-restricted latent topic model for person re-identification. Pattern Recognition (2012)

Online Friends Recommendation Based on Geographic Trajectories and Social Relations

Shi Feng[1,2], Dajun Huang[1], Kaisong Song[1], and Daling Wang[1,2]

[1] School of Information Science and Engineering, Northeastern University, China
[2] Key Laboratory of Medical Image Computing (Northeastern University),
Ministry of Education, Shenyang 110819, China
{fengshi,wangdaling}@ise.neu.edu.cn,
{huangdajun,songkaisongabc}@126.com

Abstract. With the rapid development of GPS-enabled mobile devices, people like to publish online data with geographic information. The traditional online friend recommendation methods usually focus on the shared interests, topics or social network links, but neglect the more and more important geographic information. In this paper, we focus on users' geographic trajectories that consisting of a series of positions in time order. We reduce the length of each trajectory by clustering the points and normalize every trajectory according to its positions and time in the trajectory. The similarity between trajectories is computed based on the distance of each corresponding point pair in the respective trajectory and the trajectories' trends. The potential online friends are recommended based on the trajectory similarity and social network structures. Extensive experiment results have validated the feasibility and effectiveness of our proposed approach.

Keywords: Friend Recommendation, Geographic Trajectory, Social Network.

1 Introduction

One major difference between virtual Web-based social network and real life social network is, new friends in real world tend to be geographically related. With the rapid development of GPS-enabled mobile devices, people like to publish online data, such as tweets, with geographic information, which have filled the gap between the virtual cyber world and real life world.

The traditional online friend recommendation methods usually focus on the shared interests, topics or social network links, but neglect the more and more important geographic information. There are potential similarities between users underlying in geographic information. For examples, people who have worked in a series of same cities or shared the same travel routes may be potential good friends.

How to apply users' geographic position information for friend recommendation is an important problem. Because a user does not always stay in one place, his (her) geographic position should change over time and form a trajectory. Recently, several papers have been published for computing the similarity between two users based on

H. Motoda et al. (Eds.): ADMA 2013, Part I, LNAI 8346, pp. 323–335, 2013.
© Springer-Verlag Berlin Heidelberg 2013

their GPS data or positions [1,2,3]. However, most previous literatures only considered the different positions in the social network, but not the sequence of the positions with time labels. That means the trajectory (A, B, C) is regarded as the same as (C, B, A), when the two trajectories belong to two users respectively. As a matter of fact, the trajectory in time order may reflect certain personal habits that the position set neglects.

In this paper, we take the temporal order into account in user geographic position information and propose an approach of online friend recommendation method based on geographic trajectories and social relations. We regard a user's geographic trajectory just like a time sequence, consider the shape between time-ordered points as the trend of the trajectory or sequence, and compute the similarity of two time sequence trajectories by comparing both the distance between them and the trend of each sequence. The potential friends are recommended considering the trajectory similarity and the social network structures.

The rest of the paper is organized as follows. Section 2 describes geographic trajectory reduction method. Section 3 gives the algorithm of trajectory similarity based on distance and trend. Section 4 presents and discusses the experimental results. Section 5 introduces the related work, and the conclusion is given in Section 6.

2   Geographic Trajectory Reducing

2.1   Problem Definition

We first give out the following definitions for introducing the method of reducing geographic trajectory in detail.

Definition 1 (user position set). A user position set PS is a set consisting of geographic positions P, i.e. it is the position record of the user's traveling during a period of time. Here we denote it as $PS=(P_1, P_2, ..., P_m)$.

Definition 2 (user geographic trajectory). For a user position set, if every position in the set is in chronological order, we say the user positions form a geographic trajectory. For PS in Definition 1, we say PS is a user geographic trajectory $T=(<P_1, t_1>, <P_2, t_2>, ..., <P_i, t_i>, ..., <P_m, t_m>)$, if the user stayed at P_i at time t_i, stayed at P_j at time t_j, and t_i is earlier than t_j ($i<j$ and $i, j=1, 2, ..., m$).

Definition 3 (trajectory dividing domain). Given the distance threshold and a certain period of time, the domain which is constituted by some positions within range is called trajectory dividing domain, and it is denoted as TDD.

Definition 4 (trajectory center point). The point which can represent the distribution of all points in the TDD it belongs to is called the trajectory center point, and it is denoted as TCP. In this paper, a TCP is the mean point of all points in the TDD it belongs to.

Definition 5 (trajectory trend vector). A user trajectory is a trajectory time sequence according to Definition 2. Here we define trajectory trend as the shape or style distributed on time of the trajectory or sequence. A user's trajectory trend vector's elements are composed of the slopes of two successive positions, denoted by $TTV=(slope_1, slope_2, \ldots, slope_{m-1})$, where $slope_i = \frac{P_{i+1}.lngt - P_i.lngt}{P_{i+1}.lat - P_i.lat}$ and $P_i.lat$, $P_i.lngt$ are the latitude and longitude of P_i, respectively.

2.2 Process of Geographic Trajectory Reducing

The purpose of reducing geographic trajectory is to compress a long geographic trajectory with many positions into a short geographic trajectory with relatively less positions by clustering the nearest positions in the geography. In this paper, each user's positions are represented by a time sequence. For each user's all positions, we cluster them by considering not only their latitude and longitude, but also the occurring time. The main idea of our method is shown in Figure 1.

position	latitude	longitude	time
P_1	lat_1	$lngt_1$	t_1
P_2	lat_2	$lngt_2$	t_2
...	...	...	...
P_{12}	lat_{12}	$lngt_{12}$	t_{12}

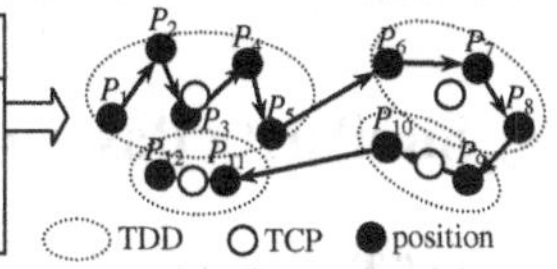

Fig. 1. Method of clustering geographic positions by latitude, longitude, and time

In Figure 1, if we do not consider the time, P_1 to P_5, P_{11} and P_{12} are clustered into one TDD because of their close latitude and longitude. However, in Figure 1, P_{11} and P_{12} are not clustered together with P_1, P_2, to P_5, because their time interval is too long. In next section, we measure the similarity between trajectories consisting of a series of positions according to their distance and trend. The trend is relevant to time, so we should consider the influence of time.

2.3 Algorithm of Geographic Trajectory Reducing

According to section 2.2, we consider not only the geographic positions but also the time of these positions during the clustering process. In detail, for a trajectory time sequence, we set a distance threshold α, then orderly compare these positions from the first position for clustering the sequence into some TDDs based on α. Finally, we use TCP to represent the corresponding TDD for every TDD. We give the algorithm of reducing geographic trajectory in Algorithm 1.

In Algorithm 1, a user's trajectory is represented with positions $(P_1, P_2, \ldots, P_n)$, where P_i $(i=1, 2, \ldots, n)$ is denoted by a triplet $(t_i, lat_i, lngt_i)$, and the elements represent the time, latitude, and longitude respectively. They can be used to compute the distance between two positions in line 5). Resultant $(TCP_1, TCP_2, \ldots, TCP_k)$ is k TCPs of k TDDs, where k is not given beforehand as k-means, but is generated in the process of the reduction.

Algorithm 1: Geographic Trajectory Reducing

Input: A user's trajectory presented with $(P_1, P_2, ..., P_m)$, radius threshold α;
Output: Refined the user's trajectory presented with $(TCP_1, TCP_2, ..., TCP_k)$;
Description:
  1) $i=1$;   \\ select the first position in trajectory
  2) $k=0$;
  3) Repeat
  4) $j=i+1$;   \\ select the next position for comparing
  5)   while $j \leq m$ and $|lat_i - lat_j| \leq \alpha$ and $|lngt_i - lngt_j| \leq \alpha$ do $j=j+1$;
  6) $k=k+1$;
  7) $TDD_k = \{P_i, P_{i+1}, ..., P_j\}$;   \\ build kth cluster TDD_k
  8)   compute TCP_k from TDD_k;   \\ compute the mean TCP_k of TDD_k
  9) $i=j+1$;   \\ start a new clustering process
  10) Until $j>m$;

3   Trajectory Similarity Based on Distance and Trend

3.1   Normalization of Trajectory

After reducing every user trajectory, we discover that the number of each user's TCPs is not the same. In order to facilitate calculation, we have to normalize the number of all of users' TCPs as a fixed number. Assume that the number of each user's position before reducing is $(p_1, p_2, ..., p_n)$, and after reducing is $(k_1, k_2, ..., k_n)$, where n is the number of users. Our algorithm requires the number of two users' positions should be the same, so we can revise α according to the following iteration formulas until the number of each user's TCP equals $\bar{k} = \left\lfloor \dfrac{1}{n} \sum\limits_{i=1}^{n} k_i \right\rfloor$.

For user i, his (her) number of position before reducing and TCP after reducing are p_i and k_i respectively, the jth iteration process is:

$$\hat{k}_i^j = Num(TDDs^j) \tag{1}$$

where $TDDs^j$ is the set of $TDDs$ at jth iteration, and $Num(TDDs^j)$ is the number of TDDs at jth iteration.

Let

$$\alpha^{j+1} = (\hat{k}_i^j / \bar{k}) \cdot \alpha^j \tag{2}$$

back to (1) for iteration, until $\hat{k}_i^j = \bar{k}$.

Note that it is difficult to get an ideal result straightly according to the above iteration formulas. In other words, we can't make the number of each user's TCPs reach $\bar{k}$. We try to consider the following two particular situations.

(1) When $\alpha^i = \alpha^j$ occurs during the process of iteration, i.e. the threshold α of the ith iteration (namely α^i) is equal to the jth iteration (namely α^j). Here we assume $i<j$, then according to Formula (2) we have:

$$\alpha^j = (\hat{k}_i^j / \bar{k}) \times \alpha^{j-1} = (\hat{k}_i^{j-1} / \bar{k}) \times (\hat{k}_i^{j-2} / \bar{k}) \times \cdots \times (\hat{k}_i^i / \bar{k}) \times \alpha^i \qquad (3)$$

Since $\alpha^j = \alpha^i$, so we have:

$$(\hat{k}_i^{j-1} / \bar{k}) \times (\hat{k}_i^{j-2} / \bar{k}) \times \cdots \times (\hat{k}_i^i / \bar{k}) = 1 \qquad (4)$$

We can say that it is a period of j-i in the process of iteration. This will lead the iteration to an infinite loop and never get the results what we want. In order to solve this problem, we let a tag array to tag whether the iteration has been the state of infinite loop and a threshold β. After each iteration, we should determine whether Formula (4) is supported. If not, we use Formula (2) to continue our iteration. When Formula (4) is supported, we should use Formula (5):

$$\alpha^{j+1} = (k_i^j / \bar{k}) \cdot \alpha^j + \beta \qquad (5)$$

(2) Based on the situation that not all the final iteration of each user's TCPs can reach $\bar{k}$, we have to settle for the suboptimal results which is the closest to $\bar{k}$. In order to normalize the results and reduce error as far as possible, we put forward an additional iterative condition: the result of each user reaches to $\bar{k}$ from the *right side* of itself. Specifically, on the one hand, if the result of the user can precisely reach $\bar{k}$, the iteration will stop immediately. On the other hand, if the result of the user is higher than $\bar{k}$ and next iteration it will be smaller than $\bar{k}$, we can use it to tentatively represent the result. By the end of the iteration, we should have truncated those results which are higher than $\bar{k}$ and made them equal to $\bar{k}$.

3.2   Similarity Measure of Distance and Trend

The reason why we choose the distance and trend as the measures of two users' geographic trajectory similarity has two aspects. On the one hand, according to Definition 3, in a user's trajectory, the relation of context positions can infer the tendency of the trajectory sequence. On the other hand, for two users' trajectories, the distance of the corresponding position can intuitively reflect how near or far the two trajectories are. For two users i and j, there are two extreme situations shown in Figure 2.

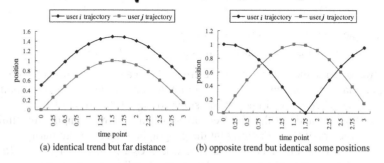

(a) identical trend but far distance   (b) opposite trend but identical some positions

Fig. 2. Two extreme situations of distance and trend for two user geographic trajectories

In Figure 2, it's obvious that considering a single of any situation is not comprehensive. In Figure 2(a), the corresponding elements of the two trajectory trend vectors are identically same but they are far apart, while some corresponding elements of the two trajectory trend vectors are opposite but their distance can be close in Figure 2(b). So in this paper, we consider that combining the distance with trend for computing the similarity between trajectories.

Firstly, we discuss the similarity of distance. As we know, we can use distance to measure the similarity between two users. Some frequently-used distance includes absolute distance, Huasdorff distance, Minkowski distance, Mahalanobis distance, etc. Huasdorff distance is often used in collection which ignores the order of elements. Obviously, it's not suitable for this paper because we need to consider the time order. Minkowski distance (when $p=1$ it's Manhattan distance. When $p=2$ it's Euclidean distance. When $p=\infty$ it's Chebyshev distance) also has such drawback. Simplicity, for example, we put the factor of time aside temporarily and only take latitude and longitude into consideration. Assume that there are three people $A(20,110)$, $B(30,110)$ and $C(20,120)$. No matter what's p equal to, the Minkowiski distance between A and B is equal to the Minkowski distance between A and C. In fact, $10°$ of latitude is not equal to $10°$ of longitude. The applicative condition of Mahalanobis distance seems too harsh, so it's also not suitable in this paper. In conclusion, the traditional methods of measuring the similarity of uses' geographic trajectory are not appropriate, thus we try to measure the similarity by the distance of latitude and longitude.

We can do some trigonometric transformation for the distance formula of longitude and latitude. Assume A (latitude: φ_1, longitude: θ_1), B (latitude: φ_2, longitude: θ_2), then we have $\cos(A,B)=A_1 \times A_2 + B_1 \times B_2 + C_1 \times C_2$, where $A_1 = \cos\varphi_1 \times \sin\theta_1$, $B_1 = \cos\varphi_1 \times \cos\theta_1$, $C_1 = \sin\varphi_1$, $A_2 = \cos\varphi_2 \times \sin\theta_2$, $B_2 = \cos\varphi_2 \times \cos\theta_2$, $C_2 = \sin\varphi_2$. Moreover, $dist(A,B) = R \times arcos(A,B) \times P_i/180$, where R is the radius of the earth. Since we only want to compare the similarity, the $\cos(A,B)$ can be the measurement. Assume that the geographic trajectory sequence of user i can be shown as $T_i=\{P_{i1}, P_{i2}, ..., P_{in}\}$, where $P_{ij}=<t_{ij}, lat_{ij}, lngt_{ij}>$ ($j=1, 2, ..., n$) and $n=\bar{k}$, lat denotes latitude and $lngt$ denotes longitude, then the distance similarity $simofdist$ between two trajectories of user i and j (their trajectories are denoted as T_i and T_j respectively) can be shown as:

$$simofdist\,(T_i, T_j) = \frac{1}{\bar{k}} \sum_{k=1}^{\bar{k}} \cos\Delta_k \qquad (6)$$

where $\cos\Delta_k = \cos lat_{ik} \times \sin lngt_{ik} \times \cos lat_{jk} \times \sin lngt_{jk}$
$\qquad + \cos lat_{ik} \times \cos lngt_{ik} \times \cos lat_{jk} \times \cos lngt_{jk} + \sin lat_{ik} \times \sin lat_{jk}$

Next we discuss the selection of the trend. We can treat them as 2 dimension vectors, i.e. (latitude, longitude). The way to implement the above idea is mapping the TCPs to plane. Then we can choose the slope to depict the trend of the geographic trajectory sequence and get the slope eigenvector. At last, we compute the similarity of the slope vectors of users to represent the geographic trajectory similarity. It's easy to get the eigenvector of user i, where $C_i=(l_{i,1}, l_{i,2}, ..., l_{i,n-1})$, $l_{i,j}=(lngt_{ij+1}-lngt_{ij})/(lat_{ij+1}-lat_{ij})$. For facilitating calculation, we handle with the initial eigenvector as follows:

$$C_i' = (l_{i,1}', l_{i,2}', \cdots, l_{i,n-1}'), \text{ where } l_{i,k}' = \begin{cases} 1 & \text{if } l_{i,k}' > 0 \\ 0 & \text{if } l_{i,k}' = 0 \\ -1 & \text{if } l_{i,k}' < 0. \end{cases}$$

Then the trend similarity *simoftrend* of user i and j is:

$$simoftrend'(T_i,T_j) = \frac{\sum_{k=1}^{\bar{k}-1} l'_{i,k} \times l'_{j,k}}{\sqrt{\sum_{k=1}^{\bar{k}-1}(l'_{i,k})^2} + \sqrt{\sum_{k=1}^{\bar{k}-1}(l'_{j,k})^2}} \tag{7}$$

The range of values of the trend similarity that gets from Formula (7) is [-1, 1], which is not consistent with the traditional distance formulas, because the range of values of the traditional is [0, 1]. In fact, the result is just a right feedback to our method of handling with the location data. As we said before, we regard users' position data as vectors, so the results may not only be positive, but also be negative. It seems entirely reasonable. Specifically, the positive indicates the tendency of the geographic trajectory of the two users' are in the same direction, while the negative indicates the opposite direction. In order to combine with the similarity of distance better, we can also standardize the similarity of the characteristic.

$$simoftrend(T_i,T_j) = \frac{1 + simoftrend'(T_i,T_j)}{2} \tag{8}$$

We have obtained the similarity of the distance and trend respectively. Next we need to introduce a parameter λ as weight to represent how the two kinds of similarity prorate in the geographic trajectory similarity. We will give the value of λ by experiment. In a word, the geographic trajectory similarity of the user i and j is given as follows:

$$sim(T_i,T_j) = (1-\lambda)simofdist(T_i,T_j) + \lambda simoftrend(T_i,T_j) \tag{9}$$

According to Formula (6), (7), (8), and (9), we give the algorithm of computing geographic trajectory similarity based on distance and trend as Algorithm 2.

Algorithm 2: Measuring Trajectory Similarity

Input : two user trajectories reduced: T_i, T_j;
Output: similarity between T_i and T_j: $sim(T_i,T_j)$;
Description:
1) For every user trajectory T_i and T_j
2) {Repeat; \\ normalize trajectories reduced, i.e. T_i and T_j
3)   adjust the threshold α according to Formula (2);
4)   If there is any cycle Then take Formula (5);
5)   back to Formula (1);
6)   Until $\hat{k}_i(\hat{k}_j)$ is nearest to $\bar{k}$ and $\hat{k}_i(\hat{k}_j) \geq \bar{k}$;
7)   compute distance similarity $simofdist(T_i,T_j)$ with Formula (6);
8)   compute trend similarity $simoftrend(T_i,T_j)$ with Formula (7) and (8);
9)   compute trajectory similarity $sim(T_i,T_j)$ with Formula (9);
10) }

For online friend recommendation, firstly we consider the social network structures to find the potential friends. Then the trajectory similarity is calculated to locate the most relevant friends of a candidate user.

4   Experiment

In this paper, the experimental data comes from a certain position service community users and Sina microblog users, where Sina microblog's data is the geographic positions of "second degree" friends of 100. Here "second degree" friends mean friends of friends. We choose 100 "second degree" friends for each user. Here we use the data of the position service community to verify that our algorithm is feasible and use the data of Sina microblog to show its practical significance. Choosing the geographic positions of "second degree" friends is to weaken the influence of the social network structure. If directly choose the user's friends, it will be inevitably blend the subjective factors into the results and reduce the credibility of the experimental results. The core of the traditional algorithm of friend recommendation is whether he/she is friend's friend. It shows that the user's "second degree" friends are the potential friends. Thus choosing the geographic positions of "second degree" friends in this paper not only takes the potential factor, but also weakens the impact of the social network structure as far as possible.

4.1   Results of User Trajectory Reduction and Normalization

In setting threshold α, on the one hand, if we set α a too high value, the number of some especial users' (their positions are almost exactly the same no matter at any time, i.e. they belong to the "Otaku") positions will possibly reduce to 1. Obviously, this is infeasible. On the other hand, if we set α a too small value, then the number will change slightly and it can't achieve the result of reducing the data. In the experiment of reduction, we set threshold $\alpha=1$. After reducing, the geographic trajectory information of a user' 100 "second degree" friends is shown as Figure 3.

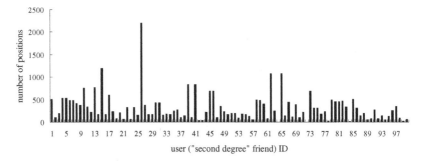

Fig. 3. Position number in 100 user trajectories after reducing the trajectories

We find that the number of each "second degree" friend's geographic positions is very different. In fact, this can be explained by the fact that different people have different routine patterns. Some people are active while others are male/female "otaku". For example, the number of geographic locations of user 26 still reaches up to 2199, while the user 72 has only 1. This can show that user 26 is more active than user 72 during this period of time. We mine this kind of information to determine whether the user is active or not. Obviously, here we can say that the user 26 is an active user and the user 72 is a typical male/female "otaku".

Next we normalize the trajectories. Here we set β=0.001. The position number (i.e. TCP number in Definition 4) of a user' 100 "second degree" friends after normalizing is shown as Figure 4.

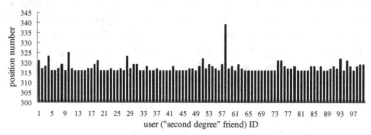

Fig. 4. Position number in 100 user trajectories after normalizing the trajectories

From Figure 4 we find that most of the number of the positions of "second degree" friends is equal to 316. For those which are not equal to 316, we only need truncate the portion which is higher than 316. Of course, we admit that we could face with the errors.

4.2   Similarity Computation Methods Comparison

In order to verify the feasibility of the algorithm proposed in this paper, we compare the proposed algorithm with the algorithm of use cosine formula to compute the similarity after DBSCAN clustering. Experimental data comes from the 100 users' locations in a certain location service community, and the number of each user's geographic location is greater than 3410.

To compute the accuracy of the algorithm we adopt 10-cross validation method which divides equally 100 users' location data into 10 portions (1~10, 11~20, ..., 91~100) and treats 9 of 10 as a training set, the rest one as a test set, then uses the mean of all of the results to estimate the accuracy of the algorithm. The main purpose of the training model is to determine the value of the parameter, here, we will use the Least Squares (LS) to determine the value of λ.

$$\arg\min_{\lambda} \sum_{i=1}^{90} \sum_{j=i+1}^{90} [sim(T_i,T_j) - simofDBSCAN(T_i,T_j)]^2$$

$$= \arg\min_{\lambda} \sum_{i=1}^{90} \sum_{j=i+1}^{90} [(1-\lambda)simofdist(T_i,T_j) + \lambda simoftrend(T_i,T_j) - simofDBSCAN(T_i,T_j)]^2$$

(10)

where $SimofDBSCAN(T_i,T_j)$ is the similarity between trajectory T_i and T_j for user i and j by the algorithm of using DBSCAN clustering. Then, we calculate the accuracy of our algorithm, the formula is as follows:

$$Accuracy = 1 - \frac{1}{\sum\limits_{i=1}^{10}\sum\limits_{j=i+1}^{10} j} \sum_{i=1}^{10} \sum_{j=i+1}^{10} | sim(T_i,T_j) - simofDBSCAN(T_i,T_j) | \qquad (11)$$

The value of $simofdist(T_i,T_j)$, $simoftrend(T_i,T_j)$ and $simofDBSCAN(T_i,T_j)$ of partial users in a certain process of training the model is shown as Figure 5.

From Figure 5, we can find that in addition to the individual points, the values of the $simofDBSCAN(T_i,T_j)$ are basically between the values of the $simofdist(T_i,T_j)$ and the values of the $simoftrend(T_i,T_j)$. This shows that our algorithm is feasible. A more in-depth mining, we find that the values of the $simofDBSCAN(T_i,T_j)$ are closer to the values of the $simofdist(T_i,T_j)$. Table 1 gives the result of each cross validation based on Formula (10) and the accuracy based on Formula (11). The average of λ is 0.31 and the average of the accuracy is 85.927%.

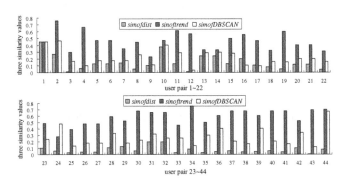

Fig. 5. Comparison of simofdist(T_i,T_j), simoftrend(T_i,T_j) and simofDBSCAN(T_i,T_j)

Table 1. λ and accuracy in every cross validation

Training ID	1	2	3	4	5	6	7	8	9	10
λ	0.29	0.33	0.21	0.39	0.27	0.37	0.41	0.28	0.35	0.20
Accuracy	0.8907	0.8473	0.8701	0.8284	0.8674	0.8143	0.8521	0.8365	0.8745	0.9114

4.3   Online Voting Activity

We launch an online voting activity and say "Now there will be an activity in your city, please choose the top-10 *second degree* friends whom you most want to invite to go with you". The voter determine whom will be invited by observing the each of "second degree" friend's personal information and the geographic locations recently emerged in Sina Weibo (It does not need to observe the latitude and longitude directly, but need to visually see the information of nation, province/district, street, etc.). Here we give the formula of computing the precision.

$$precision = \frac{NumofTop[1,10]}{10} + 0.3\frac{NumofTop[11,40]}{10} + 0.2\frac{NumofTop[41,70]}{10} \quad (12)$$
$$+ 0.1\frac{NumofTop[71,100]}{10}$$

The reason why we definite the precision with Formula (12) is that the expectation of adoption rate in the case that voting 10 from 100 users is 0.1 if and only if when the voted top-10 "second degree" friends are just the numbers of the range from 71 to 100. We know that the similarity between the voter and his/her "second degree" friends will be determined when maximizing the precision. In our experiment, what we want to get is the value of the parameter λ. We can obtain the information that which kind of similarity plays a more important role in our voting activity.

Our experiment chooses 100 voters. We can compute each user's precision and the average precision of all 100 voters whenever the parameter λ takes a specific value. Figure 6 plots the average precision of all 100 voters as the parameter λ changes.

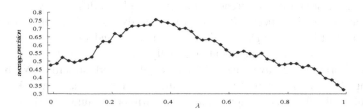

Fig. 6. Average precision of all 100 voters as the parameter λ changes

We find that the average precision up to the maximum value when λ is about 0.38. That is, when the average precision up to the maximum value, the distance similarity is more important than the trend similarity. In fact, the result is acceptable, because our experiment of voting does not impose strict restrictions on the order of the location and the voters consider more about the factor of distance.

5   Related Work

Researching on the geographic position information is very popular in recent years. Knowledge of users' positions can help improve large scale systems, such as cloud computing [12], content-based delivery networks [4], and location-based recommendations [5, 6, 7]. Additionally, the research work based on GPS data of Microsoft Research Asia has made achievements which can fix academia's eyes. The proposed methods include several stages: 1) Use GPS data to do some simple mining, such as the traveling ways [8], mining users' similarity based on trajectory [9, 10], understanding users' behaviors, and mining interesting location [11]. 2) Provide LBS service based on cloud by combining with external information [12]. 3) Achieve the recommendation of locations and activities by combining collaborative filtering algorithm [6, 13]. 4) GPS data is close to real life and also service real life[14].

Our work is quite different from above work. Firstly, we reduce user trajectories according to the order of time and unify them. Secondly, when computing the similarity between trajectories, we consider not only the distance between the trajectories but also the trend of every trajectory.

6   Conclusion

In this paper, we propose an online friend recommendation approach based on geographic trajectory similarity and social relations. We regard the trajectory consisting of a series of positions as a time sequence. We reduce the trajectory based on position and time order, and use the precise distance of latitude and longitude as the measure of the distance similarity and the slope as the measure of trend similarity. Finally, the similarity between trajectories is the result of weighted combination on distance similarity and trend similarity.

The distance similarity and trend similarity will play a different weighted role in different applications respectively. We propose a dynamic similarity formula. It's feasible to identify the weights of distance similarity and trend similarity through practical application and experiment. Certainly, for verifying the validity of the proposed algorithm, we have weakened the social network structure. We believe if combining the geographic position information with the social network structure, more ideal result will be got in applications. This will be our future work.

Acknowledgments. This research is supported by the State Key Development Program for Basic Research of China (Grant No. 2011CB302200-G), State Key Program of National Natural Science of China (Grant No. 61033007), National Natural Science Foundation of China (Grant No. 61100026, 61370074), and the Fundamental Research Funds for the Central Universities (N120404007, N100704001).

References

1. Yu, X., An, A., Tang, L., Li, Z., Han, J.: Geo-Friends Recommendation in GPS-based Cyber-physical Social Network. In: ASONAM 2011, pp. 361–368 (2011)
2. Ye, Y., Zheng, Y., Chen, Y., Feng, J., Xie, X.: Mining Individual Life Pattern Based on Location History. In: Mobile Data Management, pp. 1–10 (2009)
3. Ying, J., Lu, E., Lee, W., Weng, T., Tseng, V.: Mining user similarity from semantic trajectories. In: GIS-LBSN, pp. 19–26 (2010)
4. Leighton, T.: Improving Performance on the Internet. ACM Queue (QUEUE) 6(6), 20–29 (2008)
5. Hao, Q., Cai, R., Wang, C., Xiao, R., Yang, J., Pang, Y., Zhang, L.: Equip tourists with knowledge mined from travelogues. In: WWW 2010, pp. 401–410 (2010)
6. Zheng, V., Zheng, Y., Xie, X., Yang, Q.: Collaborative location and activity recommendations with GPS history data. In: WWW 2010, pp. 1029–1038 (2010)
7. Zheng, Y., Zhang, L., Xie, X., Ma, W.: Mining interesting locations and travel sequences from GPS trajectories. In: WWW 2009, pp. 791–800 (2009)

8. Zheng, Y., Liu, L., Wang, L., Xie, X.: Learning transportation mode from raw gps data for geographic applications on the web. In: WWW 2008, pp. 247–256 (2008)
9. Zheng, Y., Chen, Y., Li, Q., Xie, X., Ma, W.: Understanding transportation modes based on GPS data for web applications. TWEB 4(1) (2010)
10. Li, Q., Zheng, Y., Xie, X., Chen, Y., Liu, W., Ma, W.: Mining user similarity based on location history. In: GIS 2008, vol. 34 (2008)
11. Xie, X., Zhang, Y.: Understanding User Behavior Geospatially (November 30, 2008), http://research.microsoft.com/apps/pubs/?id=74370
12. Yuan, J., Zheng, Y., Xie, X., Sun, G.: Driving with knowledge from the physical world. In: KDD 2011, pp. 316–324 (2011)
13. Herlocker, J., Konstan, J., Terveen, L., Riedl, J.: Evaluating collaborative filtering recommender systems. ACM Trans. Inf. Syst. (TOIS) 22(1), 5–53 (2004)
14. Yuan, J., Zheng, Y., Zhang, C., Xie, W., Xie, X., Sun, G., Huang, Y.: T-drive: driving directions based on taxi trajectories. In: GIS 2010, pp. 99–108 (2010)

The Spontaneous Behavior in Extreme Events: A Clustering-Based Quantitative Analysis

Ning Shi[1], Chao Gao[1,*], Zili Zhang[1,2], Lu Zhong[1], and Jiajin Huang[3]

[1] College of Computer and Information Science, Southwest University, Chongqing, China
[2] School of Information Technology, Deakin University, VIC 3217, Australia
[3] International WIC Institute, Beijing University of Technology, Beijing, China
cgao@swu.edu.cn

Abstract. Social media records the pulse of social discourse and drives human behaviors in temporal and spatial dimensions, as well as the structural characteristics. These online contexts give us an opportunity to understand social perceptions of people in the context of certain events, and can help us improve disaster relief. Taking Twitter as data source, this paper quantitatively measures exogenous and endogenous social influences on collective behaviors in different events based on standard fluctuation scaling method. Different from existing studies utilizing manual keywords to denote events, we apply a clustering-based event analysis to identify the core event and its related episodes in a hashtag network. The statistical results show that exogenous factors drive the amount of information about an event and the endogenous factors play a major role in the propagation of hashtags.

1 Introduction

It is urgent and useful for improving the efficiency of disaster management to understand collective behaviors in an extreme event [1]. These results can help government monitor the damages and effects of an event and prepare for the disaster relief [2]. Traditional questionnaire-based empirical studies in collective behaviors can reveal which factors will impact on human behavior from a few of statistical data [3, 4]. However, they cannot reflect the dynamic changes of human behaviors over time. Particularly, there is no the same scenario in two extreme events. Therefore, it is difficult for us to predict human behaviors utilizing previous empirical studies. With the popular of social media, the spontaneous emergence of online human behaviors can be tracked by Micro-blogging systems such as Twitter [5, 6], which contains some hashtags standing for events or objects in its tweets [6–8]. So, it is practicable for us to perceive people's feelings or reactions towards different social or natural events that may emerge suddenly in the real world based on these textual information [9]. Specifically, there are two research questions for us to perceive human responses towards a certain event from social media.

The first one is event detection from social media that is a very popular topic in recent years [3, 10, 11]. In their studies, an event is extracted by some keywords (e.g., hashtags in Twitter) [6–8]. However, an event is often accompanied by a series of episodes [12],

---

* Corresponding author.

H. Motoda et al. (Eds.): ADMA 2013, Part I, LNAI 8346, pp. 336–347, 2013.

which are associated subevents within the entire event [13]. And these episodes may also draw different public concerns. That is to say, people will perceive different potential risks and ongoing effects of an event due to the uncertainty of an event. For example, there are different dynamic changes of emotional reaction toward nuclear crisis and natural disaster during Japanese earthquake in 2011 [14]. Moreover, some irrelevant contents may affect the quality of the results. For example, tsunami may be inputted as 'tsunmi', 'Gaddafi' and 'Qaddafi' may mean the same person in the Libya crisis. Therefore, how to accurately identify an event and its related episodes is an important problem. In this paper, we use a clustering-based analysis to distinguish event-related hashtags from a whole hashtag network that is consisted by hashtags appearing in one tweet. Taking advantages of clustering analysis, our method also can overcome typos in social media (e.g., 'Gaddafi' and 'Qaddafi' are grouped together in Fig.2(a)), which is one of the most difficult problems in NLP.

The other is identifying the effect of social influences on human behaviors that is the other popular topic in recent years. Although some researches have subdivided the social influences into global and local signals [9], they just provide experimental analysis about how two social influences affect human behaviors through modeling such effects [15–17], rather than measuring the effects of social influences from empirical data. Based on the statistical results in our dataset and some existing researches on social media [9, 15], some nonlinear characteristics of social media can be observed (as shown in Fig. 1). That is to say, human behaviors in the social media can be seen as a complex system [5]. Therefore, some methodologies in the filed of complex system can be used to measure social influences. In this paper, we use standard fluctuation scaling (FS) method, which is a typical method to measure exogenous and endogenous influences on a complex system [5, 18], to measure the effects of two social influences on human behaviors when people face different types of events.

The remainder of this paper is organized as follows: Section 2 introduces the related work and Section 3 describes the characteristics of dataset used in this paper. Section 4 proposes a method to identify an event and its related episodes from a hashtag network, which is extracted and constructed from all tweets. Section 5 quantitatively measures spontaneous collective behaviors in different events by a standard fluctuation scaling method. Section 6 highlights our major contributions.

2 Related Work

In a real world, an extreme event is composed of a series of episodes, e.g., tsunami may be associated with earthquake. It is difficult and inadequate for us to use manual keyword (e.g., one hashtag) to describe an event and its effects [19]. Currently, people often use different hashtags in one tweet to describe an event and its related subevents [6–8]. Based on the relationship about two hashtags appearing in one tweet, a hashtag-based network can be constructed. Taking advantages of a clustering method for network classification, we can categorize hashtags into different groups based on the co-occurrences of hashtags in one tweet. Therefore, an event can be denoted as a set of closely connected hashtags (i.e., a cluster), rather than only one hashtag. Based on such clusters, we can study the dynamic changes of collective attention on an event and its related subevents.

During an extreme event, it is a difficult task for emergent management to monitor collect human behaviors [1]. With the development of social media, it provides us a microscopic perspective to analyze human behaviors [2]. There are already some researches utilizing social media to analyze collective behaviors in many domains, e.g., political elections [20], financial markets [21], and interpersonal health [22]. Especially, human beings, in extreme events, may adjust their behaviors based on perceived senses from their friends and public news [23, 24]. That is to say, the social influence plays an important role in human decision making process [16]. For example, social influence can be applied to alert disasters [25] and reduce anxious emotion of our society [14]. Existing studies subdivide these influences into endogenous and exogenous factors [9, 15, 17]. How to identify the effects of both influences on human collective behaviors is the important problem as mentioned in Sec.1.

Up to now, two typical methods are used to analyze human collective behaviors in social media: empirical studies and modeling. Based on the empirical data from social media, recent researches have analyzed the influences of exogenous and endogenous factors on collective behaviors from a complex system perspective [8, 26]. They have found that a burst of collective activities is followed by a power-law relaxation [9, 15, 26, 27]. In order to further explain this finding, researchers use population-based [17, 27] and individual-based [24] models to simulate human collective behaviors. Although existing two methods have got some conclusions about the reason of collective behaviors in an event, both of them don't address the classification of events. Specially, existing methods don't address the dynamic changes of exogenous and endogenous influences over time.

Taking advantages of Twitter data in 2011 and a standard fluctuation scaling method, this paper aims to measure the effects of two social influences on collective spontaneous behaviors in different events, which are identified by a clustering method from a hashtag network.

3   Data

Because social media reflects individual spontaneous behaviors, it is desirable for us to timely and accurately capture collective social behaviors by analyzing social media. We crawled some tweets from March 8 to 31 in 2011. For the identification of events, we extracted all the hashtags through matching the tweets content to the pattern '*#*'. Table 1 lists the general statistics of dataset. Each tweet includes textual information (e.g., content, the time at which it was posted), author's information (e.g., the location, registration time) and the forwarding source if any.

Currently, lots of researches have proved the complexity of social media and human behaviors. First, the temporal characteristics of human collective activities in events are discussed in [6, 9, 26]. They have found that the spontaneous behaviors are followed by a power-law relaxation [9, 15]. Then, the structural characteristics of social media and the complexity of spatial characteristics of human responses in events are plotted in Fig. 1(a) and Fig. 1(b), respectively. According to existing finding, we apply a standard fluctuation scaling method to quantitatively measure these endogenous and exogenous influences from a complex system perspective [5, 18], which can further reveal how social influences affect human collective behaviors.

Table 1. General statistics about the database

Total number of tweets	513441
Total number of tweets with hashtags	215227
Total number of hashtags	4984
Total number of user	512473

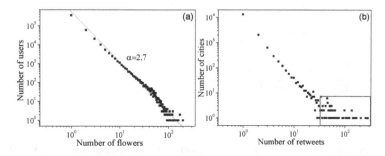

Fig. 1. The statistical results about (a) the relationship of users based on each flower following how many users, which illustrates the structural characteristic of social media. (b) the distribution of cities based on the amount of tweets released from these cities, i.e., which cities will forward a certain tweet. This result shows the spatial characteristic of human behaviors.

4 Methods

This section provides a semantic analysis for extreme events based on a clustering method. Based on the occurrences of hashtags appearing in a tweet, a hashtag network is emerged as shown in Fig. 2. This result shows that the hashtag network, consisted of an event and its related episodes, has a core-periphery structure. That is to say, an event consists of a few of core hashtags and many border hashtags. These border hashtags, as weak ties, build a connection between two events. Due to previous researches about collective behaviors are based on manual keywords (e.g., a certain hashtag), it cannot adequately describe an event. Meanwhile, it is also difficult to identify the typos and abbreviations by manual selections. In order to overcome the incomplete and biased hashtags manually selected and the typos in social media, this section proposes a clustering-based method to categorize hashtags into different groups, which presents a more precise description about an event and its related episodes, with the following five steps.

Step 1 Pretreatment
    Extracting all tweets with hashtags from dataset.
Step 2 Constructing a hashtag network
    A hashtag network, as shown in Fig. 2, is denoted as $G=<N, E>$ that has N nodes and E edges. Based on the occurrences of two hashtags in a tweet, we measure the weight e_{ij} of an edge. The larger e_{ij} an edge has, the closer two hashtags appear.

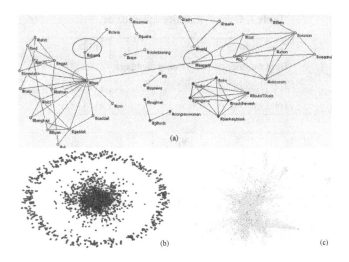

(a)

(b)   (c)

Fig. 2. (a) An example of a hashtag network based on tweets released in March 8, 2013. Red nodes are core nodes with many edges. And black nodes are border nodes with a few edges. These border nodes are bridges linking two events at times. (b) An illustration of the hashtag network G that has 4984 nodes and 8849 edges. (c) The giant component network G^* that has 471 nodes and 2837 edges in our dataset.

Step 3 Extracting the giant component network from a hashtag network

For simplicity, we remove those isolated nodes and nodes whose concurrence are less than 10 times from G as shown in Fig. 2(b). After that, we extract a giant component network G^* from G as a target, as shown in Fig. 2(c).

Step 4 Clustering analysis

In order to find out what kinds of hashtags are clustered together to describe episodes of an event, we should select an effective algorithm to separate hashtags in G^* into different communities based on their connectivity.

Currently, one of the most classical and popular community detection algorithms is the modularity-based algorithm proposed by Newman [28] which has been cited more than 2000 times since 2004. Due to the higher computational efficiency exhibited by this algorithm, we can apply it for online analysis. The modularity Q is defined as $Q = \sum_k (c_{kl} - a_k^2)$, where c_{kl} is the fraction of edges that connect hashtags in group C_k to others in group C_l, and $a_k = \sum_l c_{kl}$. This greedy algorithm aims to maximize Q at each step through merging different small clusters into a big one, i.e., maximizing $\triangle Q = 2(c_{kl} - a_k a_l)$ at each step. When Q reaches to the maximum value, the best division is obtained. Each hashtag is labeled with a cluster ID and grouped into a different cluster C_k based on the similarity of two hashtags.

Step 5 Measuring social influences based on standard fluctuation scaling

Based on the above clustering, an event can be denoted as a set of hashtags. For a hashtag i, there is an aggregate time series $n_i(t)$ that denotes there are $n_i(t)$ tweets containing i at time t. The activity of i is defined as $f_i(t) = n_i(t) - n_i(t-1)$ that corresponds to the increment of i between two adjacent time scales [5]. Each new tweet containing i, in addition to increasing the overall user base of the posting such hashtag and thus its global signal, also generates a local signal, through which a user posting i may in turn influence the future behavior of this friends. Therefore, each tweet thus acts as a microscopic social stimulus and creates a positive feedback for whole system [5].

In order to study the extent to which the behavior of an individual is related to the behavior of others, we use fluctuation scaling (FS) to divide a key signature of the system's behavior purely on the basis of the above aggregate data [18]. First, there are two measurements to estimate the time series $n_i(t)$: temporal average ($<f_i>$) and temporal standard deviation (σ_i), defined in Eq. (1) and Eq. (2), respectively.

$$< f_i >\equiv \mu_i = \frac{1}{T_i} \sum_{t=1}^{T_i} f_i(t) \tag{1}$$

$$\sigma_i = (\frac{1}{T_i - 1} \sum_{t=1}^{T_i} [f_i(t)- < f_i >]^2)^{1/2} \tag{2}$$

where T_i is the time series length [5], which reflects the fact that different hashtags are posted at different times. Then, two measurements are related based on FS by the relationship $\sigma_i \sim \mu_i^\alpha$. The fluctuation scaling exponent α lies in the rather narrow range [1/2,1] [18]. If the collective behaviors about an event is independent of the external influences $\alpha=1/2$. Whereas if the collective behaviors is fully correlated with external influences $\alpha=1$ [5, 18]. Based on this threshold, we can quantitatively measure how exogenous and endogenous influences affect human behaviors.

5   Results and Discussions

In this section, we analyze endogenous and exogenous influences on collective behaviors by measuring the fluctuation scaling exponent. Specifically, Sec. 5.1 measures emerged collective behaviors based on all tweets in one month. Sec. 5.2 further analyzes human collective behaviors in different events through categorizing event-related tweets into different groups from all tweets. Based on clustering results, we can reveal the different responses mechanisms of social media in different types of events.

5.1   Global Collective Behaviors Analysis in Social Media

For all hashtags in G^*, the relationship between $log\mu_i$ and $log\sigma_i$ is shown in Fig. 3. It reveals that posting behaviors in the social network can be seen as a complex system. Besides that, the exponent α marking by blue line in Fig. 3 is close to 1/2, which indicates that individual behaviors are independent and the propagation of these hashtags is influenced by endogenous factors. However, the exponent in the upper half of Fig. 3 equals to 0.8, which indicates the propagation of these hashtags is influenced by exogenous factors.

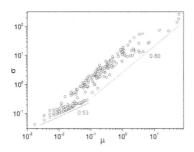

Fig. 3. The whole distribution of spontaneous emergence of human behaviors in extreme events. The figure shows that posting behaviors in the social network can be seen as a complex system. There exists an inflection point. Hashtags above inflection point are popular hashtags, hashtags below inflection point are rare hashtags.

Figure 3 illustrates that a mix of exogenous and endogenous influences have an impact on the whole collective behaviors. Our goal is to analyze social behaviors corresponding to these influences in different events described by a set of hashtags. These hashtags in our study can be divided into two parts: popular hashtags (i.e., the core nodes as shown in Fig. 2(a) highlighted with red color) and rare hashtags (i.e., the border nodes as shown in Fig. 2(a) highlighted with black color). Based on the statistical result, we find that most of core nodes, on behalf of different events, appear more than 80 times in our datasets. In contrast, border nodes are used to express the profiles of this event and most of them appear less than 30 times. Sometimes, the border nodes are bridges linking two events. The hashtag network is so complicated that we need a method to distinguish associated hashtags in order to clearly classify events.

5.2 Clustering-Based Collective Behaviors Analysis in Different Events

5.2.1 Event Detection Based on the Clustering Analysis

Based on the clustering analysis at step 4 in Sec. 4, each hashtag is labeled with a class ID. Fig. 4 plots three structures of hastags network composed of different events i.e., natural disasters (Quake and Tsunami), nuclear crisis in Japan and social crisis in Libya. We find that some hashtags about event episodes (e.g., Japan, Nuclear, Fukushima) and their corresponding impacts will cluster together, respectively. Specially, some hashtags, which seem like irrelevant to an event literally, will be grouped together based on their underlying relationships. For example, 'Australia' and 'Quake' are clustered together due to the close co-occurrence relationship in the Christchurch 2011 earthquake. Therefore, traditional analysis based on manual keyword may not reflect the real situation.

In real world, the description of event is composed of a set of related hashtags in which some core hashtags are surrounded by other border hashtags. Taking advantage of clustering method, our study aims to further analysis how social influences impact on collective behaviors in different events. Especially, we try to find out which influence plays an important role in collective behaviors through identifying the classification of hashtags in Fig. 3.

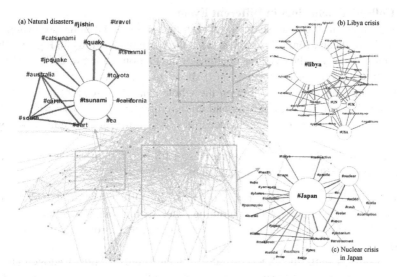

Fig. 4. Three events and their related episodes in G^* as shown in Fig. 2(c). Size of nodes and the thickness of lines in the subfigure denote the degree of a node and the weight between two nodes, respectively. An event is composed of a few core nodes and many border nodes. The core nodes describe the name and type (properties) of an event, and the border nodes illustrate the effects (relation) of an event.

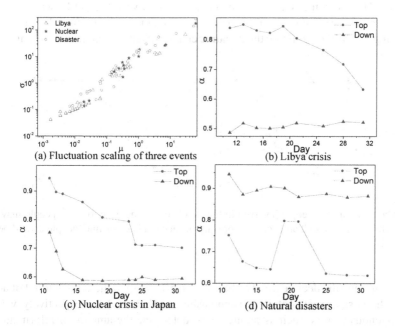

Fig. 5. (a) Fluctuation scaling of three events and their dynamic changes over time in (b)(c)(d)

5.2.2   Collective Behaviors in Different Events

Based on the clustering analysis mentioned in Sec. 4, some hashtags are grouped together to illustrate an event. This section takes three events, as shown in Fig. 4(a)(b)(c), as examples to analyze the dynamic changes of fluctuation scaling of collective behaviors in different types of events. Fig. 5(a) shows the distribution of FS when facing three different types of events based on the whole data. As discussed in Sec. 5.1, the FS exponent has two regimes: the top and bottom regions (divided by $\sigma{=}10^0$) indicate the exogenous and endogenous influences play important role on human behaviors, respectively.

Figures 5(b)-(d) plot the dynamic changes of FS in different events over time. By comparing three subfigures, we conclude that:

1. For the same event, collective behaviors are mainly affected by the exogenous influence (e.g., public news) at the beginning time. Such influence declines down over time. And endogenous influences slowly play important role in collective behaviors.
2. For different events, the effects of social influences are different. Endogenous influences always trigger more discussion among people during social crisis (e.g., Libya and Egypt disturbance).

There exits the role reversal between two influences: exogenous influences provide an alert about the event for people, and endogenous influences arouse the propagation of events. For natural disaster, people can perceive doubtless risk about the event. Therefore, the endogenous influence is weak (i.e., the value of FS is larger than 0.5 in Fig. 5(d)). However, the uncertainty of nuclear risk is very high [14], which triggers more information exchange through social media about this event. In order to estimate this finding, we further compare the dynamic change of active people in different events.

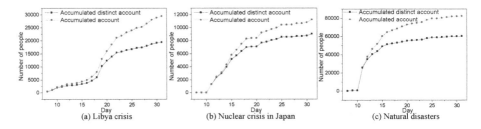

Fig. 6. The accumulated people involved in the events over time. Due to some people may post event-related tweets more than once, we compare the accumulated distinct people and all people respectively.

By comparing the decline trend of FS in Figs. 5(b)-(d) and the growth trend of active people who post a tweet containing such hashtag in Figs. 6(a)-(c) respectively, we find that exogenous factors, such as public news, determine the amount of information of an event; and endogenous factors, such as social influences, determine the duration of propagation of an event. When there is more and more people taking part in the topic, the value of FS is on a declining curve. The more the people increases, the more FS decreases. This demonstrates that the degree of social involvement (endogenous stimuli) determines the duration of propagation of an event. The number of tweets increases sharply when a network media publishes tweets on twitter, which results in the rise of FS value. The participation of network media, as an exogenous factor, determines the amount of information about an event.

6 Conclusion

Existing studies have shown that people's emotions and behaviors may be affected by each other. Different from traditional researches in the domain of social psychology based on offline data (e.g., questionnaires), the social media provides a complementary perspective for us to analyze the dynamic changes of social influences on spontaneous behaviors. In this paper, we take Twitter as example to explore the effects of social influences on collective behaviors in events. Specifically, we propose a network-based clustering method to identify events and their related episodes. And then, fluctuation scaling (FS) is used to quantitatively measure the effects of exogenous and endogenous influences on collective behaviors, as well as their dynamic changes in different events.

Different from previous work, our method can (1) eliminate the noises of typos and abbreviation in the social media; (2) identify different types of events based on the co-occurrences of hashtags, and can detect its core-periphery structural characteristic in which the hashtag network is mainly composed of a few core nodes and many border nodes; (3) analyze the dynamic changes of human collective behavior based on a fluctuation scaling method from a complex system perspective. More importantly, we reveal the underlying effects of social influences by analyzing propagation of hashtags and the amount of information of an event. We find that exogenous and endogenous influences play distinct roles in collective behaviors. Specifically, exogenous and endogenous influences determine the amount of information of an event and the duration of propagation of an event, respectively. Furthermore, taking data in one month as an example, we find that social influences containing public news and social media can be a useful resource to analyze online spontaneous emergence in an event, so as to improve effectiveness in emergence planning and response (e.g., information releasing about certain aspects of an event by monitoring human concern on the event).

Acknowledgment. Thanks to Dr. Fabian Abel for providing the related data. This project was supported by the Fundamental Research Funds for the Central Universities (No.XDJK2012B016, XDJK2012C018), SRFDP 20120182120016, the PhD fund of Southwest University (SWU111024, SWU111025) and CQ CSTC (cstc2012jjA40013, cstc2013jcyjA40022, cstc2012jjB40012), and in part by the National Training Programs of Innovation and Entrepreneurship for Undergraduates (201310635068).

References

1. Yin, J., Lampert, A., Cameron, M., Robinson, B., Power, R.: Using social media to enhance emergency situation awareness. IEEE Intelligent Systems 27(6), 52–59 (2012)
2. Adam, N.R., Shafiq, B., Staffin, R.: Spatial computing and social media in the context of disaster management. IEEE Intelligent Systems 27(6), 90–96 (2012)
3. Jones, J.H., Salathe, M.: Early assessment of anxiety and behavioral response to novel swine-origin influenza A(H1N1). PLoS ONE 4(12), e8032 (2009)
4. Oh, O., Kwon, K.H., Rao, H.R.: An exploration of social media in extreme events: rumor theory and twitter during the Haiti earthquake 2010. In: Proceedings of 31st International Conference on Information Systems (ICIS 2010), paper 231 (2010)
5. Onnela, J.P., Reed-Tsochas, F.: Spontaneous emergence of social influence in online systems. Proceedings of the National Academy of Sciences 107(43), 18375–18380 (2010)
6. Lehmann, J., Goncalves, B., Ramasco, J.J., Cattuto, C.: Dynamical classes of collective attention in Twitter. In: Proceedings of the 21st World Wide Web Conference (WWW 2012), pp. 251–260 (2012)
7. Signorini, A., Segre, A.M., Polgreen, P.M.: The use of Twitter to track levels of disease activity and public concern in the U.S. during the influenza a H1N1 pandemic. PLoS ONE 6(5), e19467 (2011)
8. Sasahara, K., Hirata, Y., Toyoda, M., Kitsuregawa, M., Aihara, K.: Quantifying collective attention from Tweet stream. PLoS ONE 8(4), e61823 (2013)
9. Crane, R., Sornette, D.: Robust dynamic classes revealed by measuring the response function of a social system. Proceedings of the National Academy of Sciences of the United States of America 105(41), 15649–15653 (2008)
10. Sakaki, T., Okazaki, M., Matsuo, Y.: Tweet analysis for real-time event detection and earthquake reporting system development. IEEE Transactions on Knowledge and Data Engineering 25(4), 919–931 (2013)
11. Sarma, A.D., Jain, A., Yu, C.: Dynamic relationship and event discovery. In: Proceedings of the 4th ACM International Conference on Web Search and Data Mining (WSDM 2011), pp. 207–216 (2011)
12. Mendonca, D., Wallace, W.A.: A cognitive model of improvisation in emergency management. IEEE Transactions on Systems, Man, and Cybernetics- Part A: System and Humans 37(4), 547–561 (2007)
13. Lindell, M.K., Prater, C.S., Perry, R.W.: Fundamentals of Emergency Management. Federal Emergency Management Agency Emergency Management Institute, Emmitsburg (2006)
14. Gao, C., Liu, J.: Clustering-based media analysis for understanding human emotional reactions in an extreme event. In: Chen, L., Felfernig, A., Liu, J., Raś, Z.W. (eds.) ISMIS 2012. LNCS, vol. 7661, pp. 125–135. Springer, Heidelberg (2012)
15. Figueiredo, F., Benevenuto, F., Almeida, J.: The tube over time: characterizing popularity growth of youtube videos. In: Proceedings of the 4th International Conference on Web Search and Data Mining (WSDM 2011), pp. 745–754 (2011)
16. Bassett, D.S., Alderson, D.L., Carlson, J.M.: Collective decision dynamics in the presence of external drivers. Physical Review E 85, 036105 (2012)
17. Myers, S., Chenguang, Z., Leskovec, J.: Information diffusion and external influence in networks. In: Proceedings of the 18th ACM SIGKDD International Conference on Knowledge Discovery and Data Mining (KDD 2012), pp. 33–41 (2012)
18. de Menezes, M.A., Barabasi, A.-L.: Separating internal and external dynamics of complex systems. Physical Review Letters 93(6), 068701 (2004)
19. Thelwall, M., Wilkinson, D., Uppal, S.: Data mining emotion in social network communication: Gender differences in MySpace. Journal of the American Society for Information Science and Technology 61(1), 190–199 (2010)

20. O'Connor, B., Balasubramanyan, R., Routledge, B.R., Smith, N.A.: From tweets to polls: linking text sentiment to public opinion time series. In: Proceedings of the 4th International AAAI Conference on Weblogs and Social Media (ICWSM 2010), pp. 122–129 (2010)
21. Bollen, J., Mao, H., Zeng, X.: Twitter mood predicts the stock market. Journal of Computational Science 2(1), 1–8 (2011)
22. Christakis, N.A., Fowler, J.H.: The spread of obesity in a large social network over 32 Years. The New England Journal of Medicine 357(4), 370–379 (2007)
23. Durham, D., Casman, E.: Incorporating Individual health-protective decisions into disease transmission models: a mathematical framework. Journal of The Royal Society Interface 9, 562–570 (2012)
24. Moran, J., Cordaro, J.: Understanding the hit-rate dynamics of a large website with an agent-based model. In: Processing of 8th International Conference on Autonomous Agents and Multiagent System (AAMAS 2009), pp. 105–109 (2009)
25. Sakaki, T., Okazaki, M., Matsuo, Y.: Earthquake shakes twitter users: real-time event eetection by social sensors. In: Proceedings of the 19th International World Wide Web Conference (WWW 2010), pp. 851–860 (2010)
26. Sano, Y., Yamada, K., Watanabe, H., Takayasu, H., Takayasu, M.: Empirical analysis of collective human behavior for extraordinary events in the blogosphere. Physical Review E 87(1), 012805 (2013)
27. Matsubara, Y., Sakurai, Y., Prakash, B.A., Li, L., Faloutsos, C.: Rise and fall patterns of information diffusion: model and implications. In: Proceedings of the 18th ACM SIGKDD International Conference on Knowledge Discovery and Data Mining (KDD 2012), pp. 6–14 (2012)
28. Newman, M.E.J.: Fast algorithm for detecting community structure in networks. Physical Review E 69(6), 066133 (2004)

Restoring: A Greedy Heuristic Approach Based on Neighborhood for Correlation Clustering

Ning Wang and Jie Li

School of Computer and Information Technology, Beijing Jiaotong University, Beijing, China
{nwang,11120451}@bjtu.edu.cn

Abstract. Correlation Clustering has received considerable attention in machine learning literature due to its not requiring specifying the number of clusters in advance. Many approximation algorithms for Correlation Clustering have been proposed with worst-case theoretical guarantees, but with less experimental evaluations. These methods simply consider the direct associations between vertices and achieve poor performance in real datasets. In this paper, we propose a neighborhood-based method called Restoring, in which we argue that the neighborhood around two connected vertices is important and two vertices belonging to the same cluster should have the same neighborhood. Our algorithm iteratively chooses two connected vertices and restores their neighborhood. We also define the cost of keeping or removing one non-common neighbor and identify a restoring order based on the neighborhood similarity. Experiments conducted on five sub datasets of Cora show that our method performs better than existing well-known methods both in results quality and objective value.

Keywords: Correlation Clustering, Neighborhood Similarity, Entity Resolution.

1 Introduction

Clustering is the problem of partitioning a set of data records into clusters, which is one of the most important problems in data mining and has applications in numerous fields. Correlation Clustering is a recent clustering formulation, introduced by Bansal et al. [1], which partitions records based on complete but possibly inconsistent pairwise information. Formally, given a complete graph whose edges are labeled either + (similar) or − (dissimilar), the goal is to produce a clustering that minimizes the number of disagreements, *i.e.*, the number of negative edges within clusters and positive edges between clusters. In the weighted case, each edge may have both positive and negative weights. The goal is to minimize the total weights of disagreements. This formulation provides us a method of clustering a set of records into the optimal number of clusters, without specifying that number in advance.

Correlation Clustering has many applications, *e.g.* Entity Resolution (ER) [2, 3] (see [4] for a tutorial of Entity Resolution), Topic Segmentation [5]. As an illustration, we take an example from ER, which will run throughout our paper. Fig. 1(a) shows a graph built based on an ER result, where each record is represented by a vertex and each matching record pair is connected by an edge. Ideally, each component in Fig. 1

H. Motoda et al. (Eds.): ADMA 2013, Part I, LNAI 8346, pp. 348–359, 2013.
© Springer-Verlag Berlin Heidelberg 2013

(a) should be a complete graph. However, the presence of inconsistent judgments made by ER algorithm makes them non-complete. Fig. 1(b) and (c) show two clustering solutions to eliminate these inconsistencies. Fig. 1(b) simply performs a transitive closure analysis on the graph shown in Fig. 1(a) and groups these vertices into two clusters. The number of disagreements made by Fig. 1(b) is 69. Fig. 1(c) groups them into five clusters with 8 disagreements. Obviously, the latter partitioning solution produces fewer disagreements with the ER result than the former. Thus, the latter can be more likely to be a Correlation Clustering solution than the former.

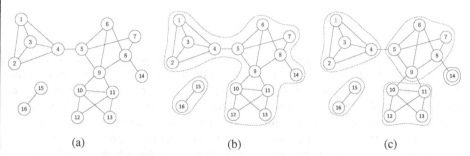

(a)   (b)   (c)

Fig. 1. (a) A graph built based on an ER result; (b) A partition solution by simply performing transitive closure analysis on the graph in (a); (c) One possible Correlation Clustering solution

The aim of Correlation Clustering is to find a clustering that eliminates inconsistencies meanwhile producing minimum disagreements with the input. Unfortunately, finding such a clustering is NP-Hard [1]. Various greedy algorithms for Correlation Clustering have been developed such as [1, 6-10]. These algorithms typically focus on providing worst-case theoretical guarantees, but experimental evaluations. However, these methods achieve poor performance in real datasets because they tend to be simple and only consider the direct associations between vertices.

In this paper, we argue that the neighborhood around two connected vertices is also important and two vertices belonging to the same cluster should have the same neighborhood. Thus, we propose a greedy method based on neighborhood, called Restoring, which iteratively chooses two connected vertices and restores their neighborhood. Refer again to the example in Fig. 1(a). Suppose the vertex pair (1, 4) belongs to the same cluster. We can restore their neighborhood by either removing the vertex 5 from the neighborhood of the vertex 4 or introducing it to the vertex 1. Which one operation we take depends on their cost. Additionally, we do not know which vertex pair will fall into the same cluster in advance. So, we need to define a measure to determine the restoring order.

The major contributions are summarized as follows:

- To our best knowledge, we are the first that considers the neighborhood in Correlation Clustering.
- We define the cost of keeping or removing one non-common neighbor when restoring a given edge and propose a restoring order based on neighborhood similarity, which has been proved effective in evaluations.

- We propose and implement a greedy heuristic neighborhood-based algorithm for Correlation Clustering, and evaluate its performance using extensive experiments.

The rest of the paper is organized as follows. Section 2 briefly reviews the related work for Correlation Clustering. Some preliminaries for the problem are given in section 3. Then our method Restoring is proposed in section 4. Empirical experiments to evaluate the proposed approach are presented in section 5. And, this paper is concluded in section 6.

2   Related Work

Correlation Clustering has been considered frequently in the machine learning community. As Correlation Clustering is hard to solve in polynomial time [1], many previous work focused on two approximation approaches. One approach is based on minimization of the disagreement while the other is on maximization of agreement. In [1], the authors focus on complete graphs and provide a constant factor approximation algorithm for minimizing disagreements and a polynomial-time approximation scheme (PTAS) for maximizing agreements. Demaine et al. [7] extended the minimizing disagreements problem to general weighted graphs where they give an $O(\log n)$ approximation algorithm based on rounding a linear program using the region growing techniques. Charikar et al. [6] gave a factor 4 approximation algorithm for minimizing disagreements on complete graphs, and also a factor $O(\log n)$ approximation for general graphs. Ailon et al. [9] presented a constant factor 2.5 approximation for minimizing disagreements, which is currently the best known factor. Swamy [8] gave a 0.76-approximation algorithm for maximizing agreements in general graphs using semi-definite programming.

Recently, many variants of Correlation Clustering have been proposed. Giotis and Guruswami [11] consider Correlation Clustering when the number of clusters is given. Bonchi et al. extend Correlation Clustering to allow overlaps in [12] and propose a generalized Correlation Clustering, in which the edge labels can be multiple but not limited to binary in [13].

An important application of Correlation Clustering is Consensus Clustering [9, 14, 15, 16], which is a methodology for combining different partitions of the same data into a single representative. These partitions could be obtained either from different clustering algorithm or from multiple runs of non-deterministic clustering algorithms. Each partition provides observations of graph vertices to co-occur in a cluster. The Consensus Clustering problem can be mapped to Correlation Clustering by combining these observations to estimate the (dis)similarity among vertices in the graph.

3   Preliminaries

Before introducing our method, we first give some preliminaries about the Correlation Clustering problem and neighborhood. Let $G = (V, E)$ be an undirected graph with n vertices, where each edge $e = (i, j)$ is labeled either + or − depending on whether vertices i and j have been deemed to be similar or dissimilar. The goal is to produce a partition that agrees as much as possible with the edge labels. In practice, each edge (i, j) in the graph has a probability $p_{i,j}$ reflecting our belief as to whether i and j can be grouped into the same cluster. In this case, each edge is assigned two real valued weights $w_{i,j}^+$ and $w_{i,j}^-$, where $w_{i,j}^+$ called positive weight represents the cost of cutting an edge whose probability is $p_{i,j}$ and $w_{i,j}^-$ called negative weight represents the cost of keeping it. Then, the goal is to find a clustering minimizing the total negative weight within clusters and positive weight between clusters.

Let $x_{i,j}$ be a binary indicator variable, where $x_{i,j} = 1$ if vertices i and j are assigned to the same cluster and $x_{i,j} = 0$ if not. Thus, the Correlation Clustering objective can be written as

$$min \sum_{ij:i<j} x_{i,j} w_{i,j}^- + (1 - x_{i,j}) w_{i,j}^+. \tag{1}$$

Naturally, we assign $w_{i,j}^+ = p_{i,j}$ and $w_{i,j}^- = 1 - p_{i,j}$. We can view the un-weighted case as weighted case by taking $p_{i,j} = 1$ if edge (i, j) is labeled $+$ and $p_{i,j} = 0$ otherwise. Additionally, we take $w_{i,j}^{\pm} = w_{i,j}^+ - w_{i,j}^-$. In the rest of this paper, we refer to an edge with $w_{i,j}^{\pm} > 0$ as positive edge and negative edge otherwise. For computational convenience, we set $w_{i,i}^+ = 1$ and $w_{i,i}^- = 0$.

The neighborhood of a vertex includes all the vertices connected to it by a positive edge. We write the neighborhood of vertex i as $\mathcal{N}(i)$, where

$$\mathcal{N}(i) = \{i\} \cup \{j : w_{i,j}^{\pm} > 0\}. \tag{2}$$

It is understandable to union the vertex itself $\{i\}$ to its neighborhood if we assume there exists one positive edge between the vertex with itself.

For a positive edge (i, j), we define its common neighborhood as the set of common neighbors of i and j, and define the non-common neighborhood of i to j as the set of non-common neighbors of i to j, denoted by $\mathcal{CN}(i, j)$ and $\mathcal{NCN}(i, j)$. Formally

$$\mathcal{CN}(i, j) = \mathcal{N}(i) \cap \mathcal{N}(j) \tag{3}$$

$$\mathcal{NCN}(i, j) = \mathcal{N}(i) - \mathcal{N}(j). \tag{4}$$

Note that the definition of $\mathcal{CN}(i, j)$ is symmetric but $\mathcal{NCN}(i, j)$. Take the example in Fig. 1(a). Consider the vertex pair $(1, 4)$ connected by a positive edge. We have the following observations: $\mathcal{N}(1) = \{1, 2, 3, 4\}$, $\mathcal{N}(4) = \{1, 2, 3, 4, 5\}$; $\mathcal{CN}(1,4) = \{1, 2, 3, 4\}$; $\mathcal{NCN}(1, 4) = \{\}$, $\mathcal{NCN}(4,1) = \{5\}$.

4   Our Method: Restoring

4.1   Framework of Restoring

Suppose we know the optimal clustering solution for the data shown in Fig. 1(a). We modify the negative edges within clusters to positive edges and meanwhile modify the positive edges between clusters to negative edges. We see these operations as a process of restoring the edge labels to their true state. As such, the final result produced by restoring is that any vertex's neighborhood will fall in the same cluster with the vertex itself. In other words, each vertex pair (i, j) connected by a positive edge in the final result share the same neighborhood. In our method, what we do is a reversal process. The details of our method Restoring are described in Algorithm 1.

Algorithm 1. Restoring(G)

> **Input:** $G = (V, E)$: a complete graph, in which each edge $e(i, j)$ is assigned
> two weights: $w_{i,j}^+$ and $w_{i,j}^-$
>
> **Output:** $\mathcal{C}$: a clustering on V
>
> 1 **begin**
> 2 $E^+ = \{(i, j) \in E \mid w_{i,j}^{\pm} > 0\}$;
> 3   Initialize all the vertices' neighborhood based on E^+;
> 4   Pick up an edge $e(i, j)$ from E^+ with $\mathcal{N}(i) \neq \mathcal{N}(j)$;
> 5   **while** $e(i, j)$ is not null **do**
> 6 $E_{added}, E_{removed}$ = RestoringEdge(e);
> 7 $E^+ = E^+ - E_{removed}$;
> 8 $E^+ = E^+ \cup E_{added}$;
> 9   Update the neighborhood;
> 10   Go to step 4;
> 11 $\mathcal{C} = \{V' \mid G' = (V', E')$ is a connected component of $G^+ = (V, E^+)\}$;
> 12 **end**

As shown in Algorithm 1, the Restoring algorithm iteratively picks up one vertex pair that is connected by a positive edge but does not share the same neighborhood and restores it. After one vertex pair restored, we update the neighborhood. Our algorithm terminates when all vertex pairs in E^+ share the same neighborhood. The final clustering solution can be obtained by simple outputting the connected components of the graph $G^+ = (V, E^+)$.

4.2   Cost of Keeping or Removing One Non-common Neighbor

Now, we discuss how to restore the neighborhood of one vertex pair. As mentioned earlier, for the vertex pair $(1, 4)$, we have two ways to make vertices 1 and 4 share the same neighborhood: 1. keeping the edge $(4, 5)$ and introducing vertex 5 to the neighborhood of vertex 1; 2. removing vertex 5 from the neighborhood of vertex 4. Which way we adopt depends on its cost. So, we need to define the cost of keeping or removing an edge. Suppose the current restoring edge is (i, j). Without loss of generality, let k be a neighbor of i and not a neighbor of j. We view the two vertices i and j as one vertex $v(i, j)$ which has the neighborhood $\mathcal{CN}(i, j)$. Then we define the keeping cost as the minimum total weight required to transform the non-common neighborhood to common neighborhood between $v(i, j)$ and k, and define the removing cost as the minimum total weight of disconnecting the common neighborhood of $v(i, j)$ and k, denoted by $cost_{keep}^{i,j}(k)$ and $cost_{remove}^{i,j}(k)$ respectively. Formally,

$$cost_{keep}^{i,j}(k) = \sum_{x \in \mathcal{CN}(i,j)-\mathcal{N}(k)} min\left\{ \frac{w_{x,i}^+ + w_{x,j}^+}{2}, w_{x,k}^- \right\} + \sum_{x \in \mathcal{N}(k)-\mathcal{CN}(i,j)} min\left\{ \frac{w_{x,i}^- + w_{x,j}^-}{2}, w_{x,k}^+ \right\} \quad (5)$$

$$cost_{remove}^{i,j}(k) = \sum_{x \in \mathcal{N}(i,j) \cap \mathcal{N}(k)} min\left\{ \frac{w_{x,i}^+ + w_{x,j}^+}{2}, w_{x,k}^+ \right\} \quad (6)$$

We keep the edge (i, k) and introduce k to j, if $cost_{remove}^{i,j}(k) > cost_{keep}^{i,j}(k)$; and remove the edge (i, k) otherwise. The pseudo code of restoring one single edge is shown in Algorithm 2.

Algorithm 2. RestoringEdge(e)

    Input: $e(i, j)$: the current restoring edge
    Output: E_{added}, $E_{removed}$: added edges set and removed edges set
1   **begin**
2   **for** each $k \in \mathcal{NCN}(i, j)$ **do**
3   **if** $cost_{remove}^{i,j}(k) > cost_{keep}^{i,j}(k)$ **then**
4   Add $e(j, k)$ into E_{added};
5   **else**
6   Add $e(i, k)$ into $E_{removed}$;
7   **for** each $k \in \mathcal{NCN}(j, i)$ **do**
8   **if** $cost_{remove}^{i,j}(k) > cost_{keep}^{i,j}(k)$ **then**
9   Add $e(i, k)$ into E_{added};
10   **else**
11   Add $e(j, k)$ into $E_{removed}$;
12 **end**

4.3 Restoring Order

We may have noticed that the restoring order would influence the later determination. So, in this section, we aim to identify a restoring order that makes as few negative effects on the later as possible. Recall the process of restoring one vertex pair in section 4.2, during which we view the two vertices as one vertex. That is, we have made an assumption that two vertices belong to the same cluster. Although we do not know the optimal clustering upfront, we can find a measure to quantify each pair's likelihood that they belong to the same cluster and restore vertex pairs based on the decreasing order of their likelihood.

In this paper, we argue that two vertices are assigned to a cluster according to how they share neighborhood. For example, consider the vertex pairs (1, 4) and (4, 5). The vertex pair (1, 4) more likely belong to the same cluster than (4, 5) since former share more neighbors. But, simply taking the size of common neighborhood of two vertices as their likelihood will be not fair for those vertices that have fewer neighbors. So, to compensate for the effect of neighborhood size, we normalize the number of common neighborhood by the geometric mean of the two neighborhoods' size as neighbors sharing degree, denoted by $\mathcal{D}(i,j)$. Formally,

$$\mathcal{D}(i,j) = \frac{|\mathcal{CN}(i,j)|}{\sqrt{|\mathcal{N}(i)||\mathcal{N}(j)|}} \tag{7}$$

A neighbors sharing degree of 1 corresponds to that two vertices have the same neighborhood whereas a neighbors sharing degree of close to 0 corresponds to a poor connectivity of two vertices.

The disadvantage of neighbors sharing degree is that it cannot capture the information of edge weights. In the weighted case, we prefer using another notion, called neighborhood similarity, which is the generalization of neighbors sharing degree. Before giving the notion, we need to define neighborhood vector ahead.

Definition: (Neighborhood Vector)
For each vertex $i \in V$, we define its neighborhood vector as $\mathcal{NV}(i) = <p_{i,1}, \dots, p_{i,n}>$, where the components of this vector are the edge probability between v and its neighborhood, and n is v's neighborhood size.

We align two vertices' neighborhood vector and fill the gaps using the corresponding probability. We employ the cosine similarity of two vertices' neighborhood vector after alignment to measure the neighborhood similarity of two vertices i and j, denoted by $Sim(i,j)$. Formally,

$$Sim(i,j) = \frac{\mathcal{NV}(i) \cdot \mathcal{NV}(j)}{\|\mathcal{NV}(i)\|\|\mathcal{NV}(j)\|} \tag{8}$$

Obviously, the neighbors sharing degree is a special case of neighborhood similarity if we set the probability 1 to positive edges and 0 otherwise.

5   Experimental Evaluation

5.1   Experimental Setup

Dataset. We use a typical Entity Resolution dataset to evaluate our method. CORA[1], is a collection of 1295 citations of 112 different research papers in computer science. It has $1295 \cdot (1295\text{-}1)/2 = 837, 865$ record pairs in total. Each record in this dataset has a class label. Records that have the same label belong to the same cluster. We randomly remove 0, 100, 200, 300, 400 records from Cora to create five sub datasets which we experiment with, to which we refer by 0, -100, -200, -300, -400, respectively.

---

[1]   CORA is provided by McCallum,
(http://www.cs.umass.edu/mccallum/data/cora-refs.tar.gz)

Edge Labels and Edge Weights Settings. In this paper, we use Jaccard similarity as the similarity function. We adopt the following ways to transform the similarity into the probability. Fig. 2 shows a simple experiment. We divide all the record pairs into 20 buckets according to their similarity values. For each bucket, we randomly select 10% record pairs as the training data. We check the proportion of the truly matching record pairs and take the value as the probability of all record pairs that fall into the bucket. For example, we have 1509 record pairs in the similarity range [0.55, 0.6]. We randomly choose 150 record pairs from them, where 135 are truly matching. Thus, it seems reasonable to assign $\frac{135}{150} = 0.9$ as the probability of these record pairs located in range [0.55, 0.6]. Obviously, the relatively smaller similarity range we set, the more accurate estimation on the matching probability we can get. Hence, in our experiment, we evenly divided the record pairs into 100 buckets.

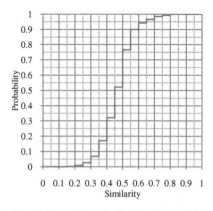

Fig. 2. Transform similarity to probability

Once we have the probability of each record pair, we can add the edge labels and the edge weights. We set the threshold of positive edges 0.5, which means only the edges whose probability are not less than 0.5 are considered as positive edges. 0.5 can produce relatively sound edge labels and weights which will provide sufficient and high quality evidences to the clustering algorithm to get better results.

Three Greedy Algorithms. We choose following three greedy algorithms for comparison in our experiments, which have been compared previously in [17].

- Pivot [9]: The algorithm repeatedly takes an un-clustered vertex i as pivot and creates a new cluster containing all un-clustered vertices j with $w_{i,j}^{\pm} > 0$.
- Best [2]: The algorithm adds each vertex i to the cluster with the vertex j which has the strongest $w_{i,j}^{\pm} > 0$, or to a new singleton if no vertex has $w_{i,j}^{\pm} > 0$.
- Vote [18]: The algorithm adds each vertex i to the cluster c which has the minimum $\sum_{j \in c[c]} w_{i,j}^{\pm}$, or to a singleton if no total is positive.

All these algorithms step through the vertices according to a vertex permutation, while our method partitions vertices according to an edge permutation. In order to make the results comparable, we propose two edge permutations, neighborhood-based permutation and probability-based permutation, which sort the positive edges by decreasing order of neighborhood similarity and probability respectively. We transform the edge permutation to vertex permutation by taking the vertices' appearance order in the edge permutation. Additionally, we will compare these methods both in weighted and un-weighted cases. So, we have four cases to compare. For weighted case, we label them "Sim" and "Prob+W". For un-weighted case, we label them "Degree" and "Prob".

Evaluation Measures. In our experiments, we take precision, recall, F-measure, Adjusted Rand Index (ARI) [19, 20] and the objective value of Correlation Clustering as the measures to evaluate the performance of these algorithms.

5.2   Experimental Results

Agreement with the Expected Clustering. The agreement of different clustering algorithms with the expected clustering under different permutations was measured using precision, recall, F-measure and ARI. Due to the limited space, we only report the result with dataset 0.

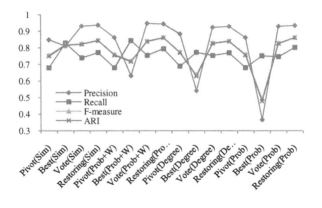

Fig. 3. Agreement with the expected clustering with dataset 0

The results shown in Fig. 3 demonstrate Restoring outperforms others in most cases. More detailed, Vote and Restoring can achieve high precision, and Vote is even slightly better. However, Vote performs a lover recall than Restoring does. The most obvious reason for that is a wrong judgment made by Vote on one vertex's cluster would lead to more wrong judgments and finally split original cluster into two parts. This will influence the recall. Unlike Vote, Restoring determines one vertex's cluster until all the positive edges' neighbors sharing degree become 1. In the process of the Restoring, any wrong judgment may be corrected in the later. Fig. 3 shows Restoring can get a stable good result both in F-measure and ARI under any one of permutations. Pivot (Prob+W) has the same accuracy result with Pivot (Prob) since Pivot only involves vertex permutations and edge labels. It is noticeable that the performance of Best is relatively poor among these algorithms, especially under the probability-based permutation because it only considers the direct associations between vertices.

Agreement with the Correlation Clustering. We use the objective values of four algorithms to evaluate their agreement with the Correlation Clustering. Minimum objective value means higher agreement with the Correlation Clustering. We first compare four algorithms under the neighborhood-based permutation. The results are given in Fig. 4. We see that Restoring can achieve better results both in weighted and un-weighted cases. Then, we compare four algorithms under the probability-based permutation, in which we add the data obtained by Restoring under the neighborhood-based permutation. The results shown in Fig. 5 demonstrate Restoring would not get the minimum objective values under the probability-based permutation, but Restoring (Sim) and Restoring (Degree) are still the better one.

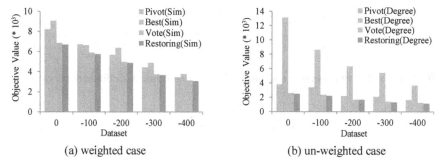

(a) weighted case   (b) un-weighted case

Fig. 4. The objective values of four algorithms under the neighborhood-based permutation

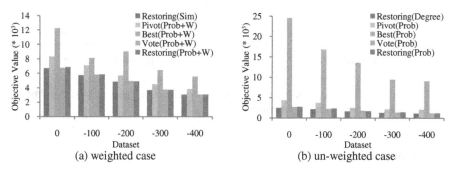

(a) weighted case   (b) un-weighted case

Fig. 5. The objective values of four algorithms under the probability-based permutation, in which the data of Restoring(Sim) and Restoring(Degree) come from Fig. 4

Effects of the Permutations on the Performance of Algorithms. As described in section 5.1, the Best algorithm adds each vertex to the cluster with the strongest positive weight connecting to it. In the same weight settings, the Best algorithm turns into one which will group each vertex i with the first vertex that has the positive edge with i in the permutation into the same cluster. So, for the Best algorithm, the permutation determines the performance of the Best algorithm. Fig. 6 shows the values of F-measure and ARI of the Best algorithm both under the neighborhood-based and probability-based permutations. We see that the better results can be obtained by adopting the neighborhood-based permutation in the Best algorithm. Similar dramatic improvement on the objective value of Best under the neighborhood-based permutation can be illustrated in Fig. 7. These great improvements imply that the neighborhood-based permutation can be more telling to determine whether two records belong to the same cluster.

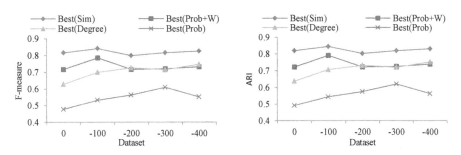

Fig. 6. The values of F-measure and ARI of the Best algorithm under different permutation

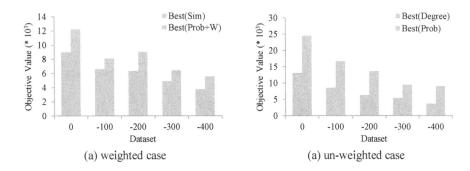

(a) weighted case   (a) un-weighted case

Fig. 7. The illustration of dramatic improvement on the objective value of Best algorithm by adopting the neighborhood-based permutation

6   Conclusion

In this paper, we proposed a neighborhood-based method for Correlation Clustering. The main idea is that two vertices belonging to the same cluster should have the same neighborhood. We defined the cost of keeping or removing one non-common neighbor when restoring a given edge. We found that the restoring order has a significant effect on the result quality. So, we proposed a restoring order based on the neighborhood similarity we designed, which proved effective in our experiments. The experiments conducted on five sub datasets of Cora showed that the proposed method performs better that the existing well-known algorithms.

Acknowledgements. This work is supported by National Natural Science Foundation of China (No.61370060), Jiangsu Provincial Natural Science Foundation of China (No.BK2011454).

References

1. Bansal, N., Blum, A., Chawla, S.: Correlation clustering. Machine Learning 56(1-3), 89–113 (2004)
2. Ng, V., Cardie, C.: Improving machine learning approaches to coreference resolution. In: Proceedings of the 40th Annual Meeting on Association for Computational Linguistics, pp. 104–111. Association for Computational Linguistics (2002)
3. Cohen, W.W., Richman, J.: Learning to match and cluster large high-dimensional data sets for data integration. In. In: Proceedings of the Eighth ACMSIGKDD International Conference on Knowledge Discovery and Data Mining, pp. 475–480. ACM (2002)
4. http://www.cs.umd.edu/~getoor/Tutorials/ER_VLDB2012.pdf (2012)
5. Malioutov, I., Barzilay, R.: Minimum cut model for spoken lecture segmentation. In: Proceedings of the 21st International Conference on Computational Linguistics and the 44th Annual Meeting of the Association for Computational Linguistics, pp. 25–32. Association for Computational Linguistics (2006)
6. Charikar, M., Guruswami, V., Wirth, A.: Clustering with qualitative information. Journal of Computer and System Sciences 71(3), 360–383 (2005)
7. Demaine, E.D., Emanuel, D., Fiat, A., Immorlica, N.: Correlation clustering in general weighted graphs. Theoret. Comput. Science 361(2–3), 172–187 (2006)
8. Swamy, C.: Correlation clustering: maximizing agreements via semi definite programming. In: Proceedings of the Fifteenth Annual ACM-SIAM Symposium on Discrete Algorithms, pp. 526–527. Society for Industrial and Applied Mathematics (2004)
9. Ailon, N., Charikar, M., Newman, A.: Aggregating in consistent information: rank in gand clustering. Journal of the ACM (JACM) 55(5), 23 (2008)
10. VanZuylen, A., Williamson, D.P.: Deterministicpi voting algorithms for constraine dranking and clustering problems. Mathematics of Operations Research 34(3), 594–620 (2009)
11. Giotis, I., Guruswami, V.: Correlation clustering with a fixed number of clusters. In: Proceedings of the Seventeenth Annual ACM-SIAM Symposium on Discrete Algorithm, pp. 1167–1176. ACM (2006)
12. Bonchi, F., Gionis, A., Ukkonen, A.: Overlapping correlation clustering. In: 2011 IEEE 11th International Conference on Data Mining (ICDM), pp. 51–60 (2011)
13. Bonchi, F., Gionis, A., Gullo, F., Ukkonen, A.: Chromatic correlation clustering. In: Proceedings of the 18th ACMSIGKDD International Conference on Knowledge Discovery and Data Mining, pp. 1321–1329. ACM (2012)
14. Bertolacci, M., Wirth, A.: Are approximation algorithms for consensus clustering worth while? In: SDM (2007)
15. Goder, A., Filkov, V.: Consensus Clustering Algorithms: Comparison and Refinement. In: ALENEX, vol. 8, pp. 109–117 (2008)
16. Gionis, A., Mannila, H., Tsaparas, P.: Clustering aggregation. ACM Transactions on Knowledge Discovery from Data (TKDD) 1(1), 4 (2007)
17. Elsner, M., Schudy, W.: Bounding and comparing methods for correlation clustering beyond ILP. In: Proceedings of the Workshop on Integer Linear Programming for Natural Langauge Processing, pp. 19–27. Association for Computational Linguistics (2009)
18. Elsner, M., Charniak, E.: You Talking to Me? A Corpus and Algorithm for Conversation Disentanglement. In: ACL, pp. 834–842 (2008)
19. Meilă, M.: Comparing clusterings—an information based distance. Journal of Multivariate Analysis 98(5), 873–895 (2007)
20. Hubert, L., Arabie, P.: Comparing partitions. Journal of Classification 2(1), 193–218 (1985)

A Local Greedy Search Method for Detecting Community Structure in Weighted Social Networks[*]

Bin Liu[1] and Tieyun Qian[2]

[1] Computer School, Wuhan University, Wuhan, 430071, China
binliu@whu.edu.cn
[2] State Key Lab of Software Engineering, Wuhan University, Wuhan, 430071, China
qty@whu.edu.cn

Abstract. In this paper, we give a new definition of community which is composed of two parts: community core and the periphery. Community core consists of highly densely connected nodes. And we propose LGSM (Local Greedy Search Method) for discovering community structures in social networks. LGSM sorts node according to weighted degree. For each node, LGSM derives a maximal weighted clique as a seed cluster. Then, LGSM adds new nodes into the seed cluster until the weighted edge density is smaller than the threshold value. After all community cores are detected, LGSM allots isolated nodes to the detected cores, and optimizes the community structure based on modularity. Our method is an integrative method, which is applicable not only to discovering overlapping communities, but also to discovering non-overlapping community. Experiments illustrate that LGSM can achieve good community structure on synthetic and real-world networks and the time complexity is O(|E||g(|V|)).

Keywords: overlapping, community core, community structure.

1 Introduction

Nowadays, researchers have found that many real-world networks possess community structure, such as large-scale social networks, Web graphs, and biological networks. This implies that the network are naturally partitioned into groups of nodes with dense internal connections while sparse connections among groups [1-5]. For example, communities in biological networks may imply functional modules [2]; communities in a citation network might indicate related papers on a research topic [2] [6], and communities in social networks represent people with common interest or background [2] [5]. Identifying these sub-structures within a network can provide insight into the network's function and interaction among communities. In real social networks, every individual typically belongs to more than one community, such as the community of his family, the community of his joining club, the community of his co-workers.

Figure 1 is a piece of research collaboration network and its community structure which is divided into two communities symbolized by circle and square. Each edge is

---

[*] This research is supported by NSFC Projects (61070011 and 61272275).

H. Motoda et al. (Eds.): ADMA 2013, Part I, LNAI 8346, pp. 360–371, 2013.
© Springer-Verlag Berlin Heidelberg 2013

assigned a nonnegative real value to evaluate the strength of the collaboration. And we assume the collaboration is closer, the value is greater. Node 3 is engaged in inter-disciplinary research. In community C1, we find that node 6 is only connected with node 2, and the subgraph consisted of node 0, 1, 2, 3, 4 is highly densely connected. We regard the subgraph as the core of C1, and Node 6 as the community periphery. Node 3 should belong to C1, and C2 community.

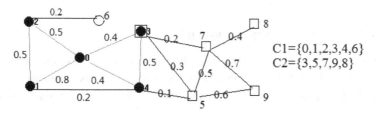

Fig. 1. A weighted network G

Our main contributions are summarized as follows.

We propose community a new definition that community equals community core and community periphery which characterizes different role of nodes in the community.

We design LGSM (local greedy search method) to find the community structure in social networks. LGSM is an integrative method, which is applicable not only to discovering overlapping communities, but also to discovering non-overlapping community.

Experimental results show that LGSM algorithm outperforms the most recent, effi-cient technique, towards both community accuracy and efficiency when the communi-ty structure is well known. By further experiments on synthetic networks, the results also show that LGSM method has high scalability on the graph size.

The rest of the paper is organized as follows. In section 2 we formulize the concep-tions used in LGSM. In section 3, we describe the algorithms in detail. In section 4, LGSM is applied to different benchmark networks and compare its performance with several baseline methods. Section 5 introduces the related works. Finally, we sum-marize our conclusions and suggest future work in section 6.

2   Preliminaries

A social network can be modeled by a weighted graph $G = (V, E, \omega)$, where V is node set, E is edge set. The cardinality of V and E are |V| and |E| respectively. An edge between nodes u and v is represented by e_{uv}. Edge weight can be represented as a function $\omega: E \rightarrow R$ that assigns each edge $e_{uv} \in E$ a value $\omega(e_{uv})$. In this paper, higher $\omega(e_{uv})$ value means high linking strength between u and v. In Figure 1, the weight of edge <2, 1> is 0.5.

2.1   Weighted Degree

The weighted degree of node v, $wd(v)$, is defined as the sum of the weights of its incident edges [9], represented in (1). $wd(0)$ is 2.1 in Figure 1.

$$wd(v) = \sum_{<u,v>\in E} \omega(e_{uv})$$ (1)

2.2   Weighted Edge Density of Subgraph

A graph $G' = (V',E',\omega)$ is a subgraph of the graph G if $V' \subseteq V$ and $E' \subseteq E$. The cardinality of V' is $|V'|$. The weighted edge density of G' is calculated by (2) [9]. For example, C is a subgraph containing node 0, 1, 2, 3, 4, $WED(C)$ equals 0.33.

$$WED(G') = \frac{2 \times \sum_{<u,v>\in E} \omega(e_{uv})}{|V'| \times (|V'|-1)}.$$ (2)

2.3   Local Subset, Boundary Subset and Peripheral Subset

For a given subgraph G' of graph G, nodes in G can be partitioned into three parts by G': Local Subset L, Boundary Subset B and Peripheral Subset U, defined in (3).

$$L = \{v \mid v \in G'\}$$
$$B = \{v \mid (\exists u)(\exists v)(u \in G') \wedge (v \in G) \wedge (v \notin G') \wedge (e_{uv} \in E)\}$$ (3)
$$U = G - L - B$$

For subgraph C, L= {0, 1, 2, 3, 4}, B= {5, 6, 7} and U= {8, 9}

2.4   Internal Weighted Degree

The internal weighted degree of node v to G', $k_v^{in}(G_{sub})$, is defined as the sum of weights of edges between v and the nodes in G', as shown in (4). For subgraph C, and node 3, 5, $k_3^{in}(C) = 0.9$, and $k_5^{in}(C) = 0.4$

$$k_v^{in}(G') = \sum_{u \in G'} \omega(e_{uv})$$ (4)

2.5   Community Core

The community core is a subgraph whose weighted edge density is greater than a given threshold. For a threshold α and a subgraph $C = (V',E',\omega)$, C is a community core if it satisfies (5).

$$WED(C) \geq \alpha \wedge WED(C\{v\}) < \alpha \quad s.t. <uv> \in E \wedge u \in C \wedge v \notin C$$ (5)

If we set α 0.3, $WED(C)$ is 0.33 which is greater than α. And if we add node 5 in-to C, $WED(C \cup \{5\})$ is 0.25. The WED is smaller than α. So $C \cup \{5\}$ is not a community core. It is NP-hard problem to find all subgraphs with weighted edge density greater than α. LGSM will adopt a heuristic search strategy to find them.

3   LGSM Method

3.1   Obtaining Community Core

LGSM chooses the seed nodes from the social network and uses local search strategy to mine community cores from those seed nodes. Seeds are very important for LGSM. A clique has been shown to be a better alternative over an individual node as a seed [6], [7].

Firstly, LGSM employs weighted degree to sort nodes. After choosing a node v, LGSM derives the max weighted clique from v and its neighbors as seed subgraph. Then all nodes in the remainder network are split into three subsets: L, B and U.

The second step is to expand the seed subgraph to obtain a community core with its weighted edge density greater than a given α. Here we adopt two heuristic search rules, (6) and (7), to expand the seed cluster by selecting the appropriate node v from B and adding it into the L.

$$k_v^{in}(L) \geq \forall k_u^{in}(L)   v, u \in B \tag{6}$$

$$(1 + \beta)k_v^{in}(L) \geq k_v^{ex}(L) \tag{7}$$

Rule (6) makes the edge density of community core may be greater than α. Rule (7) makes some nodes to become overlapping nodes if β is greater than 0. If β equals 0, LGSM can mine non-overlapping community structure.

When a community core is found and cannot be enlarged any more. LGSM will choose another seed node and repeats above procedure until all community cores are discovered. In LGSM, the seed node cannot be regarded as the overlapping node.

Here we introduce Edge Density Lemma to prove that the searching method is effective.

Lemma 1. For a given subgraph $G' = (V', E', \omega)$ and its boundary subset B, $v \in B$, if $k_v^{in}(G') \geq |V'| \times WED(G')$, then $WED(G' \cup \{v\}) \geq WED(G')$

The pseudo code of LGSM is shown in PROGRAM LGSM. When LGSM chooses the node having maximal internal weighted degree from subset B and adds it into subset L during local search, LGSM compares current $WED(C)$ with α. The result includes three cases: 1) If $WED(C)$ is greater than α. LGSM repeats the above procedure until $WED(C)$ is smaller than threshold α (from 4 to 10 lines). 2) If $WED(C \cup \{v\})$ is smaller than α but greater than $WED(C)$. LGSM repeats the above procedure until $WED(C)$ reaches maximum value (from 12 to 17 lines). If the maximum value of $WED(C)$ is greater than α, LGSM will repeat the procedure until $WED(C)$ is smaller than α (from 18 to 25 lines). 3) If $WED(C \cup \{v\})$ is smaller than α and also smaller than $WED(C)$. LGSM will stop searching and choose next seed to repeat above procedure.

```
Program LGSM
input:
    Seed Node v
    α: the threshold of weight edge density
    β: overlapping parameter
Output:
    Community Core C
(1) begin
(2)     C= MaxWeightClique(v);
(3)     Initialize B, U;
(4)         if (WED(C)≥α) then
```

(5) $\quad$ v=max($k_v^{in}(C)$);

(6) $\quad$ **while** $(WED(C \cup \{v\}) \geq \alpha)$ **do**

(7) $\quad$ C = C∪{u};

(8) $\quad$ **update** B and U;

(9) $\quad$ v=max($k_v^{in}(C)$);

(10) $\quad$ **end**

(11) $\quad$ **else**

(12) $\quad$ v=max($k_v^{in}(C)$);

(13) $\quad$ **while** $(WED(C \cup \{v\}) \geq WED(C))$ **do**

(14) $\quad$ C = C∪{v};

(15) $\quad$ **update** B and U;

(16) $\quad$ v=max($k_v^{in}(C)$);

(17) $\quad$ **end**

(18) $\quad$ **if** $(WED(C) \geq \alpha)$ **then**

(19) $\quad$ v=max($k_v^{in}(C)$);

(20) $\quad$ **while** $(WED(C \cup \{v\}) \geq \alpha)$ **do**

(21) $\quad$ C = C∪{v};

(22) $\quad$ **update** B and U;

(23) $\quad$ v=max($k_v^{in}(C)$);

(24) $\quad$ **end**

(25) $\quad$ **end**

(26) $\quad$ **end**

(27) $\quad$ **return** C

(28) **end**

Figure 2 illustrates the process of obtaining a community core. In step 1, LGSM finds three cliques: C1= {0, 1, 2}, C2= {0, 3, 4}, C3= {0, 1, 4}. But the weight of C1 is 1.8 greater than C2, and C3.   So LGSM applies C1 as the seed subgraph.

In step 4, although node 5 satisfies (6) and (7), the edge density is 0.24 after adding node 5. LGSM cannot add it into the community core and stop search.

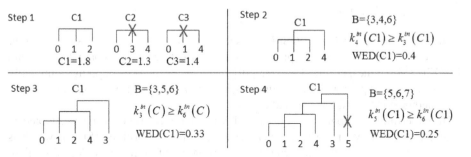

Fig. 2. Obtaining a community core from node 0 in Figure1 with $\alpha =0.3$, $\beta =0.2$

3.2   Allotting Isolated Nodes

After all communities are discovered, each isolated node, which does not belong to any community core, is allotted to the core which the isolated node has the maximal internal weighted degree to. We get the community structure of the network which is called the initial partition.

3.3   Optimizing Community Structure

LGSM intends to split large community into small communities because LGSM only chooses the node matching (6) and (7) during the expanding process. We select the modularity [2] [5] [7], Q, as the object function to optimize community structure because of its practical effectiveness and efficiency. Suppose the initial partition contains s communities, and C_i and C_j (1<=i, j<=s) are two communities, modularity can be calculated by defining modularity matrix e with s dimension. The diagonal elements e_{ii} equal to the sum of weight of edges which fall within C_i. And e_{ij} is one-half of the weight sums of the edges between nodes in C_i and nodes in C_j. Let $a_i = \sum_{j=1}^{s} e_{ij}$,be the fraction of edges attached to nodes in C_i. Q can be calculated by (8).

$$Q = \sum_{i=1}^{i=s}(e_{ii} - a_i^2) \qquad (8)$$

If there are no edges between C_i and C_j, merging them can never increase Q. We need only consider those pair communities having edges between them. The gain in Q upon merging two communities is given by (9).

$$\Delta Q = e_{ij} + e_{ji} - 2a_i a_j = 2(e_{ij} - a_i a_j) \tag{9}$$

LGSM merges two communities which can achieve the maximal gain of Q until Q reaches its maximal value.

4   Experiments

In this section, we evaluate LGSM by using synthetic benchmark datasets and four real-world datasets. In overlapping community structure, we compare LGSM with EAGEL [8], Game [9], and CFinder [10] methods. In non-overlapping community structure, we compare with our method with several representative community detection methods: DA [14], SM [3], and FM [4].

In experiment comparison, Normalized Mutual Information (NMI) is adopted to evaluate the quality of clusters generated by different algorithms [5] [7], which is currently widely used in measuring the performance of community detection algorithms. Given two community structures A and B of the same network G, A is the real and B is the detected. Suppose N is the confusion matrix whose element N_{ij} is the number of nodes in both C_i of A and C_j of B, then NMI(A, B) is defined in (10), where $N_{i.}$ is the sum over row i of N and $N_{.j}$ is the sum over column j of N.

High value of the NMI(A, B) indicates that the detected partition has high similarity with the real one. The two partitions are exactly equivalent if NMI(A,B)=1 while the two partitions are definitely different if NMI(A,B) = 0.

$$\mathrm{NMI}(A,B) = \frac{-2\sum_{i=1}^{k}\sum_{j=1}^{k} N_{ij}\log(N_{ij}|V|/N_{i.}N_{.j})}{\sum_{i=1}^{k} N_{i.}\log(N_{i.}/|V|) + \sum_{j=1}^{k} N_{.j}\log(N_{.j}/|V|)} \tag{10}$$

LGSM has two parameters α and β. For a given social network G, and a community core C, $WED(C)$ should be greater than $WED(G)$ because G is a sparse network. In the following experiments, α is set 5 times of $WED(G)$. The parameter of β controls the numbers of overlapping nodes. β is given a default value 0.2.

4.1   Experiments on Synthetic Networks with Overlapping Community Structure

Lancichinetti-Fortunato-Radicchi (LFR) algorithm is used to generate benchmark graphs [5][6] [7]with overlapping community structure. Some important parameters of the benchmark networks are listed in Table 1.

Two type weighted networks are generated with the number of node |V|=1000 and |V|=5000 [6]. By varying the parameters of the networks, we can analyze the behavior of the algorithms in detail. The mixing parameter μ is taken from the range {0.1, 0.3}. The average degree is <k> = 10, while the maximum degree is maxk = 50.

Table 1. Important Parameters of LFR algorithm

Parameters	Meaning
|V|	number of nodes
$<k>$	average degree of the nodes
maxk	maximum degree
μ	mixing parameter, each node shares a fraction$<k>$ of its edges with nodes in other communities
minc	minimum for the community sizes
maxc	maximum for the community sizes
O_N	fraction of overlapping nodes of the whole network
O_m	number of memberships of the overlapping nodes

And community sizes vary between minc = 20 and maxc = 100. We set ON to be 10% of the total number of nodes, Om to vary from 2 to 8 indicating the diversity of overlapping nodes. For each network, we generated 10 instantiations. We set α= 0.15 and apply LGSM to find the overlapping community structure.

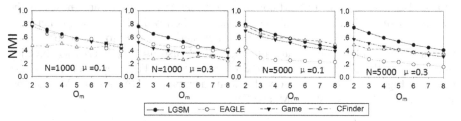

Fig. 3. Comparison with EAGLE, Game, and CFinder on computer-generated networks. Each point corresponds to an average over 100 graph realizations.

Figure 3 shows the comparison result. LGSM is better than EAGLE [8], Game [9], and CFinder [10] in both type networks. As the Om value increases, a node belongs to more communities. Then the community structure becomes fuzzy. This makes that the algorithm to detect community structure accurately is becoming more and more difficult. The accuracy of the methods is reducing with the Om value increasing except CFinder, which is not stable in \|V\|=1000.

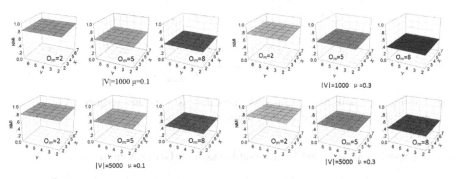

Fig. 4. The influence of α and β on prediction accuracy of LGSM. The x-axis varies from 2 to 7 which is the result of α divided by 0.05.

Then we analyze the influence of α and β on prediction accuracy of the LGSM. The value of α is set from 2 times of $WED(G)$, to 7 times of $WED(G)$. And β is set 0.1 to 0.35. The experiment results are shown in Figure 4.From Figure 4, when μ is fixed, the value of α and β has little effect on the prediction accuracy of LGSM. The prediction accuracy of LGSM is affected by μ more.

4.2   Experiments on Synthetic Networks with Non-overlapping Community Structure

In this section we analyze the performance of LGSM in detecting non-overlapping community structure. If the value of αis set 0, a node can only belong to one community. The community structure is non-overlapping.

In order to compare with existing algorithms better, LFR algorithm is applied to generate two weighted undirected networks with the number of nodes |V|=1000 and |V|=5000. For each network, three individual networks are generated with 15, 20 and 25 as average node degree, with μ varying from 0.1 to 0.6 with a span of 0.1.

DA [14], SM [3], and FM [4] community detection algorithms are chosen as the baseline methods. DA is a global optimization method which employs modularity as objection function. SM is a classical spectral clustering method to detect community structure. FM is a local optimization method which also uses modularity as objection function. All these algorithms are free of parameters. The comparison results are shown is Figure 5. It can be observed that the accuracy of DA and LGSM is almost same, and they are better than SM, and FM.

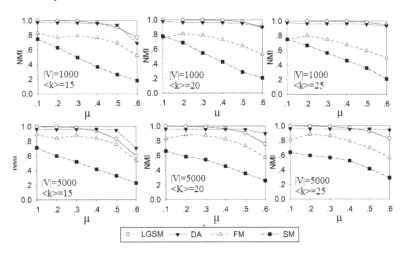

Fig. 5. Comparison with FM, DA, and SM algorithms on computer-generated networks and each point corresponds to an average over 100 graph realizations

4.3   Experiments on Real-World Networks with Ground Partition

In this section, LGSM is applied on four real-world networks: "Karate Club Network", "Dolphin Social Network", "American College Football", and "Books about

US Politics" [2] [3], which community structure is all known. Table 1 shows the comparison result of LGSM, and existing algorithms such as FM, DM, and SA. As can be seen, DA gets best results in most of cases. For "Karate Club", "American Football network", and "Books about US Politics", the community structure detected by LGSM is as good as that detected by DA.

Table 2. Comparison LGSM, with representative community detection algorithms and each cell is is a NMI value corresponding to the deteced community structure and the ground truth

Algorithm	Karate Club	American College Football	Dolphins	Books about US Politics
LGSM	0.71	0. 87	0.58	0.49
FM	0.69	0.71	0.53	0.53
DA	0.69	0.88	0.62	0.56
SM	0.69	0.70	0.48	0.49

4.4   Running Time Complexity

Finally, we analyze the time complexity of our algorithm LGSM. The complexity of computing degree is $O(|V| < k >)$ where <k> is the average neighbor nodes of each node. LGSM applies heap sort to rank nodes, so the time complexity of heap sorting is $O(|V| \lg(|V|))$.

Since the network contains |V| nodes, |V| is the maximal number of seed node. When a seed is identified, LGSM derive a clique from its one-order-neighbor-subgraph. Although the time complexity of finding a clique is high, but LGSM only find the clique from one-order-neighbor-subgraph. The real time cost actually low and ignored here.

The next step is to build the boundary subset and choose the nodes according to their internal weight degree. Since the seed node has $< K >$ neighbors and each neighbor node has $< k >$ neighbors, the boundary subset contains $< k >^2$ nodes at most. The complexity of choosing node is $O(< k >^2)$.

After LGSM updating subset B, the complexity is $O(< k >^2)$. Because the average diameter of a social network is $\lg(|V|)$ [1], the average iterative step of LGSM is $\lg(|V|)$. The total complexity is $O(2|V| < k >^2 \lg(|V|) + |V| \lg(|V|))$. Since |V|*<k>=2|E| and the network is a sparse network, the complexity approximates $O(|E| \lg(|V|))$.

Table 3. Time Complexity

Algorithm	Time Complexity
LGSM	$O(|E| \lg(|V|))$
FM	$O(|V|^2)$ or $O(|V|(|V|+|E|))$
DA	$O(|V|^2)$ or $O(|V|(|V|+|E|))$
SM	$O(|V|^3)$

To illustrate the running time of the proposed algorithms, we generate five net-
works with the number of nodes |V| ranging from 1,000 to 5,000 with <k> as 20. Fig-
ure 8 shows the running time. We observe that LGSM is faster than DA, FM, and SM.
Furthermore, we generate larger synthetic networks with the number of nodes |V|
ranging from 10,000 to 50,000 with <k> as 15, 20 respectively. We can find that
LGSM can process the network of 50,000 nodes within 800 seconds.

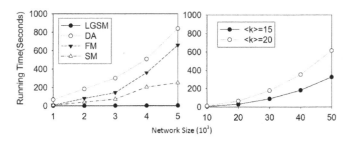

Fig. 6. Runing time on synthetic networks

5   Related Works

Community detection has become a challenge task which has received a great deal of
attention in recent years [5] [6] [7]. And many algorithms are put forward, which can
be divided into two types: non-overlapping, and overlapping.

Non-overlapping methods. Agglomerative methods merge nodes into a cluster
according to some criterion such as node similarity [11], while divisive methods remove edges from the network until the network is split into clusters based on edge or
node properties such as betweenness [2] [3]. These methods need to set a condition to
stop them [3].

Modularity, denoted as Q, is a benefit function that measures the quality of a particular division of a network into communities which is put forward by Girvan and
Newman [3] [7]. And many optimization approaches are proposed to discover communities in a network [12] [13] [14]. Modularity-based methods suffer from time
complexity [3] [7].

Overlapping methods. Chen proposed a game-theoretic framework to find overlapping community structure, in which a community is associated with a Nash local
equilibrium [9]. Palla designed CFinder method based on the clique percolation [10].
CFinder begins by identifying all cliques of size k in a network. And all k-cliques,
sharing k-1 nodes, are regarded as overlapping communities. However it also fails to
terminate in many large social networks. EAGLE uses the agglomerative framework
to produce a dendrogram [8]. First, all maximal cliques are found and made to be the
initial communities. Then, the pair of communities with maximum similarity is
merged. The optimal cut on the dendrogram is determined by the extended modularity
[7]. And also its time complexity is high.

6   Conclusion and Future Work

In this paper, we partition community into two parts: community core and community periphery according to different roles which the nodes play in a community. And we propose a method, LGSM, to detect community structure in weighted social networks. This method is not only suitable for overlapping community detection but also for non-overlapping community detection. Experiments on synthetic networks and real-world networks show that LGSM can get better performance and has lower time complexity than the benchmark community detection algorithms. Our future work will apply LGSM to investigate the local communities in large-scale online networks and to use our method to analyze complex networks in various applications.

References

1. Newman, M.: Communities, modules and large-scale structure in networks. Nature Physics 8, 25–31 (2012)
2. Newman, M.: Modularity and community structure in networks. PNAS 103(23), 8577–8582 (2006)
3. Girvan, M., Newman, M.: Community structure in social and biological networks. PNAS 99(12), 7821–7826 (2002)
4. Newman, M.: Fast algorithm for detecting community structure in networks. Phys. Rev. E 69, 066133 (2004)
5. Lancichinetti, A., Fortunato, S.: Community detection algorithms: A comparative analysis. Phys. Rev. E 80(5), 056117 (2009)
6. Xie, J., Kelley, S., Szymanski, B.K.: Overlapping Community Detection in Networks: The State of the Art and Comparative Study. ACM Computing Surveys 4 (2013)
7. Fortunato, S.: Community detection in graphs. arXiv:0906.0612 (2009)
8. Shen, H., Cheng, X., Cai, K., Hu, M.-B.: Detect overlapping and hierarchical community structure. Physica A 388, 1706 (2009)
9. Chen, W., Liu, Z., Sun, X., Wang, Y.: A game-theoretic framework to identify overlapping communities in social networks. Data Min. Knowl. Discov. 21, 224–240 (2010)
10. Palla, G., Derényi, I., Farkas, I., Vicsek, T.: Uncovering the overlapping community structure of complex networks in nature and society. Nature 435, 814–818 (2005)
11. Jianbin, H., Heli, S., Jiawei, H., Hongbo, D., Yizhou, S., Yaguang, L.: SHRINK: A Structural Clustering Algorithm for Detecting Hierarchical Communities in Networks. In: CIKM, pp. 219–228 (2009)
12. Clauset, A., Newman, M., Moore, C.: Finding community structure in very large networks. Phys. Rev. E 70, 066111 (2004)
13. Medus, A., Acuna, G., Dorso, C.O.: Detection of community structures in networks via global optimization. Physical A 358, 593–604 (2005)
14. Duch, J., Alex Arenas, A.: Community detection in complex networks using extremal optimization. Phys. Rev. E 72, 027104 (2005)

Tree-Based Mining for Discovering Patterns of Reposting Behavior in Microblog

Huilei He[1], Zhiwen Yu[1], Bin Guo[1], Xinjiang Lu[1], and Jilei Tian[2]

[1] School of Computer Science, Northwestern Polytechnical University, Xi'an, China
{nwpuhhl,ramber1836}@gmail.com,
{zhiwenyu,guobin.keio}@nwpu.edu.cn
[2] Nokia, China
jilei.tian@nokia.com

Abstract. Discovering behavior patterns is important in online human interaction understanding (e.g., how information is shared through reposting, what roles do people play in a conversation). As reposting has become the key mechanism for information propagation in social media (e.g. microblog) and contributes a lot to users' participation in online events, it is important to explore how repost works. Different from previous studies, we make two contributions in this work: firstly, we analyze the patterns of reposting behavior from the perspective of microblog user and employ a special mining method which successfully find interesting results; secondly, our analysis is based on the Sina Weibo, which has different characteristics with Twitter. Specifically, information flow for a certain message in Weibo is represented as a tree. Tree-based pattern mining algorithm is presented to extract a number of interesting patterns which are useful for understanding information diffusion in the Weibo network.

Keywords: Information propagation, Reposting behavior, Microblog, Tree-based pattern mining.

1   Introduction

Along with the development of Web 2.0 applications, social network services have gained ever-increasing popularity as a result of people's growing communication demand as well as Internet's permeation into everyone's daily life. These services have profoundly changed the way people acquire knowledge, share information and interact with one another on a societal scale.

Microblog is one of the most important types of social media services and has become a popular communication tool among Internet users. As microblog services have gained wide popularity, users have applied them for many purposes, such as sharing news, promoting political views, showing off, marketing, and tracking real-time events [1,3]. In microblog, short messages of a maximum of 140 characters are posted by users, which are called "tweet" or "post". A user can retweet or repost a message posted by others. Users may "follow" one another to receive all up-to-date

H. Motoda et al. (Eds.): ADMA 2013, Part I, LNAI 8346, pp. 372–384, 2013.
© Springer-Verlag Berlin Heidelberg 2013

messages published by interested users and get "followed" by other users to spread his messages. Those who follow a user are called his or her *followers* and those whom the user follows are called his or her *followees*. Since following someone does not necessarily mean that they will follow you back, the Followees-Followers network is a directed graph [5]. One can also use "@" to mention others and address messages to them directly. The ease of usage of tweets has made possible the swift propagation of news and messages in Twitter network [3].

Among various microblog systems, Twitter is the first microblog site which has been well studied. But in contrast, Chinese social media has not been well-studied. Although Chinese social media services are much younger than Twitter, but the number of users, real time content distribution and influence it produces are tremendous. China's unique cultural and social environment also suggest that individuals' behaviors online might be different than that in Western counterparts. Yu et al. [12] found that the effect of reposting is much larger in Sina Weibo where users are more likely to learn about a particular topic through reposts. In this paper, we examine in detail a fascinating online environment: Sina Weibo [10]. Reposting is one of the most important features in Weibo, which relays a post that has been written by another user. When a user finds an interesting post published by someone and wants to share it, s/he can simply repost the message. Reposting can apparently be considered as an efficient way of information propagation since the original tweet is propagated to a new set of consumers, namely the followers of the reposter. With its rapid growth, Sina Weibo plays an important role in the social networking world and thus studying the patterns of reposting behavior in Weibo is significant.

Since reposting has become the key mechanism for information propagation in microblog and contributes a lot to users' participation in online events, there have been some studies devoted to explore reposting behaviors. While some works [1, 2, 7] have been carried out in studying the retweet behavior, they mainly focused on the global factors, i.e., factors related to retweeting behaviors or from the perspective of information spreading [4, 8]. However, none of them has answered the questions that what are the patterns of the reposting behavior (e.g., is it true that a celebrity is more likely to be reposted and reasons behind) on the microblog network. It has also been proved that many special features in Sina Weibo differ from Twitter [9]. In our work, we conduct an in-depth study on the patterns of reposting behavior in Sina Weibo.

To discover the frequent patterns of reposting behavior in Weibo is non-trivial. The challenges are two folds: *i) How to accurately represent the information diffusion process for a certain original post*. After an original post is sent out on the microblog network, it will spread in a complex cascaded way and the mutual relationship between the involved users is intricate. It is difficult to obtain the complete structure for a certain reposting process (e.g., how a piece of information is spread to a user from her/his followees and then s/he reposts it to her/his followers), especially for those famous posts which have been reposted thousands or even tens of thousands of times. *ii) What data mining techniques could be employed to solve this problem.* Unlike the problem of associations mining or sequences mining, we focus on mining frequent patterns of the reposting behavior in Weibo, which is presented as heterogeneous collection of ill-structured data like forest or graph and thus is more difficult to analyze. Some algorithms for discovering the tree-like patterns basically adopted a straightforward generate-and-test strategy [15,16], which is not applicable here because the mining algorithm should depend on the size of the dataset. In our work, the information flow for a certain message in Weibo is represented as a repost tree. We treat repost trees as representations of information diffusion structure. In order to find out the most frequent patterns of reposting behavior, we label the nodes in the repost tree according to the corresponding users' followers' number and formulate the problem of mining subtrees in a forest of rooted, labeled, and ordered trees. We investigate data mining techniques to detect and analyze frequent reposting behavior patterns and hope to discover various types of new knowledge on information propagation.

The rest of this paper is organized as follows: In Section 2, we discuss previous studies related to our work. Section 3 introduces the repost behavior modeling. The pattern mining method is presented in Section 4. Section 5 presents the dataset and experimental results. Finally, we conclude in Section 6.

2   Related Work

Microblog has attracted much attention in the research community since it became an important social network service. Twitter has been well studied. Kwak et al. [4] conducted a large-scale study to analyze the topological characteristics of Twitter and its power as a new medium of information sharing. Java et al. [3] provided initial analysis on the topological and geographical properties of the twitter social graph along with observations on what type of content people used to tweet. Cha et al. [5] developed a framework to measure and model an individual's influence on twitter, and found that a high follower count does not necessarily lead to many retweets. In general, the above-mentioned studies aim at analyzing the basic properties of the twitter network, while our work focuses on higher level knowledge of the microblog to discover frequent patterns of reposting behavior.

As retweeting has become the key mechanism for spreading information in Twitter network, there have been a number of studies on retweeting to explore how information is diffused. The propagation graph and statistics are studied in [4]. Boyd et al. [1] have investigated retweeting as a conversational practice, such as how authorship, attribution, and communicative fidelity are negotiated in various ways. Suh et al. [2] investigated a number of tweet features that have potential relationship with the retweetability of tweets. Yang et al. [7] found that almost 25.5% of the tweets posted by users are actually retweeted from friends' blog spaces and proposed a factor graph model to predict users' retweeting behaviors. Yang et al. [8] studied the underlying mechanism of the retweeting behaviors. These studies mainly focus on analyzing the related factors of retweeting from the information spreading perspective. However, since the information flows in social network carry rich information about user behaviors, it is still an open question that what are the patterns of reposting behavior from the users' perspective (e.g., how does the message propagate between users with different influence) and why the information spreads in that way on microblog network?

Until now, quite few studies on Sina Weibo have been conducted. The differences between Sina Weibo and Twitter were studied in [9] wihich gain an in-depth understanding of what kinds of users are more active and what kinds of content are favored most for reposting. Qu et al. [11] studied the roles played by microblog systems in response to major disasters. Our work mainly focuses on finding out the patterns of the reposting behavior to reveal how information spread between diffetnt users in Weibo microblog.

3   Weibo Repost Behavior Modeling

A common practice on Sina Weibo is reposting, or rebroadcasting someone else's messages to one's followers. This section we present how repost behavior should be modeled.

3.1   Weibo Repost Tree Definition

When a piece of information is generated on the microblog network, it will spread in a complex cascaded way, which forms an information flow tree. Here, we represent the information flow for a certain post as repost tree. The root node of the repost tree stands for the author of the original post and other nodes stand for users that have reposted the original post onward. A node can have many child nodes, which means that the post has been reposted by many other users. All repost trees are subgraphs of the Weibo network.

Definition 1 (Weibo Repost Tree). *A tree is used to represent an information flow in Weibo. In this paper, trees are rooted, directed, sequential, and labelled. A tree is denoted as $T = (V, E)$, where $V = \{0,1,...,n\}$ is the set of vertices representing each retweeting Weibo users, and $E = \{ (v_i, v_j) | v_i, v_j \in V, v_i \neq v_j \}$ is the set of edges to connect the retweeting users. One distinguished vertex $v_r \in V$ is designated the root. If $(v_i, v_j) \in E$, then v_i is the parent of v_j or v_j is a child of v_i, denoting that v_j retweet from v_i. Each child has one and only one parent but a parent may have multiple children. Further, $l: V \rightarrow L$ is a labelling function mapping vertices to a set of Labels $L = \{l_1, l_2 ...\}$; for any node $v_i \in V$, $L (v_i)$ is the label of v_i. Edges are not labelled. The children of each vertex are ordered, i.e., left vertex represents that the retweeting behaviour occurs earlier than the right ones.*

Ma et al. [17] conduct a study on modeling the popularity of microblogs considering the number of followers. Intuitively, a tweet from a user with millions of followers is much more influential than that from a user with tens of followers. Furthermore, a retweet by a popular user, who has many followers, may increase the popularity of the original tweet. Specifically, we adopt the number of followers as the label criterion to describe a user's popularity, where $L = \{a, b, c, d, e\}$ indicates a Weibo user's number of followers is larger than 100K, between 10K-100K, between 1K-10K, between 100-1K, and between 1 to 100, respectively.

3.2   Weibo Repost Tree Construction

Users may "follow" one another to receive all up-to-date messages published by interested users and get "followed" by other users to spread his messages. Consequently, users in microblog sphere receive all messages which are published by the followees and aggregated in a single reverse-chronologically order. Thus we construct the repost tree based on the follower/followee relationship of the involved users in the ordered repost list. Original posts are represented as the root node, and the boosting reposts are placed as the children or grandchildren of the root. The child nodes in each depth of the tree are ranked in the chronological order. The relationship of Weibo users change dynamically, as it is common for users to add or remove his/her followees and due to the limitation of Sina Weibo APIs, it is difficult to obtain the complete structure of a certain reposting process. We formulate the Weibo repost tree construction in Algorithm 1 to get the optimal approximation of the information diffusion path.

For each user in a given repost list RL, the algorithm firstly labels each user using the corresponding user's number of followers as mentioned above and sorts the repost list in chronological order (Steps 1-2). The initialization is then processed (Steps 3-5). For each user u in RL, we iterate to select the appropriate parent node p from P^i and then add u to the tree (Steps 6-11). We start the construction of next tree level by adding I to i (Step 12) until breaking the loop. If RL is not empty, the remaining users in RL are added as child nodes of the root (Steps 13-14). These users are not followers of any users in the repost tree, they may repost from other approaches, such as trending topics and search engine. Finally, we transform the generated repost tree to an xml format file for storage (Step 15).

Algorithm 1. RTC (RL) (Repost Tree Construction)
Input: repost list for a given original post, RL
Output: repost tree in xml format
Procedure:
(1)   for each repost user u in RL, do
      label u using the labelling function l
(2)   sort RL using the repost time from earlier to late
(3)   init root node of the tree as r using author of the post
(4) $i \leftarrow I$
(5)   add r to the parent nodes set P^i
(6)   while $RL \neq \emptyset$ and $P^i \neq \emptyset$
(7)   for each repost user u in RL, do
(8)   for each candidate parent node p in P^i, do
(9)   if p is the followee of u && u repost from p
(10)   add u as the right most child of the p
(11) $P^{i+1} \leftarrow P^{i+1} \cup u$, then remove u from RL
(12) $i \leftarrow i + I$
(13)   if $RL \neq \emptyset$
(14)   add the remaining users in RL as child nodes of r
      ranking by the repost time
(15)   transform the repost tree to xml format file

3.3   An Example of Repost Tree

To illustrate the reposting function on Sina Weibo and the process of the repost tree construction, we present an example which is a piece of news about "The disappearance of iPhone5's magic", as shown in Fig. 1. Sina Weibo provides convenient retweet buttons for users to retweet a tweet easily. The equivalent of a repost on SinaWeibo is shown as two amalgamated entries: the original entry and the current user's actual entry which is a commentary on the original entry. In this example, the original author (refered to as *A*) was reposted 65 times totally while the current user (refered to as *B*) contributed once to the reposting. There are a total of 66 nodes in this repost tree and depth of the tree is 4. Most of the repost users appear at depth 2 and five of them are reposted by their followers further.

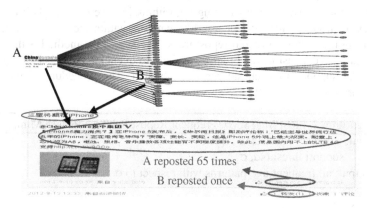

A reposted 65 times

B reposted once

Fig. 1. An example of a repost on Sina Weibo

4   Tree-Based Pattern Mining

In this section, based on the repost tree structure, we present the frequent pattern mining algorithm. With the tree constructed for each repost list in our dataset, we built the tree sets consisting of the generated repost trees. To mine the frequent patterns, we firstly give the definitions for patterns, tree set size, and support for determining patterns. Afterwards, we present the mining algorithm. Table 1 shows the symbols we use.

Table 1. Symbols

Notation	Description
TS	A tree set consists of repost trees
t	A tree
t^k	A subtree with *k* nodes, i.e., *k-subtree*
C^k	A set of candidates with *k* nodes
F^k	A set of frequent *k-subtrees*
σ	A support threshold *minsupp*

Definition 2 (Patterns). *Patterns are defined as frequent SubTrees in the tree set.*

Definition 3 (Tree Set Size). We *use |TS| to represent the numbers of nodes in a tree set named TS.*

Definition 4 (Support). *Given a subtree T and a tree set TS, the support of T is defined as:*

$$supp(T) = \frac{number\ of\ occurrences\ of\ T}{total\ number\ of\ vertices\ in\ TS}$$

If the value of *supp(T)* is more than a threshold value *minsupp* (e.g., 1%), *T* is called a "frequent subtree". Given a minimum support threshold σ, we aim to find all the subtrees that appear at least $\sigma \times |TS|$ times in the set.

Based on the frequent pattern mining algorithms [6], we employ the tree mining techniques to discover all frequent tree-like patterns within the tree set, which is a large collection of labeled ordered repost trees. The key of the method is the concept of the rightmost expansion. In order to get all frequent subtrees according to the *minsupp*, the support for each subtree is calculated. The procedure of the mining process is presented in algorithm 2.

Algorithm 2. FRSPM *(TS, σ)* (Frequent Repost SubTree Pattern Mining)

Input: a tree set consists of repost trees, *TS*
        support threshold, σ
Output: all frequent tree patterns with respect to σ
Procedure:
(1) $i \leftarrow 1$
(2)   scan *TS*, calculate the support for each labelled node
(3)   select the nodes whose support are larger than σ to form F^i
(4)   while $F^i \neq \varnothing$
(5)   for each tree t^i in F^i, do
(6)   *expandResult* ← Right_Most_Expand (t^i)
(7) $C^{i+1} \leftarrow expandResult \cup C^{i+1}$
(8)   for each pattern $T \in C^{i+1}$, do
(9)   if supp(T) > σ
             $F^{i+1} \leftarrow F^{i+1} \cup T$
(10) $i \leftarrow i + 1$
(11) output all frequent subtrees in $F_k (1 \leq k < i)$ whose supporting values
      are larger than σ

The algorithm firstly calculates the corresponding support for nodes with distinct label from *L*, and then selects the nodes whose support are larger than σ to form the set of frequent nodes, F^1 (Steps 1-3). It then calls the Right_Most_Expand for each existing frequent *i-subtrees* from F^i to generate the set of candidate subtrees with i + 1 nodes C^{i+1} (Steps 5-7). If there are any trees whose supports are larger than σ, it selects them to form F^{i+1} (Steps 8-9). Then repeatting the procedure by adding *1* to *i* (Step 10) until breaking the loop. Finally, it outputs all the subtrees whose supports are larger than σ from F^i to F^{i-1} (Step 11). Step 6 calls a subprocedure (Right_Most_Expand) to get the expanding result of t^i. The subprocedure, Right_Most_Expand, is executed for right most expansion for a pattern, which is used to grow a tree by attaching new nodes only on the rightmost branch of the tree. This subprocedure is presented as below. It firstly gets the right most branch of t^k (Step 1). Then, for each node r in the branch, it adds a new node with different label as the right most child of r (Steps 2-4). In Step 5, each new generated pattern c^{k+1} is then added to S^{k+1} (all expansion results of t^k). It returns the candidate patterns (Step 6).

Subprocedure. Right_Most_Expand (t^k)
Input: a tree pattern of k nodes, t^k
Return: candidate frequent subtree patterns, each including k+1 node
(1) $rmb \leftarrow$ the right most branch of t^k
(2)   for each node r in rmb, do
(3)   for each l in L, do
(4)   create a new node n labeled as l, then
$c^{k+1} \leftarrow$ add n as the right most child of r
(5) $S^{k+1} \leftarrow c^{k+1} \cup S^{k+1}$
(6)   return all the generated candidate patterns S^{k+1}

We define number of labels in L as $|L|$ and depth of the right most branch for T as $|D|$, then there would be $|L| \times |D|$ right most expansion results for T. Here we give an example and illustrate the expansion process in Fig. 2. Suppose that $L = \{a, b\}$. As we can see, there are 4 different expansion results for the given pattern tree. There are two nodes in the right most branch, which are the gray nodes in Fig. 2(a) and Fig. 2(b). In Fig. 2(a), new nodes labeled as a and b are added as the right most child of the gray node a respectively. The same process is showed in Fig. 2(b) for the gray node b.

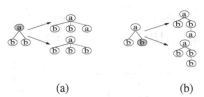

(a)   (b)

Fig. 2. Example of the Rightmost Expansion for a Pattern Tree

5   Experiments and Discussion

In this section, we firstly describe the dataset used in this paper, and then present the experiments and findings of the popular repost and pattern mining results.

5.1   Dataset

To trace the dissimilation of posts, we implemented a crawler tool and collected 241 message cascades (more than 1 million reposts totally) based on Sina Weibo open API[1]. We also collected the repost list of these messages and the corresponding follower/followee relationships of related Weibo users with Sina Weibo's open API[2,3]. To ensure that our social graph contains most of the follower/followee relationships, we collected the followeeIDs of the users who are involved in the retweeting process.

We analyze the repost tree depth distribution. As shown in Fig. 3, we observe that the tree depth varies from 2 to 11, and the number of repost trees for each depth decrease progressively from small to large. A deeper repost tree means that the information spreads more widely in the Weibo sphere and maybe reposted by more users. In other words, there is a high correlation between depth of repost tree and total number of retweets (with a correlation coefficient of 0.81).

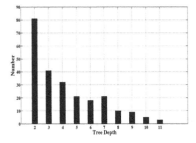

Fig. 3. Repost tree level distribution   **Fig. 4.** Tree set nodes distribution

Table 2. TreeSet

TreeSet	Number of Trees	Number of Nodes	Reposts Number Range
C1	111	1907	1-100
C2	22	6494	100-1k
C3	50	149069	1k-10k
C4	58	865376	More than 10k
C5 (C1 + C2 + C3 + C4)	241	1022846	More than 1

---

[1] http://open.weibo.com/wiki/2/statuses/user_timeline
[2] http://open.weibo.com/wiki/2/statuses/repost_timeline
[3] http://open.weibo.com/wiki/2/friendships/friends/ids

5.2   Pattern Mining Results and Findings

In order to study the frequent patterns of the repost tree, we first generate five different tree sets according to the number of reposts, as shown in Table 2. As nodes in the tree set are labeled (refer to Section 3.1), we analyze the nodes' label distribution. From Fig. 4, we can see that nodes of *d* and *e* account for a greater proportion than the others, which means that there are a lot of users with small number of followers in our tree set.

With the frequent repost subTree pattern mining algorithm mentioned in Section 4, we set the support threshold σ as 0.3 percent and discover the frequent patterns of *C5*. As showed in Fig. 5, we select the top 10 frequent patterns, which are ranked by their support values. Fig. 5(1) and Fig. 5(2) show the most frequent patterns contain only one node (*e* and *d*), which indicates that users on Sina Weibo expressing their attitudes by reposting are often the ones with fewer followers, and even their neighbors seldom take any action to response them. Therefore the most of the information propagation processes are interrupted and sliced. In Fig. 5(3), one user labeled as *a* is reposted by another user labeled as *e*. The pattern shown in Fig. 5(4) indicates that one user labeled as *a* is reposted by two users, which are labeled as *d* and *e* respectively. And the left node (*d*) is prior to the right node (*e*) for the repost behavior.

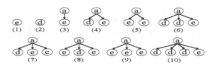

Fig. 5. The top 10 frequent patterns for TreeSet *C5*

For different TreeSets, we ignore the patterns with only one node and compare the generated frequent patterns in Fig. 6, which depicts the top three patterns for a certain number of nodes (from 2 to 5) ranked by the value of support. Comparing the generated patterns of the different TreeSets, it can be observed that information flows are all started with users labeled as *a* and messages tend to propagate in some certain paths. Some of them are similar to traditional media such as newspapers, while the others are inhered in the online social network structure. Several interesting findings are presented as follows:

- *Information tends to propagate from advanced-users to low-level users.* It indicates that the advanced-users get a lot of reposts and promote the information spreading to a deeper level in the Weibo sphere. For patterns with two nodes, we discover that pattern 2-1, pattern 2-2, and pattern 2-3 in TreeSet *C1* also exist in other TreeSets of *C2, C3* and *C4*, except the pattern 2-3 (*a* → *a*) in TreeSet *C3*. This pattern may indicate the situation that the user labeled as *a* reposts him/herself a lot during the information propagation process. Further study is needed to explain this pattern precisely.

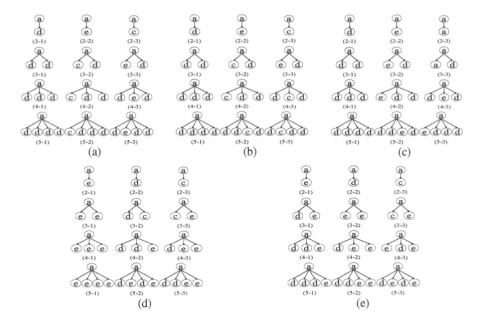

Fig. 6. Top three frequent patterns for two, three, four and five nodes for different TreeSet (The numbers under the pattern indicate the number of nodes and the rank, e.g., pattern 3-2 is the second rank of three-node patterns)

- *Frequent patterns tend to be wider rather than deeper.* It means that the possibility of reposting gets lower when the information spreads to a deeper level For all patterns in Fig. 6, they are all trees of depth two and with the number of nodes increased, they become wider not deeper.
- *The information flows of a → e and a → d are the most frequent in all TreeSets.* It means that users labeled as *e* and *d* repost a lot of messages from those influential users in Weibo sphere. For example, there are a total of 30 edges involved in each TreeSet. And the number of flows of $a \rightarrow e$ and $a \rightarrow d$ in TreeSet *C1, C2, C3, C4,* and *C5* is 26, 24, 30, 27, and 28 respectively.
- *The users labelled as b don't exist in all extracted frequent patterns.* We find that all advanced-users are labelled as *a* while the most frequent low-level users in descending order are *e, d* and *c* respectively, which indicate that the information flows of $a \rightarrow b$ and $b \rightarrow d$ (or *e, c*) are less frequent.
- *For TreeSet C5, consisted of all repost trees in our dataset, the most frequent patterns with two nodes are a → e, a → d and a → c.* Users at end of an arrow with less followers are more likely to repost, which is in accordance with the percentage distribution of the three nodes: $e > d > c$, as is shown in Fig. 6.

5.3   Discussion

As social media has revolutionized the nature of influence and the role of influential people, our findings have provided evidence to support the influence topology theory.

Based on Edelman's topology of influence (TOI) [13], Tinati et al. [14] developed a model based upon the Twitter message exchange which enables to analyze conversations around specific topics and identify different communicator roles within a Twitter conversation. As is shown in Fig. 6, most of the users in Sina Weibo have a less number of followers and they tend to follow and repost from those influence individuals with a large number of followers, such as movie stars, grassroots celebrities, and public news media. Users with fewer followers are more likely to be the information viewers while users with a lot of followers are the idea starter or the information amplifier. The idea starter and information amplifier all have a large network of follower/followee connections and most of the information viewers repost from them.

6   Conclusion

In this study, we use the data collected from Sina Weibo to study the patterns of reposting behavior in microblog services. In order to gain insights into the patterns of reposting behavior the microblog network, each message cascade is represented as a repost tree based on its reposting process. The patern mining results indicate that messages tend to propagate in some certain manners. Advanced users have a higher influential and can get a lot of reposts to promote the information spreading to a deeper level in the Weibo sphere. For example, information flows of a (users with more than 100K followers) to e (users with 1-100 followers) or d (users with 100-1K followers) are the most frequent in all patterns. In the future, we plan to study the relationship between the patterns of reposting behavior and the media modalities (i.e., picture, video, and URL) embodied in the content.

Acknowledgments. This work was partially supported by the National Basic Research Program of China (No.2012CB316400), the National Natural Science Foundation of China (No. 61222209, 61103063), the Program for New Century Excellent Talents in University (No. NCET-12-0466), the Specialized Research Fund for the Doctoral Program of Higher Education (No. 201126102110043), and the Natural Science Basic Research Plan in Shaanxi Province of China (No. 2012JQ8028).

References

[1] Boyd, D., Golder, S., Lotan, G.: Tweet, tweet, retweet: Conversational aspects of retweeting on twitter. In: Proc. of HICSS 2010 (2010)

[2] Suh, B., Hong, L., Pirolli, P., Chi, E.H.: Want to be retweeted? Large scale analytics on factors impacting retweet in Twitter network. In: Proc. of SocialCom 2010 (2010)

[3] Java, A., Song, X., Finin, T., Tseng, B.: Why we twitter: An analysis of a microblogging community. In: Zhang, H., Spiliopoulou, M., Mobasher, B., Giles, C.L., McCallum, A., Nasraoui, O., Srivastava, J., Yen, J. (eds.) WebKDD 2007. LNCS, vol. 5439, pp. 118–138. Springer, Heidelberg (2009)

[4] Kwak, H., Lee, C., Park, H., Moon, S.: What is twitter, a social network or a news media? In: Proc. of WWW 2010 (2010)

[5] Zhou, Z., Bandari, R., Kong, J.S., Qian, H., Roychowdhury, V.: Information resonance on twitter: Watching Iran. In: Proc. of SOMA 2010 (2010)

[6] Asai, T., Abe, K., Kawasoe, S., Arimura, H., Sakamoto, H., Arikawa, S.: Efficient substructure discovery from large semi-structured data. In: Proc. of SIAM 2002 (2002)

[7] Yang, Z., Guo, J., Cai, K., Tang, J., Li, J., Zhang, L., Su, Z.: Understanding retweeting behaviors in social networks. In: Proc. of CIKM 2010 (2010)

[8] Yang, J., Counts, S.: Predicting the speed, scale, and range of information diffusion in twitter. In: ICWSM 2010 (2010)

[9] Wang, C., Guan, X., Qin, T., Li, W.: Who are active? An in-depth measurement on user activity characteristics in sina microblogging. In: GLOBECOM (2012)

[10] Sina weibo, http://en.wikipedia.org/wiki/SinaWeibo

[11] Qu, Y., Huang, C., Zhang, P., Zhang, J.: Microblogging after a major disaster in China: a case study of the 2010 Yushu earthquake. In: Proc. of CSCW 2011 (2011)

[12] Yu, L.L., Asur, S., Huberman, B.A.: Artificial Inflation: The True Story of Trends in Sina Weibo. In: J. arXiv preprint arXiv:1202.0327 (2012)

[13] Bentwood, J.: Distributed influence: Quantifying the impact of social media. Edelman (2008)

[14] Tinati, R., Carr, L., Hall, W., Bentwood, J.: Identifying communicator roles in twitter. In: Proc. of MSND 2012 (2012)

[15] Miyahara, T., Shoudai, T., Uchida, T., Takahashi, K., Ueda, H.: Discovery of frequent tree structured patterns in semistructured web documents. In: Cheung, D., Williams, G.J., Li, Q. (eds.) PAKDD 2001. LNCS (LNAI), vol. 2035, p. 47. Springer, Heidelberg (2001)

[16] Wang, J.T.L., Shapiro, B.A., Shasha, D., Zhang, K., Chang, C.Y.: Automated discovery of active motifs in multiple RNA seconary structures. In: Proc. KDD 1996 (1996)

[17] Ma, H., Qian, W., Xia, F., et al.: Towards modeling popularity of microblogs. J. Frontiers of Computer Science (2013)

An Improved Parallel Hybrid Seed Expansion (PHSE) Method for Detecting Highly Overlapping Communities in Social Networks

Ting Wang, Xu Qian, and Hui Xu

School of Mechanical Electronic & Information Engineering,
China University of Mining & Technology, Beijing
wangting33184@gmail.com, {xuqian,xuh}@cumtb.edu.cn

Abstract. It is still undeveloped in the domain of detecting a *"highly"* overlapping community structure in social networks, in which networks are with high overlapping density and overlapping nodes may belong to more than two communities. In this paper, we propose an improved LFM algorithm, Parallel Hybrid Seed Expansion (PHSE), to solve this problem. In order to get nature communities, the local optimization of the fitness function and greedy seed expansion with a novel hybrid seeds selection strategy are employed. What's more, to get a better scalability, a parallel implementation of this algorithm is provided in this paper. Significantly, PHSE has a comparable performance than LFM on both synthetic networks and real-world social networks, especially on LFR benchmark graphs with high levels of overlap.

Keywords: overlapping community detection, social networks, PHSE.

1 Introduction

Community structure is one of the specific structural features in complex networks, besides power-law degree distribution in small-world networks or scale-free networks. A general definition of community is that a network divides naturally into clusters of nodes that are densely connected internally and sparsely connected otherwise. Empirical research for community detection (CD) algorithms of real-world networks such as social networks has become one of the hottest research activities in this field in recent years, especially overlapping community detection algorithms. The detection results of overlapping communities indicate that one node can belong to more than one community. Several methods for community finding have been proposed, such as the minimum-cut method, the hierarchical clustering method, the Girvan-Newman algorithm, the modularity maximization, and the recently proposed surprise maximization approach [1].

Also, a number of local based optimization methods have been developed utilizing seed expansion to grow natural communities [2]. LFM [3] recursively expands a randomly selected seed node (not belonging to any already accepted community) to a community until the fitness function reaches a local maximum. However, experiments are not repeatable due to randomly node selection.

H. Motoda et al. (Eds.): ADMA 2013, Part I, LNAI 8346, pp. 385–396, 2013.
© Springer-Verlag Berlin Heidelberg 2013

A *clique* is the most intensively researched unit structure in networks. Either two nodes are connected in one clique. Clique based methods, such as Clique Percolation Method (CPM) [4], build up communities as percolation maximum union of k-cliques by finding adjacent cliques that share k-1 nodes. But this kind of method needs to preset k value which varies in different kind of networks, with the most appropriate value of 4 or 5 [5].

Seed selection is critical in local expansion methods. Cliques as seeds have shown better performance than nodes as seeds in many algorithms. Greedy Clique Expansion (GCE) [6] takes all maximum cliques as initial seeds to greedily expand the fitness function to find overlapping communities. There are many optimization methods for cutting computing complexity based on various distance strategies. But, the initial seeds got from the maximum cliques of the networks are still numerous to compute.

Order Statistics Local Optimization Method (OSLOM) [7] optimizes locally the statistical significance information of a cluster with respect to random fluctuation with Extreme and Order Statistics. It can deal with weighted, directed edges, overlapping communities, hierarchies and dynamic communities.

In this paper, we propose an improved Parallel Hybrid Seed Expansion (PHSE[1]) method to find overlapping community structure in social networks. This method is aimed to get the seed's natural community by greedy optimizing the fitness function. Firstly, we design a novel seed selection strategy based on the Node Betweenness that can represent the node that is the key position in the network. Secondly, we solve the problem in LFM that experiments cannot be repeatable by improving the algorithm. Thirdly, consider the overlapping degree of communities as similarity to merge the highly similar communities. At the end, we make a parallel version of our algorithm to meet the larger networks. Most of the algorithms didn't give the parallel version, due to the graph structure we deal with, it could be very difficult when processing a large graph with one PC, and so our algorithm is adjusted to a parallel version to fit the single PC environment. This makes it easy to deal with a large graph from local and it can be parallel.

Our research provides unique theoretical contributions to community detection algorithms through accurately detecting the highly overlapping community structure when nodes belong to several communities and implementing a parallel implementation to increase the scalability of the algorithm due to its local nature.

The paper is organized as follows. The proposed method is described detailed in Section 2. Section 3 gives the experiments and results on synthetic networks and real-world social networks, and discussions & conclusions in Section 4.

2   Method

2.1   Hybrid Seed Expansion (HSE) Algorithm

There are some vertex-specify metrics centrality measures, such as degree centrality, betweenness centrality, closeness centrality and clustering coefficient in social network analysis. Degree centrality and betweenness centrality are the two metrics of centrality we considered in this paper. Degree centrality of a node v is the sum of

---

[1] Our Python implementation of PHSE is available at
https://github.com/sumnous/PHSE

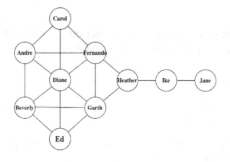

Fig. 1. Kite Network

Table 1. Vertex Metrics in Kite Network

Nodes Name	Degrees	Betweenness
Andre	0.444	0.023
Beverly	0.444	0.023
Carol	0.333	0.000
Diane	**0.667**	0.102
Ed	0.333	0.000
Fernando	0.556	0.231
Garth	0.556	0.231
Heather	0.333	**0.389**
Ike	0.222	0.222
Jane	0.111	0.000

links between other nodes and v (only undirected and unweighted graphs are considered). Betweenness centrality of a node v is the sum of the fraction of all-pairs shortest paths that pass through v, also called Node Betweenness in this paper.

See Kite Network developed by David Krackhardt [8] in Fig. 1. Node Betweenness means the importance of the node's location in the network; nodes with higher betweenness control the transfer of information. In Table 1, take the node Heather for example, its degree is low, but it is the broker in this network. Without Heather there would not be communication between Ike and Diane's cluster. Once we find these nodes like Heather, we would not lose any information or links of one community. Therefore, we chose Node Betweenness as node metric and use Ulrik Brandes [9] algorithm to compute the shortest-path betweenness centrality for nodes in this paper.

Cliques as seeds show better performance than nodes as seeds. But finding all the maximum cliques in network is time-consuming and pointless. To achieve balance on two aspects, we can just find the important nodes and get their maximum cliques in network as seeds depending on the Node Betweenness metrics. The metric with higher value means the node is more important, as well as the maximum clique including this node. The way to get the maximum clique of a specific node n in network is following. First, put n and all neighbor nodes of n as a graph g. Then find the maximum clique in graph g. That is our seed selection strategy.

Hybrid Seed Expansion (HSE) algorithm is an improved LFM algorithm to detect overlapping communities based on the seed selection strategy aforementioned.

Algorithm 1. HSE

1. Sort the nodes by betweenness as a candidate list indicating that the node with large betweenness is in an important location in the network.
2. Take the maximum clique of the first node in the candidate list as seed.
3. Get the natural community of the seed.
4. Get rid of the nodes of the community in the candidate list, and repeat calculating the maximum clique of the first node in the list, getting the nature community and getting rid of the nodes already in community, until there is no node in the candidate list.

2.2   Graph Fitness

During getting the nature community of the seed, we use fitness function [3] to reach a local maximum, which is a property function for maximization finding community. Graph Fitness is the fitness value of a particular graph. We see graph G as a middle graph of the final community. Here is the fitness function:

$$Fitness(G) = \frac{k_{in}}{(k_{in}+k_{out})^\alpha} \qquad (1)$$

where k_{in} and k_{out} are the total internal and external degree of the graph G, and parameter α determines the community size. We set $\alpha = 1$ in our experiments.

Node Fitness Contribution to Community (*NFCC*) is what decides whether add a node to the graph G or not; it means the increasing part of fitness contribution to community of one neighbor node of graph G, that is the gap between the fitness of graph G with node A, $F_{\{G+A\}}$ and the fitness of Graph without node A, F_G. Let F represents the fitness function, *NFCC* is:

$$NFCC = F_{\{G+A\}} - F_G \qquad (2)$$

If *NFCC* is over zero, it means that node A possibly belonged to graph G. In the algorithm, we calculate all *NFCC* of the neighbor nodes of graph G, chose the nodes having the maximum fitness contribution (the value of *NFCC* is largest) to add in graph G as a new graph G', and continue to compute until there is no expanding node, that's we get one nature community C.

2.3   Similarity of Communities

The detection results of communities are always having the same nodes between two communities. If their majority members of communities are the same, probably the two communities are one community and we should merge them into one. That's why we shall compute the similarity of the two communities to make decision about merging them or not. The similarity of communities could be obtained through the Community Overlapping Degree (*COD*) measure:

$$COD = \beta * \frac{|C_1 \cap C_2|}{|C_1 \cup C_2|} + (1 - \beta) * \frac{|N_1 \cap N_2|}{|N_1 \cup N_2|} \qquad (3)$$

In this formula, we are concerned with not only the effect of the overlapping degree of the two communities C_1 and C_2 (determining the similarity of communities), but also the overlapping degree of the two communities neighbors N_1 and N_2. The overlapping degree of two communities' neighbors is important: If a person has many common friends with another, the possibility of these two people being friends is high. That is why for two communities, not only their degree of overlap, but also their neighbors' degree of overlap is relevant and will contribute to the similarity of these two communities. We introduce the similarity threshold γ, that is, if *COD* is higher than the similarity threshold, we can consider the two communities are actually one community; if *COD* is lower than the similarity threshold, the two communities are not supposed to be merged. We set $\beta = 0.6$ and $\gamma = 0.59$ in our experiments; the value can be tuned to obtain the highest detection accuracy for different networks.

2.4   Parallel-HSE Algorithm

We develop a new feature to make it parallel in our algorithm rather than the non-parallel LFM. That is Parallel Hybrid Seed Expansion (PHSE) method. We shall choose some nodes as a candidate list, whose betweenness is above average and eliminate the nodes connected to previous nodes in the candidate list. The goal is to get the maximum clique of each node as seed in the network and to compute in parallel the natural community of each seed.

Algorithm 2. PHSE

1.   Input the network as a Graph.
2.   We select nodes above average in terms of their node betweenness, and choose those that are not connected.
3.   Get the maximum clique of each node, and neglect the maximum clique with the size smaller than threshold k ($k = 4$) [10]. The rests are as seeds.
4.   Parallel computing the natural community of each seed.
    4.1   Put a seed as original community C.
    4.2   Calculate $NFCC$ of each neighbor of C, get the nodes having maximum fitness contribution, if the value is over zero, then add them to original community, as C'.
    4.3   Continue step 4.2, until there is no node fitness contribution over zero. Then we get the natural community of the clique.
    4.4   If there are nodes that are not in the communities we get above, pick the node with the largest degree, go to step 4.1, until there is no single node left.
5.   Compute the Community Overlapping Degree (COD) of each two communities, sort COD value from large to small, if the value is over γ (the parameter we give, here is 0.59), then merge the two corresponding communities, until there is no $COD > \gamma$.
6.   We get the final detecting result of communities in this network.

In this way, the parallel algorithm makes itself extensional, less time-consuming and full CPU utilization than the scenario we need to wait until the detection of one nature community is complete and then to begin a new one in LFM.

2.5   Further Optimization

When conducting the experiments, we found step 4.2 cost more time than expected. The $NFCC$ means fitness contribution to community of the node. It is a huge work to calculate the $NFCC$ of each neighbor of C. To solve this problem, we make some more optimizations to get the maximum fitness contribution nodes FAST in step 4.2.

Therefore, we optimize this part of our algorithm to enhance the effectiveness of the process of recursively getting nature communities through fitness function. We use $CN = V_{con} / V_{nei}$ to evaluate the community C's neighbors' contribution to community first. V_{nei} is the number of neighbors of the node. V_{con} is the number of the edges connected with C. The larger the CN value means that the node is closer to community C. If $CN_i > CN_j$, then $NFCC_i > NFCC_j$ except when $V_{coni} < V_{conj}$ and $V_{neii} < V_{neij}$. This will get the maximum fitness quickly without adding each neighbor node in community C for calculating $NFCC$, only the situation when $V_{coni} < V_{conj}$ and $V_{neii} < V_{neij}$, we need to add them to C and computing $NFCC$ and comparing.

Compared with LFM, PHSE is more stable and repeatable with consideration of the neighbors' contribution to the similarity with two communities, without the premier knowledge of the number of communities. Furthermore, PHSE is a parallel algorithm, and thus readily scalable.

3   Experiment and Evaluation

The undirected and unweighted networks are considered in our experiments. All experiments are run on a PC with 1.7 GHz dual-core processor and 4 GB memory. We take Normalized Mutual Information (NMI) [3,11] as the standard evaluation figure to evaluate the communities detected by algorithms. The experiment results can be divided into two parts according to different network data sets including different kind of synthetic networks and real-world social networks.

3.1   Experiments on Synthetic Networks

The LFR benchmark [5,12] is adopted to generate networks of different community size and node degree, which are governed by power law distributions with exponents τ_1 and τ_2, respectively. But there is a drawback that the number of memberships of each node is the same in LFR benchmark graph. Two parameters need to be specified when generating overlapping communities-On, the fraction of overlapping nodes and, Om, the number of memberships of the overlapping nodes. The parameters in LFR benchmark can be set to control the network topology, including the mixing parameter μ, the average degree k, the maximum degree k_{max}, the maximum community size C_{max}, and the minimum community size C_{min} [13,14].

Table 2. Parameters for Synthetic Benchmarks

Parameters	Description	Fig. 2.	Fig. 3.	Fig. 4.	Fig. 5.	Fig. 6.
N	number of nodes	1k/5k	1k/5k	1k	1k	5k
k	average degree	20	20	20	20	20
k_{max}	maximum degree	50	50	50	50	50
μ	mixing parameter	0.05-1.0	0.05-1.0	0.1	0.1/0.3	0.3
τ_1	minus exponent for the degree sequence	-2	-2	-2	-2	-2
τ_2	minus exponent for the community size distribution	-1	-1	-1	-1	-1
C_{min}	minimum community size	10/20	10/20	10/20	10/20	10/20
C_{max}	maximum community size	50/100	50/100	50/100	50/100	50/100
On	fraction of overlapping nodes	0	0.1	0.1-0.9	0.1	0.5
Om	number of memberships of the overlapping nodes	1	2	2	2-8	2-8

To construct LFR synthetic benchmark graphs, we need to set ten particular parameters shown in Table 2. The mixing parameter μ indicates the proportion of the random edges end in random communities to one node's total edges; it means when $\mu = 0.3$, 70% of each node's edges end in that node's communities, the rest 30% of edges randomly end in some communities. The community size is another indicator influenced the performance of CD algorithms. Normally, two different types of community size are chosen in a network: *Small* = (10 to 50 nodes) and *Big* = (20 to 100 nodes). The parameter *On* refers to the percentage of the overlapping nodes, and *Om* indicates to the number of communities that each overlapping node belonged to.

Non-overlapping Communities
We first work on non-overlapping communities so that one node can only belong to one community. The mixing parameter μ varies from 0.05 to 1.0, and four kinds of networks are involved: 1000 nodes with 10 to 50 nodes in each community (1000-*Small*), 1000 nodes with 20 to 100 nodes in each community (1000-*Big*), 5000 nodes with 10 to 50 nodes in each community (5000-*Small*) and 5000 nodes with 20 to 100 nodes in each community (5000-*Big*).

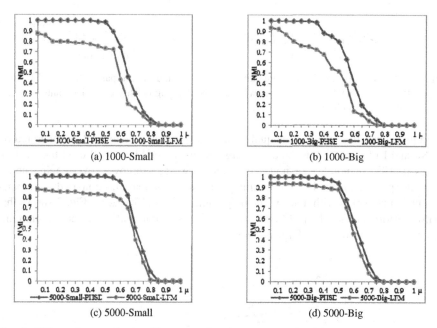

(a) 1000-Small   (b) 1000-Big

(c) 5000-Small   (d) 5000-Big

Fig. 2. Effect of network size N, community size range, and mixing parameter μ on non-overlapping communities

We compare the results we get from our method PHSE with LFM that inspired our algorithm. Fig. 2 (a-d) suggest that the NMI decreased with the mixing parameter μ increased. Consequently, the network constituted with small communities can be detected more accurately than those one constituted with large communities. We got slightly better performance as we increased the network size from 1000 to 5000.

Overlapping Communities

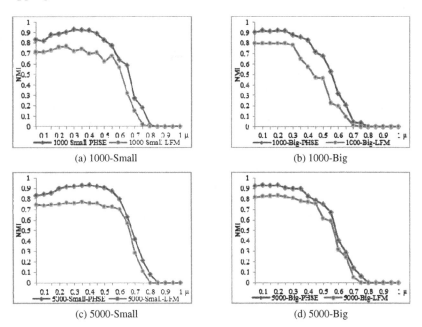

Fig. 3. Effect of network size *N*, community size range, and mixing parameter μ on LFR networks. Plots show NMI's for networks with low overlapping density $On = 10\%$ and the numbers of memberships $Om = 2$.

In this subsection, we test the quality of the overlapping communities generated by PHSE and LFM on the four kinds of networks mentioned before. We set the fraction nodes to 10% of number of nodes for each network. Moreover, overlapping nodes alters the performance when the mixing parameter is increased, compared with experiments on networks with non-overlapping communities [15,16,2,6]. **Fig. 3** shows that PHSE performs better than LFM no matter in sparse communities or dense communities with small networks or large networks.

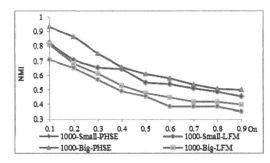

Fig. 4. Effect of community size range and overlapping density *On* on LFR networks. NMI as a function of the overlapping density *On*. Plots show NMI's for networks with network size $N = 1000$, mixing parameter $\mu = 0.1$ and the numbers of membership $Om = 2$.

Fig. 4 shows that the NMI value for networks with large community size range is typically higher than those for networks with small community size range. The bigger the number of community nodes is, the higher the agglomeration degree of the community is. That's why the detection of large communities is easier, lending more accuracy to clique based method.

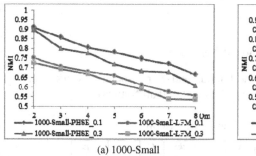
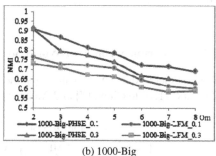

(a) 1000-Small   (b) 1000-Big

Fig. 5. Evaluation of overlapping community detection on LFR networks with low overlapping density $On = 10\%$. NMI as a function of the number of memberships Om. Results for small community size range are shown on the left, and results for large community size range are shown on the right. Plots show NMI's for networks with network size $N = 1000$ and the mixing parameter $\mu = 0.1$ and 0.3.

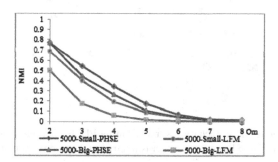

Fig. 6. Evaluation of overlapping community detection on LFR networks with high overlapping density $On = 50\%$. NMI as a function of the number of memberships Om. Plots show NMI's for networks with large network size $N = 5000$ and the mixing parameter $\mu = 0.3$.

Considering that in real-world social networks, a node may belong to multiple communities as an overlapping node; this kind of nodes is important as hubs to communities. It is hard to identify those overlapping nodes when the number of memberships is large. For this case, we conducted a group of experiments to study effects of the number of memberships Om on LFR networks with low overlapping density $On = 0.1$ (Fig. 5) and high overlapping density $On = 0.5$ (Fig. 6).

In Fig. 5, the PHSE algorithm gets high NMI even the Om is set to 8 (i.e. when $N = 1000$-Big, $\mu = 0.1$ and $Om = 8$, the NMI of our algorithm is 0.690).

Furthermore, we conduct experiments on *highly* overlapping community structure networks with overlapping density On increased to 50%, network size N is 5000, and the mixing parameter $\mu = 0.3$. From the detection results in Fig. 6, we can find that PHSE can still get highly accuracy detection results compared with the LFM method. The performance of detection results in both algorithms significantly drops in the case when On increased to 50%. But PHSE gets better performance than LFM when each node has multiple memberships. Each overlapping node belonged to 2 communities, NMI can reach to 0.7691, 0.7679 respectively with larger or small community size by PHSE.

3.2   Experiments on Real-World Social Networks

We conducted our experiments on some famous real-world social networks, such as Zachary's Karate Club [17] and American College Football Network [18].

Table 3. Basic Statistics of Real Networks

Network	NMI		N	E	$<k>$	C	$<s>$	$<m>$
	PHSE	LFM						
Karate	0.708	0.690	34	78	4.59	2	19.5	1.147
Football	0.803	0.754	115	159	10.7	12	9.45	1.00

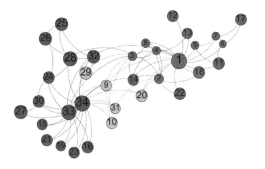

Fig. 7. The detection result of Zachary's Karate Club by PHSE. (For interpretation of the references to color in this figure legend, the reader is referred to the web version of this article.)

In Table 3, we list the analysis of Karate and Football networks with some basic statistics obtained from the application of PHSE to detect their community structure. From left to right, we list the NMI detected by PHSE and LFM, number of vertices N and edges E, the average degree $<k>$, along with the number of clusters C, the average cluster size $<s>$, the average number of memberships per vertex $<m>$ all detected by PHSE.

In the case of Zachary network [17], including 34 nodes, Node 34 represents the administrator of the club and Node 1 represents the club's instructor who had disagreement with the administrator, then left with some original members and started a new club. In Fig. 7, we detect two main communities that surrounded Node 1 and Node 34 in our test. PHSE detects two communities, matches the fact from the observation. Besides that, the nodes [9, 10, 20, 29, 31] are the overlapping nodes, meaning they may hesitate to choose side or want to be the good men to connect to both sides. The NMI is 0.708 by PHSE more than 0.690 by LFM.

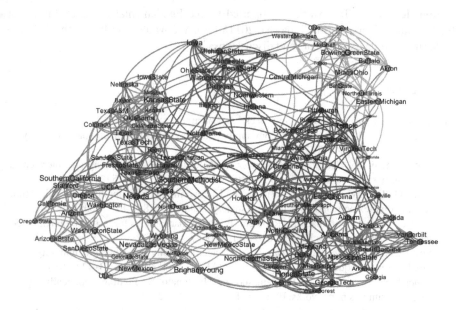

Fig. 8. The detection result of American College Football Network by PHSE. (For interpretation of the references to color in this figure legend, the reader is referred to the web version of this article.)

American College Football Network [18] represents the schedule of games between American college football teams in the 2000 season. The network constructed by 115 teams is divided into 12 groups or "conferences" containing around 8 to 12 teams each, with intra-conference games being more frequent than inter-conference games. As Fig. 8 reveals, 102 of 115 teams are classified correctly and exactly twelve communities are detected by the proposed method. Six communities are detected correctly, including Atlantic Coast, Big Ten, Big Twelve, Mountain West, Pacific Ten and Southeastern. The NMI of the proposed method is 0.803, higher than the LFM algorithm, which detected communities with only four correct communities.

4   Conclusion

This paper developed an improved and parallel algorithm, PHSE, for detecting overlapping community structure in social networks. It is getting nature community through the local optimization of the fitness function with an improved hybrid seed expansion strategy. Compared with LFM algorithm, PHSE performed competitively on both synthetic networks and real-world social networks. Especially PHSE can accurately detecting communities on synthetic networks with high overlapping degree both in the fraction of overlapping nodes ($On = 50\%$) and the membership of nodes (Om varies from 2 to 8). In the future work, we will improve our algorithm to meet the need of detecting highly overlapping communities in very large real-word networks.

Acknowledgment. This work is supported by the Fundamental Research Funds for the Central Universities of China (No.2011YJ09). We would also like to thank Jingui Li for his help on experiments and discussions.

References

1. http://en.wikipedia.org/wiki/Community_structure
2. Lancichinetti, A., Fortunato, S.: Community detection algorithms: a comparative analysis. Phys. Rev. E 80, 056117 (2009)
3. Lancichinetti, A., Fortunato, S., Kertész, J.: Detecting the overlapping and hierarchical community structure of complex networks. New J. Phys. 11, 033015 (2009)
4. Palla, G., Derényi, I., Farkas, I., Vicsek, T.: Uncovering the overlapping community structure of complex networks in nature and society. Nature 435, 814–818 (2005)
5. Lancichinetti, A., Fortunato, S., Radicchi, F.: Benchmark graphs for testing community detection algorithms. Phys. Rev. E 78, 046110 (2008)
6. Lee, C., Reid, F., McDaid, A., Hurley, N.: Detecting highly overlapping community structure by greedy clique expansion. In: Proc. SNAKDD Workshop, pp. 33–42 (2010)
7. Lancichinetti, A., Radicchi, F., Ramasco, J.J., Fortunato, S.: Finding statistically significant communities in networks. PLoS ONE 6(4), e18961 (2011)
8. Hansen, D., Shneiderman, B., Smith, M.A.: Analyzing social media networks with NodeXL: Insights from a connected world. Morgan Kaufmann (2010)
9. Brandes, U.: A faster algorithm for betweenness centrality. Journal of Mathematical Sociology 25(2), 163–177 (2001)
10. Shen, H., Cheng, X., Cai, K., Hu, M.-B.: Detect overlapping and hierarchical community structure. Physica A 388, 1706 (2009)
11. Danon, L., Daz-Guilera, A., Duch, J., Arenas, A.: Comparing community structure identification. Journal of Statistical Mechanics, P09008 (2005)
12. Lancichinetti, A., Fortunato, S.: Benchmarks for testing community detection algorithms on directed and weighted graphs with overlapping communities. Physical Review E 80(1), 016118 (2009)
13. Xie, J.: Overlapping Community Detection in Networks: The State of the Art and Comparative Study 45(4), 1–37 (2013)
14. Xie, J., Szymanski, B.K.: Towards linear time overlapping community detection in social networks. In: Proc. PAKDD Conf., pp. 25–36 (2012)
15. Wu, Z., Lin, Y., Wan, H., Tian, S., Hu, K.: Efficient Overlapping Community Detection in Huge Real-world Networks (2011)
16. Gregory, S.: Finding overlapping communities in networks by label propagation. New J. Phys. 12, 10301 (2010)
17. Zachary, W.W.: An information flow model for conflict and fission in small groups. Journal of Anthropological Research 33, 452–473 (1977)
18. Girvan, M., Newman, M.E.J.: Community structure in social and biological networks. Proc. Natl. Acad. Sci. USA 99(12), 7821–7826 (2002)

A Simple Integration of Social Relationship and Text Data for Identifying Potential Customers in Microblogging

Guansong Pang[1,*], Shengyi Jiang[2], and Dongyi Chen[2]

[1] School of Management, Guangdong University of Foreign Studies, Guangzhou 510006, China
pangguansong@163.com
[2] School of Informatics, Guangdong University of Foreign Studies, Guangzhou 510006, China
{jiangshengyi,dongyi_chen}@163.com

Abstract. Identifying potential customers among a huge number of users in microblogging is a fundamental problem for microblog marketing. One challenge in potential customer detection in microblogging is how to generate an accurate characteristic description for users, i.e., user profile generation. Intuitively, the preference of a user's friends (i.e., the person followed by the user in microblogging) is of great importance to capture the characteristic of the user. Also, a user's self-defined tags are often concise and accurate carriers for the user's interests. In this paper, for identifying potential customers in microblogging, we propose a method to generate user profiles via a simple integration of social relationship and text data. In particular, our proposed method constructs self-defined tag based user profiles by aggregating tags of the users and their friends. We further identify potential customers among users by using text classification techniques. Although this framework is simple, easy to implement and manipulate, it can obtain desirable potential customer detection accuracy. This is illustrated by extensive experiments on datasets derived from Sina Weibo, the most popular microblogging in China.

Keywords: identifying potential customers, user profiling, social relationship, text data, text classification, microblog marketing.

1 Introduction

Microblog is characterized by a huge number of users, fast message propagation and a broad range of influence, so that microblog marketing has been made as one of the most important portions of social media marketing in many companies. Nowadays, microblog marketing activities are mostly random. As a result, they are blocked and even denounced by most users. This does not produce any positive effects on companies but mostly severe negative impacts. Identifying potential customers is a fundamental problem for microblog marketing, which can enable marketers to conduct cost-effective marketing activities.

---

[*] Corresponding author.

H. Motoda et al. (Eds.): ADMA 2013, Part I, LNAI 8346, pp. 397–409, 2013.
© Springer-Verlag Berlin Heidelberg 2013

One challenge in potential customer detection in microblogging is how to generate an accurate characteristic description for users, i.e., user profile generation. Some methods have been proposed to deal with this issue recently. The most relevant work is [1], where the authors focused on utilizing user's default profile features, posting behavior features, linguistic content features and social network features to construct user profiles. The gradient boosted decision trees were then applied to perform user classification tasks, including prediction of potential followers (i.e., potential customers) of Starbucks. Since this method involves a wide range of analysis over various features and is a task-specific framework, it is not easy to manipulate in real-world applications (e.g., extensive parameter tuning needed for a new task). Moreover, the best F_1-measure value in the prediction of Starbucks's potential followers was about 76%, which could be improved. User profiles can also be generated from conversational data in social media [2]. Other related work focused on some other user classification tasks, e.g., demographic attribute value prediction and communication role identification, by making use of a variety of features mentioned above [3-7]. We focus on potential customer detection by employing microblog users' self-defined tag features only.

Intuitively, the preference of a user's friends (i.e., the person followed by the user in microblogging) is of great importance to capture the characteristic of the user. Also, a user's self-defined tags are often concise and accurate carriers for the user's interests. In this paper, for identifying potential customers in microblogging, we propose a method to generate user profiles via a simple integration of social relationship and text data. In particular, our proposed method constructs self-defined tag based user profiles by aggregating tags of the users and their friends. We further identify potential customers among users by using text classification techniques such as K Nearest Neighbors (KNN), Naive Bayes (NB), Rocchio, centroid-based classification (Centroid) and Support Vector Machines (SVM). The effectiveness of our proposed framework has been illustrated by experiments on datasets derived from Sina Weibo, the most popular microblogging in China.

Our contribution is to propose a simple yet effective framework for identifying potential customers in microblogging. Although this framework is simple, easy to implement and manipulate, it can obtain desirable potential customer detection accuracy. With the aid of the fruitful social relationship in microblogging, our proposed framework can not only provide potential customer detection with a new perspective and method, but also present important reference to many classic Customer Relationship Management (CRM) problems such as customer churn.

2   Related Work

Identifying potential customers in microblogging, which is essentially a target-specific binary user classification task, is an emerging research field. Previous work has explored user profile generation methods and their application on user classification in microblogging. In [1], the authors constructed user profiles based on users' default profile features (e.g., user name, location, number of followers and friends), posting behavior features (e.g., total number of microblogs posted, average number of hashtags and URLs per microblog), linguistic content features (e.g., prototypical words and hashtags, sentiment words, generic and domain-specific topics generated via Latent Dirichlet Allocation [8]) and social network features (e.g., number and fraction of intersected friends between users). The user profiles were then used for user classification tasks, including a potential customer detection task. This method defined user profiles from a range of dimensions and required somewhat complicated parameter tuning for different classification tasks. Other related work is [2], which aimed at extracting terms and concepts that can reflect users' interests from social media conversional data, in order to target advertisements or conduct other commercial activities. Instead of generating user profiles from a variety of features or complex conventional data, our proposed framework focuses on making use of users and their friends' self-defined tag features only, which is easy to implement and manipulate. There have been many methods proposed for demographic attribute value prediction and communication role identification [3–7]. These methods can be further improved to use for commercial utilities, e.g., identifying potential customers, while our proposed method is especially designed for potential customer detection.

A range of previous work has aimed at identifying potential customers based on a customer database [9–12]. These work normally built potential customer detection models based on customers' demographic information. Compared to these work, we investigate potential customer detection methods in a different application context, i.e., microblogging. Moreover, we capture customers' interests from social and personal perspectives rather than personal demographic attributes. Much work has been made to develop text classification algorithms [13, 14]. In general, these work usually focused on devising more effective or efficient algorithms, while our framework is to use text classification algorithms to make better use of microblog user data for identifying a company's potential customers.

3 Identifying Potential Customers in Microblogging

A company's potential customers are conventionally defined as anyone who has a particular need for some products of the company. In this work, we define the potential customers of a company in microblogging as users who share similar interests to followers of the company's authorized microblog account, since these followers tend to be the company's customers or potential customers. Alternatively, identifying potential customers of a company in microblogging is equivalent to potential follower prediction of the company's microblog account.

3.1 Our Proposed Framework

This work aims to make use of the self-defined tags of a particular company's followers and their friends to generate tag-based user profiles, such that each user can be denoted by a tag vector, where each tag represents a dimension. Identifying potential customers among a large number of users is then transformed into a text classification problem. Our proposed framework is shown in Fig. 1. Given a company, we first collect tags of the company's followers and their friends to construct user profiles, and then build a model based on the user profiles using text classification algorithms and classify a given user into either the potential customer class or the general user class.

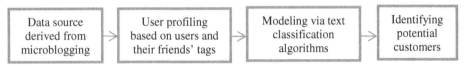

Fig. 1. Our proposed framework for identifying potential customers in microblogging

Compared to traditional methods with customer databases, our framework possesses three strengths: (i) there are abundant publicly accessible customer data in microblogging; (ii) the users' interests are denoted by self-defined tags of the users' and their friends, which are likely to reflect the users' characters from various perspectives; (iii) marketers can directly design and conduct marketing activities in microblogging other than emails or mass media.

3.2   User Profile Generation

After registering an account in microblogging, e.g., Sina Weibo, user can write multiple tags to describe their personal interests. These tags, such as "Post-80s", "data mining", "football", "housewife", "foodie" and "faculty", contain critical information for a user's personality, common interests or occupation. This enables visitors to understand the user better and further help the user socialize. Since these tags are normally self-defined, they are often concise and accurate carriers for users' interests. Thus, we focus on using tags to construct user profiles.

In general, a user's personal interests or purchase behaviors tend to be affected by the user's friends, or someone who has closed relationship with; in turn, these fellow's interests also reflect the user's interests to some extent. Thus, in microblogging, the person who has a bi-direction relation with or is followed by a user (i.e., the user's friends in microblogging) is very likely to share some similar interests to the user. We therefore devise a method to construct a user's profile by a sum of tag vectors of the user and his or her friends, as detailed in Eq.(1).

$$\mathbf{up}_i = \sum_{j=1}^{s} \mathbf{friend}_j + \mathbf{user}_i \tag{1}$$

where $\mathbf{user}_i$ is a user tag vector, $\mathbf{friend}_j$ denotes the tag vector of a friend of the user $\mathbf{user}_i$ among s friends, $\mathbf{up}_i$ is the user profile vector (where each tag represents a dimension and its dimension value is the frequency occurred in the $\mathbf{user}_i$ and all the $\mathbf{friend}_j$). Each user profile vector is regarded as a document vector and weighted by a general $TF \times IDF$ method [15] (TF and IDF are short for Term Frequency and Inverse Document Frequency respectively). The weighted user profile vectors are then used for modeling and detection. Our experiments showed that the user profile will become less effective, when we do not use the tags of the user's friends, or when we discriminate the tags of the user and the user's friends. This is why we sum tag vectors of the user and his friends with equal weights.

3.3   Modeling and Detection

We investigate various text classification algorithms for potential customer modeling and detection, including KNN, NB, Centroid, Rocchio and SVM. We briefly describe the modeling and classification processes of these five classifiers below (More detailed descriptions refer to [13, 14]).

KNN: The process of KNN text classification can be described as follows: given a test document, find the nearest neighbors for the test document among the training document set, and score class candidates for the test document based on the classes of the neighbors. KNN then assigns the class with the highest score to the test document.

NB: NB uses the Bayes Rule to determine the class of a given test document. Its independent assumption assumes term features to contribute independently for modeling and classification. Maximum likelihood estimation is usually used to compute the prior and conditional probabilities.

Centroid: In Centroid, each class is represented by its class centroid, which is the average of document vectors within the class, and a test document is assigned to the class label of its closest centroid.

Rocchio: Rocchio functions similarly to Centroid except that it uses prototype vectors to classify documents. The prototype vectors are generalized from class centroids, calculated as Eq. (2).

$$\mathbf{pv}_k = \alpha \frac{1}{|C_k|} \sum_{d_m \in C_k} \mathbf{d}_m - \beta \frac{1}{|D - C_k|} \sum_{d_n \in D - C_k} \mathbf{d}_n \tag{2}$$

where $\mathbf{pv}_k$ denotes the prototype vector of the class C_k, D is the entire training document set. Parameters α and β adjust the relative importance of positive and negative documents. Rocchio is equivalent to Centroid with $\alpha = 1$ and $\beta = 0$.

SVM: The main principle of SVM is to determine support vectors that maximize margins of separation between classes in a hyperplane. SVM performs well on linearly separable space, and it can use kernel methods such as polynomial and *RBF* kernels, to adapt to non-linearly separable data space. As shown in [13, 16], compared to non-linear kernels, the linear kernel can enable SVM to achieve better or very comparable text classification accuracy and perform more efficiently.

A critical step at this stage is to determine the training sets of positive and negative instances. In general, given a potential customer detection task for a specific company, the followers of the company's microblog account can be regarded as positive instances, while other users can be regarded as negative instances. However, this general rule is not without problems. This is because (i) those who do not follow the company may be the followers of the microblog accounts of this company's rivals, and those users are more appropriate to be used as positive instances rather than negative instances, because followers of companies within the same sector tend to have similar characters and interests; (ii) there are a huge number of users who do not follow the microblog accounts of the company and its rivals, so it is a problem about how to determine the compositions of negative instances. For example, negative instances can be composed of followers of a single company from a distinct sector or multiple companies from various sectors. After determining the positive and negative instances for the given task, we then can use a text classification algorithm to construct a potential customer model and identify potential customers.

4   Experimental Results

4.1   Datasets

Our experimental datasets derived from Sina Weibo, which is the most popular microblog platform in China. A large number of companies have registered in this microblogging, amounting to 130 thousand. These companies come from different sectors such as healthcare, education, tourist, personal services, foods and e-commerce. Our datasets consist of 8 small or medium sized companies distributed through 4 sectors, i.e., healthcare, education, tourist and personal services, as shown in Table 1. We define an ID for each company, and the columns "No. of Followers" and "No. of Friends" are the number of followers and the total number of the followers' non-overlapping friends respectively.

Table 1. A summary of the eight companies used for our experiments

Sectors	ID	No. of Followers	No. of Friends
Tourist	A1	3,397	703,607
	A2	4,308	1,006,333
Healthcare	B1	2,141	569,505
	B2	4,022	894,985
Education	C1	3,976	588,906
	C2	4,999	686,545
Personal services	D1	4,130	734,886
	D2	1,170	389,202

We focus on small or medium sized companies for two main reasons. One is that we can only gather at most 5,000 followers' information of one particular microblog account due to limitations of microblogging service providers. The number of followers of small or medium sized companies is often less than 5,000, so that we can collect all the followers of these companies. Therefore, hopefully, this kind of data can well cover the characters of the companies' customers. Another reason is that, compared to large-sized companies, it is more demanding for smaller sized companies to identify potential customers.

It is not uncommon that numerous microblogging users have been registered automatically by machines, also known as "zombie users". This kind of user is actually noise instances, since they can be acted as "zombie followers" of many companies. In general, normal users are likely to form bi-direction relations with their friends. While "zombie users" build a huge number of one direction relations, i.e., randomly follow numerous users, yet they do not have any followers or very few followers (e.g., less than 10). In addition, normal users are likely to build a personalized user domain using their names like "http://weibo.com/" plus "user's name", while "zombie users" do not set this domain. Based on these observations, we removed the "zombie users" by filtering out the users who have less than 10 bi-direction relations and do not have a personalized user domain. The remaining data is detailed in Table 2.

Table 2. Basic information about the eight companies after removing the "zombie users"

ID	No. of Followers	No. of Friends	ID	No. of Followers	No. of Friends
A1	2080	386845	C1	2716	381613
A2	1232	337241	C2	2287	328240
B1	1726	321319	D1	1213	290805
B2	1297	295139	D2	609	213267

4.2   Parameter Settings

The parameters of the five text classifiers used, i.e., KNN, NB, Rocchio, Centroid and SVM, are set as follows:

KNN: For KNN, the parameter tuning issue is how to decide an appropriate K value [17]. Various K values has been adapted into different application domains and scenarios [13], we set $K = 50$ in our experiments after a pilot study.

NB: NB is a non-parametric classifier. We used the Laplace smoothing method to deal with the "zero probability" issue.

Rocchio: Following [18], we set $\alpha = 2\beta = 1$ after a pilot study, which enables the Rocchio classifier to obtain stable classification performance.

Centroid: Centroid is a non-parametric classifier, and it therefore does not require any parameter tuning.

SVM: We applied a widely used SVM implementation package, called LibSVM [19]. Linear kernel is used in our experiments with other settings default [13, 20].

4.3   Performance Evaluation

The datasets are split into training and test sets with the ratio 2:1. A widely used measure, F_1 measure, is used as the classification accuracy metric. F_1 is a combination measure of precision and recall, i.e., $F_1 = \dfrac{2 \times precision \times recall}{precision + recall}$, where

$precision = \dfrac{TP}{TP + FN}$ and $recall = \dfrac{TP}{TP + FP}$. TP, FN and FP are short for True Positive, False Negative and False Positive respectively. In our experiments, the average F_1 value of two classes has a similar trend to the F_1 value of an individual class. Since only the effectiveness of identifying positive instances, i.e., potential customers, is of our interest, we only report the F_1 value on the positive class.

4.4   Results

We examined our proposed framework on various contexts. In particular, since the composition of positive and negative instance sets is very flexible, but is a critical step for potential customer detection, we conducted extensive experiments with varying compositions of negative or positive instances.

Negative Instances from Multiple Sectors. Microblogging consists of users with a wide range of background and character, so the variety of negative instances is important for examining the model's generalization. In this subsection, for each sector we chose one company (i.e., A1, B1, C1 and D1) as the target company, so positive instances were followers of the selected company, while negative instances were randomly selected from companies of other three sectors evenly. We employed the five classifiers to conduct the task, and each classifier runs five times for each company. Table 3 shows the average F_1 values of the classifiers and the baseline. The baseline is calculated by randomly identifying potential customers without using any classifiers. Since the datasets are made up of positive and negative instances evenly, all the baselines are 0.5.

Table 3. F1 values of five classifiers on four companies from different sectors

	KNN	NB	Rocchio	Centroid	SVM	Baseline
A1	0.7637	0.8749	0.7883	0.8310	0.9264	0.5000
B1	0.8183	0.9101	0.8924	0.8974	0.9445	0.5000
C1	0.6206	0.9472	0.8477	0.9008	0.9758	0.5000
D1	0.8581	0.8694	0.8535	0.7859	0.9272	0.5000
Avg.	0.7652	0.9004	0.8455	0.8538	0.9435	0.5000

The results show that all the classifiers outperform the baseline. SVM consistently outperforms other four classifiers with a very high averaging F_1 value, 0.9435, followed by NB, Centroid and Rocchio. The worst performance among these classifiers goes to KNN due to the fact that the datasets used have a number of noisy tags and KNN is sensitive to these noisy tag features [21].

Negative Instances from Single Sector. Apart from choosing negative instances from multiple sectors, negative instances can also derive from only one sector. We first used

Centroid to conduct classification on every selected company, with one of the remaining companies as negative class. This could help us decide which company would be optimal for being the negative class for the target company. The results are shown in Table 4. The best and the worst performance are bold and underline respectively. It is clear that Centroid obtains the worst performance when the positive and negative classes come from the same sector. This is because instances of the classes from the same sector are likely to distribute closely in the feature space, as they bear a number of common features. Centroid tends to achieve promising performance when the sectors of the two classes differ significantly, such as healthcare (i.e., B1 or B2) vs. education (i.e., C1 or C2). This result is consistent with the general classification intuition that instances of diversified classes are distributed separately and easier to be classified. Therefore, the company, which can enable Centroid to obtain the best classification performance, was chosen to be the negative class for a specific target company (e.g., A1-vs-B2 and A2-vs-D2). The other four classifiers were then applied to perform classification on each dataset. The results are illustrated in Table 5.

Table 4. Centroid's classification results on eight companies with one of the remaining companies as negative class

	A1	A2	B1	B2	C1	C2	D1	D2	Avg.
A1	-	0.6814	0.8931	**0.9296**	0.8635	0.8453	0.9004	0.9237	0.8624
A2	0.6390	-	0.8662	0.8795	0.8182	0.8414	0.8456	**0.9063**	0.8642
B1	0.8720	0.8919	-	0.6017	**0.9600**	0.9323	0.9015	0.9209	0.8296
B2	0.8801	0.8553	0.6357	-	0.8951	0.8568	0.8995	**0.9216**	0.9140
C1	0.8899	0.8944	**0.9746**	0.9560	-	0.6729	0.9318	0.9533	0.8682
C2	0.8748	0.9026	**0.9505**	0.9325	0.7340	-	0.9173	0.9489	0.8485
D1	0.7738	0.7766	0.8196	**0.8700**	0.8076	0.7994	-	0.7373	0.7715
D2	0.7263	0.7861	0.7857	**0.8393**	0.7990	0.7969	0.6565	-	0.8624

Table 5. F1 values of five classifiers on eight target companies

	KNN	*NB*	*Rocchio*	*Centroid*	*SVM*
A1	0.8464	0.9523	0.8899	0.9296	0.9660
A2	0.8550	0.9050	0.8997	0.9063	0.9166
B1	0.8449	0.9603	0.9643	0.9600	0.9826
B2	0.8678	0.9481	0.9318	0.9216	0.9660
C1	0.8888	0.9754	0.9774	0.9746	0.9890
C2	0.7874	0.9498	0.9601	0.9505	0.9727
D1	0.7725	0.9605	0.9174	0.8700	0.9789
D2	0.5338	0.8962	0.8626	0.8393	0.9231
Avg.	0.7996	0.9435	0.9254	0.9190	0.9619

It can be seen from Table 5 that the average F_1 values of all the five classifiers outperform their counterparts in Table 3. It also shows that SVM achieves very promising F_1 values over all the datasets, outperforming the rest of classifiers. NB, with 0.9453 in the average F_1 value, comes in second. It is worthwhile to note that all the five classifiers' F_1 values decrease substantially on the dataset with D2 as positive class. This is mainly due to the fact that the classifiers lack sufficient instances to make predictions on this dataset.

Increasing Positive Instances with Followers of the Rival Company. One obvious advantage for identifying potential customers in microblogging is that it is easy to augment positive instances. This can be done by collecting information of followers of the rival's microblog account. Augmenting positive instances is very likely to improve classification accuracy. We examined this strategy in this subsection. The classification task is the same as the one in the first subsection with the only exception that we increased the positive instances with the instances of another company within the same sector in Table 2. The F_1 values of the five classifiers are shown in Table 6. The "origin" column directly derives from Table 3, and the "augment" column denotes the classifiers' results on each dataset with augmented positive instances. On average, NB, Rocchio, Centroid and SVM with augmented positive instances slightly outperform that with original positive instances. Specifically, these four classifiers with the augmented strategy obtain improvements on the first two datasets. KNN with the augmented positive instances only performs well in the first dataset. It indicates that the augmented positive instances do not always help improve the classifiers' performance. The main reason for this result, we conjecture, is that the instances within A1 and A2, B1 and B2 are closed to each other, while the instances within C1 and C2, D1 and D2 tend to be diversified. This can be observed from Table 4, where the averaging F_1 values of C1-vs-C2 and D1-vs-D2 (0.7035 and 0.6969 respectively) are much larger than that of B1-vs-B2 and A1-vs-A2 (0.6187 and 0.6602 respectively), indicating that the class pairs A1, A2 and B1, B2 are more likely to share an underling latent feature space than the pairs C1, C2 and D1, D2.

Table 6. Classification results with augmented positive instances

	KNN		NB		Rocchio		Centroid		SVM	
	Origin	Augment	Origin	Augment	Origin	Augment	Origin	Augment	Origin	Augment
A1	0.7637	**0.7780**	0.8749	**0.8906**	0.7883	**0.7885**	0.8310	**0.8405**	0.9264	**0.9373**
B1	0.8183	0.7956	0.9101	**0.9281**	0.8924	**0.9211**	0.8974	**0.9191**	0.9445	**0.9489**
C1	0.6206	0.5997	0.9472	0.9415	0.8477	0.8297	0.9008	**0.9125**	0.9758	0.9733
D1	0.8581	0.8269	0.8694	0.8584	0.8535	0.8436	0.7859	0.7489	0.9272	0.9254
Avg.	0.7652	0.7501	0.9004	**0.9047**	0.8455	**0.8457**	0.8538	**0.8553**	0.9435	**0.9462**

Computation Time Comparisons. Table 7 shows the averaging running time of 5 runs of the five classifiers on the datasets. All the results were obtained from one computer with configuration: Intel (R) Core (TM) 2 Duo CPU 3.00GHz 1.85GB. NB achieves the best classification efficiency, followed by Centroid, Rocchio, SVM and KNN. NB runs about 28-fold and 37-fold faster than SVM and KNN. SVM runs faster than KNN, and Centroid runs slightly faster than Rocchio.

Table 7. Running time (in seconds) of the five classifiers on the four datasets

	KNN	*NB*	*Rocchio*	*Centroid*	*SVM*
A1	1091	19	75	70	685
B1	605	10	63	59	532
C1	1623	62	109	103	1105
D1	335	8	37	35	422
Avg.	914	25	71	67	686

5   Conclusions

We proposed a framework of identifying potential customers in microblogging. Specifically, we first devised a user profile generation method via a simple integration of users' self-defined tags based on their social relations. This generated a tag vector to represent a user. We further proposed a potential customer detection method by using text classification algorithms. The promising performance of our proposed framework has been illustrated by a series of experiments on datasets with varying compositions of positive or negative instances. In particular, negative instances derived from a single diversified sector enable classifiers to perform more accurately than negative instances from multiple sectors. However, it should be noted that, in the former case, models are likely to overfitting. Augmented positive instances from a company's rivals can improve the detection accuracy, when customers of the company and its rivals share a large number of features, i.e., an underlying latent feature space.

Our proposed framework is simple and easy to implement and manipulate, which hopefully can be mastered by marketers quickly. Our results also showed SVM is the best choose in terms of accuracy, while NB is preferable to the other four classifiers when taking both effectiveness and efficiency into consideration.

We continue to further investigate a combination of dimension reduction methods with our proposed framework. We are also interested in improving our proposed framework to deal with other CRM issues.

Acknowledgements. We thank the anonymous reviewers for their fruitful suggestions. This paper was completed when Guansong Pang was a visiting student in the Web Sciences Center at University of Electronic Science and Technology of China. He wishes to thank his supervisor Prof. Mingsheng Shang in the Web Sciences Center for his support on this work. This work was supported in part by the Natural Science Foundation of China under Grant No. 61070061 and No. 61202271, and by the Social Science Foundation of China under Grant No. 13CGL130, and by the National Key Technologies R&D Program Project under Grant No. 2012BAH02F03.

References

[1] Pennacchiotti, M., Popescu, A.: Democrats, republicans and starbucks afficionados: user classification in twitter. In: Proceedings of the 17th ACM SIGKDD International Conference on Knowledge Discovery and Data Mining, pp. 430–438 (2011)

[2] Konopnicki, D., Shmueli-Scheuer, M., Cohen, D., Sznajder, B., Herzig, J., Raviv, A., Zwerling, N., Roitman, H., Mass, Y.: A statistical approach to mining customers' conversational data from social media. IBM Journal of Research and Development 57(3/4), 11–14 (2013)

[3] Burger, J.D., Henderson, J., Kim, G., Zarrella, G.: Discriminating gender on Twitter. In: Proceedings of the Conference on Empirical Methods in Natural Language Processing, pp. 1301–1309 (2011)

[4] Fink, C., Kopecky, J., Morawskib, M.: Inferring Gender from the Content of Tweets: A Region Specific Example. In: Proceedings of the 6th International AAAI Conference on Weblogs and Social Media (2012)

[5] Rao, D., Yarowsky, D., Shreevats, A., Gupta, M.: Classifying latent user attributes in twitter. In: Proceedings of the 2nd International Workshop on Search and Mining User-generated Contents, pp. 37–44 (2010)

[6] Tinati, R., Carr, L., Hall, W., Bentwood, J.: Identifying communicator roles in twitter. In: Proceedings of the 21st International Conference Companion on World Wide Web, pp. 1161–1168 (2012)

[7] Yu, S., Kak, S.: A Survey of Prediction Using Social Media. arXiv preprint:1203.1647 (2012)

[8] Blei, D.M., Ng, A.Y., Jordan, M.I.: Latent dirichlet allocation. The Journal of Machine Learning Research 3, 993–1022 (2003)

[9] Drew, J.H., Mani, D.R., Betz, A.L., Datta, P.: Targeting customers with statistical and data-mining techniques. Journal of Service Research 3(3), 205–219 (2001)

[10] Kim, Y., Street, W.N.: An intelligent system for customer targeting: a data mining approach. Decision Support Systems 37(2), 215–228 (2004)

[11] Kim, Y., Street, W.N., Russell, G.J., Menczer, F.: Customer targeting: A neural network approach guided by genetic algorithms. Management Science 51(2), 264–276 (2005)

[12] Berry, M.J., Linoff, G.S.: Data mining techniques: for marketing, sales, and customer relationship management. Wiley Computer Publishing (2004)

[13] Yang, Y., Liu, X.: A re-examination of text categorization methods. In: Proceedings of the 22nd Annual International ACM SIGIR Conference on Research and Development in Information Retrieval, pp. 42–49 (1999)

[14] Aggarwal, C.C., Zhai, C.: A survey of text classification algorithms. Mining Text Data, pp.163–222. Springer US (2012)

[15] Salton, G., Wong, A., Yang, C.: A vector space model for automatic indexing. Communications of the ACM 18(11), 613–620 (1975)

[16] Kim, H., Howland, P., Park, H.: Dimension reduction in text classification with support vector machines. The Journal of Machine Learning Research 6, 37–53 (2005)

[17] Guo, G., Wang, H., Bell, D., Bi, Y., Greer, K.: Using kNN model for automatic text categori-zation. Soft Computing 10(5), 423–430 (2006)

[18] Pang, G., Jiang, S.: A generalized cluster centroid based classifier for text categorization. Information Processing & Management 49(2), 576–586 (2013)

[19] Chang, C., Lin, C.: LIBSVM: a library for support vector machines. ACM Transactions on Intelligent Systems and Technology (TIST) 2(3), 27 (2011)

[20] Guan, H., Zhou, J., Guo, M.: A class-feature-centroid classifier for text categorization. In: Proceedings of the 18th International Conference on World Wide Web, pp. 201–210 (2009)

[21] Han, E., Karypis, G., Kumar, V.: Text categorization using weight adjusted k-nearest neigh-bor classification. In: Proceedings of the 5th Pacific-Asia Conference on Knowledge Discovery and Data Mining, pp. 53–65 (2001)

An Energy Model for Network Community Structure Detection

Yin Pang and Kan Li

Intelligent Information Technology, School of Computer Science Technology,
Beijing Institute of Technology, Beijing, 100081, China
Beijing Institute of Tracking and Telecommunication Technology,
Beijing, 100094, China

Abstract. Community detection problem has been studied for years, but no common definition of community has been agreed upon till now. Former modularity based methods may lose the information among communities, and blockmodel based methods arbitrarily assume the connection probability inside a community is the same. In order to solve these problems, we present an energy model for community detection, which considers the information of the whole network. It does the community detection without knowing the type of network structure in advance. The energy model defines positive energy produced by attraction between two vertices, and negative energy produced by the attraction from other vertices which weakens the attraction between the two vertices. Energy between two vertices is the sum of their positive energy and negative energy. Computing the energy of each community, we may find the community structure when maximizing the sum of these communities energy. Finally, we apply the model to find community structure in real-world networks and artificial networks. The results show that the energy model is applicable to both unipartite networks and bipartite networks, and is able to find community structure successfully without knowing the network structure type.

Keywords: community detection, energy model, unipartite community, bipartite community.

1 Introduction

Finding community structure and grouping vertices in the complex network, is the key to learning a complex network topology, understanding complex network functions, finding hidden mode, linking prediction, and evolution detection [2–7]. However, community detection seems like a chicken and egg problem [1], because the task is to discover the actual communities which we dont know; as the real communities are unknown, it is difficult for us to determine the community detection method is good or not. Depending on the properties of the network and the quality measure used, an appointed algorithm may be "better" than others. In other words, algorithms claimed they have good community results based on different understanding of community and no common definition of community

H. Motoda et al. (Eds.): ADMA 2013, Part I, LNAI 8346, pp. 410–421, 2013.
© Springer-Verlag Berlin Heidelberg 2013

has been agreed upon. In this paper, to make the method and experiment results easy to understood, we give our illustration of community through comparing two typical different former definitions. Measuring the relationship between vertices or edges is a key to community detection. Generally speaking, as no vertex pair is isolate in a connected network except a network with only two vertices or an isolate vertex, we should not only consider the affection between the pair of two vertices, but also the affection from others on the two of them in the whole network at the same time, while measuring the relationship between the two vertices. Many former theories have been proposed and each of them has its own theoretical foundation. Some consider that connected vertices are more likely to be in the same community such as spectral clustering [8], or edges sharing a common vertex may be in the same community [1], etc. In these methods, the relationship between two vertices only using the degree of vertices or edges may lose the information of the affection on the two vertices from other ones. Some are the theory that vertices in the same community (block) should have same connection probability, called blockmodels [9, 10], which may be arbitrary as we will explain in related work. Different from methods above, we propose an energy model using energy to measure the relationship between any two random vertices in this paper. The energy of each vertex pair is computed through considering not only the attraction between the two vertices but also the attraction from other vertices in the whole network. The energy model is as follows: Lets consider the network as an undirected graph with n vertices and m edges. Some factors may make two vertices tend to be in same community, like connections, vertex similarity etc. based on different theoretical foundation. To minimizing the loss of information of the network, we use innerproduct as the theoretical foundation in our energy model. We suppose that for each vertex pair there is an attraction between them. The attraction between the two vertices produces positive energy. On the other hand, attraction from other vertices will weaken the attraction between the two vertices. The attraction from other vertices produces negative energy. Vertices with bigger energy tend to be in the same community. Most former methods need to know the network type is unipartite or bipartite in advance to find community structure. However, in many cases, researchers have no prior knowledge of the network structure. For example, when we get a network which consists of peoples relationships in schools, the type of network may not be sure. This is because that if links are between students only, the network will be a unipartite network; or if links are between students and teachers, the network will be a bipartite one. An effective method which can be used for finding community structure only with the network topology only is needed. The energy model can recognize the network is a unipartite one or a bipartite one. In the energy model, given a network, it will separate the two parts of the network if it is bipartite without knowing the type and do not lose any connecting information. Moreover, it can do the community detection in a mixed network with some parts are organized as unipartite while some are bipartite. Mixed networks exist extensively in protein-protein interaction networks. We compare the community detection results with some successful methods: modularity [11], degree corrected blockmodel (we called blockmodel for short in the following) [12], infomap [13], and BRIM [14]. We will analyze these methods in related works.

2   Related Work

The most popular method for finding community structure is the modularity matrix method proposed by Newman et al. If the type of the network structure is known, the method is able to find community structure in both unipartite and bipartite networks by maximum or minimum eigenvalue separately [11]. Then, some researchers have sought to detect the community in bipartite networks. BRIM [14] proposed by Barber and his col-leagues can determine the number of communities of a bipartite network. Furthermore, Barber and Clark use the label-propagation algorithm (LPA) for identifying network communities [15]. However, the pre-vious work cannot be used without knowing the type of network or a mixed network. Hierarchical clustering is used frequently in finding community structures. In a hier-archical network, vertices are grouped into communities that further subdivide into smaller communities, and so forth [16]. Clauset, Moore and Newman proposed HRG [17] using the maximum likelihood estimation to predict the probability of connections between vertices. Hierarchical methods perform remarkably in clear hierarchical network, but not so impressive under contrary circum-stance. Moreover, a hierarchical method always has high computational complexity. Since being proposed in 2008, the Infomap algorithm [12] is becoming accepted as one of the best and most accurate node partitioning methods [20]. A random walk model is adopted to simulate the process of information diffusion on networks, with an objective to minimize the length of the information described by a geographical cod-ing schema. In this way, the task of community detection is turned into a coding problem: finding the partitions of networks that can minimize the description length of an infinite random walk process. Infomap manages to optimize such a length through a greedy search method combined with a simulated annealing strategy, which usually converges slowly. In 2009, Roger Guimera and Marta Sales-Pardo pro-posed a stochastic block model based on HRG [18]. Different from traditional concept which divides network by principle of inside connection dense outside sparse in previous work, the probability that two vertices are connected depends on the blocks to which they belong. However, the assumption that vertices in same blocks have same connection probability is not accurate. Recently, Karrer and Newman also proposed a stochastic block model which considers the variation in vertex degree [19]. This stochastic block model solves the heterogeneous vertex degrees problem and got a better result than other previous researches without using degree correction. It can be used in both types of networks, but different types of networks should be dealt separately. Furthermore, block model needs the number of communities in the beginning which is hard to be fixed. In 2010, Yong-Yeol Ahn et al. proposed an approach to find communities in hierarchical networks with overlaps based on edges [1]. Moreover, edges may lose, wrong or overlapped in a network. Nevertheless, it seems that most of these methods use the relationship of edges or vertices in the network instead of both of them in the community. They also need the prior knowledge of the network structure to deal with unipartite network and bipartite network separately.

3 An Energy Model

We propose an energy model in which energy is produced by the attraction between any two vertices, and attraction makes two vertices tend to be in the same community. There are two kinds of energy in the energy model: "positive energy" produced by the attraction between two vertices; negative energy produced by the attraction from other vertices which weakens the attraction between the two vertices. Positive energy can be defined using other theoretical foundation. Here we use inner product to measure the positive energy between vertices, which includes the vectors and the degrees, and the angle between two vertices. Vertex pair with larger energy tends to be in the same community, so the community structure is found when energy inside communities are maximized and outside are minimized. The energy model is independent of the type of network, so it is a unified method for both unipartite, bipartite networks and mixed networks. In the graph, we define the adjacency matrix A which is a symmetric matrix with element A_{ij}. If there is an edge joining vertices i and j, $A_{ij} = 1$; otherwise, $A_{ij} = 0$. In energy model, a_i is the label of vertex i, detonated by the ith vector of A. Let k_i be the degree of vertex i,

$$k_i = |a_i|^2, K = \sum_{i=1}^{n} a_i = (k_1, k_2, ..., k_n)$$

where $|a_i|$ is the module of a_i.

We define the positive energy pe_{ij} produced by the direct attraction from vertex j to vertices i as

$$pe_{ij} = a_i \cdot a_j = |a_i||a_j| \cos \theta = \sqrt{k_i}\sqrt{k_j} \cos \theta \qquad (1)$$

$\theta \in [0, \pi]$ is the angle between vectors a_i and a_j. The larger θ is, the smaller pe_{ij} will be. The total positive energy of the network is

$$Tpe = \sum_{i,j=1}^{n} pe_{ij} = \sum_{i,j=1}^{n} a_i \cdot a_j = K \cdot K \qquad (2)$$

As a result, the probability of the positive energy between vertices i and j is

$$pep_{ij} = \frac{pe_{ij}}{Tpe} \qquad (3)$$

When vertices i and j attract each other, other vertices in the graph may attract i and j at the same time. We believe that the attraction acting on i and j from other vertices in the graph will weaken the attraction and reduce the positive energy of between vertices i and j. It is viewed that, relative to the positive energy from vertex j to i, a vertex k in the graph which has the attraction with vertex i, produces negative energy. The negative energy of vertex i is

$$ne_i = -\sum_{k=1}^{n} pe_{ik} = -\sum_{k=1}^{n} a_i \cdot a_k = -a_i \cdot K \tag{4}$$

The total negative energy is

$$Tne = \sum_{i=1}^{n} ne_i. \tag{5}$$

The total negative energy of j is ne_j, so the probability of the negative energy of vertex j acting on other vertices is

$$nep_j = \frac{ne_j}{Tne}.$$

We define ne_{ij} as the negative energy from vertex j to vertex i through all vertices of the network.

$$ne_{ij} = ne_i nep_j = a_i \cdot K \frac{a_j \cdot K}{K \cdot K} \tag{6}$$

Obviously, for two random vertices i and j, $ne_{ij} = ne_{ji}$. The energy E_{ij} between vertices i and j is the sum of positive energy and negative energy, E_{ij} is normalized as

$$E_{ij} = \frac{1}{K \cdot K}(pe_{ij} + ne_{ij}) = [\frac{a_i \cdot a_j}{K \cdot K} - (\frac{a_i \cdot K}{K \cdot K})(\frac{a_j \cdot K}{K \cdot K})] \tag{7}$$

The energy model is given as follows,

$$\begin{aligned}
Q &= \sum_{k=1}^{c} \sum_{i,j \in c_k} E_{ij} \\
&= \sum_{k=1}^{c} \sum_{i,j \in c_k} [\frac{a_i \cdot a_j}{K \cdot K} - (\frac{a_i \cdot K}{K \cdot K})(\frac{a_j \cdot K}{K \cdot K})] \\
&= \sum_{k=1}^{c} \frac{d_k \cdot d_k}{K \cdot K} - (\frac{d_k \cdot K}{K \cdot K})^2
\end{aligned} \tag{8}$$

where Q is the energy within a community, $d_k = \sum_{i \in c_k} a_i$. When Q is maximized, we will get the community structure. Equation (8) is a function that divides the network into groups, with larger values indicating stronger community structure. Our task to find largest energy inside communities is changed to find $Max(Q)$. In the graph, E is the energy matrix. What we should notice is that total energy of the network is $E = \sum_{i,j} E_{ij} = 0$, It shows that the positive energy in a graph is equal to the negative energy. In a real complex network, vertices and edges change diversely. According to the law of conservation of energy, energy can only be transferred and cannot be produced. In the graph, no matter how it changes, E is constant which complies with the law of conservation of energy. Vertices within the same community have larger energy and vertices between communities have smaller energy. The sum energy of vertices in the same community is non-negative.

4 Performance Evaluation

Without knowing the type of the network in advance, our energy model can find the communities, and recognize the network is unipartite or bipartite. In this section, we compare the energy model in unipartite networks with the modularity model, infomap and Degree correction block model, and in bipartite networks with the modularity model and BRIM model; then, we evaluate the accuracy of energy model on large scale artificial networks [20] generated by LFR benchmark for finding communities with three properties: (1) The edges inside and outside communities. From the explanation of non-overlapping community, unipartite networks have small outside edges and bipartite networks have small inside edges. The first property is denoted as Edges (In/Out) in unipartite network, and Edges (Out/In) in bipartite networks.(2) Value of modularity which we maximum in unipartite networks and minimize in bipartite networks (denoted as edges(Q-Modularity)).(3) $I_{n}orm$ to compare with other algorithms. $I_{n}orm$ [21] is the normalized mutual information. When the found communities match the real ones, we have $I_{n}orm = 1$, and when they are independent of the real ones, we have $I_{n}orm = 0$. The normalized mutual information allows us to measure the amount of information common to two different grouping types.

4.1 Finding Communities in Actual Networks with Real Grouping Results

Five famous actual unipartite networks are used, of which are so classical that we have the real grouping results of each of them: Dolphin social network [23] with 62 bottlenose dolphins ties between dolphin pairs established by direct observation over a period of several years; Karate club network [23] contains the network of friendships between 34 members of a karate club at an American university. This club was by chance split into two smaller ones due to the divergence of opinions about the club fees; Football network [24] have 115 vertices representing the football teams while an edge divided into 12 conferences and, except for one conference; Politics books network (Krebs unpublished) contains 105 books about US politics indicates by the customers who bought this book also bought these other books feature on Amazon; Politics blogs network with 1409 vertices and 16715 edges [25] is based on incoming and outgoing links and posts around the time of the 2004 presidential election. Two famous actual networks are used: Southern women network [26]: describes the grouping of 18 women in 14 social events; Scotland network [27] in early twentieth century, the data set is characterized by 108 Scottish firms and totaling 136 individuals. All the three measurements are used in this part. Edges and Q-Modularity which depend on different properties of the network are not definitely correct. But I_{norm} is always correct, because it describes the difference between the result and the real grouping of communities. We calculate Edges and Q-Modularity of the real grouping of communities. Energy model has the most approximate value of the two measurements with the real grouping.

Comparison results among energy model and other three networks in the five actual networks are shown in Table.1. The GBLL algorithm is used for grouping community. It shows that, compared with real networks and other typical methods in unipartite networks, the performance of the energy model is the best in the dolphin network, karate network, and polbook network. Although modularity of Newman performs a little better than energy model in football network and blockmodel performs a little better than energy model in poblog network, the community structure found by energy model is quite similar with the actual community structure.

Given a network, the energy model does the community detection directly using Equation (7). In Table.2, it shows that, compared with other typical methods, energy model performs well in all three properties. The community structure found by energy model is better than other models. I_{norm} of BRIM is extremely small, because it tends to group networks into too many communities. For example, it may find more than 50 communities in Scotland which may be meaningful, which is much bigger than that of community number that most people accepted.

Table 1. The results in the unipartie networks

Networks	Methods	Edges inside/outside communities	Q-Modularity	I_{norm}
Dolphins	real	151/8	0.356	1
	energy	152/7	0.375	0.886
	Modularity of Newman	150/9	0.386	0.752
	HRG	38/121	0.093	0.208
	blockmodel	117/42	0.125	0.533
Karate	real	68/9	0.383	1
	energy	68/9	0.383	1
	Modularity of Newman	57/20	0.430	0.679
	HRG	24/53	0.372	0.467
	blockmodel	68/9	0.383	1
football	real	394/219	0.554	1
	energy	439/174	0.597	0.855
	Modularity of Newman	424/189	0.601	0.927
	HRG	55/158	0.232	0.494
	blockmodel	336/227	0.448	0.781
Polbook	real	371/70	0.415	1
	energy	399/42	0.523	0.550
	Modularity of Newman	376/65	0.520	0.510
	HRG	215/226	0.093s	0.247
	blockmodel	384/57	0.499	0.460
Polblog	real	15140/1575	0.405	1
	energy	15472/1243	0.424	0.710
	Modularity of Newman	15382/1333	0.430	0.640
	HRG	1110/15605	0.006	0.168
	blockmodel	15501/1214	0.426	0.745

Table 2. The results in the bipartie networks

Networks	Methods	Edges inside/outside communities	I_{norm}
Southern women	energy	0/93	1
	BRIM	53/40	0.520
	HRG	21/72	0.412
	blockmodel	0/93	1
Scottland	energy	0/358	0.846
	BRIM	220/138	0.553
	HRG	34/324	0.525
	blockmodel	4/354	0.279

4.2 Finding Communities in Artificial Networks

It is hard for us to get the real network community struc-ture to evaluate our model. As a result, we use LFR benchmark to generate artificial network for finding com-munities instead of real large networks. We test our algorithm on the LFR benchmark with 1000 vertices and 5000 vertices. For each class of benchmarks, the algorithm is run 100 times. As the number of communities of LFR benchmark is un-known at the beginning, blockmodel cannot be used in this section. We compare energy model with Modularity and infomap in this part.

LFR benchmark will lose community structure as the mixing parameter keeps growing. We compare energy model with BGLL and blockmodel (see Fig.1). It shows that the energy model and BGLL get perfect result when networks have strong community structure (mixing parameter < 0.5) while blockmodel per-forms relatively not well. The energy model has similar performance with BGLL in networks with big communities. As communities become smaller, the accu-racy of energy model declined a little fast-er than BGLL when mixing parameter > 0.5. However, when mixing parameter ≥ 0.5, the network becomes a weak community structure one, it is hard to evaluate the result is good or not.

4.3 Recognize Bipartite Networks

If we have the topology of a bipartite network with the two parts are mixed. Energy model can be used to distin-guish the two parts.

If the network is bipartite, energy model finds the two parts of the network accurately, as shown in Table.3. It shows that energy model can recognize a bipartite network. Comparison results among energy model and other five models in two actual bipartite networks are shown in Table.3.

We use LFR to make bipartite networks with $\mu \geq 0.5$. As the mixing param-eter grows, there are more edges between two parts. If we set the artificial net-work into two parts, the network will be a bipartite network. Among the method mentioned above, blockmodel and BRIM can be used in bipartite networks while

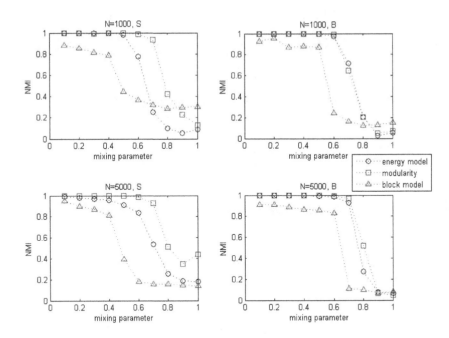

Fig. 1. Performances of the energy model, modularity, and blockmodel on the LFR benchmark as a function of the mixing parameter of the benchmark graphs. The different curves for each method refer to different system sizes (1000 and 5000 vertices) and community size ranges [(S) = from 10 to 50 vertices, (B) = from 20 to 100 vertices]. NMI means "I_{norm}".

Table 3. Result of the energy model to find the two parts of bipartite networks

Networks	Methods	Edges inside/outside communities	I_{norm}
Southern women	real	0/93	1
	energy	0/93	1
Scottland	real	0/358	1
	energy	0/358	1

BRIM. To group networks into many communities instead of separate the two parts, blockmodel needs to know community number in advance. We compare energy model with blockmodel (see Fig.2), it is shown that both models can detect bipartite structure accurately.

4.4   Finding Communities in Protein-Protein Interaction (PPI) Networks

PPI networks are always mixed networks with unipartite and bipartite parts in them. Former methods will no longer useful in a PPI network, because they

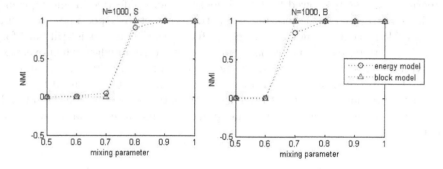

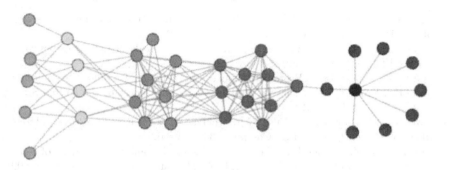

Fig. 2. The network is divided into two parts. As the mixing parameter grows, there are more edges between the two parts. It is shown that the enery model can recognize bipartite structure accurately. NMI means "I_{norm}"

Fig. 3. Results of a mixed network where vertices of different colours are in different communities

need the type of the network at the beginning in which a PPI network is uncertain. Fig.3 is a a PPI network model, which shows that energy model can find meaningful communities in a PPI network.

5 Conclusion

In many cases, we cannot know that network type is bipartite or unipartite. In order to solve the problem, we propose an energy model to find community structure in the paper. The network is viewed as the energy system and it includes positive energy between two vertices and negative energy produced by other vertices attracting the two vertices. The energy model is to make that energy inside communities largest, and then divides the network into communities without knowing the network type. Finally, we use our model in the actual networks and artificial networks generated by LFR benchmark to find communities.

Experimental results showed that the Energy model is effective in finding community structure without knowing the network structure type. There are still some problems that we should solve in the future work. First, it is still hard for us to illustrate unipartite networks from the perspective of edges when overlapping vertices exist in the network. Second, artificial networks have its limitation and could not replace actual networks completely. Moreover, it is hard for us to know the real community structure of the large actual networks. We should find an effective method to evaluate the energy model using large actual networks in the future work.

Acknowledgments. The research was supported in part by Natural Science Foundation of China (No.60903071).

References

1. Ahn, Y.Y., Bagrow, J.P., Lehmann, S.: Link communities reveal multiscale complexity in networks. Nature 466, U761-U711 (2010)
2. Newman, M.E.J.: Communities, modules and large-scale structure in networks. Nature Physics 8, 25–31 (2012)
3. Spirin, V., Mirny, L.A.: Protein complexes and functional modules in molecular networks. Proceedings of the National Academy of Sciences of the United States of America 100, 12123–12128 (2003)
4. Flake, G.W., Lawrence, S., Giles, C.L., Coetzee, F.M.: Self-organization and identification of web communities. Computer 35, 66 (2002)
5. Dourisboure, Y., Geraci, F., Pellegrini, M.: Extraction and Classification of Dense Implicit Communities in the Web Graph. ACM Transactions on the Web 3 (2009)
6. Moody, J., White, D.R.: Structural cohesion and embeddedness: A hierarchical concept of social groups. American Sociological Review 68, 103–127 (2003)
7. Wellman, B.: The development of social network analysis: A study in the sociology of science. Contemporary Sociology-A Journal of Reviews 37, 221–222 (2008)
8. Barnes, E.R.: An Algorithm for Partitioning the Nodes of a Graph. Siam Journal on Algebraic and Discrete Methods 3, 541–550 (1982)
9. Doreian, P., Batagelj, V., Ferligoj, A.: Generalized blockmodeling of two-mode network data. Social Networks 26, 29–53 (2004)
10. Reichardt, J., Bornholdt, S.: Statistical mechanics of community detection. Physical Review E 74, 14 (2006)
11. Newman, M.E.J.: Finding community structure in networks using the eigenvectors of matrices. Physical Review E 74 (2006)
12. Karrer, B., Newman, M.E.J.: Stochastic blockmodels and community structure in networks. Physical Review E 83 (2011)
13. Rosvall, M., Bergstrom, C.T.: An information-theoretic framework for resolving community structure in complex networks. Proceedings of the National Academy of Sciences of the United States of America 104, 7327–7331 (2007)
14. Barber, M.J.: Modularity and community detection in bipartite networks. Physical Review E 76 (2007)
15. Barber, M.J., Clark, J.W.: Detecting network communities by propagating labels under constraints. Physical Review E 80 (2009)

16. Newman, M.E.J., Girvan, M.: Finding and evaluating community structure in networks. Physical Review E 69 (2004)
17. Clauset, A., Moore, C., Newman, M.E.J.: Hierarchical structure and the prediction of missing links in networks. Nature 453, 98–101 (2008)
18. Guimera, R., Sales-Pardo, M.: Missing and spurious interactions and the reconstruction of complex networks. Proceedings of the National Academy of Sciences of the United States of America 106, 22073–22078 (2009)
19. Karrer, B., Newman, M.E.J.: Stochastic blockmodels and community structure in networks. Physical Review E 83 (2011)
20. Lancichinetti, A., Fortunato, S., Radicchi, F.: Benchmark graphs for testing community detection algorithms. Physical Review E 78 (2008)
21. Danon, L., Diaz-Guilera, A., Arenas, A.: The effect of size heterogeneity on community identification in complex networks. Journal of Statistical Mechanics-Theory and Experiment (2006)
22. Lusseau, D., Schneider, K., Boisseau, O.J., Haase, P., Slooten, E., Dawson, S.M.: The bottlenose dolphin community of Doubtful Sound features a large proportion of long-lasting associations - Can geographic isolation explain this unique trait? Behavioral Ecology and Sociobiology 54, 396–405 (2003)
23. Zachary, W.W.: An information flow model for conflict and fission in small groups. Journal of Anthropological Research 33, 452–473 (1977)
24. Girvan, M., Newman, M.E.J.: Community structure in social and biological networks. Proceedings of the National Academy of Sciences of the United States of America 99, 7821–7826 (2002)
25. Adamic, L.A., Adar, E.: Friends and neighbors on the Web. Social Networks 25, 211–230 (2003)
26. Davis, B.B., Garder, M.R.: Deep South. University of Chicago Press, USA (1941)
27. Scott, J., Hughes, M.: The anatomy of Scottish capital: Scottish companies and Scottish capital. The anatomy of Scottish capital: Scottish companies and Scottish capital (1980)

A Label Propagation-Based Algorithm for Community Discovery in Online Social Networks

Yitong Wang, Yurong Zhao, Zhuoxiang Zhao, and Zhicheng Liao

School of Computer Science, Fudan University, Shanghai, China
{yitongw,12210240048}@fudan.edu.cn, dygfzzx@gmail.com,
zchliao@fudan.edu.cn

Abstract. With the rapid development of Internet and Web 2.0 applications, many different patterns of online social networks become fashionable all over the world. These sites help people share and exchange information, as well as maintain their social relations on the Internet. Therefore, it is very important to study the structure of communities in online social network.

Most of existed community discovery algorithms are very costly. Moreover, the behavior of users in online social networks is rather dynamic. We first investigate Label Propagation Algorithm (LPA), which has near linear time complexity and discuss some limitations of LPA. Then, we propose a new algorithm for community discovery based on label influence vector (LIVB), an improved variation of LPA. In this algorithm, we abstract several types of nodes corresponding to different kinds of entities such as users, posts, videos as well as comments. Different types of relations between nodes are also taken into account. A node will update its label by calculating its label influence vector. We conduct experiments on crawled real data and the experimental results show that communities discovered by LIVB algorithm have more concentrative topics. The quality of the communities is improved and LIVB algorithm remains a near linear time complexity.

Keywords: Community discovery, label propagation, social network, label influence value, label influence vector.

1 Introduction

With the emergence and prosper of Web 2.0, there become some new research issues and challenges in social network analysis. A social network is always abstracted as a directed or undirected graph, a node in which represents a person and an edge in which represents a relationship between people. Related studies are mainly focused on the structure of social network and its evolution, as well as the influence and effect of the topology on these dynamic social systems [1, 2].

Generally speaking, a community in a social network consists of a group of similar nodes. Admittedly, there is no specific definition for community that could be generally accepted [3, 4], however, it could not be denied that there exist such implicit community structures in social networks. We adopt the definition

H. Motoda et al. (Eds.): ADMA 2013, Part I, LNAI 8346, pp. 422–433, 2013.

of community proposed in [5], which is composed of a group of nodes while edges within the community are much denser than edges between communities.

Study on community structure has many meaningful applications such as public opinion analysis, focusing on target group, finding a better business environment and so forth. Community discovery/extraction has been extensively studied and many algorithms have been proposed in literature. However, most of these algorithms are very costly and cannot be adapted to current online social networks, which consist millions of users. With the advent of "big data", how to extract meaningful communities has posed great challenges for current research. Moreover, comparing with traditional social network, current online social works are with huge volume, more dynamic, composed of different types of entities and different types of relationships between entities. Most existed community discovery algorithms are not suitable to online social networks both in term of efficiency and effectiveness. In this paper, we propose a new algorithm for community discovery based on label influence vector, which is an improved variation of LPA. The experimental results indicate that the proposed algorithm could identify the high quality communities efficiently.

In section 2, we review the priori related work. Details of Label-Influence-Vector-Based (LIVB) algorithm are described in section 3. We present the experimental results on real data in section 4 In section 5, we give the conclusion and the future work.

2 Related Work

2.1 Related Algorithms

Most algorithms for community discovery can be categorized into three types [6]. The first method is based on graph theory, such as random walk methods and physics-based methods [7, 8], spectral methods [9, 10] and heuristic graphs partitioning methods [11]. The second type are hierarchical methods, such as agglomerative methods based on structural similarity metrics [12, 13, 14] and divisive methods based on betweenness metrics, like *Girvan* and *Newman's* algorithm[5] and *Radicchi's* method[15]. The third method is for web community discovery based on hyperlinks, such as HITS algorithm [16] and spreading activation energy algorithm [17].

Pons and *Latapy* proposed a community discovery algorithm based on random walk [7], it starts from a community with a single node and merges adjacent communities repeatedly to minimize the mean of squared distances between each node and its community. The distances between communities will be updated dynamically in each step. The distance between nodes is the directional transition probability between them in a random walk process. *Girvan* and *Newman* proposed a divisive method [5] based on Edge Betweenness Centrality which is the number of shortest paths among all pairs of nodes in the network passing through that edge. The main idea is that the edge betweenness centrality for edge between communities is higher than that lie within a community. The network is divided into some disjoint connected components by removing edges with highest betweenness values. If there are n nodes and m edges in the network, the time complexity for each step is $O(mn)$. The mentioned algorithms focus on partitioning the network structure with a comparatively higher time complexity and ignore the propagating property of social networks.

Raghavan et al. proposed an algorithm for community discovery based on label propagation in [18], based on which we propose a novel algorithm for community discovery in social networks. The proposed algorithm takes near liner time to discover high-quality communities with a better stability.

2.2 Community Discovery Based on Label Propagation (LPA)

The main idea of the label propagation algorithm (LPA) [18] can be described as follows:

1. Allocate a unique label (like an integer) to each node in the graph as an identifier of a community.
2. Update the labels of all nodes in the graph iteratively. At each iteration, the label of a node x is updated by replacing it by the label used by the greatest number of its neighbors. If there is more than one candidate label, it will choose a label randomly from the candidates. After several iterations, the label changes become stable
3. All nodes with the same label form a community

Fig.1 shows a label propagating process. At first, each node has a unique label and finally, all nodes will have the same label to form a community.

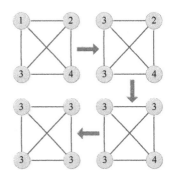

Fig. 1. Label propagation process

However, the algorithm cannot always guarantee the convergence. For example, when we update all nodes synchronously, some sub-graphs in the network that are bipartite or nearly bipartite may lead to oscillations of labels. Therefore, *Raghavan et al.* propose to update the nodes asynchronously in [17] to avoid label oscillations. This algorithm has a near liner time complexity $O(m + n)$. One drawback is that the number of iterations is difficult to predict. *Leung et al.*

made a detail comparison between synchronous updating and asynchronous updating in [19] and found that asynchronous updating need less iterations but also less stable. *Gergely et al.* modeled the label influence process [20] and found it is equivalent to zero-temperature kinetic Potts-model. The main drawback of LPA is that the final results are very unstable due to random selection of candidate label during its iterations.

2.3 Measurement of Community Quality

Most work use modularity measure Q in [21] as the metric to evaluate the quality of communities. Q is a good measurement for the modularity of a graph. However, there are many different types of entities/nodes and different types of relationships in our graph. So, instead of modularity, we use some other metrics such as distributions of communities, topic coverage of communities and tag overlapping to measure the quality of communities in online social networks.

3   Label-Influence-Vector-Based Algorithm

3.1   Main Idea

Large online social networking sites are increasingly becoming an important part of people's life. The social networking site allows a variety of user behaviors to share and manage different kinds of entities, such as uploading and sharing photos, posting blogs, share bookmarks and so forth. The behaviors of a user are related to the behavior of other users and other entities, as well as the historical behavior of the user. Therefore, we think that the relationships between users are not only explicit (such as adding friends) but also implicit which is linked by many common user behaviors (such as two users commented on the same article, shared the same label and had a common interest or theme). If there exists many such implicit contacts between two users, they may be active at the same potential social circle.

We regard both users and other entities as vertices in the graph. All entities will participate into the label propagation process so that the propagation process contains more information. As shown in Fig.2, circle represents user, triangle represents videos, star represents post, and square represents comment. We regard all this entities as nodes in the graph to participate into the label propagation process.

3.2   Label-Influence-Vector-Based Algorithm for Community Discovery

A social network can be abstracted as a graph $G = (V, E)$, in which V represents the node set and E represents the edge set. For each $e \in E$, there is a real number $w(e)$ represents the weight of edge e. Every label is attached with a label influence vector (Static Value, Dynamic Value), in which they represent static influence

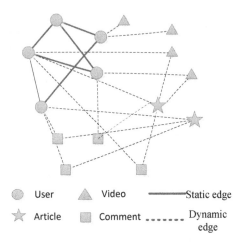

Fig. 2. Social network graph with different entities

value and dynamic influence value respectively. While updating the label of a node, we will calculate a label vector for every label in its neighbors. Then we will update the node with the label having the largest influence vector.

LIVB algorithm can be described as follows:

1. Allocate a unique label (like an integer) to each node in the graph as an identifier of a community.
2. Update the labels of all nodes in the graph iteratively. At each iteration, the label of a node x is updated by replacing it by one label in its neighbors. We calculate a label vector for each existed label and choose one with the greatest influence.
3. All nodes with the same label form a community.

Calculating Label Influence Vectors. We will calculate static value and dynamic value respectively while updating the label of a node i. The notations used in the paper have been listed in the Table 1.

We calculate the Static Value as follows:

$$StaticValue(l) = \sum Weight(e_S^l) + (1 - e^{-Degree(l_S)}) \qquad (1)$$

In formula (1), when we compare two label influence values, the sum of weights will be a decisive factor if their values of $Weight(e_S^l$ are different. Otherwise, the label will be decided by the sum of neighbor nodes degrees.

We calculate the Dynamic Value as follows:

$$DynamicValue(l) = \sum Weight(e_D^l) * (1 + \frac{1}{Count(l_D)}) \qquad (2)$$

Table 1. Notations used in formulas

Notation	Meaning
l	a given label of node $i's$ neighbor nodes
L	all labels of node $i's$ neighbor nodes
$e_S^l(e_D^l)$	a static or dynamic edge which connect a node with label l
$l_S(l_D)$	the set of nodes (except the node i)which connected by e_S^l or e_D^l
$Count(l_D)$	the cardinality of set l_D
$Weight(e_S^l\|e_D^l)$	the weight of e_S^l or e_D^l
$Degree(l_S)$	the sum degree of nodes in l_S

In formula (2), DynamicValue is almost the same with StaticValue, but the label will be decided by the average weight while their values of $Weight(e_D^l)$ are the same. We calculate the label vector for each label according to two formulas.

In the paper, we assume that the weight of static edge is larger than the one of dynamic edge, which has been turn out to be consistent with the reality.

Comparisons between Label Influence Vectors. When we updating a node's label, we need to find the most influence label and sign the label to the node. Before that, we can use some normalization methods to operate these static values and dynamic values. We use the sum of $StaticValues$ and $DynamicValues$ to normalize them respectively.

If the each $StaticValue$ is 0, then the most influence label is the one with the highest $DynamicValue$;

If the each $DynamicValue$ is 0, then the most influence label is the one with the highest $StaticValue$;

Else then, we use the formula(3) to normalize $staticvalue$ and $dynamicvalue$.

$$Sta(l) = \frac{StaticValue(l)}{\sum_{l \in L} StaticValue(l)}, Dyn(l) = \frac{DynamicValue(l)}{\sum_{l \in L} DynamicValue(l)} \quad (3)$$

If we want to combine the effects of static value and dynamic value, then can use two parameters to operate, but it is difficult to train the parameters. Therefore, we use formula (4) to compare two labels' influence, and the label l_1 has more important influence than l_1 if and only if the inequality (4) is established.

$$V_1 > V_2 : \frac{Sta(l_1) - Sta(l_2)}{max(Sta(l_1), Sta(l_2))} + \frac{Dyn(l_1) - Dyn(l_2)}{max(Dyn(l_1), Dyn(l_2))} > 0 \quad (4)$$

This comparison is transitive, which means if $V1 > V2$ and $V2 > V3$ then $V1 > V3$.

LIVB Algorithm for Community Discovery. After running LIVB algorithm(as shown in Table 2), nodes with the same label form a community. Obviously, the proposed LIVB algorithm is suitable for current online social networks and could overcome the drawbacks of LPA algorithm.

Table 2. The pseudo-code of LIVB algorithm

Input	$G(V, E), k$
(1)	allocate a unique label for each node
(2)	$t = 1$;
(3)	Update label for all nodes, choose a label with a greatest label vector
(4)	if($t > k$) END; Else $t = t + 1$, goto step (3);

3.3 Time Complexity of LIVB Algorithm

Suppose that we have n nodes and m edges in graph G. The time complexity of LIVB algorithm is analyzed as follows: (1) Label initializing for all nodes: $O(n)$; (2) One propagating iterative process in worst case: $O(m + n)$; (3) Find all communities: $O(m + n)$. Obviously, LIVB algorithm has near-liner time complexity.

4 Experiments for LIVB Algorithm

4.1 Data Description

We chose a social network dataset collected by *Choudnury*[23] from Digg Website. Contents in Digg are all produced by users, who can dig a story or comment in a story to change the rank of stories. As shown in Fig. 3, there are three kinds of nodes and relations between users are static and other relations are dynamic.

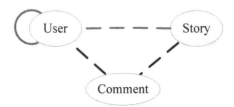

Fig. 3. Relations between nodes

4.2 Experiments

We conduct two experiments. The first one is to compare LIVB algorithm and LPA algorithm on the entire graph in which stories include many topics. The second one is to compare LIVB and LPA algorithm in a partial graph in which stories are focusing on one general topic-"food".

1)Experimental Results on an Entire Graph. In this section, the dataset of the entire graph consists of 295388 nodes and 1682183 edges. Table 3 and Table 4 show detail description of this dataset.

Table 3. Different types of nodes

Node type	Node count
User	9583
Story	44005
Comment	241800
Total	295388

Table 4. Different types of edges

Relations	Edge count
user-user	56440
user-story	1157529
story-comment	241800
user-comment	226414
Total	1682183

As can be seen in Fig. 4, we have got large amount of small communities and only a few large communities with pocket distribution when we ran the original LPA algorithm. When we ran LIVB algorithm, we got a smaller amount of communities with more uniform distribution and more meaningful communities.

In order to analyze the result more from social aspects, we take out a community and exclude all non-user nodes to analyze its topic concentration from LVIB and LPA respectively. Community discovered by LPA (5301): 233, 4536, 5301, 6643; Community discovered by LIVB (233): 233, 993, 1549, 1915, 1998, 2071, 4536, 5301 6643.

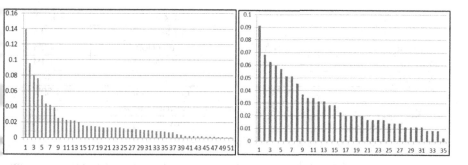

Fig. 4. Topic coverage for community discovered by LPA

Fig. 5. Topic coverage for community discovered by LIVB

We can see that community discovered by LPA has 4 members, which are all included in the community discovered by LIVB. We will analyze the performance in detail based on the topic coverage for each community. Topic coverage have two meanings that the times of topic appeared in different communities at the same time and the number of topics discovered by algorithms.

Fig.4 and Fig.5 show the topic coverage for two communities. The Topics amount in the community discovered by LPA is 35 while the community discovered by LIVB is 51. Fig.6 and Fig.7 show the coverage summation of Top-8 topics in two communities respectively.

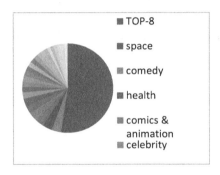

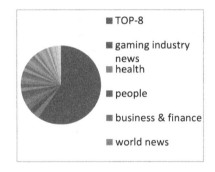

Fig. 6. Top-8 Topic coverage summation in community discovered by LPA

Fig. 7. Top-8 topic coverage summation in community discovered by LIVB

We can see Top-8 topics covers 48.72% and 57.04% in LPA and LIVB respectively. The latter improves 8.32% compared with the former so that the concentration in LIVB is higher. As shown in Table 5 and Table 6, LIVB community involves much more topics and has a higher topic concentration compared with LPA community.

Table 5. Top-8 topic coverage in community discovered by LPA

Table 6. Top-8 topic coverage in community discovered by LIVB

linux/unix	9.12%
us election	6.84%
tech industry new	6.27%
movies	5.98%
software	5.70%
food	5.13%
apple	5.12%
gaming industry news	4.56%

linux/unix	13.92%
apple	9.59%
us election	8.02%
tech industry news	7.64%
space	5.44%
general science	4.39%
political news	4.19%
movies	3.85%

In summary, LIVB can discover more meaningful communities and each community has higher concentration in terms of topics (story, comments) as users in the community has more similar topics. We have the reason to believe that community discovered by LIVB has higher quality from viewpoint of sociology.

2)Experiments Results and Analysis in a Partial Graph. In this section, we run the LIVB algorithm in a partial graph which is related to a concrete topic. We choose the data about the topic of "food", which cover a big part in original dataset. It consists of 30146 nodes and 136237 edges. Table 7and Table 8 show details of the dataset.

3)Experiment for comparison of LIVB and LPA. For the same story about the general topic "food", different users will view it from different aspects

Table 7. Different types of nodes

Node type	Node count
User	9583
Story	461
Comment	20102
Total	30146

Table 8. Different types of edges

Relations	Edge count
user-user	56440
user-story	54706
story-comment	5681
user-comment	1941
Total	136237

and then will use different tags to comment and reply this story. Therefore, there exist many-to-many relationships between user entity and story entity. Based on this observation, we expect that the set of tags described the food stories that are clustered into the same community will have a higher overlapping. Overlapping means the intersection part between some sets, but we assume it means the number of tags appeared more than once in a community. Since the numbers of tags for different communities are on different scale, we have to normalize them for final comparison.

We use $tags$ to measure the distribution of tags in detected communities. Note that $tags = \frac{\#tag - \#tag^*}{\#tag}$, $\#tag$ is the number of tags in a community and $\#tag^*$ is the number of unique tags by removing overlapping tags. Obviously, the larger the $tags$, the degree of overlapping of tags in a community is also higher.

The tag overlapping for LIVB and LPA are shown in Fig.8 and Fig.9. In Fig.8, 2 of 30 communities detected by the LIVB algorithm have a value 0(that means $\#tag = \#tag^*$).In Fig.9, 38 of 120 communities detected by the original LPA algorithm have a value 0. From the macroscopic view, tag overlapping of each community in the Fig.8 is much higher, and the maximum overlapping value is more than 0.7, but the tag overlapping of each community in the Fig.9 is much less. So, from Fig.8 and Fig.9, it is indicated once more that the quality of community detected by LIVB is higher.

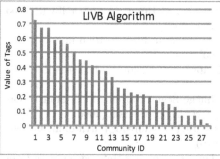

Fig. 8. Distribution of tag overlapping in the communities detected by LIVB

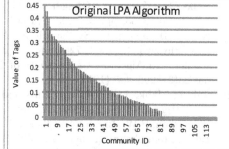

Fig. 9. Distribution of tag overlapping in the communities detected by original LPA

5 Conclusion

In this paper, we consider the problem of community discovery/extraction for online social networks, which are very complex including millions of users and various types of entities as well as different relationships between entities. Existed community discovery algorithms are not suitable both in terms of efficiency and effectiveness. We propose a new algorithm based on label influence vector, an improved variation of label propagation. Label-Influence-Vector-Based (LIVB) algorithm considers different types of nodes in the graph. It calculates static and dynamic components of a label influence vector for each label. We conduct two experiments on a crawled real data. The experimental results demonstrate that communities discovered by LIVB algorithm are more realistic and have a better quality.

We plan to test our current study on directed graph and signed graph in future work. Overlapping community discovery is also under consideration.

References

1. Newman, M.E.J.: The Structure and Function of Complex Networks. SIAM Review 45(2), 167–256 (2003)
2. Albert, R., Barabsi, A.L.: Statistical Mechanics of Complex Networks. Reviews of Modern Physics 74(1), 47 (2002)
3. Fortunato, S.: Community Detection in Graphs. Physics Reports 486(3), 75–174 (2010)
4. Alba, R.D.: A Graph–Theoretic Definition of a Sociometric Cique. Journal of Mathematical Sociology 3(1), 113–126 (1973)
5. Girvan, M., Newman, M.E.J.: Community Structure in Social and Biological Networks. Proceedings of the National Academy of Sciences 99(12), 7821–7826 (2002)
6. Yang, B., Cheung, W.K., Liu, J.: Community Mining from Signed Social Networks. IEEE Transactions on Knowledge and Data Engineering 19(10), 1333–1348 (2007)
7. Pons, P., Latapy, M.: Computing communities in large networks using random walks. In: Yolum, p., Güngör, T., Gürgen, F., Özturan, C. (eds.) ISCIS 2005. LNCS, vol. 3733, pp. 284–293. Springer, Heidelberg (2005)
8. Kim, D.H., Jeong, H.: Systematic Analysis of Group Identification in Stock Markets. Physical Review E 72(4), 46133 (2005)
9. Fiedler, M.: Algebraic Connectivity of Graphs. Czechoslovak Mathematical Journal 23(2), 298–305 (1973)
10. Shi, J., Malik, J.: Normalized Cuts and Image Segmentation. IEEE Transactions on Pattern Analysis and Machine Intelligence 22(8), 888–905 (2000)
11. Kernighan, B.W., Lin, S.: An Efficient Heuristic Procedure for Partitioning Graphs. Bell System Technical 49(1), 291–307 (1970)
12. Danon, L., Daz-Guilera, A., Arenas, A.: The Effect of Size Heterogeneity on Community Identification in Complex Networks. Journal of Statistical Mechanics: Theory and Experiment 2006(11), P11010 (2006)
13. Burt, R.S.: Positions in Networks. Social Forces 55(1), 93–122 (1976)
14. Wasserman, S., Faust, K.: Social Network Analysis: Methods and Applications. Cambridge University Press (1994)

15. Radicchi, F., Castellano, C., Cecconi, F., et al.: Defining and Identifying Communities in Networks. Proceedings of the National Academy of Sciences of the United States of America 101(9), 2658–2663 (2004)
16. Kleinberg, J.M.: Authoritative Sources in a Hyperlinked Environment. Journal of the ACM (JACM) 46(5), 604–632 (1999)
17. Pirolli, P., Pitkow, J., Rao, R.: Silk from a Sow's Ear: Extracting Usable Structures from the Web. In: Proceedings of the SIGCHI Conference on Human Factors in Computing Systems: Common Ground, pp. 118–125. ACM (1996)
18. Raghavan, U.N., Albert, R., Kumara, S.: Near Linear Time Algorithm to Detect Community Structures in Large-Scale Networks. Physical Review E 76(3), 036106 (2007)
19. Leung, I.X.Y., Hui, P., Li, P., et al.: Towards real-time community detection in large networks. Physical Review E 79(6), 066107 (2009)
20. Tibly, G., Kertsz, J.: On the Equivalence of the Label Propagation Method of Community Detection and a Potts Model Approach. Physica A: Statistical Mechanics and its Applications 387(19), 4982–4984 (2008)
21. Newman, M.E.J., Girvan, M.: Finding and Evaluating Community Structure in Networks. Physical Review E 69(2), 026113 (2004)
22. Wang, Y., Feng, X.: A Potential-Based Node Selection Strategy for Influence Maximization in a Social Network. In: Huang, R., Yang, Q., Pei, J., Gama, J., Meng, X., Li, X. (eds.) ADMA 2009. LNCS, vol. 5678, pp. 350–361. Springer, Heidelberg (2009)
23. Leskovec, J., Kleinberg, J., Faloutsos, C.: Graph Evolution: Densification and Shrinking Diameters. ACM Transactions on Knowledge Discovery from Data (TKDD) 1(1), 2 (2007)
24. De Choudhury, A., et al.: Social Synchrony: Predicting Mimicry of User Actions in Online Social Media. In: International Conference on Computational Science and Engineering 2009, CSE 2009, pp. 151–158. IEEE CPS, Vancouver (2009)

Mining Twitter Data for Potential Drug Effects

Keyuan Jiang and Yujing Zheng

Department of Computer Information Technology & Graphics
Purdue University-Calumet, Hammond, Indiana, USA
jiang@purduecal.edu, hzyh64160@gmail.com

Abstract. Adverse drug reactions have become one of the top causes of deaths. For surveillance of adverse drug events, patients have gradually become involved in reporting their experiences with medications through the use of dedicated and structured systems. The emerging of social networking provides a way for patients to describe their drug experiences online in less-structured free text format. We developed a computational approach that collects, processes and analyzes Twitter data for drug effects. Our approach uses a machine-learning-based classifier to classify personal experience tweets, and use NLM's MetaMap software to recognize and extract word phrases that belong to drug effects. Our results on 5 medications demonstrate the validity of our approach, and its ability to correctly extract potential drug effects from the Twitter data. It is conceivable that social media data can serve to complement and/or supplement traditional time-consuming and costly surveillance methods.

Keywords: Big Data, Drug Effect Identification, Natural Language Processing, Pharmacovigilance, Social Media Mining, Twitter.

1 Introduction

A pharmaceutical product, before it can be brought to market, must go through rigorous testing, in clinical trials, for short-term safety and efficacy on a very limited number of subjects which were carefully chosen using restrictive inclusion and exclusion criteria. Once on the market, a drug is legally set free for consumption by the general public [1] and is frequently for off-label uses [2]. Adverse drug reactions (ADRs) have already become a top cause of deaths in developed nations such as USA and UK [4-7], contributing to more than 100,000 estimated deaths each year in USA [8]. And more than half of all new serious adverse reactions are not identified until seven or more years after a drug was approved by the FDA [3]. Yet, the progress towards solving the problem has yet to be made [9]. It is important and necessary to collect information on the use of a drug by the public and risks caused by the product in order to ensure the safety of the medication.

Traditional methods of reporting adverse drug effects include clinical trials and spontaneous reporting which has been an effective solution for post-marketing surveillance of approved medications in normal clinical use, detecting many important signals. Recently, mining electronic medical records (EMRs) has been

H. Motoda et al. (Eds.): ADMA 2013, Part I, LNAI 8346, pp. 434–443, 2013.
© Springer-Verlag Berlin Heidelberg 2013

studied in detecting ADRs. Not only do these methods rely heavily on caregivers such as medical doctors and nurses, but also they are time-consuming and costly.

Patients play the most important role in pharmacovigilance, the science of detection, assessment, understanding and prevention of adverse effects of medicines, as the actual benefits and harm of any particular medication are only known to patients who take it. A healthcare professional may only provide the interpretation of the description provided by the patient. Involving in patients in reporting drug related problems may improve post-marketing surveillance of a drug and overcome shortcomings due to the sole reporting from healthcare professionals. Patient reporting could identify new adverse drug effects sooner than that by healthcare professionals alone [10]. This type of patient reported outcomes (PROs) is typically conducted through the use of dedicated computer software systems or Web sites where data are more structurally organized.

Recently, thanks to the development of social networking on Internet, patients are able to communicate with others and share their personal experiences in a way much easier than many other communication platforms. Instead of using structured approaches, many choose to use non-health-related, general purpose social networking sites such as Facebook and Twitter to share their responses to medications they took in free text. This kind of activities generates a collection of useful and valuable patient reported information (PRI) related to the use of pharmaceutical products. This set of unstructured free-text social media data could potentially be valuable in discovering signals of adverse drug effects.

Presented in this work is our effort in developing an automated method to extract potential drug effects from the Twitter data. Unlike the data collected during a clinical trial, from the electronic medical records, spontaneous reporting systems, and health-specific websites or social networking site, Twitter data are very dirty and noisy with most of the posts being unrelated to the medication. To obtain the Twitter data needed for this study, we designed and followed the following steps. We started with retrieving tweets related to the drug name and its brand name(s). The collected tweets were cleaned and preprocessed before being analyzed. The cleaned and preprocessed tweets were filtered to only include the tweets that describe author's feeling and/or experience with the drug. And the last step is to identify any drug effects in the remaining tweets. To assess the validity of our approach, we compared the discovered effects with what has been reported or known for the same medications.

2   Related Work

Very few efforts have been made on detecting drug effects using social networking sites, most of which are dedicated, and more often patients were aware that they were submitting their postings to the health related site/group.

Knezevic and colleagues created a FaceBook group that invited public to submit the information of the drugs they took and their adverse experiences with the drugs [16]. In a period of 7 months, the group had a total of 1,034 members, among which 21 members reported 21 different ADRs for 19 drugs and supplements such as multivitamins. This is probably the very first effort of detecting drug effects using a general-purpose, non-health-related social network site. However, besides the manual approach to determine the drug effects, the study shows a very low rate of participation in reporting – only 2% of the group members reported ADRs.

A group at Arizon State University mined, using its own-developed lexicon, user comments posted to the DailyStrength health-related social network, and discovered that those posts do contain information related to adverse drug reactions of 6 drugs studied. The discovered ADRs correlate well with those previously documented [17], although the frequencies of each ADRs do not match well. In addition to noticeable errors in its results, the group's text-mining method only showed marginally acceptable performance (78.3% precision, 69.9% recall, and 73.9 F-measure).

Bian and colleagues [19] studied 5 investigational cancer drugs that were used in ongoing clinical trials by retrospectively mining a set of 2 billion Twitter posts. Their data processing pipeline starts with classifying drug users, retrieves their timelines (the collection of all Twitter posts submitted by a Twitter user), and identifies drug side effects in those tweets. They developed two similar classifiers using the Support-Vector-Machine-based machine learning algorithm, and each used both textual and semantic features. The semantic features were derived from the Unified Medical Language Systems (UMLS) semantic types from the result of executing the National Library of Medicine's (NLM's) MetaMap software [22]. Out of 2 billion Tweets they studied, 239 potential drug users were found, and 72 of them were identified as drug users. In the tweets posted by those 72 drug users, 27 users' tweets were discovered to contain a report of adverse event. Their results only show the performances of both classifiers without demonstrating the drug effects were actually detected. They used the same 72 drug users' tweets for testing and training the classifiers, but did not seem to actually use the classifiers to classify any more raw Tweets. The choice of using investigational drugs is of a big concern because they can lead false detections of side effects as placebos are commonly used in clinical trials and the study participants do not know whether they were given the actual medicines or placebos.

3   Method

We developed a data processing pipeline that automatically extracts and detects potential drug effects from the Twitter data:

1. Collecting and pre-processing drug-related tweets,
2. Classifying personal experience tweets through the use of a machine-learning-based classifier, and
3. Identifying drug effects from tweet text with the help of National Library of Medicine's MetaMap software.

3.1   Selection of Medications

To query Twitter data, we needed to have particular drug names. In this pilot study, we chose 5 drugs to test if any related effects could be found in the Twitter data, and they were chosen based upon the following. First, they should have been on market for a number of years so that sufficient tweets would exist for effects to be reported, and second, they should not be taken by patients whose conditions kept them from posting tweets regularly. Table 1 shows the drugs we chose to be investigated in this research.

Table 1. The Drugs Chosen to be Studied[1]

#	Drug Name	Brand Names	Prescribed for	First Approval Year
1	Duloxetine	Cymbalta, Duloxetine Hydrochloride	Depression and Generalized Anxiety Disorder	2004
2	Gabapentin	Gabapentin, Gralise, Horizant, Neurontin	Epilepsy Seizure	1993
3	Baclofen	Baclofen, Gablofen, Kemstro, Lioresal,	Spinal Cord Nerves and Muscle Spasms	1977
4	Glatiramer	Copaxone	Multiple Sclerosis	1996
5	Pregabalin	Pregabalin, Lyrica	Neuropathic pain	2004

3.2   Collecting and Preprocessing Drug-Related Tweets

Twitter data were collected through the use of Twitter API 1.0 which only allowed searching for tweets posted within about past 2 weeks. We continuously queried Twitter with both drug and brand names shown in Table 1 as keywords in a period of 80 days (from January 29, 2013 to April 18, 2013), collecting a total of 6,829 tweets related to the 5 study drugs. Any non-English tweets and retweets (RTs) were removed. Any tweets whose author's name contains a drug name were also removed. This was intended for disambiguation between a drug and a name with some other senses. For example, Lyrica is a brand name for Pregabalin, and is also widely used on Twitter for the name of singers, music, and so forth. Although this treatment is conservative, it improves the relevance of the drug-related tweets.

The collected drug-related tweets were preprocessed before being analyzed. The collect Tweets were normalized to expand condensed words and phrases, abbreviations and acronyms to the normal format. In addition, any mentions, URLs, and hastags were converted to USERNAME, URL, and SOMETOPIC.

3.3   Classifying Personal Experience Tweets

Effects are physical or mental signs and conditions shown on patients who take the medication. Not all the drug-related tweets collected were related to drug effects.

---

[1]   Sources: Drugs @ FDA [http://www.accessdata.fda.gov/scripts/cder/drugsatfda/] and MedlinePlus Drug Information [http://www.nlm.nih.gov/medlineplus/druginformation.html]

Tweets that describe the author's experience and reactions to the medication were mostly relevant in our study, and are called "personal experience" tweets. Personal experience tweets are those that describe the author's or another person's own experience or feeling toward the drug. It is necessary to distinguish personal experience tweets from non-personal-experience tweets. Examples of "personal experience" tweets are shown in Table 2.

Table 2. Examples of Tweets with Personal Drug Experience

Tweets	Drug	Effects
Feeling dizzy every time I took pregabalin so I google-d the side effects of it #TOPIC :))	Pregabalin	Dizzy
my mom taking this pills.. during medication she got swelling in her fingers,,, is this because of this #lyrica	Lyrica	Swelling fingers
Just starting lyrica, tho it reduced the pain, i cant sleep at the night	Lyrica	Can't sleep
I am anxious than before, thats how cymbalta should work?	Cymbalta	Anxious

No classification methods existed to distinguish "personal experience" tweets (or sentences). However, there were a few efforts dealing with personal tweets and sentences. Verma et al. consider personal tweets as that the author injects him or herself into the situation in some way, whereas impersonal ones as that the author displays a sense of emotional distance from the event/situation [18]. Elgersma and colleague observed that personal blogs are likely to contain more personal pronouns than others such as news or political blogs [19], and they used pronouns as one of the features to classify personal and non-personal blogs in their experiment. In classifying sentiment, Li et al. distinguished personal sentences from impersonal sentences by defining a personal sentence as having a person subject (using a first person nominative pronoun) and an impersonal sentence as having a non-person subject [20].

Table 3. Classifier Performance Summary

Classifier	Precision	Recall	F-Measure
Naïve Bayes	0.858	0.827	0.835
Support Vector Machine	0.856	0.810	0.820
Maximum Entropy	0.866	0.842	0.848

Upon analyzing tweets we collected, we observed that personal experience tweets tend to use more personal pronouns than those that are not related to personal experiences. This is consistent with (or similar to) the aforementioned works. In addition, personal experience tweets seem to have higher sentiment than those that are not. Based upon our observations, we decided to develop a supervised machine learning (ML) classifier to classify personal experience tweets. Occurrences of personal pronouns and sentiment which was derived from the Natural Language

Toolkit (NLTK) [21], were used as features fed to the machine learning (ML) methods. Three ML methods were tested: Naïve Bayes, Support Vector Machine, and Maximum Entropy, and they demonstrated slightly different accuracies and performances. In our experiment, we used 300 personal experience tweets and 300 non-personal experience tweets as the training data, and another 285 tweets for testing. The Maximum Entropy (MaxEnt) classifier demonstrated the best performance as shown in Table 3 and was subsequently used to classify personal experience tweets related to the five drugs.

3.4   Recognizing Drug Effects

Recognizing drug effects is to identify word phrases in the tweet text that are considered as effects which are signs, symptoms, syndromes, or diseases. To perform this computationally, it involves using natural language processing (NLP). Word phrases in the tweet text are separated, and are classified by looking up in a domain lexicon while taking into the consideration of word sense disambiguation.

Table 4. Drug Effects and Their Semantic Types

Drug Effect	UMLS Semantic Types
Anxiety	Mental or Behavioral Dysfunction
Bruises	Injury or Poisoning
Depressed	Finding
Depression	Finding
	Mental or Behavioral Dysfunction
Dizziness	Finding
Exhausted	Sign or Symptom
Fat	Organism Attribute
Frequent Urination	Organism Function
Inflammation	Pathologic Function
Insomnia	Finding
Low Libido	Finding
Numbness	Sign or Symptom
Pain	Sign or Symptom
Restless	Disease or Syndrome
Sweating	Finding
Swelling	Pathologic Function
Tired	Sign or Symptom

MetaMap [22] developed by the US National Library of Medicine was used to extract from the preprocessed tweets word phrases that are signs, symptoms, syndromes, or diseases. Primarily designed for classifying biomedical text and widely used for text mining, classification, and knowledge discovery, MetaMap includes a number of NLP features: tokenization, lexical lookup, shallow parsing, semantic type mapping, and word sense disambiguation. MetaMap maps word phrases to UMLS[2] semantic types and computes a mapping score for each mapping. A mapping with a score of 1000 implies a perfect mapping, and anything with a score less than 1000 indicates an imperfect mapping. The semantic types relevant to drug effects include *Disease or Syndrome, Finding, Injury or Poisoning, Mental or Behavioral Dysfunction*, and *Sign or Symptom*[3]. In this study, an effect is anything that is mapped to any of the above semantic types. For example, the word phrase "neuropathic pain" maps to the UMLS semantic type of *Sign or Symptom*. Table 4 lists examples of drug effects and their corresponding semantic types.

3.5   Verification

To determine the validity of our approach, we compared effects discovered using our method with those reported known effects of the same medications. The reported known effects were retrieved from two sites: one is the authoritative NLM MedLinePlus Drug Information site[4] and another is the health-specific social networking site (PatientsLikeMe). Combining the two known effect lists, we generated a list of known drug effects for each medication, which was then used for our comparison.

4   Results

In this study, we focused more on the consistency between the effects discovered with our method and those known or reported. For each of 5 medicines studied, 20 or so effects were discovered in the Twitter data, and most of the effects match the known effects. Table 5 below shows the statistics of matches. Not only did we include perfect mappings whose score is 1000, but also we included any mappings with a score of 850 or higher. This relaxation allowed for partial mappings which were manually examined for accuracy. The Discovered Effects column lists the number of effects discovered in the Twitter data using our computational method, and the figures in the Matched Effects column indicate the number of discovered effects that match the known effects. The Matching Rate, as a measure of consistency, is the percentage of the discovered effects that match the known effects – it will not make much sense measure if we try to find how many known effects that are found in the discovered effects because not all the known effects are necessarily reported in Twitter posts. In addition, it is interesting to note that there are a few unmatched effects for each of the study drugs.

---

[2] Unified Medical Language System

[3] http://semanticnetwork.nlm.nih.gov/SemGroups/SemGroups.txt

[4] http://www.nlm.nih.gov/medlineplus/druginformation.html

Table 5. Statistics of the Discovered Effects

Drug	Discovered Effects	Matched Effects	Unmatched Effects	Matching Rate
Duloxetine	21	18	3	86%
Gabapentin	22	19	3	86%
Baclofen	17	15	2	88%
Glatiramer acetate	19	14	5	74%
Pregabalin	23	19	4	83%

5 Discussion

Given the sheer volume of users and postings on various topics, it is interesting to note that there exist a plenty of postings on drug effects on the general purpose social media platform such as Twitter. This is an indication that social media may serve as a valuable data source to identifying potential drug effects. Our study only collected Twitter data for a period of 80 days. It is our hypothesis that there will be more drug effect related tweets if the Twitter data can be collected for a longer period of time. To use Twitter data for long-term drug safety surveillance, tweets can be collected continuously, and processed and analyzed periodically.

In our study of 5 medications, we have found a number of drug effects for each, and among the discovered effects, most of them match the known effects with matching rates ranging from 74% to 88% as shown in Table 5. This demonstrates the effectiveness of our approach. For unmatched effects, some may be the underlying condition being treated, and others may be the potentially new effects that have not been reported. The confirmation of the latter unmatched effects warrants a further rigorous clinical investigation.

The computational approach we developed in this study is effective, by which we were able to identify and extract a number of drug effects that match the known effects for each drug investigated. Our Maximum Entropy classifier achieves a relatively good accuracy in classifying personal experience tweets. However, we noticed that there is a room to improve the algorithm for classifying personal experience tweets as the classifier misidentified a number of non-personal experience tweets as personal experience ones.

Leveraging the power of the MetaMap tool has helped achieve a high accuracy in identifying drug effects. The tool itself contains a domain specific lexicon and implements word sense disambiguation. Mapping tweet word phrases to the UMLS semantic types simplified the task of recognizing drug effects.

It is known that every medication has effects which can be beneficial or adverse. It is conceivable that our approach can assist in finding beneficial drug effects – any discovered effects that do not match the reported effects could be beneficial effects or the underlying condition being treated.

Twitter is a short text microblogging social networking platform. The limitation of 140 characters restricts the amount of information that can be posted in each tweet.

Many pieces of information related to the authors of tweets are typically unavailable, rendering the difficulty in collecting quantitative data from the patients who take the medications. Therefore, unlike other rigorous drug safety surveillance methods, it is unlikely that high-quality, accurate quantitative results can be derived from the Twitter data.

One major issue related to use social media as a data source for scientific research is the inability to verify the source and author, and accuracy of the reported information, leading to trustability of the information posted. Regardless, social media data may provide some useful hints as to the signals of drug-related adverse events [2].

6   Conclusion

In this study, we developed a feasible yet effective method for extracting potential drug effects from the Twitter data, through the use of machine learning and natural language processing. To our best knowledge, this is the first attempt of this kind. It appears that text data of general purpose social medial such as Twitter may serve as a valuable source of drug effect signals, complementing and/or supplementing other existing drug safety surveillance methods. Any findings of potential drug effects in social media should be further verified and confirmed by more rigorous clinical studies.

Acknowledgement. Authors wish to thank reviewers' constructive comments on the first draft of this manuscript.

References

1. World Health Organization. Pharmacovigilance: ensuring the safe use of medicines. Geneva, Switzerland: World Health Organization (2004)
2. Dal Pan, G.J.: Monitoring the safety of medicines used off-label. Clin. Pharmacol. Ther. 91(5), 787–795 (2012)
3. Edwards, B.J., Gounder, M., McKoy, J.M., Boyd, I., Farrugia, M., Migliorati, C., et al.: Pharmacovigilance and reporting oversight in US FDA fast-track process: bisphosphonates and osteonecrosis of the jaw. Lancet Oncol. 9(12), 1166–1172 (2008)
4. Pirmohamed, M., James, S., Meakin, S., Green, C., Scott, A.K., Walley, T.J., et al.: Adverse drug reactions as cause of admission to hospital: prospective analysis of 18 820 patients. BMJ 329(7456), 15–19 (2004)
5. Lazarou, J., Pomeranz, B.H., Corey, P.N.: Incidence of adverse drug reactions in hospitalized patients: a meta-analysis of prospective studies. JAMA 279(15), 1200–1205 (1998)
6. Moore, T.J., Cohen, M.R., Furberg, C.D.: Serious adverse drug events reported to the Food and Drug Administration, 1998-2005. Arch. Intern. Med. 167(16), 1752–1759 (2007)
7. Levinson, D.R., General, I.: Adverse events in hospitals: national incidence among Medicare beneficiaries. Department of Health & Human Services (2010)

8. Landow, L.: Monitoring adverse drug events: the Food and Drug Administration MedWatch reporting system. Regional Anesthesia and Pain Medicine 23(6), 190–193 (1998)

9. Edwards, I.R.: Pharmacovigilance. Br. J. Clin. Pharmacol. 73(6), 979–982 (2012)

10. Egberts, T.C., Smulders, M., de Koning, F.H., Meyboom, R.H., Leufkens, H.: Can adverse drug reactions be detected earlier? A comparison of reports by patients and professionals. BMJ 313(7056), 530–531 (1996)

11. Chunara, R., Andrews, J.R., Brownstein, J.S.: Social and news media enable estimation of epidemiological patterns early in the, Haitian cholera outbreak. Am J. Trop. Med. Hyg. 86(1), 39–45 (2010)

12. Stone, B.: https://blog.twitter.com/2009/whats-happening2009 (cited 2013)

13. Signorini, A., Segre, A.M., Polgreen, P.: The use of Twitter to track levels of disease activity and public concern in the U.S. during the influenza A H1N1 pandemic. PLoS One 6(5), e19467 (2011)

14. Heaivilin, N., Gerbert, B., Page, J.E., Gibbs, J.L.: Public health surveillance of dental pain via Twitter. J. Dent. Res. 90(9), 1047–1051 (2011)

15. Corley, C.D., Mikler, A.R., Singh, K.P., Cook, D.J.: Monitoring influenza trends through mining social media. In: International Conference on Bioinformatics & Computational Biology (2009)

16. Knezevic, M.Z., Bivolarevic, I.C., Peric, T.S., Jankovic, S.M.: Using Facebook to increase spontaneous reporting of adverse drug reactions. Drug Saf. 34(4), 351–352 (2011)

17. Leaman, R., Wojtulewicz, L., Sullivan, R., Skariah, A., Yang, J., Gonzalez, G.: Towards internet-age pharmacovigilance: extracting adverse drug reactions from user posts to health-related social networks. In: Proceedings of the 2010 Workshop on Biomedical Natural Language Processing. Association for Computational Linguistics (2010)

18. Verma, S., Vieweg, S., Corvey, W.J., Palen, L., Martin, J.H., Palmer, M., et al.: Natural Language Processing to the Rescue?: Extracting 'Situational Awareness' Tweets During Mass Emergency. In: Proc ICWSM (2011)

19. Elgersma, E., de Rijke, M.: Personal vs non-personal blogs: Initial classification experiments. In: Proceedings of the 31st Annual International ACM SIGIR Conference on Research and Development in Information Retrieval. ACM (2008)

20. Li, S., Huang, C.-R., Zhou, G., Lee, S.: Employing personal/impersonal views in supervised and semi-supervised sentiment classification. In: Proceedings of the 48th Annual Meeting of the Association for Computational Linguistics. Association for Computational Linguistics (2010)

21. Bird, S.: NLTK: The natural language toolkit. In: Proceedings of the COLING/ACL on Interactive Presentation Sessions. Association for Computational Linguistics (2006)

22. Aronson, A.R., Lang, F.M.: An overview of MetaMap: historical perspective and recent advances. J. Am. Med. Inform. Assoc. 17(3), 229–236 (2010)

23. Bian, J., Topaloglu, U., Yu, F.: Towards large-scale twitter mining for drug-related adverse events. In: Proceedings of the 2012 International Workshop on Smart Health and Wellbeing, pp. 25–32. ACM (2012)

Social-Correlation Based Mutual Reinforcement for Short Text Classification and User Interest Tagging

Rong Li and Ya Zhang

Shanghai Key Laboratory of Multimedia Processing and Transmissions
Shanghai Jiao Tong University, Shanghai, China
rliemail688@gmail.com

Abstract. Short text such as micro-blog messages is becoming increasingly prevalent in China. Due to the sparseness of the features associated with short text, accurately classifying short text and tagging user interest have become important and challenging tasks. Many recent studies have focused on utilizing external data to address the data sparsity issue but fail to leverage the social-correlation which is expected to help improve the accuracy of short text classification. In this paper, we present a new method using a semi-supervised coupled mutual reinforcement framework based on social-correlation to simultaneously classify short text and tag user interest. Specifically, our method requires relatively few labeled examples to initialize the training process. More importantly, experimental results have demonstrated that our method can achieve 100% accuracy in classifying certain categories and significantly improve the accuracy of classifying the other categories. Meanwhile, the experiments show that our model is effective in user interest tagging.

Keywords: mutual reinforcement, user interest tagging, short text classification.

1 Introduction

In recent years, micro-blog sites such as Sina Weibo have developed rapidly and attracted much attention. According to a report, Sina Weibo has accumulated over 330 billion users and hundreds of messages have been posted every minute all over the world [1]. The database of Sina Weibo becomes a huge treasure house of information, which can provide abundant resource for research needs.

In leveraging information generated on mocroblog platforms for various purposed such as trending topic discovery, personalization and recommendation, two of the fundamental research issues are to classify microblogging messages and tag user interest. While document classification and user interesting tagging have been widely investigated in the Internet domain, on mocroblog site, the two tasks face some unique challenges. Different from traditional document, the characteristics of microblogging messages are as follows: 1) shortness: the microblogging messages are limited in length (e.g. up to 140 characters for twitter messages), making the textual features too sparse to perform an accurate classification; 2) noisiness: there are quite

H. Motoda et al. (Eds.): ADMA 2013, Part I, LNAI 8346, pp. 444–455, 2013.
© Springer-Verlag Berlin Heidelberg 2013

few meaningless bubbles and the terms used contain quite a few Internet slangs; and 3) relationships: microblogging sites allow users to follow each other which forms a large social network, and the social relationship could partially reflect users' interests.

For classifying short text, one topical approach is to efficiently and iteratively select unlabeled data from external domain as subset to train a transfer classifier with other labeled training data [2]. A framework is proposed to utilize external knowledge from Wikipedia and MEDLINE to discover hidden topics, and then build a classifier on both labeled training dataset and a large set of hidden topics to achieve high accuracy [3]. However, there are often conditions under which no external data can be acquired for some categories. Therefore, "high-quality" topic clusters are discovered by grouping similarity topics together on the training data [4]. Then not only the raw words of short text but also the topic similarities between topic clusters are used to extend the word feature vector. In addition, other studies attempt to expand the feature domain using the author's profile [5].

Unfortunately, to the best of our knowledge, existing studies on classifying short text require either large amounts of training examples or external information to enhance the feature domain. On the microblogging platform, the social relationship among the users may be leveraged to improve classification accuracy. It is observed that 1) user interests and their messages are highly correlated; 2) users tend to share similar interests with their social relationships, forming a coupled mutually reinforcing relationship. Building on the above observations, we propose a framework for simultaneously predicting user interests and classifying microblogging messages as short text. We model the relationship between user interest, social-correlation and short text as an undirected bipartite graph $Ge(U, E, T)$ where U denotes user, T denotes short text, and E denotes the undirected adoption links between U and T. We also assume a social graph $Gs(U, S)$, where U denotes the same set of users, and S denotes the social links between users. Both Ge and Gs are un-weighted graph (i.e. the edges either exists or does not exist) to keep our model simple.

We develop a machine learning approach that starts with a set of known labels for short text, and utilize the mutual reinforcement between the connected entities in two graphs to respectively predict user interest and classify short text simultaneously, iteratively refining the labels for the unlabeled entities. We make the following contributions in this paper:

- We propose a social-correlation based mutual reinforcement framework to simultaneously predict user interests and classify short text.
- We present a novel approach to model the social-correlation between users as a function of latent feature factors through our framework named as social-correlation mutual reinforcement Model (SCMM) that required relatively few amount of labeled data.
- Through extensive experiments on real world datasets (Sina Weibo), the proposed model has led to promising results in terms of high accuracy of user interest tagging and short text classification.

2   Related Work

Generally speaking, traditional methods on text-classification focus on classifying documents [6-9], but with the development of Internet-enabled devices and the popularity of social network, short text is becoming prevalent on the web which boosts the studies on it. However, the classification of short text is different from long text. The most significant difference is its length and sparsity [10]. Related work can roughly be classified into several types of method of text categorization (also known as text classification).

For the problem of text categorization, traditional approaches on short text classification are to train a classifier with labeled datasets, and then use it to predict the categorization of test data. One of the most important training steps is how to obtain the sufficient feature of texts. Meng et al. [11] leverage public search engines to extract additional features from the search results that are relevant to the short text. With the feature expansion process, they can obtain enough corpuses for feature vector to address the problem of short text sparseness.

At the same time, many researches in machine learning find that the combination of unlabeled data and labeled data can largely boost the performance of text categorization. Several relevant approaches are proposed in [13-14]. It has demonstrated that combing of labeled training dataset and a corpus of unlabeled background knowledge can significantly decrease error rates [13]. The method, called "second-order", uses unlabeled corpus as "background knowledge" to aid labeled training examples in classifying unknown test examples. A PAC-style framework is proposed that provides a way to use the unlabeled dataset to augment the labeled dataset for significant benefits in practice [14].

Traditional approaches to classify short text generally require a large number of labeled training sets to achieve high accuracy of classification. However, labeled sets of texts may not be available in some cases but unlabeled sets are often easy to obtain [15-16]. [12] proposes a novel method for unlabeled short text classification based on the semantic relations between concepts which define the categories. The classification process depends on proximity relations utilizing the ability of an extension of the standard Prolog language, which allows flexible matching of documents and knowledge representation.

Existing works attempt utilizing external data either to expand the corpus of feature terms or to calculate the similarity of feature vectors. Most of them are shown to be successful in solving the sparsity issue of short text and can greatly improve the accuracy of text categorization both in precision and recall. Borrowing the idea of mutual reinforcement framework [17], we exploit social relationships and propose a social-correlation based mutual reinforcement framework for simultaneous user interest tagging and short text classification.

3   Short Text Categorization and User Interest Tagging

3.1   Problem Description

There are two different types of entities: messages as and users in micro-blog platforms. A user may post tweets of various topics according to his interest or influenced by their friends. Specifically, we divide users into two groups: users u and his friends f. It has been observed that people with similar preference tend to post messages of same category.

Problem:

Given a dataset obtained from microblogging sites such as Sina Weibo, determine the interest of users and the categorization of short text based messages.

In the following, a detailed analysis on "social-correlation based mutual reinforcement principle" for message categorization and user interest tagging will be presented.

3.2   Social-Correlation Based Mutual Reinforcement Principle

We seek to obtain high accuracy of short text categorization and tag user interest. We now elaborate on the social-correlation based mutual reinforcement principle used in our framework to address this problem:

- A user u is likely interested in a category $\mathcal{T}$ if most of messages he posts belong to the category $\mathcal{T}$ and the majority of his friends are interested in the category $\mathcal{T}$.
- A message t is likely be classified into the category $\mathcal{T}$ if the majority of the persons who post or forward the message are interested in the category $\mathcal{T}$.

We would like to illustrate the framework underlying this principle more precisely. Considering the relationships between user and messages, we use an undirected bipartite graph with an edge between each author and the messages they post. Meanwhile, we also have a social graph between users and their friends.

We formulate the graph as $G(U, T, M^{ut}, M^{uf})$, where U is the set of users, $M^{ut} = \left| m_{ij}^{ut} \right|$ is the $|U|$-by-$|T|$ matrix indicating the relationship between users and messages, i.e., $m_{ij}^{ut} = 1$ if the user u_i posts or forwards the message t_j. Similarly, M^{uf} is a matrix represents the friendship relationship between users.

3.3   Social-Correlation Mutual Reinforcement Model

To simplify the notation, we collect the tags of user interest and the categories of microblogging messages into vectors, so the user interests inferred from one's friendship relationships is denoted as $Y_{u-f-\tau}$:

$$Y_{u-f-\tau} = M'_{uf} Y_{f-\tau} \tag{1}$$

where $Y_{f-\tau}$ is the labels of user interest which is inferred based on that of his friends.

Similarly, the interest of the user u can be inferred from the categories of these messages:

$$Y_{u-t-\tau} = M_{ut}' Y_{t-\tau} \qquad (2)$$

where $Y_{t-\tau}$ is the category of messages, and $Y_{u-t-\tau}$ is the label of user interest that is derived from the category of messages.

According to the two equations mentioned above, the final user interest tagging may be formulated as follows:

$$Y_u^{\tau} = Y_{u-t-\tau} + \gamma Y_{u-f-\tau} \qquad (3)$$

where γ is a parameter to control the weight of influence of $Y_{u-t-\tau}$ and $Y_{u-f-\tau}$ on Y_u^{τ}.

Given the tags of user interests, the category of corresponding messages $Y_{t-\tau}$ is shown as:

$$Y_{t-\tau} = M_{ut}^T Y_u^{\tau} \qquad (4)$$

The matrices M_{uf}', M_{ut}' are derived from M^{uf} and M^{ut}, respectively. M_{ut}^T is the transpose of M_{ut}'. For each $m_{ij}' \in M_{ut}'$, $m_{ij}' = \dfrac{m_{ij}^{ut}}{\sum_{j=1}^{|T|} m_{ij}^{ut}} (m_{ij}^{ut} \in M^{ut})$.

For each $m_{ij}' \in M_{uf}'$ $\quad m_{ij}^{uf} = \dfrac{m_{ij}^{uf}}{\sum_{j=1}^{|U|} m_{ij}^{uf}} (m_{ij}^{uf} \in M^{uf})$. Next we change the social-correlation based mutual reinforcement principle into a semi-supervised algorithm to classify messages and tag user interest.

3.4   Coupled Semi-supervised with Social-Correlation Based Mutual Reinforcement Principle

We use TF-IDF as the text features for messages and user interest which are denoted as $X_t = \{x_t\}$ and $X_u = \{x_u\}$, respectively.

Let $P(x_u)$ represent the probability of user interest tag and $P(x_t)$ denote the possibility of message category. We present a general method to obtain all the probabilities. Using P to represent any of $P(x_u)$ or $P(x_t)$ and x to denote the corresponding feature vector. We use logistic regression to model the log-odds of $P(x)$:

$$\log \frac{P(x)}{1 - P(x)} = \beta^T x \tag{5}$$

where β are the coefficient variables of the linear model.

When there are enough labeled examples, we can derive those coefficients through optimizing the corresponding log-likelihoods represented as equation $LL(X)$:

$$LL(X) = \sum_{x \in X} y\beta^T x - \log(1 + e^{\beta^T x}) \tag{6}$$

where $y \in \{0,1\}$ denotes the label of examples x; X represents any of U or T.

We next leverage the social-correlation based mutual reinforcement principle (Equation 3 and Equation 4) to enhance the modeling in Equation 6.

Suppose y represents the current labels for x and y' represents the new labels after updating with the social-correlation based mutual reinforcement principles. Similar to [17], we measure the conditional log-likelihood of Y using KL-divergence:

$$LL(y \mid Y) = -\sum_{i=1}^{|x|} [\ y(i) \log \frac{y(i)}{y'(i)} + (1 - y(i)) \log \frac{1 - y(i)}{1 - y'(i)} \ ] \tag{7}$$

We extend the objective function from the objective log-likelihood in equation 7 to the following:

$$L(X) = LL(X) + \alpha LL(Y \mid y) \tag{8}$$

where α is a weight parameter.

3.5   Learning Algorithm

We present a Bi-side learning algorithm for simultaneously classifying short text and tagging user interest. The bi-side learning algorithm records the propagated labels in the form of probability and fit the social-correlation based mutual reinforcement model not only on the message side but also on the side of user interest.

Algorithm 1. Bi-sides learning algorithm
Input: user feature X_u, short text feature X_t, matrix M^{ut} and M^{uf}
Output: user interest Y_u^τ, category of short text $Y_{t-\tau}$
Start with an initial prediction. Fitting the logistic regression model, for $Y_{t-\tau}$ and Y_u^τ
Begin
While $Y_{t-\tau}$ and Y_u^τ not converge do
Using equation 3 and then 4 to obtain the propagated labels of short text
Fitting the SCMM to obtain the category of short text
Using equation 4 and then 3 to obtain the propagated labels of user interest
−   Fitting the SCMM to obtain the tag of user interest
End

4   Experimental Setup

4.1   Data Collection

As most researchers chose to analyze hashtag usage in the tweets, we crawled 45903 short texts from Sina Weibo in seven months from October 2010 to April 2011. The categories of short text are pre-defined using hashtag to label the short texts automatically. Then, we label user interest using equation 2. After eliminating the messages whose length are shorter than 10 Chinese characters, and we randomly sample 5236 messages, only the users that interacts with the messages are kept in the data set. We put the messages into ten categories as shown in Table 1.

Then, we pre-process the dataset as follows: the messages are segmented into words and the words with less than 5 occurrences are eliminated. The remaining words are used as features for message classification and user interest tagging. We divided the training sets into two parts. The feature matrix of the categories 4, 5, 6, 8, 10, mostly posted by individuals, tends to be sparser than that of the categories 1, 2, 3, 7, 9 released by micro-blogging application or the third-parity websites. The number 1-10 in tables and figures represents 10 categories.

Table 1. Statistics of Our Data

Category	Proportion
1(Xiami Music)	22.9%
2(Doggy Pet)	13.5%
3(VeryCD Collection)	13.6%
4(LOLLIPOP F)	6.6%
5(WeicoLomo)	6.6%
6(iweejly)	7.0%
7(NBA)	8.1%
8(New Year Wishes)	7.9%
9(China Visual Collection)	9.95%
10(Fan Fan's wedding)	3.95%

Table 2. The Accuracy of STC and UIT of SCMM and LR

Category	STC		UIT	
	SCMM	LR	SCMM	LR
1	1	0.9958	0.8704	0.8392
2	1	1	0.9221	0.9952
3	1	0.9977	0.9709	0.8563
4	0.8932	0.8738	0.9309	0.7848
5	0.9805	0.8488	0.9306	0.7299
6	0.9178	0.8995	0.9467	0.9491
7	1	0.9922	0.9411	0.7782
8	0.9879	0.8669	0.9447	0.7509
9	1	0.9904	0.9606	0.7919
10	0.8943	0.7236	0.9827	0.7595

The main objective of this paper is to achieve high accuracy of message classification and user interest tagging with relative few manually labeled data. We compare the social-correlation based mutual reinforcement with basic logistic regression. In all experiments, we randomly select labeled examples to fit the model. In the first experiment, we set $\alpha=0.3$. $\gamma=0.5$ the training set is 40% of the total set. In other experiments, if without additional explanation, the parameters are set as experiment 1. The logistic regression-based method and the proposed method are denoted as LR and SCMM respectively. In the tables and figures, "short text categorization" and "use interest tagging" are abbreviated STC and UIT, respectively.

Table 2 reveals that SCMM can obtain higher accuracy than LR in all categories which means that SCMM is more robust than LR. The accuracy of SCMM is up to 100% for categories 1, 2, 3, 7, 9 and the accuracy of categories 4, 5, 6, 8, 10 is improved maximum by 15%. It is clear that SCMM outperforms LR in all categories.

Table 3. The Accuracy of STC and UIT of Different Density Degree

Category	STC			UIT		
	60%	80%	100%	60%	80%	100%
1	0.8623	0.9944	1	0.8745	0.8784	0.8704
2	0.9149	0.9976	1	0.9194	0.9209	0.9221
3	0.8388	0.9836	1	0.9659	0.9686	0.9709
4	0.7282	0.8155	0.8932	0.9157	0.9367	0.9309
5	0.7584	0.8780	0.9805	0.9311	0.9294	0.9306
6	0.7397	0.8584	0.9178	0.9426	0.9501	0.9467
7	0.9373	0.9765	1	0.9327	0.9356	0.9441
8	0.8505	0.9133	0.9879	0.9287	0.9383	0.9447
9	0.9519	1	1	0.9349	0.9465	0.9606
10	0.6170	0.7480	0.8943	0.9611	0.9695	0.9827

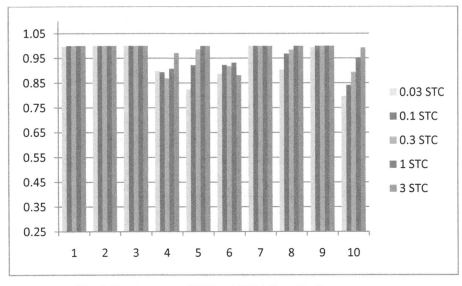

Fig. 1. The Accuracy of STC and UIT Affected by Parameter α

We run experiments on the density degree of user-short text connection to show how it affects the performance of our model. We randomly remove 40% and 20% edges of user-short text connection compared with 100% edges reserved. The accuracy of STC and UIT of different density degree is plotted in Table 3. The results indicate that the performance of our proposed method is robust in tagging user interest even though the density degree of user-short text is changed from 100% to 60%. The higher the density is, the more accuracy of STC we obtain.

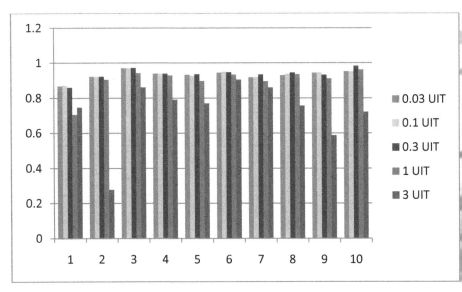

Fig. 2. The Accuracy of UIT Affected by Parameter α

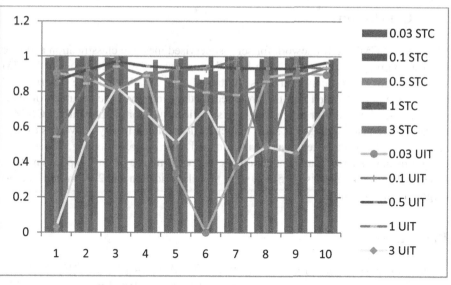

Fig. 3. The Accuracy of STC and UIT Affected by Parameter γ

For each parameter setting, we run the experiment for parameter α that denotes he degree of mutual reinforcement. The accuracy of SIT and UIT that affected by parameter α are shown in Fig. 1 and Fig. 2, respectively. Then we test the effect of parameter γ that represents the degree of social-correlation. The accuracy of SIT and UIT affected by γ are shown in Fig. 3. The lines indicate the accuracy of user interest tagging and histograms represent the accuracy of short text categorization. As we can see from in Fig. 1, when α changed from 0.03 to 3, the accuracy of STC increase until it reaches 100% except for the category 6.

From Fig. 2, we can see that the accuracy of UIT first increases and then decreases. When α fluctuates around 0.3, the highest accuracy of UIT in categories 2-8 and 10 is obtained. For categories 1 and 9, the highest accuracy of UIT is reached when α is around 0.1. The above results indicate that SCMM utilizing mutual reinforcement between users and short texts is an effective method to overcome the sparsity of short text. The reason for the declines in UIT accuracy may be that the model over-fit when α becomes large.

To show the influence of social-correlation in our model, we change γ from 0.03 to 3. From Fig. 3, we can see that the accuracy of STC increases until it reaches 100% in all categories except the category 6. The different influence between γ and α is that the precision of UIT is more sensitive to γ that also can be seen in equation 3. For the accuracy of STC, the effect of the two parameters seems to be equal that also can be derived from equation 4.

5   Conclusion

We present SCMM, a framework for semi-supervised message classification and user interest tagging, simultaneously. We have demonstrated the effectiveness of SCMM with extensive experiments on a dataset obtained from Sina Weibo. The experimental results have demonstrated that the proposed method improves the accuracy of short text classification and user interest tagging significantly. More importantly, SCMM is robust to parameter settings for short text classification and finally converges to high accuracy in both short text categorization and user interest tagging. Even some categories have sparser features than others, the performance of our proposed method is better than semi-supervised 15% in short text classification and 23% in user interest tagging by maximum.

References

1. China Internet Development Statistics Report, 第32次中国互联网络发展状况统计报告, http://www.cnnic.net.cn/hlwfzyj/hlwxzbg/hlwtjbg/201301/P0201 30724346275579709.pdf
2. Long, G., Chen, L., Zhu, X.Q., Zhang, C.Q.: TCSST: Transfer Classification of Short & Sparse Text Using External Data. In: Proceedings of the 21st ACM International Conference on Information and Knowledge Management, pp. 764–772. ACM Press, New York (2012)
3. Pan, X.-H., Nguyen, L.-M., Horiguchi, S.: Learning to Classify Short and Sparse Text & Web with Hidden Topics from Large-Scale Data Collections. In: Proceedings of the 17th International Conference on World Wide Web, pp. 91–100. ACM Press, Beijing (2008)
4. Dai, Z., Sun, A., Liu, X.-Y.: CREST: Cluster-based Representation Enrichment for Short Text Classification. In: Pei, J., Tseng, V.S., Cao, L., Motoda, H., Xu, G. (eds.) PAKDD 2013, Part II. LNCS, vol. 7819, pp. 256–267. Springer, Heidelberg (2013)
5. Sriram, B., Fuhry, D., Demir, E., Ferhatosmanoglu, H., Demirbas, M.: Short Text Classification in Twitter to Improve Information Filtering. In: Proceedings of the 33rd International ACM SIGIR Conference on Research and Development in Information Retrieval, pp. 841–842. ACM Press, New York (2010)
6. Hatzivassiloglou, V., Klavans, J.L., Eskin, E.: Detecting Text Similarity over Short Passage: Exploring Linguistic Feature Combinations via Machine Learning. In: Proceedings of the 1999 Joint SIGDAT Conference on Empirical Methods in Natural Language Processing and Very Large Corpora, pp. 203–212. Maryland (1999)
7. Li, Y.H., Mclean, D., Bandar, Z.A., O'Shea, J.D., Crockett, K.: Sentence Similarity Based on Semantic Nets and Corpus Statistics. IEEE Transactions on Knowledge and Data Engineering 18, 1138–1150 (2006)
8. Jiang, J.J., Conrath, D.W.: Semantic Similarity Based on Corpus Statistics and Lexical Taxonomy. In: Proceedings of ROCLING X, Taiwan (1997)
9. Lyon, C., Malcolm, J., Dickerson, B.: Detecting Short Passages of Similar Text in Large Document. In: Proceedings of the 2001 Conference on Empirical Methods in Natural Language Processing, pp. 118–128. Pennsylvania (2001)
10. Rafeeque, P.C., Sendhikumar, S.: A Survey on Short Text Analysis in Web. In: 2011 Third International Conference on Advanced Computing, Chennai, pp. 365–371 (2011)

11. Meng, W., Lanfen, L., Jing, W., Penghua, Y., Jiaolong, L., Fei, X.: Improving Short Text Classification Using Public Search Engines. In: Qin, Z., Huynh, V.-N. (eds.) IUKM 2013. LNCS, vol. 8032, pp. 157–166. Springer, Heidelberg (2013)
12. Francisco, P.R., Pascual, J.-I., Andres, S., Mateus, F.S., Juan, G.-C.: Classifying Unlabeled Short Texts Using a Fuzzy Declarative Approach. Language Resources and Evaluation 47, 151–178 (2013)
13. Sarah, Z., Haym, H.: Improving Short Text Classification Using Unlabeled Background Knowledge to Assess Document Similarity. In: Proceedings of the Seventeenth International Conference on Machine Learning, San Francisco, pp. 1183–1190 (2000)
14. Blum, A., Mitchell, T.: Combining Labeled and Unlabeled Data with Co-training. In: Proceedings of the Eleventh Annual Conference on Computational Learning Theory, pp. 92–100. ACM Press, New York (1998)
15. Duda, R.O., Hart, P.E.: Pattern Classification and Scene Analysis. Wiley (1973)
16. Yarowsky, D.: Unsupervised Word Sense Disambiguation Rivaling Supervised Methods. In: Proceedings of the 33rd Annual Meeting on Association for Computational Linguistics, pp. 189–196. Pennsylvania (1995)
17. Bian, J., Liu, Y.D., Zhou, D., Agichtein, E., Zha, H.Y.: Learning to Recognize Reliable Users and Content in Social Media with Coupled Mutual Reinforcement. In: Proceedings of the 18th International Conference on World Wide Web, p. 5 (2009)

Graph Based Feature Augmentation
for Short and Sparse Text Classification

Guodong Long and Jing Jiang

Centre for Quantum Computation & Intelligent Systems,
University of Technology, Sydney, Australia
{guodong.long,jing.jiang-1}@student.uts.edu.au
http://www.qcis.uts.edu.au/

Abstract. Short text classification, such as snippets, search queries, micro-blogs and product reviews, is a challenging task mainly because short texts have insufficient co-occurrence information between words and have a very spare document-term representation. To address this problem, we propose a novel multi-view classification method by combining both the original document-term representation and a new graph based feature representation. Our proposed method uses all documents to construct a neighbour graph by using the shared co-occurrence words. Multi-Dimensional Scaling (MDS) is further applied to extract a low-dimensional feature representation from the graph, which is augmented with the original text features for learning. Experiments on several benchmark datasets show that the proposed multi-view classifier, trained from augmented feature representation, obtains significant performance gain compared to the baseline methods.

Keywords: Short Text, Text Classification, Graph Based Method, Multi-view Learning, Multi-Dimensional Scaling.

1 Introduction

Text categorization, one of the most important tasks of text mining, has been studied intensively in the past decade. Due to the rapid development of information technologies and its applications, the amount of short text data is becoming rapidly available, such as micro-blogs (e.g. Twitter), web search snippets (e.g. Google), forum messages (e.g. 20Newsgroups) and product reviews (e.g. Amazon), etc. Consequently, classifying short text data with sparse information has received many research attentions.

Various methods (e.g. Naive Bayes, Maximum Entropy) and language models (e.g. Bag-of-words, N-gram) in the Natural Language Process (NLP) and in the Text Mining fields have been applied to text classification and been proved to be effective on different benchmark datasets. However, short text refers to a dataset where each text instance is short and only consists of a dozen of words to a few sentences. When these methods are applied directly to short texts classification,

H. Motoda et al. (Eds.): ADMA 2013, Part I, LNAI 8346, pp. 456–467, 2013.

their performance significantly deteriorates mainly because there are few co-occurrence words shared between two short text snippets and their similarity is hard to be determined by using the traditional document-similarity metrics.

To address this problem, one intuitive solution for short text classification is to exploit external auxiliary datasets to improve the training process. Phan [1] proposed a LDA [2] topic based method by exploiting both short texts and external auxiliary datasets to improve the classification performance. Sahami [3] measured the similarity between two short queries by involving the query results from search engine, such as Google. Vitale [4] expanded short text feature space by including annotation and link structure which come from online resources. Furthermore, several recent machine learning tools, such as transfer learning [5] and domain adaptation [6], can also be used to improve the performance of short text classification by involving auxiliary datasets. However, this type of solution depends on the assumption of existing external data source.

This paper, from another point of view, aims to solve the short text classification problem without external data source. We focus on the transductive learning scenario where there is a few labelled data and numerous unlabelled data and the learning target is to predict labels of the given unlabelled data. To improve the classification performance, we propose a multi-view classification method by learning with both the original document-term representation and a new graph based feature representation constructed from the same domain data.

The key motivation of our proposed graph based method is inspired by an intuitive idea: the traditional similarity metrics of sparse text vectors, such as Cosine Similarity and Euclidean distance, cannot be calculated without co-occurrence words, while the relationship of sparse texts can be constructed as a graph by using their shared neighbours. A simple example is shown in the Figure 1, which contains three short texts, $T0$ ("Terrible weather"), $T1$ ("I feel happy") , $T2$ ("Good mood today"), all describing the sentiment in users' micro-blog, such as Twitter. It is hard to calculate the similarity between these texts due to lack of co-occurrence words. However, we know that $T1$ and $T2$ represent positive sentiment, and the similarity between $T1$ and $T2$ should be bigger than the similarity between $T1$ and $T0$ or $T2$ and $T0$. There is another short text, $T3$ ("Good food make me feel happy"), which shared "happy" with $T1$ and shared "good" with $T2$. As a result, we can build a neighbour graph, $T1 \leftrightarrow T3 \leftrightarrow T2$, which can indicate some hidden information, such as "$T1$ and $T2$ is close in hop distance" and "$T1$ and $T2$ shared the same neighbour". The hidden information extracted from the graph can be used to measure the similarity between two short texts without shared co-occurrence words and improve the classification performance.

Graph based method can be used to calculate similarity for sparse vector. Thad etl [7] and Ramage etl [8] used Random Graph Walks model to measure the documents similarity. Yunpeng Xu [9] introduces the Markov Random Walks model to classify texts. In addition to random walks models, graph based semi-supervised learning models are introduced to propagates labels information for sentiment text by [10] and [11]. However, these graph based methods are not designed for short text, and they only utilize the information of the graph.

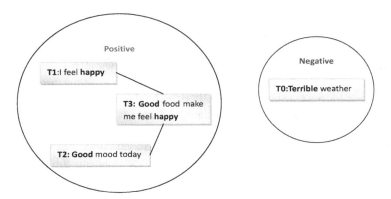

Fig. 1. A simple example of graph based short text classification

The graph and document-term representation are different views of the same source data. They both contain various important information. Therefore, we train a multi-view classifier with the augmented feature representation by exploiting both the graph and document-term representation as two independent datasets.

The main contributions of the work is summarized as follows.

- To address the problem of lack of co-occurrence words among documents, we construct the relationship of all documents as a graph by using their shared neighbours to support text classification, and further propose a novel method to improve the performance of short & sparse text classification.
- To avoid the miss of hidden information by utilizing one type of representation, we develop a multi-view classification model to incorporate different representations (views) i.e. by integrating both graph representation and document-term representation, for short text classification.

The rest of this paper is organized as follows. Section 2 discusses the definition, framework and proposed method for the **G**raph based Multi-Dimensional Scaling (MDS) feature augmentation with Multi-View learning on short and sparse text classification, **GMDS-MV**. The experiments and performance evaluation are reported in Section 3. We conclude our work in Section 4.

2   Methodology

In this section, we first define the problem of graph based feature augmentation on short and sparse text classification. Next we describe the proposed framework, followed by technical details of our method.

2.1   Definition

Consider a short text dataset D which contains n documents $< D_i, Y_i >$, where D_i is the ith document, and Y_i is the class label respectively. Set n as the number of documents in D, and v as the size of vocabulary Ψ. To simplify the illustration, we focus on binary classification problem $Y_i \in \{0, 1\}$.

Denote $G =< V, E >$ as the document graph, where vertices V are documents, and edges E represents the similarity between pair of documents. Then, a distance matrix $P = \{P_{i,j}\}$ can be inferred based on the defined distance function. Each entry $P_{i,j}$ is the distance between the ith and jth documents.

We now extend our definition to fit the multi-view learning scenario. For a short text dataset D, there are two representations:

(i) Document-term representation $D_T =< X_i^T, Y_i >$ is $n \times v$ matrix, and each row $X_i^T =< X_{i,1}^T, ... X_{i,j}^T, ... X_{i,v}^T >$ is the term vector for the ith short text. Consider the short text only contains a few terms, each term is equally important to the short text. We set $X_{i,j}^T$ as 1 while the jth term occurs in the ith document, otherwise 0.

ii) $D_G =< X_i^G, Y_i >$ is a low-dimensional representation for graph G's distance matrix P by measuring nodes relationship with shared common neighbours. Compare $P \in n \times n$, D_G is a $n \times m$ matrix with $m \ll n$. Each row $X_i^G =< X_{i,1}^G, ... X_{i,j}^G, ... X_{i,m}^G >$ represents a low dimensional representation for the document.

A multi-view classifier $H : X^T \times X^G \to Y$ will be learned by exploiting both D_T and D_G.

2.2 Framework of GMDS-MV

The framework of our proposed Graph based MDS feature augmentation with Multi-View learning framework (GMDS-MV), is shown in Figure 2. The input is a short text dataset which contains a small portion of labelled data and a large portion of unlabelled data. First, we construct a document graph G from the textual data. To reflect the relationship between documents, a distance matrix P is generated from the document graph G. This distance matrix will be transformed to a low-dimension representation D_G by utilizing MDS. After that,

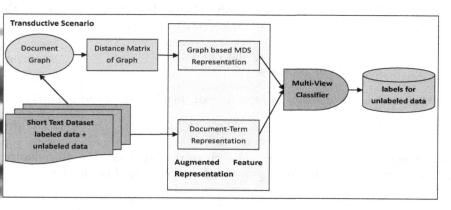

Fig. 2. Proposed Framework of GMDS-MV

we integrate the new graph based representation and the document-term representation as the new hybrid two-view augmented feature representation. Finally, we train a multi-view classifier in the transductive scenario.

2.3   Construct Document Graph

For document graph $G =< V, E >$, due to its spectral sparsity issue, a sparse graph is used to represent the document graph. There are two common ways to construct a sparse graph.

- KNN graphs: Vertex i,j are connected by an edge if i is in j's k-nearest-neighbourhood or vice versa.
- ϵNN graphs: Nodes i,j are connected by an edge, if the similarity $s(i, j) \geq \epsilon$.

Because different documents have various number of neighbours in the graph, we choose ϵNN to construct the document graph in our experiments.

Furthermore, we choose Cosine Similarity to measure the similarity for short texts.

$$Cosine(i, j) = \frac{X_i^t X_j^t}{\| X_i^t \| \ \| X_j^t \|} = \frac{| X_i^t \cap X_j^t |}{\sqrt{| X_i^t | \cdot | X_j^t |}} \tag{1}$$

where $| X_i^t |$ and $| X_j^t |$ is the number of terms in the ith and jth document respectively, and $| X_i^t \cap X_j^t |$ is the number of shared terms in the ith and jth document.

Accordingly, we summarize steps of the graph constructing as below: (i) represent each document as vertex; (ii) for each pair of vertices, if their term similarity is no less than ϵ, create an edge between them.

2.4   Distance Matrix for Graph

We consider that the pairwise documents, who share more common neighbours, are more likely to belong to the same class. Thus, the distance between two short text vertices can be measured by the number of shared common neighbours. Each document i has a neighbour vectors $N_i = \{N_{i,j} | j \in [1..n]\}$, where

$$N_{i,j} = \begin{cases} 1, & \text{if node } V_i \text{ and } V_j \text{ are neighbours, or i=j} \\ 0, & \text{otherwise} \end{cases} \tag{2}$$

The two documents' distance $P_{i,j}$ can be measured by Cosine dissimilarity.

$$P_{i,j} = 1 - Cosine(N_i, N_j) \tag{3}$$

Accordingly, P is $n \times n$ matrix, and each entries represents the dissimilarity relationship for a pairwise documents based on their shared neighbours.

2.5   Augment Feature Representation

Assume there is a latent semantic space where each document has a coordinate. MDS [12] is a representative approach to calculate the latent coordinate [13] in graph structure. MDS is able to find a low-dimensional (m) coordinate for each node while keeping the proximity or dissimilarity relationship among nodes in the graph.

Particularly MDS takes the distance matrix $P \in R^{n \times n}$ as input. Assume there is a low-dimensional matrix $S \in R^{n \times m}$ denoting the coordinates of nodes in the m-dimensional space such that S is column orthogonal, and S satisfies the equation:

$$SS^T \approx -\frac{1}{2}(I - \frac{1}{n}11^T)(P \circ P)(I - \frac{1}{n}11^T) = \widetilde{P} \tag{4}$$

where I is the identity matrix, 1 an n-dimensional column vector with each entry being 1, and $\circ$ is the element-wise matrix multiplication. Hence, the S can be obtained via minimizing the discrepancy between $\widetilde{P}$ and SS^T as follows:

$$S = \arg\min_{S} \| SS^T - \widetilde{P} \|_F^2 \tag{5}$$

Supposed that V contains the top m eigenvectors of $\widetilde{P}$ with largest eigenvalues, Λ is a diagonal matrix of top m eigenvalues $\Lambda = diag(\lambda_1, \lambda_2, ..., \lambda_m)$. The optimal S is

$$S = V\Lambda^{\frac{1}{2}} \tag{6}$$

As a result, each row in S is the document's latent coordinate in a low-dimensional semantic space.

2.6   Multi-view Classification

As demonstrated in previous section, we have two representations of the original dataset:

- D_T dataset for document-term representation.
- D_G dataset for graph based MDS representation.

The graph representation D_G describes the document's coordinate in a latent "semantic space" that consider relationship measurements - number of shared neighbours. The graph representation describe the document's coordinates in a latent "semantic space", which is calculated by the number of shared neighbours of two nodes. Accordingly, the document-term representation also can be viewed as each document's coordinate in the n-dimensional Euclidean space. Therefore, the two representations are different views for the same data source so that we use them to train a multi-view classifier.

Co-training [14] is a multi-view learning algorithm. However, Co-training suffers in the presence of view disagreement, i.e., when samples in each view do not belong to the same class due to view corruption or other noise processes [15].

Algorithm 1. Bagging-Co-Training

Input: L_A, U_A (labelled and Unlabelled data in dataset A)

L_B, U_B (labelled and Unlabelled data in dataset B)

M (# of classifiers in Bagging) , k (# of confidence samples)

Output: H_A, H_B (Bagging classifiers for dataset A and B)

1: Init $2 * M$ pools $\overline{U}_A[i], \overline{U}_B[i], i \in [0..M-1]$;
2: **repeat**
3:    Learn Bagging Classifier H_A from L_A;
4:    Learn Bagging Classifier H_B from L_B;
5:    **for** $i = 0$ **to** $M - 1$ **do**
6:       Get k confidence samples S_A from $\overline{U}_A[i]$ by H_A;
7:       Assign labels to S_A by H_A;
8:       Move S_A from $\overline{U}_A[i]$ to $L_A[i]$;
9:       Select k samples from $U_A[i]$ to replenish $\overline{U}_A[i]$;
10:      Get k confidence samples S_B from $\overline{U}_B[i]$ by H_B;
11:      Assign labels to S_B by H_B;
12:      Move S_B from $\overline{U}_B[i]$ to $L_B[i]$;
13:      Select k samples from $U_B[i]$ to replenish $\overline{U}_B[i]$;
14:    **end for**
15: **until** Reach Maximal Iterations

To address this issue, we propose a novel form of co-training, named Bagging-Co-Training. A Bagging classifier is composed of many common classifiers which we named them as unit classifier. In our algorithm, each unit classifier selects high confidence samples from different pools, consequently one noise sample only corrupt one classifier. Furthermore, the corrupted classifier can be rescued via feeding the good samples justified by a Bagging of classifiers.

We use A and B to represent two different datasets D_T and D_G which are composed of labelled part $L_A \& L_B$ and unlabelled part $U_A \& U_B$. Two Bagging classifiers H_A and H_B are learned and each of them contains M unit classifiers. Each $H_A[i]$ or $H_B[i]$ represents the ith unit classifier and owns a unique unlabelled samples pool $\overline{U}_A[i]$ or $\overline{U}_B[i]$.

The Bagging-Co-Training algorithm is presented in Algorithm 1. Firstly, we initiate a few pools which are used to hold unlabelled samples (line 1). Then, an iterative process is started by selecting high confidence samples to augment the labelled dataset. In each iteration, two initial Bagging classifiers, H_A and H_B, are learned based on two labelled training dataset (line 2-3). During this process, each pool will be updated by moving high confidence labels to labelled dataset and acquire samples to replenish itself (line 5-12). Particularly, the confidence is justified by the Bagging classifier rather than a single classifier. The iteration process will stop after a given maximal iteration times (line 15), and the final output is two Bagging classifiers.

3   Performance Study and Experiment Results

In this section we evaluate the performance of the proposed framework GMDS-MV. First, we describe the data sets used in our experiments. Then we compare the graph based representation with document-term representation by visualizing them on a 2-dimension space. Third, our proposed GMDS-MV is compared with several baseline methods. Finally, we report the results of different types of experiments.

3.1   Datasets

For the benchmark data set, we consider the well-known 20newsgroup, which is collected from 20 different newsgroups posts. We arbitrary combine two categories (computers.graphics and sci.crypt) into a binary-class dataset. Furthermore, considering the short text scenario, we only use the subject field of each document as training data, and discard other information, such as main body. We also use Twitter Sentiment dataset [16] in which each document is a Twitter message that expresses a positive or negative sentiment. To evaluate our algorithm in search snippet application, we choose a 8-class dataset which is composed of search snippets from [1]. Before utilizing the data, we execute standard pre-processing, such as removing stop words and stemming words.

3.2   Visualize the Graph Based MDS Representation

To demonstrate the different view of graph from document-term representation, we visualize the MDS results on graph representation and the Principle Component Analysis (PCA) results on document-term representation.

In Figure 3, each point represents a document and same coloured points belong to the same class. For document-term representation, we utilize PCA to project each document to a 2-dimension space, and the graph based representation are transformed to a 2-dimension vector by MDS. From the Figure 3, we can distinguish the cluster of documents obviously. The pair of subfigures (a) & (d) shows different views of the same data source. Same to the pair of subfigures (b) & (e), and (c) & (f). Particular, the red points in subfigure (c) are mixed with other color points, while they are far away from other color points in the subfigure (f). In contrast, the blue point is clearly far away from other color points in the subfigure (c) while close to others in the subfigure (f). It demonstrates that the graph based representation and document-term representation are different views for the same data source. Generally, the two-view representation can provide extra information than one-view representation, and the extra information can be useful for classification training process.

3.3   Compared Methods

To evaluate the performance on short text classification, we compare our proposed method **GMDS-MV** against the following classification methods.

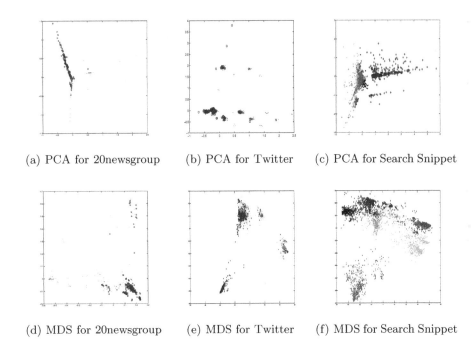

(a) PCA for 20newsgroup   (b) PCA for Twitter   (c) PCA for Search Snippet

(d) MDS for 20newsgroup   (e) MDS for Twitter   (f) MDS for Search Snippet

Fig. 3. PCA on document-term representation Vs. MDS on graph representation. Subfigure (a)(b)(c) visualize 2-dimensional PCA based on document-term represen- taiont for 20newsgroup/Twitter Sentiment/Search Snippet dataset. Subfigure (d)(e)(f) visualize 2-dimensional MDS based on document graph representation for 20news- group/Twitter Sentiment/Search Snippet dataset.

- Supervised Classification model. In particular, we choose SVM as the classifi- cation model due to its good performance on sparse data, and use SVM^{light} [17] as the tool. We refer to this method as **SVM**.
- Transductive Classification model. We choose Transductive SVM (in SVM^{light}) as the Transductive classification model for binary-class task, and refer to this method as **TSVM.**.
- Hybrid features Transductive Classification model. This method has the same settings as the Transductive classification method except that its inputs comes from the hybrid feature representation which simply joins both graph based representation and document-term representation. With the hybrid feature dataset, we choose TSVM (in SVM^{light}) to learn the binary-class classifier and refer to this method as **HTSVM**.
- Our proposed Graph based MDS feature augmentation with Multi-View classification model (**GMDS-MV**). We learn a graph based MDS represen- tation from the given labelled and unlabelled data. Then we utilize Bagging- Co-Training (Algorithm 1) to incorporate both graph representation and document-term representation. We choose SVM(in SVM^{light}) as the basic classifier for Bagging-Co-Training.

3.4   Experimental Results

Accuracy Comparison. Firstly, we compare the performance on GMDS-MV against other methods on the benchmark datasets. Particular, for search snippet data, we only choose samples, which belongs to business and computer class label, to form a binary-class dataset. We vary the number of labelled samples from 5 to 80. The results of the classifiers mentioned in Subsection 3.3 on all benchmark datasets are reported in Table 1 .

For SVM and TSVM classifiers, we use the default parameters. For GMDS-MV, the number of iterations is set to 30 and add 1 confidence sample for each class per time. The number of classifiers for Bagging is set to 20. The dimension of graph based MDS representation are set to 100. All results are the average value of 10 times experiments.

As table 1 shows that GMDS-MV is superior over others. The HTSVM is not always better than TSVM, and it demonstrates that it is necessary to design a learning framework for two representations rather than simply combining two representations.

Table 1. Accuracy Comparison on three datasets

Dataset	Size	Classifier	5 labelled samples	10 labelled samples	20 labelled samples	40 labelled samples	80 labelled samples
20News	2000	SVM	50.751%	52.392%	58.506 %	66.180%	76.014%
		TSVM	70.820%	74.871%	75.231%	78.636 %	84.485%
		HTSVM	55.888%	63.270%	71.031%	76.647%	83.220%
		GMDS-MV	**76.349 %**	**76.450 %**	**78.348 %**	**80.343 %**	**85.418 %**
Twitter	2000	SVM	70.871%	71.717%	73.890%	76.963%	83.763%
		TSVM	70.803%	78.739%	79.488%	83.125 %	86.329%
		HTSVM	69.509%	69.509%	78.861%	84.924 %	86.998%
		GMDS-MV	**77.393%**	**79.185%**	**81.224%**	**85.249%**	**88.973%**
Search	2400	SVM	58.163 %	67.616%	73.982%	80.452%	82.6891%
		TSVM	73.693%	78.391%	83.582%	86.394%	90.124%
		HTSVM	75.343%	76.437%	82.172%	88.394%	89.994%
		GMDS-MV	**80.270%**	**83.347%**	**85.258%**	**87.275%**	**92.357%**

Parameters Influence on Algorithm. To investigate the influence of different parameters in GMDS-MV method, we design a set of experiments.

- to investigate the accuracy changing with the iteration increasing, we change the maximal iterations as a variate while fixing others.
- to study the accuracy changing while increasing the number of labelled samples, we set the labelled samples as a variate while fixing others.
- to explore how the accuracy is influenced by the Bagging size, we set the Bagging size a s variate while fixing others.

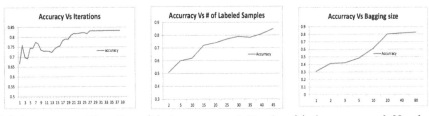

(a) Accuracy and Iterations (b) Accuracy and Number (c) Accuracy and Number
                      of labelled samples   of classifiers in Bagging

Fig. 4. Accuracy changes on various parameters in Search Snippet data

We implement our experiments on the binary-class Search Snippet dataset which only contains two classes: computers and business. We set default value to each parameter. The number of labelled data is 10, the dataset size is 2400, the maximal iteration is 30 and the Bagging size is 20. All experiments results are shown in Figure 4. In subfigure(a), GMDS-MV becomes stable approximately after Maximal Iteration reaches 25. As shown in subfigure(b), more labelled samples means higher accuracy. Subfigure(c) implies bigger Bagging size can improve accuracy.

4   Conclusions

In this paper, we proposed a novel method, GMDS-MV, which classifies short and sparse texts by using multi-view learning. GMDS-MV generates a graph of relationship among documents from original textual data and converts it to a low-dimensional representation. To integrate the graph based representation with document-term representation, GMDS-MV uses multi-view learning to leverage two representations of one domain data under Transductive learning scenario. Furthermore, we applied Bagging with Co-Training, a classical multi-view learning tool, to reduce the influence of noisy data. Experimental results demonstrate the proposed method can improve the classification performance.

References

1. Phan, X.H., Nguyen, L.M., Horiguchi, S.: Learning to classify short and sparse text & web with hidden topics from large-scale data collections. In: Proceedings of the 17th International Conference on World Wide Web, pp. 91–100. ACM (2008)
2. Blei, D.M., Ng, A.Y., Jordan, M.I.: Latent dirichlet allocation. The Journal of Machine Learning Research 3, 993–1022 (2003)
3. Sahami, M., Heilman, T.D.: A web-based kernel function for measuring the similarity of short text snippets. In: Proceedings of the 15th International Conference on World Wide Web, pp. 377–386. ACM (2006)

4. Vitale, D., Ferragina, P., Scaiella, U.: Classification of short texts by deploying topical annotations. In: Baeza-Yates, R., de Vries, A.P., Zaragoza, H., Cambazoglu, B.B., Murdock, V., Lempel, R., Silvestri, F. (eds.) ECIR 2012. LNCS, vol. 7224, pp. 376–387. Springer, Heidelberg (2012)

5. Long, G., Chen, L., Zhu, X., Zhang, C.: Tcsst: transfer classification of short & sparse text using external data. In: Proceedings of the 21st ACM International Conference on Information and Knowledge Management, CIKM 2012, pp. 764–772. ACM, New York (2012)

6. Glorot, X., Bordes, A., Bengio, Y.: Domain adaptation for large-scale sentiment classification: A deep learning approach. In: Proceedings of the 28th International Conference on Machine Learning (ICML 2011), pp. 513–520 (2011)

7. Hughes, T., Ramage, D.: Lexical semantic relatedness with random graph walks. In: EMNLP-CoNLL, pp. 581–589 (2007)

8. Ramage, D., Rafferty, A.N., Manning, C.D.: Random walks for text semantic similarity. In: Proceedings of the 2009 Workshop on Graph-based Methods for Natural Language Processing, pp. 23–31. Association for Computational Linguistics (2009)

9. Xu, Y., Yi, X., Zhang, C.: A random walks method for text classification. In: SDM (2006)

10. Zhu, X., Lafferty, J., Rosenfeld, R.: Semi-supervised learning with graphs. PhD thesis, Carnegie Mellon University, Language Technologies Institute, School of Computer Science (2005)

11. Goldberg, A.B., Zhu, X.: Seeing stars when there aren't many stars: graph-based semi-supervised learning for sentiment categorization. In: Proceedings of the First Workshop on Graph Based Methods for Natural Language Processing, pp. 45–52. Association for Computational Linguistics (2006)

12. Borg, I., Groenen, P.J.: Modern multidimensional scaling: Theory and applications. Springer (2005)

13. Tang, L., Liu, H.: Community detection and mining in social media. Synthesis Lectures on Data Mining and Knowledge Discovery 2(1), 1–137 (2010)

14. Blum, A., Mitchell, T.: Combining labeled and unlabeled data with co-training. In: Proceedings of the Eleventh Annual Conference on Computational Learning Theory, pp. 92–100. ACM (1998)

15. Christoudias, C., Urtasun, R., Darrell, T.: Multi-view learning in the presence of view disagreement. arXiv preprint arXiv:1206.3242 (2012)

16. Twitter sentiment data, http://www.sentiment140.com/

17. Joachims, T.: Making large scale svm learning practical (1999)

Exploring Deep Belief Nets to Detect and Categorize Chinese Entities

Yu Chen, Dequan Zheng, and Tiejun Zhao

Computer Science and Technology Department, Harbin Institute of Technology, China
{chenyu,dqzheng,tjzhao}@mtlab.hit.edu.cn

Abstract. This paper adapts a novel model, deep belief nets (DBN), to extract entity mentions in Chinese documents. Our experiments were designed to develop entity detection system and entity categorization system applying DBN, to complete entity extraction. Our results exhibit how the depth of architecture and quantity of unit in hidden layer of DBN influence the performance. In DBN Systems, token labels are produced independently and DBN does not concerned what the labels of surrounding token are. Viterbi algorithm is a good solution to overcome this issue. It can find the most likely probability label path to make DBN to be more suitable for entity detection. Furthermore, this paper demonstrates DBN is proper model for our tasks and its results are better than Support Vector Machine (SVM), Artificial Neural Network (ANN) and Conditional Random Field (CRF).

1 Introduction

Named Entities (NE) are defined as the names of existing objects. Named entity extraction is an important technology to provide structured information from raw text. It can be separated into two subtasks, entity detection and entity categorization. Entity detection is to identify the boundary of entity mentions whereas entity categorization is to classify the detected entity mentions into proper type.

In most situations, identifying and categorizing Chinese entity is more intricate than English. There is no space or specific boundary between Chinese characters. Many works for this task started from Chinese word segmentation (CWS). Unfortunately, all CWS systems are not perfect. They lead errors into the following systems. In addition, some long words may contain the target entities so that those entities will never be detected. In our experiments, several CWS results are combined and the fine-grain words will be selected to alleviate the problems.

Many machine learning methods have been applied to information extraction task. In this paper, we explore the usage of Deep Belief Nets (DBN), a new feature-based machine learning model for Chinese entity extraction. It is a neural network model developed under the deep learning architecture and feature learning approach. Deep architecture generates better compact representation of target function and requires less computational element comparing to shallow architecture. Feature learning approach avoids random initiation and prevents DBN from stuck in local minima. [1]

H. Motoda et al. (Eds.): ADMA 2013, Part I, LNAI 8346, pp. 468–480, 2013.

claimed DBN had capability to automatically learn high-level features from input low-level features for the complex problems. Our work is to adapt this novel model to entity extraction task and show how the depth of architecture and computational element structure influence the performance.

Entity labels for words are dependent by labels of surrounding words. However, DBN training processing and labels generation processing break this connection. Viterbi algorithm uses the transition probability, emission probability and initial probability to find a path which has the most likely probability from all paths. It guarantees that there is no illogical label sequence.

The rest of this paper is structured in the following manner. Section 2 reviews the previous work on entity extraction. Section 3 presents task definition, combination of different CWS, DBN model and the Viterbi algorithm. Section 4 provides the experimental results of entity detection and entity categorization. Finally, Section 5 gives the conclusion of this paper.

2   Previous Works

Over the past decades, immense works have been carried out to develop entity recognition in different languages and domains. Most of them can be categorized into three classes. They are rule-based approach, machine learning-based approach and knowledge-based approach.

[2] built a system for all information extraction tasks including entity recognition in MUC-7(Message Understanding Conference). They collected some corresponding grammar rules and domain model of discourse interpreter. When training examples are not available, handcrafted rule technology is preferred. The quality of rules determines the performance of the system. However, it is time-consuming to collect rules and it is lack of robustness.

Machine learning-based approach does not concern any rules but it requires large labeled data. Machine learning methods and feature selection significantly influence results. [3] proposed Hidden Markov Model (HMM) and HMM-based chunk tagger to detect and classified entities. Simple deterministic feature, semantic feature, gazetteer feature and context feature are used in this paper. They achieved F-measures 96.6% and 94.1% on MUC-6 and MUC-7 English entity extraction task.

Knowledge-based approach has been receiving much more attention over the last few years due to the rapid growth of online encyclopedia. Wikipedia is a typical database which gives important concepts and contains abundance of links of those concepts. [4] utilized local feature and global feature in Wikipedia to recognize and disambiguate entities. Local features captured the traditional features such as context similarity and TF-IDF information. Global features include the inlinks and outlinks between concepts in Wikipedia web pages.

DBN is a new feature-based approach for NLP tasks. It was reported to perform very well in many classification problems applying high dimensional feature vectors [1]. DBN had been cast to NLP task such as word disambiguation [5] and question answering [6].

3   Deep Belief Nets for Entity Detection and Entity Categorization

3.1   Task Definition

Entity extraction, added in MUC-6 and promoted by the ACE program, is a task of finding predefined entities from the texts. We used **BILOU** tagging scheme in our system to find the entities boundary. [7] claimed it could learn more expressive model and outperform **BIO** scheme in MUC-7 and CoNLL03 English shared task data.   In **BILOU** scheme: **B** means the token at the beginning of an entity; **I** means the token inside an entity; **L** means the last token of an entity; **O** means the token outside an entity and **U** means token of an unit-length entity. Systems have to give a tag to all tokens from those five tags indicating their location associated to the entities. The task can be formalized as:

$$y_i = f(x_{i-k}...x_i...x_{i+l}) \tag{1}$$

x_i is the feature vector of i-th word. k and l are small numbers indexing the word feature before and after x_i. y_i is one of the **BILOU** labels.

After identifying entities, their categorization needed to be distinguished. In ACE 2004 corpora which will be used in this paper, five predefined entities are included. They are Person, Organization, Location, GPE (a Geo-political entity) and Facility. Each entity belongs to one and only one of these five categories.

3.2   Combination of Chinese Word Segmentation

Our entity extraction system applied CWS to generate Chinese words as tokens. Although, many CWS are well developed they are still imperfect. They produce some incorrect words, or words being too long for our tasks. For example, "市政府" (Municipal Government) is a word given by a CWS but "政府" (Government) is the entity we are seeking. In this case, "政府" will never be detected.

Table 1. Number of entity segmentation errors

CWS System	Number of errors
Stanford	585
ICTCLAS	572
FudaNLP	560
Combination	282

In this paper, Stanford CWS [8], ICTCLAS and FudaNLP, which are reported to have more than 94% F-measure performance, are combined. We pick the shortest words from these three CWS. Table 1 shows the number of incorrect segments related to the entities. The results show the improvement of CWS combination.

3.3   Deep Belief Nets

DBN often consists of multiple Restricted Boltzmann Machine (RBM) layers and a Back-Propagation (BP) layer [9]. As illustrated in Figure 1, each RBM layer learns its parameters independently and unsupervisedly. RBM makes the parameters optimal for the joint distribution of its visible layer and hidden layer but not for the whole model. It detects complicated features and keeps as much information as possible when it transfers feature information to next layer. Therefore RBM can help networks to avoid local optimum. There is a supervised BP layer on top of the DBN model. The BP layer classifies its input features which is also the output of last RBM layer. Later on, the error information generated by BP layer will be back-propagated to all RBM layers to fine-tune the entire DBN in the learning process.

Deep architecture of DBN represents functions compactly. It is expressible by integrating different levels of simple functions [10]. Upper layers are supposed to represent more "abstract" concepts that explain the input data whereas lower layers extract "low-level features" from the data. In addition, none of the RBM guarantees that all the information conveyed to the output is accurate or important enough. The learned information produced by preceding RBM layer will be continuously refined through the next RBM layer to weaken the wrong or insignificant information in the input. Multiple layers filter valuable features. The final feature vectors used for classification consist of sophisticated features which reflect the structured information, promote better classification performance than direct original feature vector.

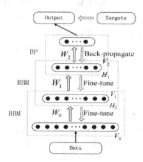

Fig. 1. The structure of a DBN

3.4   Restricted Boltzmann Machine

In this section, we will introduce RBM, which is the core component of DBN. RBM is Boltzmann Machine with no connection within the same layer. An RBM is constructed with one visible layer and one hidden layer. Each observed variable v_i in visible layer V accept raw data or the previous RBM output as input. Hidden variable h_j in the hidden layer H is output of new learned feature. To learn RBM, the optimum parameters are obtained by maximizing the joint distribution $p(v,h)$ on the training data [11]. By modifying the previous parameters with the gradient, the expected parameters can gradually approximate the target parameters as

$$W^{(\tau+1)} = W^{(\tau)} + \eta \frac{\partial \log P(v^0)}{\partial W}\bigg|_{W^\tau} \qquad (2)$$

where η is a parameter controlling the leaning rate. It determines the speed of parameters W converging to the target. log $p(v,h)$ is the log probability of the data. Traditionally, the Monte Carlo Markov chain (MCMC) is used to calculate this kind of gradient. However, MCMC typically takes a long time to converge. [9] introduced the contrastive divergence (CD) algorithm as a substitution. It is reported that CD can train the model much more efficiently than MCMC. It considers a series of distributions $\{p_n(v)\}$ which indicate the distributions in n steps. In our experiments, we set n to be 1. It means that in each step of gradient calculation, the estimate of the gradient is used to adjust the RBM weight as Equation 3. $< h^0v^0 >$ denotes the multiplication of the average over the data states and its relevant sample in hidden unit. $<h^1v^1>$ denotes the multiplication of the average over the states after 1 step of MCMC.

$$\frac{\partial \log p(v,h)}{\partial W} = \left\langle h^0v^0 \right\rangle - \left\langle h^1v^1 \right\rangle \tag{3}$$

3.5   Viterbi Algorithm

As shown in equation 1, the function that generates label y_i does not involve the information of labels y_{i-1} before y_i. In fact, y_i is limited by the label appearing before it. For example, it is obvious that the label **B** should not follow label **I**. Viterbi algorithm is a solution to make sure that illogical label sequence will not be selected. It is a dynamic programming algorithm for finding the most likely sequence of hidden states to tag sequential observed states. Supposed we are given initial probabilities s_i (i hidden states to be the beginning of sequence), transition probabilities t_{ij} (from hidden state i to hidden state j) and emission probabilities e_{ti} (from observed state t to hidden state i). The most likely hidden state sequence produced by equation 4:

$$\delta_t(i) = \max_j (\delta_{t-1}(j)t_{ji}e_{ti}) \tag{4}$$

$\delta_t(i)$ is partial probabilities of path from start to token t. The transition probabilities of illogical entity label path should be 0. For example t_{BO} is 0, the partial probabilities δ of path containing "BO" will be 0 and this path will be discarded.

3.6   Feature Sets

Table 2 summarizes the feature sets used in our experiments which are commonly used in Chinese NLP task. Our motivation is to figure out how DBN is suitable to Chinese entity extraction task so that no many sophisticated features get involved. Window size of feature sets in our experiments is 1 for all results shown in section 4. We had compared several window sizes and size 1 is the best choice.

Table 2. Feature set description

Feature	Description
Unigram	Unigram token sequences including current token and tokens of context window.
Part-of-Speech	POS of current token and context window.
Gazetteers	Various type of gazetteers like surname, location and so on.

4   Experiments and Evaluations

The experiments are conducted on the ACE 2004 Chinese entity extraction dataset, which consists of 9809 entities in 221 documents. We examined the proposed DBN model using 4-fold cross-validation. 166 documents are considered as training data and 55 documents are testing data. A mention is a reference to an entity which can be different constitution unit of extents or heads. In this paper, only head mentions will be detected and categorized. The performance is measured by precision, recall, and F-measure. They are represented by **P**, **R** and **F** in the following table.

$$F\text{-measure}=\frac{2*\text{Precision}*\text{Recall}}{\text{Precision}+\text{Recall}} \tag{5}$$

As for entity categorization, it is a typical classification task. Each detected entity should be classified into one of five pre-defined categorizations. Therefore, its recall equals precision.

4.1   Entity Detection

4.1.1   Evaluation on DBN Structure

There is no convenient method to determine unit quantity in hidden layer. Table 3 shows the results of DBN containing one RBM layer. We tried 300, 500, 1000, and 2000 units of structures in one RBM layer. The feature space is 21484, 21573 and 25824 for unigram, POS+ (unigram and POS) and Gazetteer+ (unigram, POS and Gazetteer) respectively. Results show that the F-measures are similar even though model structures are different. It means that those functions of models are all well developed. The results also indicate 500 units in hidden layer are slightly better than others on unigram and POS+. On Gazetteer+, RBM of 1000 units is better.

Table 3. Entity detection performance of DBN containing one RBM layer

Feature	Units	P(%)	R(%)	F(%)
Unigram	300	86.01	69.79	77.05
	500	85.77	70.48	**77.38**
	1000	85.49	70.08	77.02
	2000	86.86	69.44	77.18
POS+	300	87.99	75.46	81.24
	500	86.1	77.1	**81.36**
	1000	87.17	76.16	81.3
	2000	87.05	76.31	81.33
Gazetteer+	300	87.37	75.81	81.18
	500	86.88	76.46	81.33
	1000	87.3	76.46	**81.4**
	2000	87.21	76.06	81.25

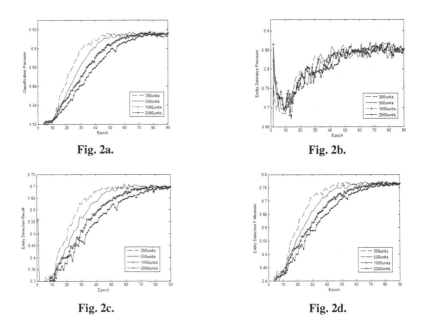

Fig. 2a.   Fig. 2b.

Fig. 2c.   Fig. 2d.

Fig. 2. Performances of DBN containing various RBM structure for entity detection. Figure 2-a is classification precision, figure 2-b is entity detection precision, figure 2-c is entity detection recall and figure 2-d is entity detection F-Measure.

Figure 2 gives the classification precision and entity detection performances by DBN using unigram feature. Classification precision is the measurement of how many words are correctly classified into one of **BILOU** labels. Figure 2-a, figure 2-b and figure 2-d indicate performances of smaller RBM structure converge faster. After enough epochs, all RBM obtain similar classification precision, detection Precision and F-Measure. All entity detection precision curves in figure 3-c oscillate strong. This is why so it is hard to tell which one is better.

4.1.2   Evaluation on Deep Architecture

Deep architecture is the one of advantages of DBN. We compared the performances of DBN containing one RBM layer to three RBM layers. We have tried to input the output of previous RBM before fine-tune into current RBM. Unfortunately, it makes performances a little worse. Therefore, we used the output of RBM after supervised learning as the input of next RBM input. We choose the best performances of DBN having deep architecture to list in table 4. The depth of RBM and the quantity of units in each layer are shown in "Unit" column. It is obvious that DBNs containing two RBM layers result in better performances which have been bold. Three RBMs does not increase the results comparing to two RBMs. Therefore more RBM layers do not ensure better performance. We need to choose proper depth for certain task. The improvement of deep architecture is not very significant. It means that the model based on word features is linear and one RBM layer has detected the sophisticated features for entity detection.

Table 4. Entity detection performance of deep architecture

Feature	Units	P(%)	R(%)	F(%)
	500	85.77	70.48	77.38
Unigram	500-1000	87.31	70.18	**77.81**
	500-1000-4000	85.96	70.38	77.39
	500	86.10	77.10	81.36
POS+	500-2000	87.97	76.06	**81.58**
	500-2000-4000	86.92	76.41	81.32
	1000	87.03	76.46	81.4
Gazetteer+	1000-500	87.67	76.46	**81.68**
	1000-500-1000	87.5	75.96	81.32

Using parameters in well-developed DBN to initiate DBN with deeper architecture gives our system three advantages. As we mentioned above, it detect high-level abstract features to improve performance. Furthermore, since shallower DBN has already been developed and has constructed proper features for entity detection, it helps the new deeper DBN to find its target function faster. This advantage can be seen from figure 3. Figure 3 describes the performance of shallow architecture and deep architecture when only unigram was used. Classification precision, entity detection recall and entity detection F-Measure rise to a stable status faster when deep architecture is applied. We can also conclude that deep architecture alleviated over-fitting according to figure 3-a and figure 3-d. When our systems used only one RBM layer, it obtained the best classification precision and entity detection F-Measure around 65 epochs. Its performances declined substantially after that. On the other hand, the performances of deep architecture do not change by much with the increase of epochs. finally, we found that two RBMs help to obtain better detection precision whereas three RBMs can obtain higher recall referring to figure 3-b and figure 3-c.

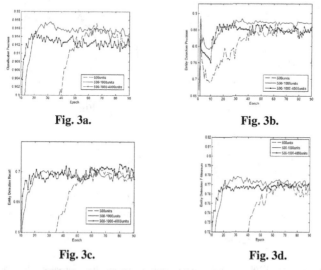

Fig. 3a.   Fig. 3b.

Fig. 3c.   Fig. 3d.

Fig. 3. Performances of DBN using shallow architecture and deep architecture for entity detection. Figure 3-a is classification precision, figure 3-b is entity detection precision, figure 3-c is entity detection recall and figure 3-d is entity detection F-Measure.

4.1.3   Evaluation on Machine Learning Methods

To prove DBN is a powerful machine learning model to detect entity, we compared performances from DBN, SVM, ANN and CRF. We wanted to note that DBN and ANN have the same structures. The difference between these two networks is that the parameters of DBN are initiated by RBM whereas ANN is randomly initiated. ANN is a typical Back-Propagation network. Performances of these models are shown in table 5. DBN's performance significantly outperforms ANN's indicating RBM overcome the local minima problem. RBM indeed constructs high-level abstracted feature. Figure 4 draws the performances of DBN and ANN when using shallow architecture. It shows us that DBN is not as good as ANN at the beginning of learning processing. However, DBN outperforms ANN after about 60 epochs on all evaluations except precision.

Table 5. Entity detection performance of various machine learning methods

Feature	Model	P(%)	R(%)	F(%)
	DBN	87.31	70.18	**77.81**
Unigram	ANN	85.17	69.19	76.35
	SVM	85.31	67.65	75.46
	CRF	86.84	66.65	75.42
	DBN	87.97	76.06	**81.58**
POS+	ANN	85.26	74.56	79.55
	SVM	86.64	73.27	79.4
	CRF	87.28	72.08	78.95
	DBN	87.67	76.46	**81.68**
Gazetteer+	ANN	86.08	74.51	79.88
	SVM	86.71	75.36	80.64
	CRF	87.44	73.12	79.64

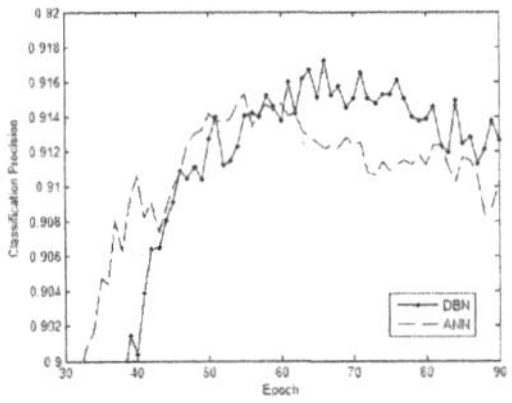

Fig. 4a.

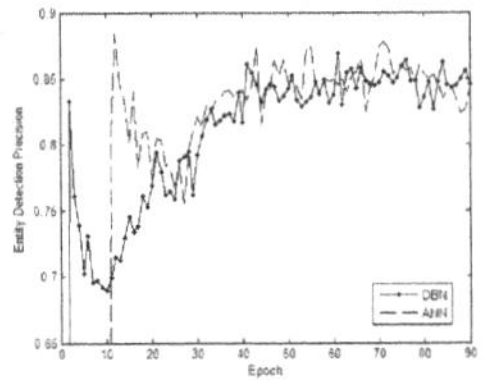

Fig. 4b.

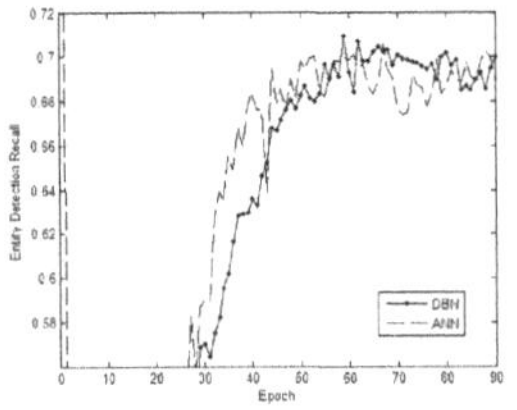

Fig. 4c.

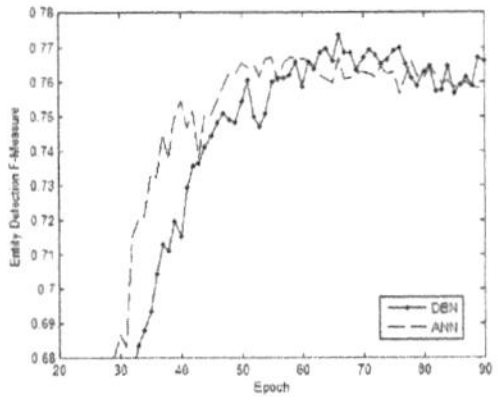

Fig. 4d.

Fig. 4. Performances of DBN and ANN using shallow architecture for entity detection. Figure 4-a is classification precision, figure 4-b is entity detection precision, figure 4-c is entity detection recall and figure 4-d is entity detection F-Measure.

SVM has been successfully applied in many classification applications. We use the LinearSVM toolkit to implement the SVM model. Linear kernel is a fast and high performance model for large data set, especially whose quantity of samples is much larger than feature spaces. The best performances of SVM are chosen in table 5. In contrast to SVM, DBN still obtains meaningful improvement.

Many researches applied CRF to extract entities and published state of the art results. We used CRF++ on our tasks to make it one of baselines. Although CRF achieves precision approaching to DBN, its recall falls behind DBN by 3.53%, 3.98% and 3.34% on the three feature sets. In addition, its F-Measure loses at least 2.04% comparing to DBN.

4.1.4   Evaluation on Viterbi Algorithm

In previous results, many illogical label paths appeared. We used Viterbi algorithm to ensure the most likely path to be found. Initial probability is the probability of each label to be the beginning of a sentence. On the other hand, it can be binary number. For example, label **O** can be the beginning of a sentence, its initial probability is set to be 1. Label **I** cannot be the beginning of a sentence, its initial probability is 0. Likewise, transition probability can be real according to corpora or binary by logic. The emission probability is produced by our entity detection models. It is the probability of each label to tag current word. Figure 5 is the performances of Viterbi algorithm and DBN having two RBMs on unigram feature set. It can easily to figure out that Viterbi algorithm using real probability is able to improve precision but it is harmful for recall. Although Viterbi algorithm using binary probability decreases precision, it increases recall significantly and improve F-Measure slightly. Table 6 exhibit the results of Viterbi algorithm. We used bold to emphasize the positive result comparing to table 5 except CRF. CRF has already comprised Viterbi algorithm and has rejected illogical label paths. When binary probabilities are applied, precision improvements of all machine learning methods are negative and recall improvements are positive. Except for DBN that use unigram feature, all F-measures resulted in higher score.

Table 6. Entity detection performance on Viterbi algorithm using binary probabilities

Model	Feature	Real Probability			Binary Probability		
		P(%)	R(%)	F(%)	P(%)	R(%)	F(%)
DBN	Unigram	-0.32	-1.94	-1.32	-3.61	**2.39**	-0.07
	POS+	-0.38	-0.15	-0.25	-2.19	**2.64**	**0.51**
	Gazetteer+	-0.55	**0.64**	**0.13**	-2.96	**2.93**	**0.28**
ANN	Unigram	**2.86**	-2.19	**0.26**	-3.06	**2.34**	**0.11**
	POS+	**0.62**	**0.35**	**0.08**	-2.32	**3.39**	**0.82**
	Gazetteer+	-0.28	-1.14	-0.78	-2.68	**2.74**	**0.33**
SVM	Unigram	**5.75**	-27.58	-19.81	-2.07	**2.09**	**0.44**
	POS+	**1.11**	-12.99	-7.93	-1.48	**1.59**	**0.28**
	Gazetteer+	**1.17**	-8.61	-4.77	-1.68	**2.39**	**0.59**

Table 7. Precision of entity categorization using DBN containing one RBM layer

Units	Unigram	POS+	Gazetteers+
500	81.48%	81.69%	82.13%
1000	81.48%	81.75%	82.91%
2000	81.48%	81.40%	82.02%

Fig. 5a.   Fig. 5b.   Fig. 5c.

Fig. 5. Performances of DBN using Viterbi algorithm. Figure 5-a is entity detection precision, figure 5-b is entity detection recall and figure 5-c is entity detection F-Measure.

4.2   Evaluation on Entity Categorization

After entity detection processing, entities should be classified into pre-defined categorizations. We used golden standard of entities in training data to train entity categorization systems. As the experiments we described above, we compared different RBM structures, shallow architecture and deep architecture, DBN and other machine learning methods on this task. In contrast to [12], we used word-based feature. Meanwhile, they classified golden standard of entities in testing data but we completed task on detected entities. Results in table 7 indicate that the quantity of units in RBM is not a problem for entities categorization on word-based features. Precisions of entity categorization are equal or extremely closed for various RBM structure. Table 8 shows that deep architecture has limited helpfulness for this task. The reason is that the internal relation among feature elements is rare. Table 9 indicates DBN is better than ANN and SVM for this classification task. In addition, SVM has slightly improvement than ANN.

Table 8. Precision of entity categorization using deep architecture of DBN

Depth	Unigram	POS+	Gazetteers+
1RBM	81.48%	81.75%	82.13%
2RBM	81.60%	81.75%	82.13%
3RBM	81.54%	81.92%	82.19%

Table 9. Precision of entity categorization using various machine learning methods

Depth	Unigram	POS+	Gazetteers+
DBN	81.60%	81.75%	82.19%
ANN	80.9%	78.88%	80.7.%
SVM	79.54%	80.93%	81.1%

4.3 Evaluation on Entity Detection and Categorization in Sequence

After entity detection and entity categorization, we combined their best results to extract entities. To extract correct entity, not only should its boundary be detected correctly but also its type should be classified into the right categorization. Table 10 shows us the performance of entity detection and entity categorization in sequence. DBN achieves the best performance on all feature sets. DBN is not much better than ANN by only using unigram. Along with the feature sets increase, DBN obtain much more improvements. Table 11 gives the performance of entity detection and entity categorization in sequence using Viterbi algorithm. Similar to the results of entity detection using Viterbi algorithm, binary probability decreases precision but increases recall and F-Measure for all methods and feature sets.

Table 10. Performance of entity detection and entity categorization in sequence

Feature	Model	P(%)	R(%)	F(%)
	DBN	81.3	65.36	**72.46**
Unigram	ANN	80.27	65.41	72.08
	SVM	79.28	62.87	70.13
	DBN	81.92	70.83	**75.97**
POS+	ANN	78.88	68.99	73.61
	SVM	80.93	68.44	74.16
	DBN	82.19	71.68	**76.58**
Gazetteer+	ANN	80.39	69.59	74.6
	SVM	81.1	70.48	75.42

Table 11. Entity extraction performance on Viterbi algorithm using binary probabilities

Model	Feature	Real Probability			Binary Probability		
		P(%)	R(%)	F(%)	P(%)	R(%)	F(%)
	Unigram	**1.12**	-0.7	**0.01**	-2.43	**3.03**	**0.8**
DBN	POS+	-0.93	-0.65	-0.77	-2.43	**2.09**	**0.09**
	Gazetteer+	-0.98	**0.2**	-0.32	-2.87	**2.39**	**0.02**
	Unigram	**0.62**	-0.5	-0.06	-2.84	**2.04**	**0.01**
ANN	POS+	**1.07**	**0.1**	**0.52**	-2.19	**3.09**	**0.7**
	Gazetteer+	-0.47	-1.25	-0.92	-2.37	**2.68**	**0.44**
	Unigram	**8.05**	-24.44	-16.76	-2.27	**1.64**	**0.08**
SVM	POS+	**1.61**	-11.75	-6.94	-1.54	**1.35**	**0.12**
	Gazetteer+	**1.47**	-7.76	-4.13	-1.84	**1.99**	**0.3**

5 Conclusions

In this paper we presented our recent work on applying a novel machine learning model, namely Deep Belief Nets, to entity detection and entity categorization. DBN is demonstrated to be effective for these tasks because of its strong representativeness and deep architecture. Experimental results clearly showed the strength of DBN which obtained better performance than other existing models such as SVM, ANN and CRF. In addition, we combined several CWS and applied Viterbi algorithm to improve performance.

References

1. Hinton, G.E., Osindero, S., Teh, Y.: A fast learning algorithm for deep belief nets. Neural Computation 18, 1527–1554 (2006)
2. Humphreys, K., Gaizauskas, R., Azzam, S., Huyck, C., Mitchell, B., Cunningham, H., Wilks, Y.: University of Sheffield: Description of the. LaSIE-II System as Used for MUC-7. In: Proceedings of 7th Message Understanding Conference, pp. 1–12 (1998)
3. Zhou, G., Su, J.: Named Entity Recognition using an HMM-based Chunk Tagger. In: Proceeding of the 40th Annual Meeting of the Association for Computation Linguistics (ACL), pp. 473–480 (2002)
4. Ratinov, L., Roth, D., Downey, D., Anderson, M.: Local and Global Algorithms for Disambiguation to Wikipedia. In: Proc. of the Annual Meeting of the Association of Computational Linguistics, ACL (2011)
5. Wiriyathammabhum, Kijssirikul, B., Takamura, H., Okumura, M.: Applying Deep Belief Networks to Word Sense Disambiguation. CoRR (2012)
6. Wang, B., Wang, X., Sun, C., Liu, B., Sun, L.: Modeling semantic relevance for question-answer pairs in web social communities. In: Proceeding of the 48th Annunnal Meeting of the Association for Computational Linguistics (2010)
7. Ratinov, L., Roth, D.: Design challenges and misconceptions in named entity recognition. In: Proceeding of the Thirteenth Conference on Computational Natural Language Learning (CoNLL), pp. 147–155 (2009)
8. Tseng, H., Chang, P., Andrew, G., Jurafsky, D.: Christopher Manning. A Conditional Random Field Word Segmenter for SighanBakeoff 2005. In: Fourth SIGHAN Workshop on Chinese Language Processing (2005)
9. Hinton, G.E.: Training products of experts by minimizing contrastive divergence. Neural Computation 14(8), 1711–1800 (2002)
10. Bengio, Y., LeCun, Y.: Scaling learning algorithms towards ai. Large-Scale Kernel Machines. MIT Press (2007)
11. Hinton, G.: Products of experts. In: Proceedings of the Ninth International Conference on Artificial Neural Networks (ICANN), vol. 1, pp. 1–6 (1999)
12. Chen, Y., Ouyang, Y., Li, W., Zheng, D., Zhao, T.: Using Deep Belief Nets for Chinese Named Entity Categorization. In: Proceeding of the 2010 Named Entity Workshop (NEWS 2010), pp. 102–115 (2010)

Extracting Novel Features for E-Commerce Page Quality Classification

Jing Wang, Lanfen Lin, Feng Wang, Penghua Yu,
Jiaolong Liu, and Xiaowei Zhu

College of Computer Science,
Zhejiang University, Hangzhou, China
{cswangjing,llf,wangfeng,yph719,jl_liu,zjuedward}@zju.edu.cn

Abstract. There're a huge amount of web pages describing the same product on e-commerce websites, while their quality varies greatly. Therefore, there is a growing need for automated, accurate and efficient quality classification methods. Several link-based, click-based and content-based approaches have been proposed to evaluate the quality of pages for general search engines. However, these methods only consider the surface features of the html documents. What's more, features like link relations have drawbacks when dealing with e-commerce pages, because the hypothesis that links mean endorsements is not always right in the environment of e-commerce. In this paper, we propose two kinds of features that can directly indicate the quality of content. We analyze pages' content structure with a corpus of labeled texts, and evaluate the property completeness with the help of ontology. Then we combine these features with other commonly used features in literature. We apply several learning methods to train and classify pages into good and bad ones. Experiments on real e-commerce pages show that the proposed novel features can greatly improve the accuracy of classification.

Keywords: E-commerce, Page quality, Semantic, Content analysis, Information extraction.

1 Introduction

Search engines are useful tools to find information that satisfies users' needs. There are two important factors to judge if the results of a search satisfy the users' needs: relevance and quality of pages. When the number of relevant pages on the whole Internet is relatively small, it's important to find as many relevant pages as possible. However, with the explosion of web pages, more and more relevant pages exist on the Internet. Especially on many e-commerce websites, sellers can publish their product pages without strict review. When customers search for a product, there are usually a huge amount of candidate web pages describing the target product published by different sellers. All of these pages are relevant to the query words while their quality varies greatly. So there is a growing need for automated, accurate and efficient quality classification methods.

H. Motoda et al. (Eds.): ADMA 2013, Part I, LNAI 8346, pp. 481–492, 2013.

Generally speaking, products that have well-designed pages tend to have better quality, which is based on the intuition that the sellers that pay more attention on the pages tend to be more professional and have better attitudes. Furthermore, a well-designed page can provide more useful information for a customer, and is organized in a more readable way, especially for a user aiming to search for the details of a product. It's very helpful if the page quality is considered when ranking the search results. In the long term, this can also encourage sellers to pay more attention on page designing and enhance the overall customer's satisfaction.

Many studies have been conducted on data quality criteria and automatic measurement of page quality. Batini et al. [1] provide systematic and comparative description of data quality assessment. Many approaches have been proposed to evaluate the pages' quality based on link structure and user click-through data. Link-based methods are based on the notion that a link from a page p to another page q can be viewed as an endorsement of q by p. However, in the context of e-commerce, a link from a page to another page doesn't necessarily mean an endorsement, yet it's more likely to be an ad or link to another product provided by the same seller. Click-based methods rely on the users' histories of behaviors, which are quite difficult to obtain and usually contain much noise. Several content-based methods have been proposed to analyze the quality of the textual contents. They are based on statistics such as the number of words and stop words, the lengths of URLs and titles. These methods don't really understand what the page is about.

The pages in e-commerce websites have many common characteristics: 1) they have similar kinds of subsections, we call them semantic blocks, such as product description, brand and manufacturer introduction, sales promotion activity, user's guide, and advertisement. A page of good quality should contain a proper length of text for each semantic block; 2) they have descriptions about the properties of a product, and a page of good quality should contain detailed properties. In addition, products of the same kind tend to have similar properties, for example, sun creams have descriptions of "SPF", "brand", "price", etc. Knowing these characteristics, we can explore novel features by analyzing pages' semantic blocks and properties. There are many other features in literature that can be used to evaluate the quality of a page. In this paper, we choose some of these features, and combine them with our proposed features to classify pages into good and bad ones. The contributions of this paper are:

1. We propose to use semantic-block level features to indicate the quality of pages' content structure, and propose a method to extract these features.
2. We propose to use property completeness as another indicator of pages' content quality, and describe a method to analyze property completeness.
3. We combine our features with other features in literature, and use a feature selection process to select the most useful features for product pages.
4. By conducting extensive experiments on pages from e-commerce websites, we demonstrate that our extracted novel features can greatly improve the accuracy of classification.

The rest of this paper is organized as follows. The next section briefly describes related work on web page quality. Section 3 presents the framework of our method. In Section 4, we proposed the semantic-block level features of web pages and described the process of how to extract them. In section 5, we analyze the property level feature of web pages. In Section 6, we evaluate the performance of our method, using the product pages from e-commerce websites. Finally, we summarize our method and discuss the future work.

2 Related Work

Existing methods of quality evaluation can be classified into three categories based on the features they use: 1) link-based methods, 2) click-based methods, 3) content-based methods.

The pages and the links between them can be represented as a graph structure where web pages comprise nodes and links denote directed edges. Link-based methods calculate pages' importance(or authority) according to this graph. Link-based methods like PageRank [2], Hits (Hub and authority) [3], and SALSA [4] are based on the assumption that a link from a page p to another page q can be viewed as a vote of q by p, and they do not explicitly take into account the actual text content of the pages. However, in the context of e-commerce, a link from a page p to another page q doesn't necessarily mean a vote, and it's more likely to be an ad or link to another product of the same seller. So the effects of link-based methods are limited.

Click-based methods, like Liu et al. [5] use the frequency of users visiting Web pages, Richardson et al. [6] use users' browsing graph to predict the importance of pages. These click-based methods usually require users to install toolbars and need users' permission to use browse behavior data, and the data contain much noise. Link-based methods and click-based methods both do not directly use the features of independent pages.

Content-based approaches use statistical data of the content (including texts and HTML tags) as features. Information-to-noise ratio (ITN) is used in [7] to measure the content quality. Measures about words and anchor text are used in [8]. Bendersky et al. [9] also use statistics of words and measures about terms, URLs, titles, anchor texts and tables. Agichtein et al. [10] use linguistic features, and other semantic features like out-of-vocabulary words, spelling mistakes, average number of syllables per word, the entropy of word lengths. However, the semantic features mentioned above are different from the semantic features in our work. They cannot actually reflect the quality of the content, for they don't understand what the pages are describing. There're also some researchers paying attention to a specific dimension of quality, such as visual quality [11] [12] and the appropriation of language [13]. Some hybrid approaches use both PageRank value and statistics of HTML document [14].

These methods only consider the surface features of the html content, without understanding what the text is about, so they cannot precisely predict the user's satisfaction about the information contained in pages. Different from previous work, our method directly analyzes the semantic features of the text content according to the characteristics of product pages.

3 Framework

The pages in e-commerce websites have many common characteristics, and these characteristics inspire us to explore novel features to evaluate pages' quality:

1. They have similar subsections, for example, sections of product description, brand and manufacturer introduction, sales promotion activities, user's guide(payment, delivery, after-sale service), and links to other products. In this paper, we use "semantic block" to name a group of sections that describe the same aspect. These semantic blocks have different importances. A page that contains little product descriptions with massive ads tends to annoy a user. Unlike the concept of "information-to-noise ratio", some information is not noise, but also not that important and should be limited in proper lengths, such as delivery or manufacturer introduction. So a page of high quality should have proper lengths of different semantic blocks.
2. They have descriptions about the properties of a product, and products of the same kind have similar properties. Different properties also have different importance. For example, a page that introduces a sun cream should contain descriptions of "SPF" (Sun Protection Factor), brand, price, production place, among which "SPF" is a very important property for a sun cream. A page of high quality should have complete description about properties, especially these important properties.

Knowing these characteristics above, we can extract new features by analyzing pages' semantic blocks and properties, and combine them with the traditional features in literature to classify pages into good and bad ones. We present the proposed framework which consists of the following steps depicted in Fig.1.

1. **Feature Generation**
   The features used in this paper include our semantic-block level features and property level, and other features chosen from conventional features.
   (a) **Semantic block recognition:** As mentioned above, different semantic blocks have different importance, so we choose the lengths of different semantic blocks as indicators of pages' quality. We split each page into several blocks and assign semantic labels to each block, thus recognize the semantic blocks in the pages. Then we use the length of each type of semantic block(blocks that have same labels) as a feature. The details of semantic block recognition will be addressed in Section 4.
   (b) **Property completeness analysis:** First, we recognize the properties of products in the pages with the help of ontology. Then we give each property a weight to indicate its importance, and then we use the weighted sum of properties as the feature. The weights of different properties can be learned from the labeled pages of good quality. The details of property completeness analysis will be addressed in Section 5.

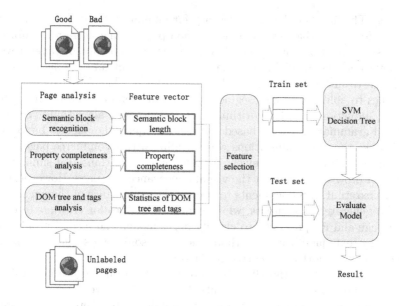

Fig. 1. Framework for page quality classification

(c) **DOM tree and tags analysis:** We summarize the DOM tree and tags related features in literature, and extract these features as parts of our initial features. The features we used will be introduced in the experiment part.

2. **Feature selection**

The features mentioned in step 1 are chosen according to the context of e-commerce by intuition, so maybe not all of these features are effective for the classification. We employ a feature subset selection process to select the subset of features from the candidate features that together have good predictive power. We use best-first searching method to search the space of all possible variable subsets[19], and evaluate the worth of a subset of features using Correlation Feature Selection(CFS)[20], and subsets of features that are highly correlated with the class while having low inter-correlation are preferred.

3. **Training and prediction**

We apply several classification methods, such as SVM, Decision Tree to train and classify pages, and conduct 5-fold cross-validation to avoid over-fitting problem.

4   Semantic Block Recognition

To extract the semantic-block level features we want, we must analyze the pages' semantic structure, and this includes two processes: splitting each page into blocks and assigning a label to each block with indicates what kind of semantic block it is. The first problem is quite easy for human, but difficult for machine, especially for pages that are designed by non-professional users and published without strict review. Sometimes they put a single sentence or several words in a single line and in different style to stress without any regulation, so it's not enough to find boundaries simply by html labels. There are many existing approaches to split pages into segments, including template based algorithms, grammar induction based algorithms and vision-based algorithms. Template based and grammar induction based algorithms require the page to be formally organized, so they have limitations when dealing with e-commerce pages which are published without strict review. Vision-based algorithms [15] simulate the behavior of browsers to present the visual appearance of a page. However, many pages use external CSS documents to control the appearance of the page, or use scripts to generate the contents, which brings difficulties for crawlers to obtain the full page and for programs to classify pages alone. The second problem can be solved by text classification methods. However, some text segments separated by labels are too short to be correctly classified.

In this paper, taking the efficiency into consideration, we use a simple but effective method to segment the semantic blocks by iteratively using text classification stages and merging stages. First, we divide the text into initial segments according to the DOM tree structure. For the segments that are too short to recognize, we use their context(former and latter segments) to improve the accuracy of recognition. Our method is easy to understand and the performance is good enough for our situation. What's more, this method has no strict requirement of the HTML documents, which is very important to analyze the pages generated by non-professional users.

Let B : $(B_0, B_1, \ldots B_n)$ be the set of semantic blocks that may appear in product pages. Our recognition method is performed as follows:

1. First, we divide the text into initial segments S: S : $(S_0, S_1, \ldots, S_n)$ according to the DOM tree structure. These segments are the smallest text nodes in the DOM tree, as shown in Fig.2. And We presume that they are the smallest independent units and each belongs to only one semantic block, which can avoid further splitting process. For the segments which are too short to recognize, we use the context to improve the accuracy.

2. For each segment S_i, calculate its probability to be assigned to semantic block B_j, $P(B_j|S_i)$. $P(B_j|S_i)$ is calculated using Bayes method.

3. For each segment S_i, find Pmax (S_i):

$$\text{Pmax}(S_i) = \arg\max_{B_j \in B} (P(B_j|S_i)) \tag{1}$$

where Pmax(S_i) denotes the max probability that segment S_i can be recognized. When Pmax(S_i) is larger than the given threshold t, S_i is labeled as the corresponding B_j, otherwise, consider merging S_i with S_{i-1}, S_{i+1} separately, and calculate Pmax $(S_{i-1} + S_i)$, Pmax $(S_i + S_{i+1})$, dealing as follows:
   (a) Pmax $(S_{i-1} + S_i)$ >   Pmax $(S_i + S_{i+1})$, and Pmax $(S_{i-1} + S_i)$ >   Pmax (S_i), merge s_i and s_{i-1}, label them as the corresponding B_j.

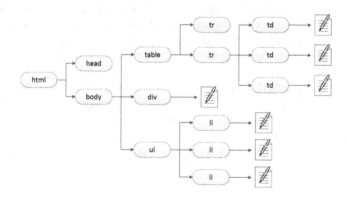

Fig. 2. Smallest text nodes in DOM tree

(b) $\text{Pmax}(S_i + S_{i+1}) > \text{Pmax}(S_{i-1} + S_i)$, and $\text{Pmax}(S_i + S_{i+1}) > \text{Pmax}(S_i)$, merge s_i and s_{i+1}, label them as the corresponding B_j.

(c) $\text{Pmax}(S_{i-1} + S_i) < \text{Pmax}(S_i)$, and $\text{Pmax}(S_i + S_{i+1}) < \text{Pmax}(S_i)$, label S_i as NULL.

After all the segments are labeled, we calculate the total length of each semantic block:

$$\text{tLen}(B_j) = \sum_{S_i \to B_j} \text{len}(s_i) \tag{2}$$

Where len(S_i) denotes the text length of segment S_i. Finally, we have the feature vector: $(\text{tLen}(B_0), \text{tLen}(B_1), \dots \text{tLen}(B_n))$, which can indicate the quality of a page's semantic structure appropriateness.

5 Property Completeness Analysis

To analyze property completeness, we first recognize the pre-specified types of properties in the texts. In this paper, we use both wrapper-based method and ontology-based method to perform the task. Wrapper-based method is suitable for the content generated automatically from internal databases and organized as tables or lists. Ontology-based method is suitable for those free style texts. These two methods can be complementary to each other. The ontology-based method can discover the properties in free style texts that cannot be processed by the wrapper-based method. While the construction of ontology is time-consuming and cannot cover all properties and new properties, the properties extracted by wrapper-based method can be used to automatically construct the ontology.

We manually write a program to extract properties for specific websites such as taobao.com(the largest e-commerce website in China). On the other hand, we construct an ontology about products, and use GATE[16] to perform the task of property recognition from free style texts according to the ontology. Gate is an open source software capable of solving almost any text processing problem. ANNIE[21] is one of many Information Extraction systems that have been developed using GATE. We use these tools to extract the properties in product pages, then we have the sets of properties:Attr $(a_1, a_2, \ldots a_n)$.

After the recognition process, we assign weights for each property, because different properties have different importances. We use the high quality pages in our labeled dataset(201 good pages in total) to learn the weights for each property. We use $W(w_1, w_2, \ldots, w_n)$ to represent the weights for properties, and count (a_i) for the times that property a_i appears in the high quality pages. We calculate the weights as follows:

$$w_i = \frac{count(a_i)}{\sum_{a_i \in Attr} count(a_i)} \tag{3}$$

Then we get the completeness of properties:

$$C = W \cdot Attr \tag{4}$$

6   Experiment

6.1   Preparation

As far as we know, we are the first to focus on the page quality of e-commerce. Existing datasets are not suitable for our experiments. So we constructed our own dataset to evaluate the performance of our method and explore the effects of our proposed features. We gathered 409 product pages of "sun cream" from taobao.com. We invited 3 users to evaluate these pages, and consider a page as good if more than two of three users think it's good. Finally we have 201 pages labeled as "good" and 208 pages labeled as "bad".

We define 5 semantic blocks in our experiment: product description, brand and manufacturer introduction, sales promotion activities, user's guide and ads. We also extracted the texts from these pages and labeled them with those semantic blocks. A part of the labeled texts are used as training corpus of text classification, and the rest are used to evaluate the performance of semantic block recognition. We also analysed the proposed features in good and bad pages, and found that the lengths of semantic blocks like product description, user's guide are usually larger in good pages than in bad ones, as shown in Fig.3.

6.2   Features

The features are organized into 3 categories. The first category includes traditional features chosen from features in literature according to the context of e-commerce. These features are listed in Table 1. The second category includes semantic-block level features proposed in this paper, namely the lengths of "product description", "brand and manufacturer introduction", "sales promotion activities", "user's guide" and "ads". The third category is property completeness - another feature proposed in this paper. In the experimental stage, a further feature selection step will be performed to find the best features. The performances of using different combinations of the three categories are compared later.

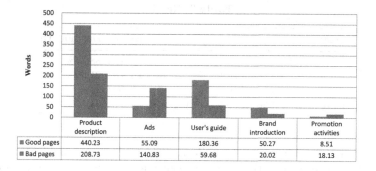

Fig. 3. Comparisons between good and bad pages

Table 1. Traditional features

Category	Features
URL	Length of URL.
Document	File size, text length, information-to-noise ratio.
DOM tree	Number of DOM elements.
Tables	Number of tables; average, median deviation of tr per table.
Lists	Number of lists; average, median deviation of li per list.
Colors	Number of unique colors.
Words	Number of words, unique words, stop-words, sentence markers.
Images	Number of images.
Relations	Number of out-links to file size, number of images to DOM elements, number of images to text length, number of words to DOM elements.

6.3 Results and Analysis

We use the Java API of Weka [17] to test these methods. Weka is a collection of machine learning algorithms for data mining tasks. We use SMO and J48 as the classifier separately, and compare the performances of using different combinations of features. We use the traditional features alone as the first group, then we add the two kinds of proposed features separately to the traditional features as the second group and third group, finally we add both of proposed features to the traditional features as the fourth group. A feature selection step is conducted on each combination of features before used to train and classify. All experiments are conducted by using 10 times of 5-fold cross-validation to avoid over-fitting problem, and the average value of the 10 times' accuracies as the final results. The results are shown in Table 2.

We can see that the accuracy of using traditional features alone is the lowest on both classifiers. Both of semantic-block level features and property-level feature can greatly enhance the accuracy. Generally speaking, it is better to use fewer features while the accuracy doesn't drop for better efficiency. The time consumption of different combinations without feature selection are almost the same because our proposed features are only a small part of the entire features.

Table 2. Results

Features	SVM	Decision Tree
Category 1	84.90%	88.37%
Category 1+ Category 2	85.82%	91.69%
Category 1+ Category 3	86.06%	91.20%
Category 1+ Category 2+ Category 3	86.55%	91.93%

It takes around 20 milliseconds to build models on different combinations of features without feature selection, while it drops down to less than 10 milliseconds using selected features. Here we list the best features selected from the fourth group (which includes the entirely features),and we can see that our proposed features are both in the selected features.

- Number of DOM elements.
- Number of unique colors.
- Number of sentence markers.
- Number of images.
- Number of images to text length.
- Number of words to DOM elements.
- Lengths of each semantic block: product description
- Weighted sum of properties

7   Conclusion

In this paper, we discuss the characteristics of pages on e-commerce websites, and propose to use the semantic-block level and property-level features in the evaluation of web pages. The lengths of different semantic blocks indicate the appropriateness of information composition, and the property-level features indicate whether the content of the page can satisfy the needs of users. These features can well model the knowledge in a page. Other features in literature focus on tags and DOM tree of HTML indicate other facets such as visual quality. Combining our features and these HTML features can evaluate the quality more comprehensively. Experimental results show that the new features can greatly enhance the accuracy of classification. Our method can also be applied to other domain where pages have common semantic blocks and properties such as music and movies. However, there are also some problems in our approach to be solved in the future:

1. We only manually labeled 409 pages for product "sun cream" to explore the effectiveness of proposed features, so there are still no sufficient labeled pages to handle other kinds of products. In the future work, we can develop an application for real world user, where users can both get score and suggestions about how to improve their pages, and users can submit their ratings and labels for pages.

2. In our experiments, the number of good and bad pages are almost equal, but in real word, the distribution of good and bad pages are different when different users label them, and we face the imbalance issue. In future work, we can explore more when the data is imbalanced and seeking for more personalized techniques.
3. We mainly focus on the problem of page quality evaluation. In future work, we can evaluate our methods in the frame of a search engine. Quantitative values instead of good and bad labels are needed to integrate with the ranking methods of search engines.

Acknowledgment. This work was supported in part by a grant from Specialized Research Fund for the Doctoral Program of Higher Education of China (No.20110101110065), National key Technology Research and Development Program of China (2011BAG05B04), and Zhejiang Key Science and Technology Innovation Team Plan of China (No. 2009R50015).

References

1. Batini, C., Cappiello, C., Francalanci, C., Maurino, A.: Methodologies for data quality assessment and improvement. ACM Computing Surveys (CSUR) 41, 1–52 (2009)
2. Brin, S., Page, L.: The anatomy of a large-scale hypertextual Web search engine. Computer Networks and ISDN Systems 30, 107–117 (1998)
3. Kleinberg, J.M.: Authoritative sources in a hyperlinked environment. J. ACM 46, 604–632 (1999)
4. Lempel, R., Moran, S.: SALSA: the stochastic approach for link-structure analysis. ACM Trans. Inf. Syst. 19, 131–160 (2001)
5. Liu, Y., Gao, B., Liu, T., Zhang, Y., Ma, Z., He, S., Li, H.: BrowseRank: letting web users vote for page importance. In: Proceedings of the 31st Annual International ACM SIGIR Conference on Research and Development in Information Retrieval, pp. 451–458. ACM, Singapore (2008)
6. Richardson, M., Prakash, A., Brill, E.: Beyond PageRank: machine learning for static ranking. In: Proceedings of the 15th International Conference on World Wide Web, pp. 707–715. ACM, Edinburgh (2006)
7. Zhu, X., Gauch, S.: Incorporating quality metrics in centralized/distributed information retrieval on the World Wide Web. In: Proceedings of the 23rd Annual International ACM SIGIR Conference on Research and Development in Information Retrieval, pp. 288–295. ACM, Athens (2000)
8. Ntoulas, A., Najork, M., Manasse, M., Fetterly, D.: Detecting spam web pages through content analysis. In: Proceedings of the 15th International Conference on World Wide Web, pp. 83–92. ACM, Edinburgh (2006)
9. Bendersky, M., Croft, W.B., Diao, Y.: Quality-biased ranking of web documents. In: Proceedings of the Fourth ACM International Conference on Web Search and Data Mining, pp. 95–104. ACM, Hong Kong (2011)
10. Agichtein, E., Castillo, C., Donato, D., Gionis, A., Mishne, G.: Finding high-quality content in social media. In: Proceedings of the 2008 International Conference on Web Search and Data Mining, pp. 183–194. ACM, Palo Alto (2008)

11. Wu, O., Chen, Y., Li, B., Hu, W.: Learning to evaluate the visual quality of web pages. In: Proceedings of the 19th International Conference on World Wide Web, pp. 1205–1206. ACM, Raleigh (2010)
12. Wu, O., Chen, Y., Li, B., Hu, W.: Evaluating the visual quality of web pages using a computational aesthetic approach. In: Proceedings of the Fourth ACM International Conference on Web Search and Data Mining, pp. 337–346. ACM, Hong Kong (2011)
13. Pun, J.C.C., Lochovsky, F.H.: Ranking search results by web quality dimensions. Journal of Web Engineering 3, 216–235 (2004)
14. Mandl, T.: Implementation and evaluation of a quality-based search engine. In: Proceedings of the Seventeenth Conference on Hypertext and Hypermedia, pp. 73–84. ACM, Odense (2006)
15. Cai, D., Yu, S., Wen, J., Ma, W.: VIPS: a vision-based page segmentation algorithm. Microsoft Technical Report, MSR-TR-2003-79 (2003)
16. Cunningham, H., Maynard, D., Bontcheva, K., Tablan, V.: A Framework and Graphical Development Environment for Robust NLP Tools and Applications. In: Proceedings of the 40th Anniversary Meeting of the Association for Computational Linguistics (ACL 2002) (2002)
17. Cen, R., Liu, Y., Zhang, M., Ru, L., Ma, S.: Web page quality estimation based on linear discriminant function. Journal of Computational Information Systems 3, 1117–1126 (2007)
18. Hall, M., Frank, E., Holmes, G., Pfahringer, B., Reutemann, P., Witten, I.H.: The WEKA data mining software: an update. SIGKDD Explor. Newsl. 11, 10–18 (2009)
19. Guyon, I., Andr, E.: An introduction to variable and feature selection. J. Mach. Learn. Res. 3, 1157–1182 (2003)
20. Hall, M.A.: Correlation-based Feature Selection for Discrete and Numeric Class Machine Learning. In: Proceedings of the Seventeenth International Conference on Machine Learning, pp. 359–366. Morgan Kaufmann Publishers Inc. (2000)
21. Cunningham, H., Maynard, D., Bontcheva, K., Tablan, V.: GATE: A framework and graphical development environment for robust NLP tools and applications. In: Proceedings of the 40th Anniversary Meeting of the Association for Computational Linguistics (2002)

Hierarchical Classification for Solving Multi-class Problems: A New Approach Using Naive Bayesian Classification

Esra'a Alshdaifat, Frans Coenen, and Keith Dures

University of Liverpool, Department of Computer Science, United Kingdom
{esraa,coenen,dures}@liv.ac.uk

Abstract. A hierarchical classification ensemble methodology is proposed as a solution to the multi-class classification problem where the output from a collection of classifiers, arranged in a hierarchical manner, are combined to produce a better composite global classification (better than when the classifiers making up the ensemble operate in isolation). A novel topology for arranging the classifiers in the hierarchy is proposed such that the leaf classifiers act as binary classifiers and the remaining classifiers (those at the root and intermediate nodes) address groupings of classes. The main challenge is how to address the general drawback of the hierarchical model, that is if a record is miss-classified early on in the classification process (near the root of the hierarchy) it will continue to be miss-classified at deeper levels too. Three different approaches, founded on Naive Bayes classification, are proposed whereby Bayesian probability values are used to indicate whether single or multiple paths should be followed within the hierarchy. Reported experimental results demonstrate that the proposed mechanism can improve classification performance, in terms of average AUC, in the context of selected data sets.

Keywords: Hierarchical classification, multi-class classification, ensemble classification.

1 Introduction

Classification is a well established element of machine learning. Classification can be categorised according to: (i) the size N of the set of classes from which class labels may be drawn ($N = 2$ for binary classification and $N > 2$ for multiclass classification) and (ii) the number of class labels that may be assigned to a single record (single-label classification or multi-label classification). The work described in this paper is directed at multi-class single-label classification, especially where N is large.

Many real-world learning problems are multi-class. The problem with multiclass classification is that when the number of classes is large the effectiveness of the classification tends to diminish. The use of an ensemble is one methodology used to solve multi-class classification problems where a set of classification models is used (as opposed to a single model) in order to obtain a better composite

H. Motoda et al. (Eds.): ADMA 2013, Part I, LNAI 8346, pp. 493–504, 2013.

global model, with more accurate and reliable classifications than obtained from using a single model [9]. The classifiers making up an ensemble can be arranged in serial, parallel or grid form. There has been much published research work with respect to serial (for example "boosting" [10]) and parallel (for example "Bagging" [2]) ensemble forms, but not on grid ensembles.

The solution proposed in this paper is to adopt ensemble classifiers that feature a hierarchy of classifiers. Classifiers at the leaves of the hierarchy feature binary classifiers, while classifiers at nodes further up the hierarchy feature classifiers directed at groupings of classes. The motivation for using ensemble classifiers arranged in a hierarchical form so as to improve the accuracy of the classification were thus: (i) the established observation that ensemble methods tend to improve classification performance [14], and (ii) dealing with smaller subsets of class labels at each node might produce better results. The research challenges are then: (i) how best to organise (group) the class labels so as to produce a hierarchy that generates the most effective classification, and (ii) how to deal with the issue that if a record is miss-classified early on in the process (near the root of the hierarchy) it will continue to be miss-classified at deeper levels of the hierarchy, regardless of the classifications proposed at lower level nodes and the final leaf nodes.

To address the first issue several techniques are proposed in this paper to organise (group) the class labels so as to produce a hierarchy that generates an effective classification. These are founded on ideas concerned with the use of clustering and splitting mechanisms to distribute the class labels. With respect to the second issue a number of alternatives are considered using Naive Bayes classifiers where by the Bayesian probability values are used to determine whether a single branch or multiple branches in the hierarchy should be followed, coupled with two alternatives for arriving at a final decision either according to some accumulated probability value, or by simply choosing the most probable outcome.

The rest of this paper is organised as follows. Section 2 gives a review of related work on multi-class classification. Section 3 describes the proposed hierarchical classification approach. Section 4 presents an evaluation of the proposed hierarchical classification mechanism as applied to a range of different data sets. Section 5 summarises the work and indicates some future research directions.

2   Literature Review

This section provides a generic overview of multi-class classification. It is widely accepted that multi-class problems can be solved in three ways: (i) using stand-alone classification algorithms, (ii) using a "chain" of binary classifiers, and (ii) using ensemble classifiers arranged in some specific form.

Starting with the standard way of solving multi-class problems using a stand-alone learning algorithm such as: decision trees [17], neural network [21], Naive Bayes [16] or Classification Association Rule Mining (CARM) [5]. Among these Bayesian classification algorithms are of interest with respect to the work described in this paper because of the Bayesian probability values generated. Naive

Bayes classification assumes that the effect of an attribute value on a given class is independent of the values of other attributes. This is the well known "class conditional independence" assumption which is made to simplify the computations involved [14]. Naive Bayes is an extremely effective, but straightforward, form of classification. Consequently it is often used as a baseline standard by which other classifiers can be measured [11]. Various comparative studies with respect to decision tree and neural network classifiers have found the operation of Naive Bayes to be comparable [7,15,3]. Bayesian classifiers have also exhibited high accuracy and speed when applied to large databases [14].

In the context of using chains of binary classifiers to solve multi-class problems; we can, from the literature, identify three commonly referenced methods: (i) One-Versus-All (OVA) [18] (ii) All-Verses-All (AVA) [20] (iii) Error-Correcting-Output-Codes (ECOC) [6]. Among these ECOC has often been found to be able to outperform the direct use of single multi-class learning algorithms [6,19].

A more specialised method directed at the solution of the multi-class classification problem is founded on the use of ensembles of classifiers arranged in some specific manner. Experimental studies have shown that ensemble learning can outperform the single classifier approach [8,12,13]. With reference to the literature, "base classifiers" within an ensemble model can be arranged into: (i) concurrent (parallel) [2], or (ii) sequential (serial) [10] form. In the context of parallel ensembles, "Bagging" [2] was the first effective method and remains one of the simplest ensemble strategies. Regarding sequential ensemble, "Boosting" [10] is the most widely used and most frequently sited method. In this paper we propose the concept of a hierarchy ensemble.

3   The Hierarchical Model Framework

This section is concerned with the creation of the suggested hierarchical model and its usage for predicting the class labels associated with new data. The section is organised as follows: Section 3.1 explains the generation of the hierarchical model, while Section 3.2 considers its operation.

3.1   Hierarchical Model Generation

The proposed hierarchical model adopts an ensemble approach founded on the idea of arranging the classifiers into a hierarchical form. Each node in our hierarchy holds a classifier. Classifiers at the leaves of our hierarchy feature binary classifiers, while classifiers at nodes further up the hierarchy feature classifiers directed at groupings of class labels. At the root we classify into two groups of class labels. At the next level we split into smaller groups, and so on till we reach classifiers that can associate single class labels with records.

Two types of classifier were considered with respect to the nodes within our hierarchy: (i) straight forward single "stand-alone" classifiers, and (ii) "bagging" ensembles (thus in effect we have ensembles of ensembles). With respect to the first approach a Naive Bayesian classifier was generated for each node in the hierarchy. With respect to bagging the data set D associated with each node was randomly divided into N disjoint partitions and a classifier generated for each (in the evaluation reported in Section 4, $N = 3$ was used).

In order to divide D during the hierarchy generation process, two different techniques were used: (i) k-means, and (ii) data splitting. Of these k-means is the most well-known and commonly used partitioning method where the input is divided into k partitions (in our model $k = 2$ was used because of the binary nature of our hierarchies). Data splitting comprises a simple "cut" of the data into two groups so that each contains a disjoint subset of the entire set of class labels.

Thus two classification styles and two types of mechanism for partitioning our data were used. Thus four different hierarchical model are considered:

1. K-means and Naive Bayes (K-means&N): The proposed approach using k-means $(k = 2)$ to group data with Naive Bayesian classifiers at each node.
2. Data splitting and Naive Bayes (DS&N): The proposed approach using data splitting to group data with Naive Bayesian classifiers at each node.
3. K-means and Bagging (K-means&B): The proposed approach using K-means $(k = 2)$ to group data with a bagging ensemble classifier at each node ($N = 3$).
4. Data splitting Bagging (DS&B): The proposed approach using data splitting to group data with a bagging ensemble classifier at each node ($N = 3$).

3.2   Hierarchical Model Classification

Section 3.1 explained the generation of the proposed hierarchical model. After the model has been generated, it can be used to classify new unseen data records. Depending on the resulting probability of the Naive Bayesian classifiers at each hierarchy node. Two different strategies are suggested: (i) Single Path and (ii) Multiple Path. In the first case the strategy is to select the class at the leaf node label by following a "single path" (the maximum probability path) within the hierarchy from the root node. However, this strategy does not address the issue that if a record is miss-classified early on in the process it will continue to be miss-classified later on in the process. The second strategy attempts to address this issue by allowing more than one path to be followed (depending on a predefined threshold σ ($0 \leq \sigma < 1$), and consequently being able to select a class label from one or more candidates. In the following two subsections the process whereby this may be achieved is explained.

Single Path. In the Single Path strategy only one path will be followed according to the classification at each hierarchy node. More specifically, the probabilities associated with class groups are used to identify a "maximum probability path". As noted earlier, during the generation process, class labels are grouped. Given the binary nature of our hierarchies, we refer to the groups associated with a node N as the left and right groups ($N.leftClassGroup$ and $N.rightClassGroup$). The Single Path algorithm is presented in Algorithm 1. The algorithm operates as follows. On each recursion the algorithm is called with two parameters: R, the record to be classified, and N a reference to the current node location in the hierarchy (at the start this will be the root node). How the process proceeds thus depends on the nature of the groups of classes or single class labels identified by the Bayesian classifier at each current node. If a class group is returned then one or other of the branches (left or right) will be followed as indicated by the associated class group probability values. If specific class label are indicated (as opposed to some grouping of labels) the class label with the highest Bayesian probability will be returned as the label to be associated with the given record R and the algorithm terminated.

Algorithm 1. SinglePath

```
INPUT :
1. R a new unseen record.
2. N a pointer to the current node in the hierarchy (root node at start).
OUTPUT :
c the predicted class label of the input record R.

C = Class label set for R with the associated probabilities generated using classifier held
at node N  (C = {N.leftClassGroup, N.rightClassGroup})

If  P(N.leftClassGroup) > P(N.rightClassGroup) Then
       If |N.leftClassGroup| = 1  Then
           Return class label c (c in N.leftClassGroup)
       Else
           Return SinglePath(R,N.leftBranch)
       EndIf
Else
       If |N.rightClassGroup| = 1 Then
           Return class label c (c in N.rightClassGroup)
       Else
           Return SinglePath(R,N.rightBranch)
       EndIf
EndIf
```

Multiple Path. Using the Multiple Path strategy more than one path may be followed as a result of the classification conducted at each current hierarchy node. More specifically the Bayesian probability P associated with individual class groups ($P_{N.leftClassGroup}$ and $P_{N.rightClassGroup}$) will be used to dictate whether one or two paths (because of the binary nature of our hierarchy) will be followed according to the predefined threshold sigma (σ). If $P_{N.leftClassGroup} > \sigma$ and $P_{N.rightClassGroup} > \sigma$ then both branches will be explored, otherwise the branch with the highest associated P value will be selected.

Two different approaches to determining the final resulting class label are also suggested: (i) best probability and (ii) accumulated best probability. Using the best probability approach the individual Bayesian probabilities associate with the identified classes at the leaf nodes are used to select a final label. Using the accumulated best probability approach the accumulated Bayesian probabilities associate with the paths that have been followed are used to select a final label.

The Multiple Path best probability algorithm is presented in Algorithm 2. The algorithm is similar to Algorithm 1; the main difference is the use of the threshold sigma (σ) to decide whether we wish to proceed down a single path or down both paths emanating from a node. The distinction between the Multiple Path best probability algorithm and the Multiple Path best accumulated probability algorithm (not presented here because of space limitations) is that accumulated probability counts are maintained for each path followed as the algorithm proceeds.

Algorithm 2. MultiPathBestProb

```
INPUT :
1. R a new unseen record.
2. N a pointer to the current node in the hierarchy (root node at start).
OUTPUT :
c the predicted class label of the input record R.

C = Class label set for R with the associated probabilities generated using classifier held
at node N  (C = {N.leftClassGroup, N.rightClassGroup})

If (P(N.leftClassGroup) > sigma and P(N.rightClassGroup) > sigma) Then
       If (|N.leftClassGroup| = 1) Then
           Add class label ci (c in N.leftClassGroup) to class list L with prob. P(ci)
       Else
           MultiPathBestPro(R,N.leftBranch)
       EndIf
       If (|N.rightClassGroup| = 1) Then
           Add class label ci (c in N.rightClassGroup) to class list L with prob. P(ci)
       Else
           MultiPathBestPro(R,N.rightBranch)
       EndIf
Else
       If (P(N.leftClassGroup) > P(N.rightClassGroup)) Then
           If (|N.leftClassGroup| = 1) Then
               Add class label ci (c in N.leftClassGroup) to class list L with prob. P(ci)
           Else
               MultiPathBestProb(R,N.leftBranch)
           EndIf
       Else
           If (|N.rightClassGroup| = 1) Then
               Add class label ci (c in N.rightClassGroup) to class list L with prob. P(ci)
           Else
               MultiPathBestProb(R,N.rightBranch)
           EndIf
       EndIf
EndIf
Process L and select class label c with highest probability
```

4   Experiments and Evaluation

In this section we present an overview of the adopted experimental set up and the evaluation results obtained. The experiments used fourteen different data sets (with different numbers of class labels) taken from the UCI data repository [1], which were processed using the LUCS-KDD-DN software [4]. Ten-fold Cross Validation (TCV) was used throughout. The evaluation measure used was average AUC (Area Under the receiver operating Curve).

In addition to the four techniques considered for hierarchical model generation (K-mean&N, DS&N, K-mean&B and DS&B) as described in section 3.1, our experiments included a stand-alone classification of the data using a single Naive Bayesian classifier (no hierarchy) and classification of the data using a bagging ensemble classifier comprised of three classifiers (again no hierarchy). The results are presented in the following sections as the follows: Section 4.1 considers the results using the Single Path strategy, Section 4.2 presents the results of using the Multiple Path strategy with the best probability approach, and Section 4.3 the Multiple Path strategy with the best accumulated probability approach. Section 4.4 then provides a comparison and discussion of the results.

4.1   Single Path Strategy

This section presents the results obtained using the Single Path strategy with respect to the fourteen different data sets and the six alternative classification methods considered in this evaluation. The results in terms of average AUC are presented in Figure 1 in the form of a collection of fourteen histograms, one for each data set considered. The best method is indicated to the top right of each histogram (in some cases two or three best methods were identified). From the figure it can be observed that by using the Single Path strategy the proposed hierarchical techniques can significantly improve the classification AUC with respect to four of the fourteen data sets considered (Nursery, Heart, Pen Digits and Chess KRvK), although in one case the AUC result was the same as that produced using stand-alone Naive Bayes as well as bagging. In the remaining ten cases the stand-alone Naive Bayesian classifiers and/or bagging produced the best result.

4.2   Multiple Path Strategy with Best Probability Approach

With respect to the Multiple Path strategy a very low threshold of $\sigma = 0.01$ was used. In other words the only time a path is not followed is when it is almost certain that it will not result in a satisfactory classification. In earlier experiments, not reported here, a range of alternative σ values were tried and it was found that $\sigma = 0.01$ produced the best performance. The results using the Multiple Path strategy, in terms of average AUC values, are presented in Figure 2, again the best method(s) with respect to each dataset is highlighted to the top right of each histogram. From the figure it can be observed that the proposed Multiple Path strategy (coupled with the best probability approach) produced the best AUC results with respect to five of the fourteen data sets considered (Wine, Nursery, Heart, Glass and Chess KRvK), although in two cases the same AUC result was produced using stand-alone Naive Bayes. Three of the "best" data sets also performed well with respect to the Single Path strategy. In the remaining nine cases out of the fourteen Naive Bayes and/or Bagging produced the best AUC result.

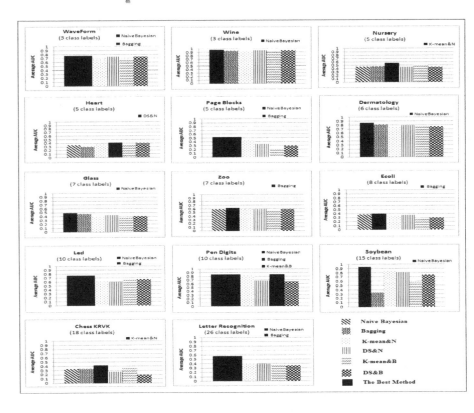

Fig. 1. Average AUC for the fourteen different data sets using the Single Path strategy

4.3   Multiple Path Strategy with Accumulated Best Probability Approach

In the case of the Multiple Path strategy coupled with the accumulated best probability approach $\sigma = 0.01$ was again used. The results in terms of average AUC are presented in Figure 3 (again the best method(s) with respect to each dataset are listed to the top right of each histogram). From the figure it can be observed that the proposed hierarchical techniques produced the best performance with respect to five of the fourteen data sets considered (Wine, Nursery, Heart, Glass and Chess KRvK) although in one case the same AUC result was produced as in the case of using stand-alone Naive Bayes. The five data sets that operated well with respect to the accumulated best probability approach were the same as those that performed well with respect to the best probability approach.

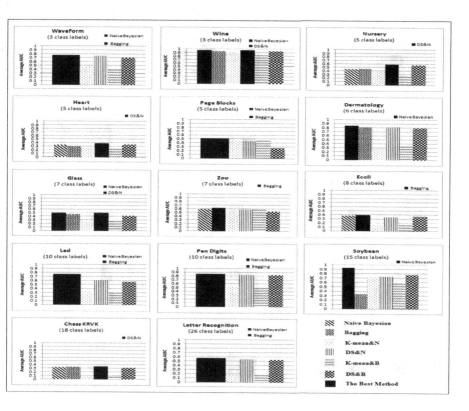

Fig. 2. Average AUC for the fourteen different data sets using the Multiple Path strategy with the best probability approach

4.4 Comparison between Single and Multiple Path Strategies

This section provides a comparison between the Single Path and Multiple Path strategies with respect to the four hierarchical model generation methods (K-mean&N, DS&N, K-mean&B, and DS&B). Figure 4 presents a set of histograms showing the average AUC values obtained in each case. From the figure it is interesting to note that K-means class distribution across nodes (K-mean&N and K-mean&B) worked best using the Single Path Strategy while the data splitting class distribution across nodes (DS&N and DS&B) worked best using the Multiple Path strategy. In the context of the Multiple Path strategy the accumulated probability approach tended to produce the best performance. The best overall result was obtained using the Multiple path strategy coupled with the best probability approach, data splitting and Naive Bayes classification.

Considering the number of class labels associated with each of the evaluation data sets it is interesting to note that our proposed techniques produced the best results for datasets with high numbers of class labels (18 in the case of the Chess

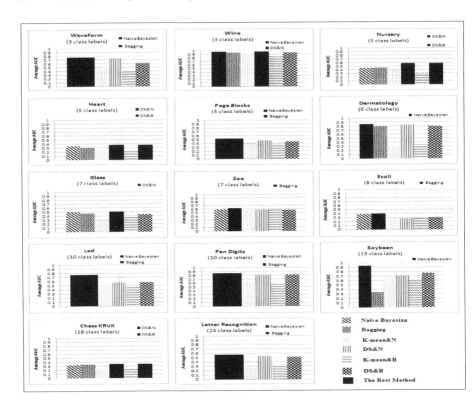

Fig. 3. Average AUC for the fourteen different data sets using the Multiple Path strategy with the best accumulated probability approach

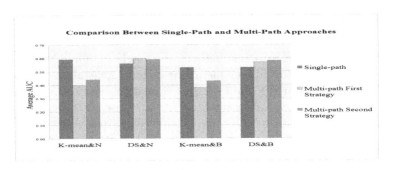

Fig. 4. Comparison between the suggested hierarchical model techniques

KRvK dataset). The reason for this is that a high number of class labels allow for the generation of a sophisticated hierarchy, while a low number of classes do not. In the case of a two class label data set the proposed hierarchical approach will generate a three node hierarchy with a classifier at the root node and class labels at the remaining two leaf nodes. In other words, given a two class label data set, no benefit can be gained from using our proposed approach. Thus the proposed approach is best suited to situations that feature a large number of class labels.

5 Conclusion and Future Work

This paper has presented a novel ensemble based techniques to achieve multi-class classification using the concept of classification hierarchies. To generate such a hierarchical ensemble two different grouping techniques were considered: (i) k-means and (ii) data splitting. We also considered two different styles of classifier at each node: (i) Naive Bayes and (ii) bagging ensemble classifiers; and two alternative classification strategies: Single Path and Multiple Path. The latter was proposed to address the hierarchical drawback that if a record is miss-classified early on in the process (near the root of the hierarchy) there is no opportunity for recovery. Two alternative approaches to determining the final classification in the case of the Multiple Path strategy were also considered: best probability and accumulated best probability.

From the reported evaluation it was noted that the proposed hierarchical classification model could be successfully used to classify data in a more effective manner than when stand-alone classifiers or even other ensemble classifiers like bagging were used in the context of some data sets, especially data sets that featured a high number of class labels. The evaluation also indicated that following more than one path in the hierarchy tended to produce a better overall performance. Best results were produced using data splitting and naive Bayesian classification using a Multiple Path strategy coupled with the best probability approach. An issue with Multiple Path strategy when coupled with the accumulated best probability approach is that the accumulated values tend to be biased according to the length of the path. To address this issue the use of other tree structures, than the binary tree structures used in this paper, is suggested. For example a Directed Acyclic Graph (DAG) structure might be adopted instead of a tree hierarchy structure so as to include a greater number of possible class label combinations at each level. In the future, we also intend to consider using Classification Association Rule Mining (CARM) at the individual hierarchy nodes; the advantage offered is that *confidence values* can be used to influence paths selection. The operation of such a system can then be compared against the Naive Bayes hierarchical model presented here.

References

1. Bache, K., Lichman, M.: UCI machine learning repository (2013), http://archive.ics.uci.edu/ml
2. Breiman, L.: Bagging predictors. Machine Learning 24(2), 123–140 (1996)
3. Cestnik, B.: Estimating probabilities: A crucial task in machine learning. In: Proceedings of the Ninth European Conference on Artificial Intelligence, pp. 147–149. Pitman, Stockholm (1990)

4. Coenen, F.: The LUCS-KDD discretised/normalised arm and carm data library (2003), http://www.csc.liv.ac.uk/~frans/KDD/Software/LUCS_KDD_DN
5. Coenen, F., Leng, P.: The effect of threshold values on association rule based classification accuracy. Journal of Data and Knowledge Engineering 60(2), 345–360 (2007)
6. Dietterich, T.G., Bakiri, G.: Solving multiclass learning problems via error-correcting output codes. JAIR (1995)
7. Domingos, P., Pazzani, M.: On the optimality of the simple bayesian classifier under zero-one loss. Mach. Learn. 29(2-3), 103–130 (1997), http://dx.doi.org/10.1023/A:1007413511361
8. Duin, R.P.W., Tax, D.M.J.: Experiments with classifier combining rules. In: Kittler, J., Roli, F. (eds.) MCS 2000. LNCS, vol. 1857, pp. 16–29. Springer, Heidelberg (2000)
9. Dunham, M.H.: Data Mining: Introductory and Advanced Topics. Prentice Hall (2003)
10. Freund, Y., Schapire, R., Abe, N.: A short introduction to boosting. Journal of Japanese Society for Artificial Intelligence 14(5), 771–780 (1999)
11. Gangrade, A., Patel, R.: Privacy preserving three-layer nave bayes classifier for vertically partitioned databases. Journal of Information and Computing Science 8(2), 119–129 (2013)
12. Giacinto, G., Roli, F.: Dynamic classifier selection. In: Kittler, J., Roli, F. (eds.) MCS 2000. LNCS, vol. 1857, pp. 177–189. Springer, Heidelberg (2000)
13. Grim, J., Kittler, J., Pudil, P., Somol, P.: Combining multiple classifiers in probabilistic neural networks. In: Kittler, J., Roli, F. (eds.) MCS 2000. LNCS, vol. 1857, pp. 157–166. Springer, Heidelberg (2000)
14. Jiawei, H., Micheline, K., Jian, P.: Data Mining: Concepts and Techniques. Morgan Kaufmann (2011)
15. Langley, P., Iba, W., Thompson, K.: An analysis of bayesian classifiers. In: Proceedings of the Tenth National Conference on Artificial Intelligence, pp. 223–228. MIT Press (1992)
16. Leonard, T., Hsu, J.S.: Bayesian Methods: An Analysis for Statisticians and Interdisciplinary Researchers. Cambridge University Press (2001)
17. Quinlan, J.R.: Induction of decision trees. Machine Learning 1(1), 81–106 (1986)
18. Rifkin, R.M., Klautau, A.: In defense of one-vs-all classification. Journal of Machine Learning Research 5, 101–141 (2004)
19. Schapire, R.E.: Using output codes to boost multiclass learning problems. In: Machine Learning: Proceedings of the Fourteenth International Conference (ICML 1997) (1997)
20. Tax, D.M.J., Duin, R.P.W.: Using two-class classifiers for multiclass classification. In: ICPR, vol. (2), pp. 124–127 (2002)
21. Zhang, G.P.: Neural networks for classification: A survey. IEEE Transactions on Systems, Man, and Cybernetics-Part C: Applications and Reviews 30(4), 451–462 (2000)

Predicting Features in Complex 3D Surfaces Using a Point Series Representation: A Case Study in Sheet Metal Forming

Subhieh El-Salhi, Frans Coenen, Clare Dixon, and Muhammad Sulaiman Khan

University of Liverpool, Department of Computer Science, United Kingdom
{hsselsal,cldixon,mskhan}@liv.ac.uk

Abstract. This paper presents an integrated framework for learning to predict geometry related features with respect to 3D surfaces. The idea is to use a training set of known prediction values to create a model founded on local 3D geometries associated with a given surfaces so that predictions with respect to a new "unseen" surfaces can be made. The local geometries are represented using point series curves. Two variations are proposed: (i) *discretised* and (ii) *real number*. To act as a focus for the work a sheet metal forming application is considered where we wish to predict the errors that are introduced as a result of applying a forming process. Given a desired surface T, the surface T' actually produced as a result of the sheet metal forming process is affected by a phenomena called *Springback* (the feature we wish to predict). The proposed process has been evaluated using two flat-topped pyramid shapes and by considering a variety of parameter settings. Excellent results have been obtained in terms of accuracy and Area Under ROC Curve (AUC).

Keywords: 3D Surface Representation, Point series curves, Classification.

1 Introduction

Pattern mining has long been studied within the field of knowledge discovery in data and machine learning. Pattern mining in the context of tabular data is well understood (see for example work on frequent item set mining). With respect to many other forms of more complex data, pattern mining remains a subject of research. The fundamental challenge is to represent such complex data in a way that facilitates meaningful pattern identification. The focus for the work described in this paper is three dimensional (3D) surfaces. More specifically we wish to identify patterns in such surfaces that are indicative of some set of feature values we wish to predict. The intention is that, once identified, these patterns can be used for prediction purposes with respect to new "unseen" 3D surfaces. The idea is to represent specific local geometries located within 3D surfaces so that we can attempt to associate specific feature values with these geometries. To this end a prediction framework is proposed whereby local geometries are represented in terms of linearisations to form a collection of point series (curves). Two variations of the linearisation are considered, a discretised linearisation and a real

H. Motoda et al. (Eds.): ADMA 2013, Part I, LNAI 8346, pp. 505–516, 2013.

valued linearisation, and a number of different local geometry "sizes". Using an appropriately defined training set prediction values, associated with some feature of interest, can be related to each linearisation. Given a new "unseen" 3D surface this can be decomposed into a collection of linearisations similar to those used with respect to the training set. By matching the new linearisations associated with the new surface to the previously generated linearisations, which have known feature values associated with them, these values can be used as predictor values with respect to the new shape.

The exemplar application, and that used to illustrate the work described in this paper, is sheet metal forming. In sheet metal forming the forming process takes as input a specification of a 3D surface to be manufactured, a shape T, to which a forming processes is applied to produce a shape T'. However, as a result of application of the process various deformations are introduced, called *springback*. As a consequence the manufactured shape T' is not equivalent to the specified shape T. Springback is caused by a number of factors of which the most significant is the geometry of the intended shape [2]. Further, springback is not distributed evenly over a given shape, in practice springback is more significant with respect to some geometries than others. Springback is therefore correlated to the nature of local geometries within the specified shape. If we can predict springback we can apply a correction to the specification T so as to minimise the springback effect.

Thus the contributions of this paper are as follows:

1. A method for representing local 3D geometries using point series.
2. A mechanism for feature value prediction with respect to local geometries present within 3D surfaces.
3. An interesting case study illustrating the significance of the proposed process.

The rest of the paper is organised as follows. In section 2 a brief overview of related work is presented. Section 3 introduces the prediction framework including the generation of the proposed point series representations. The evaluation of the proposed prediction framework, using manufactured surfaces describing flat topped pyramid shapes, is presented in Section 4 using a variety of parameters. Some conclusions are then presented in Section 5.

2   Overview of Related Work

The work described in this paper, although generally applicable, is directed at sheet metal forming as widely used in the aircraft and automotive parts manufacturing industries. A specific issue in sheet metal forming is the springback phenomenon, the elastic deformation that occurs as a result of the application of the manufacturing process. Generally speaking the nature of any resulting springback is influenced by: (i) the manufacturing parameters used and (ii) the material properties [6,15,13]. Substantial works have been conducted to characterise, analyse and predict the springback. To this end the Finite Element Method (FEM) has been extensively used for springback prediction purposes [4,14,19]. The FEM provides a simulation environment that is flexible (parameters can be easily modified) but at the same time complex. However, the application of FEM is a time consuming task [7,17,6]. Furthermore, FEM has been found to be an inaccurate prediction method due to the use of simplification assumptions with respect to the required integration calculation [3,4,15]. An alternative approach, and that advocated in this paper is to use some form of machine learning to build a predictor. More specifically to generate a generic classifier that takes as input definitions of local geometries contained on the part to be manufactured and predicts springback. To the best knowledge of the authors there is no reported work on the application of classification techniques for the purpose of predicting springback in sheet metal forming.

3 The Prediction Framework

The input to the proposed prediction framework is a grid describing some 3D shape of interest. Each grid square is defined by its centre point, the *grid point*, which is in turn defined by an *x*-*y* coordinate pair. Each grid point also has a *z* (height) value and, when used as a training set, a value associated with some feature of interest. In the case of the sheet metal forming application this will be a springback value. Thus the grid represents a mesh describing some 3D surface of interest. Using this input we wish to build a model of the 3D surface which can be used to predict values associated with new "unseen" 3D surfaces. Note that the number of grid squares required to represent a given shape will depend on the grid size *d*. Fewer grid squares will be generated if a larger *d* value is used.

Given a grid representation G, a collection of local geometries can be defined (one per grid square). Two alternative proposed point series representations are described in Sub-section 3.1 below. Once a collection of point series curves have been generated, each with a feature value of interest associated with it (a prediction value), this model can be used for prediction purposes. There are a number of mechanism where by this may be achieved; however, in the context if this paper a K-Nearest Neigbour (KNN) approach is advocated that uses the *warping distance* between local geometry curves to make predictions. This matching process is described further in Sub-section 3.2.

3.1 Point Series Representation

There are various ways that 3D local geometries can be defined. Earlier work by the authors considered a Local Binary Pattern (LBP) based representation referred to as the Local Geometry Matrix (LGM) representation [5,11,12]. In the context of sheet metal forming the authors have also considered a "distance from edge" measure (unpublished), referred to as the Local Distance Measure (LDM) representation. In this paper a point series representation is considered. The basic idea is to describe each local 3D geometry surrounding a grid point p_i using a linearisation of space. There are four issues to be considered in this respect:

1. What is the nature of the neighbourhood to be considered.
2. How best to conduct the desired linearisation.
3. How many points should we include in our point series.
4. What is the nature of the value represented by each point.

In terms of the nature of the neighbourhood this is partly related to the adopted grid size d, adoption of a large value for d will dictate larger grid squares and consequently larger neighbourhoods. The size of a neighbourhood can be simply described in terms of a Region Of Interest (ROI) surrounding each point p_i defined in terms of a $n \times n$ block of grid squares centred on p_i (note that to ensure that the ROI is symmetric about p_i the value for n should be an odd number). What the value for n should be is a matter for consideration and may also, at least to an extent, be application dependent. For the linearisation there are a number of "space filling curve" formats that could have been adopted (for example a Peano curve [16] or a Hilbert curve [8]). However, given the nature of our ROIs (see below) a straightforward spiral linearisation was adopted as this fits well with the proposed $n \times n$ ROI definition. With respect to the number of points to be considered we can include all points covered by a linearisation or a selection of "key" points. If we include all points there will be $n^2 - 1$ points per linearisation. Given a grid comprised of many grid squares this will result in a large number of point series (one per grid square) which in turn might mean that the size (length) of each point series becomes significant (from a time complexity perspective). An alternative is to consider only "corner points" and "mid-way points" in which case we will have $((n - 1)/2) \times 8$ points per linearisation instead of $n^2 - 1$. Finally, with respect to the value to be represented by each linearisation point the idea is to use the difference in the z coordinates between each neighbourhood point and the point p_i as this will clearly capture variations in local geometries. The issue is weather to use real δz values or discretised values. In the case of the LGM and LDM representations considered previously by the authors (see above) discretised values were used. This was because the classification processes adopted with respect to these representations required discrete values. However, real values may produce a better result. Thus two variations off the proposed linearisation were considered: (i) discretised linearisation and (ii) real number linearisation. Figure 1(a) shows an example linearisation using a 5×5 neighbourhood and key points instead of all points. Figure 1(b) shows the spiral linearisation and Figure 1(c) the resulting point series.

Using the above process we can create a model of a particular 3D surface application domain which can be used for prediction purposes. So that the model is as comprehensive as possible the training data should cover a wide diversity of local geometries, ideally all the potential local geometries that can occur with respect to a particular application domain.

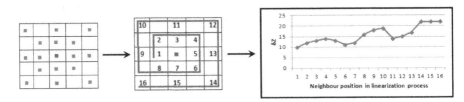

Fig. 1. Example spiral linearisation for a 5×5 neighbourhood

3.2 Prediction

As noted above a KNN classification systems is suggested to carry out the desired prediction. Given a new curve to be classified, the k most similar pre-labeled curves were identified and the most similar selected according to the warping distance between two curves. The warping distance was obtained using the established Dynamic Time Warping (DTW) process [10,18]. DTW operates as follows. Given two point series $A = \{a_1, a_2, \cdots, a_n\}$ and $B = \{b_1, b_2, \cdots, b_m\}$, we create a matrix R that has dimension $|A| \times |B| = n \times m$ is defined. The value at each matrix element $\langle i, j \rangle$ (where $0 \leq i < n$ and $0 \leq j < m$) is the the Euclidean distance between points a_i and b_j when the two curves are considered in term of a 2D reference plane. Once the matrix elements have been computed the *warping path* W is identified ($W \subset$ *elements in* R). This is the path from the bottom left corner of the matrix to the top right corner that links the matrix elements with lowest values. The most direct path will be along the leading diagonal, this will occur when curves A and B are identical. The length of the warping path, W_{dist}, is the accumulated sum of the matrix element values contained in W:

$$W_{dist} = \sum_{k=1}^{k=|W|} W_k \tag{1}$$

Given two identical curves W_{dist} will be 0.

4 Evaluation

The evaluation of the proposed prediction framework is considered in this section. To this end real data was obtained using an Asymmetric Incremental Sheet Forming (AISF) process, a sheet metal forming process used in industry [9]. Two sample surfaces (shapes) were considered referred to as the Gonzalo and Modified surfaces, both described flat topped pyramid shapes. Each was manufactured four time, twice using steel and twice using titanium. Thus eight different data sets (comprised of both before and after clouds) were available for experimentation: (i) Gonzalo Steel 1 (GS1), (ii) Gonzalo Steel 2 (GS2), (iii) Modified Steel 1 (MS1), (iv) Modified Steel 2 (MS2), (v) Gonzalo Titanium 1 (GT1), (vi) Gonzalo Titanium 2 (GT2), (vii) Modified Titanium 1 (MT1), (viii) Modified Titanium 2 (MT2). Each data set comprised two coordinate clouds, a before cloud C_{in} and an after cloud C_{out}. The pre-processing of this data is described in Sub-section 4.1 below.

The evaluation was directed at investigating the following:

1. The effect of variations in the value of d (the grid size).
2. The effect of using different sized neighbourhoods.
3. A comparison with respect to using all linearisation points versus only key points.
4. The generalisation of the proposed approach (can we build a generally applicable classifier using our proposed linearisation).

Each of these is considered in further detail in Sub-sections 4.2 to 4.5 below. The evaluation was conducted in sequence by considering pairs of data sets, the best parameter settings identified with respect to one investigation were adopted for use in following investigations. The metrics used for evaluation purposes were accuracy and AUC. A tolerance of 0.08, as suggested by BS ISO 2005 [1], was used.

4.1 Training Data Generation

As noted above the training data was derived using before and after point clouds (C_{in} and C_{out}) produced using AISF sheet metal forming. The C_{in} cloud was obtained from a CAD system and was used to describe the desired shape T. The C_{out} cloud, used to describe the produced shape T', was obtained using a GOM (Gesellschaft fur Optische Messtechnik) optical measuring tool. During pre-processing both clouds were translated into the desired a grid representation (using the same d value) to give two grids, G_{in} and G_{out} respectively. The z value associated with each p_i was obtained by averaging the z coordinates for all the points located within that grid square. The springback (error) associated with each p_i was obtained by comparing the corresponding points in G_{in} and G_{out}. The springback value in each case was obtained by calculating the distance along the normal from each point p_i in C_{in} to where it cut the C_{out} surface.

4.2 Grid Size

To determine the effete of the grid size, a range of grid sizes, $d = \{2.5, 5, 10, 15, 20\}$ mm, were considered together with a 3×3 neighbourhood using the key point linearisation. The results are shown in the Tables 1 (best result for each dataset indicated in bold font). From the table it can be seen that excellent accuracy and AUC values were recorded regardless of the shape and manufacturing material used. From the tables it can also be observed that, in general, as the grid size increases the accuracy and AUC values start to decrease, it is conjectured that this is because of the increasing coarseness of the representation. This is more evident with respect to the AUC values because of the unbalanced nature of the input data.

Table 1. The results for 3×3 neighbourhood using *key* point representation technique (Total number of key points = 8 points)

	d=2.5		d=5		d=10		d=15		d=20	
	Accuracy	AUC	Accuracy	AUC	Accuracy	AUC	Accuracy	AUC	Accuracy	AUC
GSV1	0.97	0.96	0.98	0.97	0.98	0.96	0.97	0.95	0.90	0.82
GSV2	**0.99**	0.94	0.98	0.89	0.97	0.84	0.96	0.64	0.96	0.78
GTV1	**0.99**	**0.97**	**1.00**	**1.00**	**0.99**	**0.98**	0.94	0.76	0.93	0.72
GTV2	**0.99**	0.96	0.99	0.99	**0.99**	0.97	**0.98**	0.93	0.96	**0.96**
MSV1	0.97	0.92	0.97	0.92	0.98	0.97	0.97	0.94	**0.97**	0.87
MSV2	0.96	0.94	0.98	0.97	0.98	0.93	0.97	0.92	0.92	0.71
MTV1	0.98	0.96	0.98	0.97	0.98	0.94	0.97	**0.96**	0.90	0.81
MTV2	0.97	0.96	0.96	0.94	0.93	0.90	**0.98**	0.94	0.82	0.73

4.3   Neighbourhood Size

To determine the effect of different $n \times n$ neighbourhood sizes three values of n were considered, $\{3,5,7\}$. Tables 2 and 3 show the results obtained using 5×5 and 7×7 neighbourhood sizes. Comparing these results with those presented above in Table 1 (3×3) it can be seen there there is no discernible difference between the different neighbourhood sizes although it can be noted that grid size ceases to have an impact as neighbourhood sizes increase.

Table 2. The results for 5×5 neighbourhood using *key* point representation technique (total number of key points = 16 points)

	d=2.5		d=5		d=10		d=15		d=20	
	Accuracy	AUC	Accuracy	AUC	Accuracy	AUC	Accuracy	AUC	Accuracy	AUC
GSV1	0.97	**0.96**	0.97	0.95	**0.98**	**0.97**	**1.00**	**1.00**	0.84	0.80
GSV2	**0.99**	0.94	0.97	0.89	0.96	0.75	0.94	0.64	0.92	0.73
GTV1	**0.99**	**0.96**	**0.99**	**0.99**	0.94	0.89	0.94	0.70	0.91	0.74
GTV2	**0.99**	**0.96**	**0.99**	0.98	0.96	0.85	0.95	0.90	**0.99**	**0.99**
MSV1	0.96	0.91	0.96	0.92	0.96	0.92	0.98	0.97	0.91	0.62
MSV2	0.96	0.94	0.96	0.94	0.97	0.91	0.97	0.93	0.92	0.64
MTV1	0.98	**0.96**	0.98	0.96	0.96	0.89	0.97	0.95	0.84	0.70
MTV2	0.97	**0.96**	0.96	0.94	0.93	0.92	0.96	0.93	**0.99**	0.98

Table 3. The results for 7×7 neighbourhood using *key* point representation technique (total number of key points = 24 points)

	d=2.5		d=5		d=10		d=15		d=20	
	Accuracy	AUC	Accuracy	AUC	Accuracy	AUC	Accuracy	AUC	Accuracy	AUC
GSV1	0.98	**0.97**	**0.99**	0.98	0.94	0.92	0.87	0.70	0.78	0.33
GSV2	**0.99**	0.96	0.98	0.93	0.94	0.85	0.89	0.65	0.67	0.50
GTV1	0.98	0.93	**0.99**	**0.99**	0.87	0.77	0.79	0.57	**0.81**	**0.89**
GTV2	**0.99**	0.95	0.98	0.96	0.88	0.72	0.83	0.75	0.48	0.19
MSV1	0.96	0.92	0.97	0.95	**0.99**	**0.99**	0.89	0.81	0.75	0.67
MSV2	0.97	0.95	0.97	0.97	0.96	0.95	0.92	0.86	0.67	0.50
MTV1	0.98	0.95	0.98	0.97	0.92	0.72	**0.97**	**1.00**	0.70	0.47
MTV2	0.97	0.94	0.96	0.93	0.97	0.95	0.92	0.74	0.56	0.25

4.4   All Points versus Key Points

This sub-section compares the operation of the all points linearisation with the key points linearisation using different grid size $d = \{2.5, 5, 10, 15, 20\}$. Table 4 presents the results obtained using a 5×5 neighbourhood coupled with the all point linearisation which can be compared with the results presented in Table 2 which shows the results of using a 5×5 neighbourhood and the key point linearisation. From the table it can be observed that there is no performance difference between the two linearisations. Similar results were produced using a 7×7 neighbourhood, Note that in the case of a

Table 4. The results for 5 × 5 neighbourhood using *all* point representation technique (total number of key points = 24 points)

	d=2.5		d=5		d=10		d=15		d=20	
	Accuracy	AUC	Accuracy	AUC	Accuracy	AUC	Accuracy	AUC	Accuracy	AUC
GSV1	0.96	0.95	0.97	0.95	**0.98**	**0.97**	**1.00**	**1.00**	0.84	0.80
GSV2	**0.99**	0.94	0.98	0.90	0.96	0.75	0.94	0.64	0.92	0.73
GTV1	**0.99**	0.96	**0.99**	**0.99**	0.94	0.88	0.94	0.70	0.91	0.74
GTV2	**0.99**	0.96	**0.99**	0.98	0.96	0.85	0.95	0.90	0.98	**0.99**
MSV1	0.97	0.91	0.96	0.93	0.96	0.92	0.98	0.97	0.93	0.64
MSV2	0.96	0.94	0.96	0.94	0.96	0.92	0.98	0.96	0.90	0.64
MTV1	**0.99**	**0.97**	0.98	0.95	0.96	0.91	0.97	0.96	0.83	0.67
MTV2	0.98	**0.97**	0.96	0.94	0.93	0.92	0.98	0.96	**0.99**	0.98

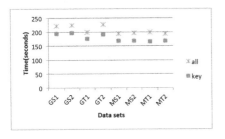

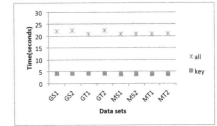

Fig. 2. The run time (s) for *all* vs *key* point representation, 5 × 5 neighbourhood, $d = 2.5$ mm   **Fig. 3.** The run time (s) for *all* vs *key* point representation, 7 × 7 neighbourhood, $d = 2.5$ mm

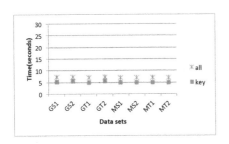

Fig. 4. The run time (is) for *all* vs *key* point representation, 5 × 5 neighbourhood, $d = 5$ mm   **Fig. 5.** The run time (s) for *all* vs *key* point representation, 7 × 7 neighbourhood, $d = 5$ mm

3 × 3 neighbourhood there is no difference between the key point and the all point linearisation. Figures 2 to 11 indicate the recorded run times using both representations and different grid size. The charts indicate that the key point linearisation is more efficient than the all point representation especially with respect to large grid sizes.

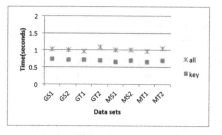

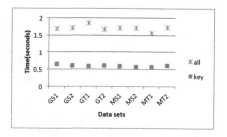

Fig. 6. The run time (s) for *all* vs *key* point representation, 5 × 5 neighbourhood, $d = 10$ mm

Fig. 7. The run time (s) for *all* vs *key* point representation, 7 × 7 neighbourhood, $d = 10$ mm

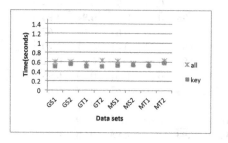

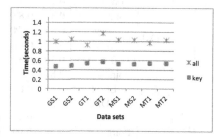

Fig. 8. The run time (s) for *all* vs *key* point representation, 5 × 5 neighbourhood, $d = 15$ mm

Fig. 9. The run time (s) for *all* vs *key* point representation, 7 × 7 neighbourhood, $d = 15$ mm

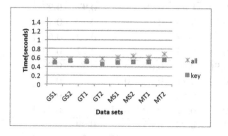

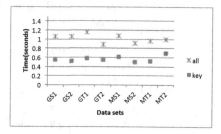

Fig. 10. The run time (s) for *all* vs *key* point representation, 5 × 5 neighbourhood, $d = 20$ mm

Fig. 11. The run time (s) for *all* vs *key* point representation, 7 × 7 neighbourhood, $d = 20$ mm

4.5 Generalisation

This section presents the results obtained from training a classifier on one shape and testing it on another. The main goal was determine whether it was possible to generate a generally applicable classifier if it was provided with a suitable shape to train on. From earlier experiments it was noted that that lower grid sizes produced better results, hence for this set of experiments $d = 5$ was used. The operation of the proposed point series representation was also compared with the LGM and LDM representations proposed previously by the authors [5]. Table 5 presents the results obtained in terms of AUC values (best results on bold). From the table it can be seen that using the proposed representation a best AUC value of 1.00 could be obtained. In terms of the the previous

Table 5. AUC results for the generic classifier

Test			Train							
			GSV1	GSV2	GTV1	GTV2	MSV1	MSV2	MTV1	MTV2
	GSV1	Point series		**0.97**	**0.94**	**0.93**	**0.96**	**0.91**	**0.96**	**0.98**
		LGM		0.66	0.59	0.70	0.44	0.52	0.48	0.52
		LDM		0.52	0.50	0.50	0.51	0.50	0.52	0.47
		LGM + LDM		0.94	0.76	0.81	0.80	0.89	0.75	0.70
	GSV2	Point series	**0.99**		**0.99**	0.92	**1.00**	**0.96**	**1.00**	**0.99**
		LGM	0.62		0.68	0.74	0.60	0.53	0.67	0.61
		LDM	0.50		0.50	0.50	0.50	0.50	0.50	0.50
		LGM + LDM	0.72		0.78	0.83	0.81	0.90	0.78	0.74
	GTV1	Point series	**0.84**	0.87		**0.95**	0.64	0.66	**0.94**	**0.94**
		LGM	0.65	0.74		0.75	0.69	0.67	0.69	0.67
		LDM	0.49	0.61		0.53	0.41	0.41	0.41	0.40
		LGM + LDM	0.70	**0.89**		0.80	**0.80**	**0.89**	0.73	0.72
	GTV2	Point series	**0.81**	0.91	**0.98**		0.65	0.61	**0.99**	**0.95**
		LGM	0.66	0.81	0.72		0.70	0.63	0.68	0.65
		LDM	0.50	0.50	0.50		0.50	0.50	0.50	0.50
Test		LGM + LDM	0.74	**0.97**	0.79		**0.83**	**0.93**	0.79	0.74
	MSV1	Point series	**0.98**	**0.99**	**0.99**	**0.98**		**0.95**	**0.98**	**0.97**
		LGM	0.61	0.57	0.70	0.76		0.82	0.76	0.77
		LDM	0.51	0.39	0.47	0.47		0.59	0.59	0.60
		LGM + LDM	0.66	0.89	0.74	0.76		0.92	0.74	0.75
	MSV2	Point series	**0.95**	**0.96**	**0.99**	**1.00**	**0.98**		**0.97**	**0.97**
		LGM	0.65	0.75	0.72	0.77	0.80		0.70	0.73
		LDM	0.51	0.39	0.59	0.47	0.59		0.59	0.60
		LGM + LDM	0.77	0.95	0.78	0.83	0.84		0.77	0.73
	MTV1	Point series	**0.85**	0.88	**0.96**	**0.94**	0.62	0.59		**0.95**
		LGM	0.62	0.74	0.72	0.71	0.75	0.73		0.76
		LDM	0.50	0.50	0.50	0.50	0.50	0.50		0.50
		LGM + LDM	0.74	**0.93**	0.76	0.81	**0.80**	**0.91**		0.71
	MTV2	Point series	**0.90**	**1.00**	**0.98**	**0.93**	0.65	0.63	**0.99**	
		LGM	0.56	0.49	0.59	0.59	0.76	0.75	0.73	
		LDM	0.50	0.39	0.47	0.47	0.59	0.59	0.59	
		LGM + LDM	0.69	0.83	0.72	0.76	**0.81**	**0.86**	0.72	
Average (Point series)			**0.90**	**0.94**	**0.98**	**0.95**	**0.79**	0.76	**0.98**	**0.96**
Average (LGM)			0.56	0.64	0.69	0.69	0.74	0.73	0.72	0.64
Average (LDM)			0.50	0.50	0.47	0.50	0.52	0.53	0.50	0.51
Average (LGM + LDM)			0.81	0.79	0.79	0.83	0.78	**0.81**	0.81	0.77

representations proposed by the authors (LGM, LDM and LGM and LDM combined (LGM+LDM)) it can be seen that, in most cases, these alternatives performed badly in comparison with the point series representation. Consequently it is argued that an effective generic classifier can be produced using the proposed representation.

5 Conclusion

This paper has presented a new representation and supporting mechanism to predict feature values associated with 3D surfaces using a point series based approach. The proposed point series representation is founded on a *linearisation* of the space describing a neighbourhood surrounding a given point where the neighbourhood in turn is defined in terms of a grid. The motivation for the work was springback prediction in sheet metal forming. Two 3D surfaces (shapes) were used to evaluate the mechanism. Various forms of linearisation were considered using all the points in a linearisation or only key (corner and midway) points as well as the effect of using different grid sizes ($d = \{2.5, 5, 10, 15, 20\}$). The experiments indicated that: (i) smaller grid sizes tended to work better, (ii) the performance using 3×3, 5×5 and 7×7 neighbourhood was almost the same and (iii) that there was no significant difference in accuracy or AUC between the representations (all and key) however the key point representation offers runtime advantages. Further experiments were conducted to determine whether the linearisation could be used to produce a generic classifier, the results indicated that this was indeed the case. Excellent results were returned, 100% in terms of AUC, indicating that the point series representation is able to capture general geometric information that can successfully be employed for prediction purposes. Overall, this is a very encouraging result. For future work the intention is to conduct further experimentation with a greater variety of surfaces (shapes). The ultimate goal is to build an intelligent process model that can predict springback errors, and suggest corrections, in the context of sheet metal forming.

Acknowledgement. The research leading to the results presented in this paper has received funding from the European Union Seventh Framework Programme (FP7/2007-2013) under grant agreement number 266208 and partially by Hashemite University in Zarqa, Jordan.

References

1. BS ISO 1101: 2005 Geometrical Product Specifications (GPS) Geometrical tolerancing Tolerances of form, orientation, location and run-out (2005)
2. Cafuta, G., Mole, N., Ltok, B.: An enhanced displacement adjustment method: Springback and thinning compensation. Materials and Design 40, 476–487 (2012),
   http://www.sciencedirect.com/science/
   article/pii/S026130691200252X
3. Chatti, S.: Effect of the Elasticity Formulation in Finite Strain on Springback Prediction. Computers and Structures 88(11-12), 796–805 (2010),
   http://www.sciencedirect.com/science/article/
   pii/S004579491000074X
4. Chatti, S., Hermi, N.: The Effect of Non-linear Recovery on Springback Prediction. Computers and Structures 89(13-14), 1367–1377 (2011)
5. El-Salhi, S., Coenen, F., Dixon, C., Khan, M.: Identification of correlations between 3d surfaces using data mining techniques: Predicting springback in sheet metal forming. In: Bramer, M., Petridis, M. (eds.) Research and Development in Intelligent Systems XXIX, pp. 391–404. Springer, London (2012)

6. Firat, M., Kaftanoglu, B., Eser, O.: Sheet Metal Forming Analyses With An Emphasis On the Springback Deformation. Journal of Materials Processing Technology 196(1-3), 135–148 (2008), http://www.sciencedirect.com/science/article/pii/S0924013607005353

7. Hao, W., Duncan, S.: Optimization of Tool Trajectory for Incremental Sheet Forming Using Closed Loop Control. In: 2011 IEEE Conference on Automation Science and Engineering (CASE), pp. 779–784 (2011)

8. Hilbert, D.: Ueber die stetige Abbildung einer Line auf ein Flächenstück. Mathematische Annalen 38(3), 459–460 (1891)

9. Jeswiet, J., Micari, F., Hirt, G., Bramley, A., Duflou, J., Allwood, J.: Asymmetric Single Point Incremental Forming of Sheet Metal. CIRP Annals - Manufacturing Technology 54(2), 88–114 (2005), http://www.sciencedirect.com/science/article/pii/S0007850607600213

10. Keogh, E.J., Pazzani, M.J.: Derivative dynamic time warping. In: First SIAM International Conference on Data Mining (SDM 2001) (2001)

11. Khan, M., Coenen, F., Dixon, C., El-Salhi, S.: A classification based approach for predicting springback in sheet metal forming. Journal of Theoretical and Applied Computer Science 6, 45–59 (2012)

12. Sulaiman Khan, M., Coenen, F., Dixon, C., El-Salhi, S.: Finding correlations between 3-D surfaces: A study in asymmetric incremental sheet forming. In: Perner, P. (ed.) MLDM 2012. LNCS, vol. 7376, pp. 366–379. Springer, Heidelberg (2012)

13. Liu, W., Liang, Z., Huang, T., Chen, Y., Lian, J.: Process Optimal Ccontrol of Sheet Metal Forming Springback Based on Evolutionary Strategy. In: 7th World Congress on Intelligent Control and Automation (WCICA 2008), pp. 7940–7945 (June 2008)

14. Narasimhan, N., Lovell, M.: Predicting Springback in Sheet Metal Forming: An Explicit to Implicit Sequential Solution Procedure. Finite Elements in Analysis and Design 33(1), 29–42 (1999), http://www.sciencedirect.com/science/article/pii/S0168874X99000098

15. Nasrollahi, V., Arezoo, B.: Prediction of Springback in Sheet Metal Components With Holes on the Bending Area, Using Experiments, Finite Element and Neural Networks. Materials and Design 36, 331–336 (2012), http://www.sciencedirect.com/science/article/pii/S0261306911007990

16. Peano, G.: Sur une courbe, qui remplit toute une aire plane. Mathematische Annalen 36(1), 157–160 (1890)

17. Tisza, M.: Numerical Modelling and Simulation in Sheet Metal Forming. Journal of Materials Processing Technology 151(1-3), 58–62 (2004), http://www.sciencedirect.com/science/article/pii/S0924013604003073

18. Xi, X., Keogh, E., Shelton, C., Wei, L., Ratanamahatana, C.: Fast time series classification using numerosity reduction. In: Proceedings of the 23rd International Conference on Machine Learning (ICML 2006), pp. 1033–1040. ACM, New York (2006)

19. Yoon, J., Pourboghrat, F., Chung, K., Yang, D.: Springback Prediction For Sheet Metal Forming Process Using a 3d Hybrid Membrane/Shell Method. International Journal of Mechanical Sciences 44(10), 2133–2153 (2002), http://www.sciencedirect.com/science/article/pii/S0020740302001650

Automatic Labeling of Forums Using Bloom's Taxonomy

Vanessa Echeverría[1], Juan Carlos Gomez[2], and Marie-Francine Moens[2]

[1] Centro de Tecnologías de Información,
Escuela Superior Politécnica del Litoral,
Km 30.5 vía Perimetral, Guayaquil, Ecuador
vecheverria@cti.espol.edu.ec
[2] Department of Computer Science, KU Leuven
Celestijnenlaan 200A, B-3001 Heverlee, Belgium
{juancarlos.gomez,sien.moens}@cs.kuleuven.be

Abstract. The labeling of discussion forums using the cognitive levels of Bloom's taxonomy is a time-consuming and very expensive task due to the big amount of information that needs to be labeled and the need of an expert in the educational field for applying the taxonomy according to the messages of the forums. In this paper we present a framework in order to automatically label messages from discussion forums using the categories of Bloom's taxonomy. Several models were created using three kind of machine learning approaches: linear, Rule-Based and combined classifiers. The models are evaluated using the accuracy, the F1-measure and the area under the ROC curve. Additionally, a statistical significance of the results is performed using a McNemar test in order to validate them. The results show that the combination of a linear classifier with a Rule-Based classifier yields very good and promising results for this difficult task.

Keywords: CSCL, Bloom's taxonomy, logistic regression classifier, Rule-Based classifier, combined classifiers.

1 Introduction

Discussion forums are considered as an application of Computer-Supported Collaborative Learning (CSCL). The goal of CSCL is to create a computer environment for assessing educational goals through a shared activity among participants of a discussion board [15]. The participants can construct their own knowledge through this social interaction by sharing ideas and negotiate their validity which leads to an active participation in this collaborative activity [8].

In order to assess the educational goals achieved in a learning environment, Bloom's taxonomy (BT) plays an important role. Benjamin Bloom created a categorization of learning objectives to evaluate learning outcomes [1], and former students of him modified this categorization to include the analysis of cognitive processes of participants [7]. The categories that Bloom's students proposed

H. Motoda et al. (Eds.): ADMA 2013, Part I, LNAI 8346, pp. 517–528, 2013.

are: *Remembering, Understanding, Analyzing, Applying, Evaluating* and *Creating*. Several studies have fostered BT to analyze the contributions of participants in the discussion forums at a cognitive level [13].

In a CSCL environment the information from forums needs to be labeled using BT. However, in large databases of information usually not all the data is labeled because the overall process is time-consuming, needing human resources with an appropiate background in education [3]. This is an important motivation for migrating from a manual to an automatic labeling by selecting automatically a level from the taxonomy and assign it to a discussion forum's response.

The automatic labeling of discussion forums with a cognitive level from BT could be considered in essence as a text categorization (TC) problem. TC is defined as a supervised task of assigning a value of true or false to a document with regard to the assignment of a certain category c_j, where $d_i \in D$ is the i-th element of the collection of documents $D = \{d_1, d_2, d_3, ...d_n\}$. Thus, each document d_i is labeled with a category $c_j \in C$, which is the j-th element of the set of categories $C = \{c_1, c_2, c_3, ...c_k\}$ [12].

Several studies have been conducted for automatic labeling of questions using BT. These questions are used for designing an effective test in order to evaluate the skills of a participant using the taxonomy. Those studies yielded systems for labeling questions using Artificial Neural Networks (ANN) [14], a Rule-Based classifier [9] and a bank of words using weights for each category [2]. However, it is important to notice that those studies addressed the automatic labeling of questions, which is not equivalent to a text containing expressions and thoughts from a person. Additionally, a study by [10] implemented a system for automatic labeling of discussion forums, but a deeper analysis of the results was not performed. As far as we know, there is not a relevant study about the automatic labeling of discussion forums.

In this paper, we conduct an analysis of several models based on supervised machine learning methods in combination with the use of several features for labeling discussion forums using BT. Such models are compared using different evaluation measures and performing statistical significance tests.

The problem presented in this paper is treated as TC task, however it presents an important difference with common TC problems resulting from the complexity for modeling the data. The goal of most TC task is to find features that can be common for a category or topic (directories, news, emails, etc.). In our case, the categories to be assigned to a forum are not topics, but levels of a cognitive process. This makes this task harder than a common TC problem, since the data that we are dealing with are not simple documents containing information related to a given category or topic, but rather information about the cognitive level of persons, which is indeed topic independent.

Three approaches are used in order to have a wide coverage of results. The first one consists on training linear classifiers, in particular Support Vector Machine (SVM) and Logistic Regression (Logit), using the following features: words, verbs, bigrams and trigrams; weighting such features using tf-idf (term frequency - inverse document frequency) [11]. Linear classifiers are well known to have good performance in TC tasks [4]. In this approach, we also perform a dimensionality reduction using Principal Component Analysis (PCA) [5] on the feature space and then use the reduced feature space with the linear classifiers. The second approach implements the creation of Rule-Based (RB) classifiers using the following features: verbs, bigrams and trigrams. Finally, the third approach comprises the combination of classifiers resulting from the first and second approaches [6].

For all the models, their performance is computed based on accuracy, F1-measure and Area Under the ROC Curve (AUC). From all the models created, at the end we select the best five models based on those measures for conducting a deeper analysis and a statistical comparison using the McNemar test.

The contributions of our work are: 1) the addressing of a novel and hard task that has not been explored in detail; 2) the good labeling results obtained; 3) the creation of highly effective but simple rules in a RB classifier; 4) the observation that a combination of classifiers, where the prior knowledge has been modeled, is a good approach to follow in order to enhance the general performance; 5) a deep experimental and error analysis of the performance of the different models, considering common measures and a statistical comparison of the results, and providing rationales for some methods performing better than others.

The remainder of this paper is organized as follows: in section 2, we provide the related work that has been carried out when implementing automatic classification of text using BT. Section 3 is dedicated to describe the framework used for finding a suitable model for this TC task, including a detailed description of the data collection and the experimental setup for each model. In Section 4 we provide the experiments and results together with a deeper analysis of them. Finally, in Section 5 we present the conclusions and future work.

2 Related Work

In the area of CSCL Several studies have been conducted for automatic labeling of examination questions from different areas (courses) using Bloom's taxonomy (BT). In [14], the aim of the study was to label a set of questions using Artificial Neural Networks (ANN). The model presented there used words to form an initial feature set, and then two feature reduction methods were performed: document frequency (DF) and category-frequency document-frequency (CF-DF), which introduces a discriminant value for features that appeared in few categories. Seven models were trained using a 3-layer feed forward neural network. The models were evaluated using precision, convergence time and error. Results showed that the use of all words reached the best performance according to the given measures.

The authors of [2] tackled the task of dealing with multiple keywords in one question. To solve the problem, they compared keywords extracted for each BT category. They stressed that the verbs in a question are the most important keywords to represent a category. To deal with shared keywords among categories, they gave weights to the keywords. Thus, in the test phase, when an unseen question was analyzed, they compared the verb with the database of keywords constructed previously and the category with higher weight was the predicted one for that example. The evaluation measure used in this study was the number of correct matched items. Moreover, the results of this study showed that the category with the label "Knowledge", which corresponds to the first cognitive level on BT, achieved a good performance (57 percent), while for other categories, the performance was lower.

In [9] the authors determined the appropriate category for exam questions using a Rule-Based approach, employing keywords and verbs. Initially, they had a collection of questions that were manually categorized by a group of experts in the programming domain. Then, to beginning the analysis of the questions, they tag each question to obtain the question structure and then, find patterns according to the tags. They also mentioned that there were certain shared patterns among categories. Therefore, to overcome this problem, they gave weights to each category. Those weights were set by experts, meaning that they did not have a good method to determine the weights of the categories; which makes the analysis expert-dependent. Finally, they created rules based on the found patterns questions and the prediction is the category with the highest weight.

The closest study related to the automatic labeling of forums is presented in [10]. There, the authors trained a Bayesian classifier over a set of 420 documents in Spanish using document frequency for feature selection. The results of their study showed that their model reached an accuracy of 51 percent for the category "Understanding", while for the other categories the performance was lower.

Although most of the previous studies label questions using BT, they do not label large texts or responses from participants in order to identify the participant's cognitive level. Furthermore, we can observe from those studies a lack of a deeper analysis regarding the employed evaluation measures. Nevertheless, we observe that most of the studies show that there exists shared information among the categories of BT.

3   Framework

In this paper we propose a framework for performing the automatic labeling of forums using the cognitive levels of BT under several models. The framework is divided in three main parts: preprocessing, indexing and learning-testing phase. Figure 1 shows the framework used for creating our models.

3.1   Document Collection

The data used for this work was gathered from nine discussion forums about several topics[1]. The discussion forums were originally written in Spanish, then

---

[1] The topics related to the forums are from two courses of a Bachelor in Computer Science: Computer Graphics and Interactive Multimedia Applications; and from an Information Security course in a Master Program of an Ecuadorian University. The data gathered is property of the CTI-ESPOL.

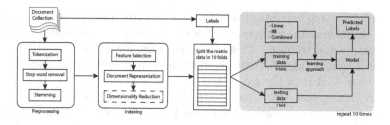

Fig. 1. Scheme for model creation

translated into English by people with an intermediate-advance knowledge of the language.

The manual tagging of forums was accomplished by three independent coders. Each entry of the forum (message) was labeled with one of the BT categories: Remembering, Understanding, Analyzing, Applying, Evaluating and Creating. An additional label *Uncodable* was used in order to track the messages that did not fit in the taxonomy.

After all coders finished the labeling of messages, a meeting between all coders was carried out and each individual result was discussed. The messages with disagreement were debated until an agreement was reached. The result of this activity is a set of messages with one category label for each message. At this point, each message in the set is treated as an individual document.

The dataset used here corresponds to a collection of 463 documents grouped in five categories (there are two categories from the taxonomy without messages). Table 1 shows the number of documents and the corresponding percentage for each category.

Table 1. Frequency of documents per category

Category	Frequency	Percent	Cum. Percent
Remembering	116	24.8	24.8
Understanding	186	39.7	64.5
Analyzing	46	9.8	74.4
Evaluating	27	5.8	80.1
Uncodable	93	19.9	100
Total	**468**	**100**	

3.2   Preprocessing and Indexing

In the preprocessing step, the document collection is analyzed document by document. Thus, we carried out a tokenization process for each document, resulting in a list of words. From this list we perform stemming and stop word removal. Hence, the result of the preprocessing part is a list of relevant words for each document.

The indexing step consists of selecting the features and representing them in machine readable data. Four types of features are identified and used: words, verbs, bigrams and trigrams. First, the verbs were identified from the returned list of words using a tagger. Then, the bigrams were created by joining each verb with the word that is to the left or to the right of it. Finally, the trigrams were created by joining the words that are to the left and right of each verb.

The document representation defines the input of the data for a given learning approach. The input for a linear classifier is a sparse vector of weights. Here we use tf-idf as a weighting factor for the vectors of features. The representation of a RB classifier is given by features related to a category; thus the words, verbs, bigrams and trigrams are used as the representation of a rule.

Dimensionality reduction is dedicated to reduce the number of features and for this work we use Principal Component Analysis (PCA), which performs a linear combination over the document representation matrix. Then, from the matrix of eigenvectors, a fixed number of components is selected and used to project the original matrix, giving a matrix with less but more relevant features. The selection of an optimal number of components q was done using a global grid search over the training set, varying the number of components from 2 to N, where N is the size of the document collection. Finally, each number of components is evaluated considering the classification performance and the best one is selected.

3.3   Training the Models

In this study, we used three types of learning approaches. First, two linear classifiers (SVM and Logit) [2]; second a RB classifier; and third the combination of the linear and RB classifiers. For each learning approach, we implement different models using variants of features or performing a dimensionality reduction.

Initially, the SVM and the Logit classifier were trained using words as features. Afterwards, only the Logit was considered for training models using verbs, bigrams and trigrams as features; and PCA with the same features. The selection of this classifier was given by analyzing the output of each classifier: the output of the Logit (a probability of membership for each category) was better suited for the combination of classifiers rather than the output of the SVM (a yes/no output). For this approach, we obtain nine trained models using the two classifiers with four types of features: SVM (words), Logit (words, verbs, bigrams, trigrams) and PCA Logit (words, verbs, bigrams, trigrams). Additionally, we accomplish for each model a global parameter optimization using a 3-fold cross validation together with a grid search over the training set. The best (regularization) parameter C of the SVM classifier was selected from the set of values {1e-6,1e-5,1e-4,1e-3,1e-2,1e-1,1,1e1}; while for the Logit classifier, the C, the tolerance (ε) and the norm were optimized using the range values [1 - 10],{1e-8,1e-7,1e-6,1e-5,1e-4,1e-3,1e- 2,1e-1,1,1e1,1e2} and {L1, L2} respectively.

---

[2] A non-linear version of SVM with a RBF kernel was also tested in the experiments, but compared to the linear SVM the performance was worse.

For the second approach, the RB classifiers, we select a set of rules for each category using verbs, bigrams and trigrams as features. First, the training set was separated by categories (giving five groups). Next, from each collection of documents per category we extract the verbs, bigrams and trigrams and compute a frequency distribution of the resulting list. Then, we select the verbs, bigrams and trigrams with a support threshold greater or equal than 1 (thr$\geq$1), 2 (thr$\geq$2) and 3 (thr$\geq$3) to compose the set of rules. However, for the verbs only a support threshold (thr$\geq$2) was selected because there were too many verbs with a threshold equal to 1 and this would affect the generalization of the classifier. In order to add meaning to the rule, the conditional probability that one verb v belongs to a category c_i given the category $P(v \mid c_i) = P(v, c_i)/p(c_i)$ was computed, where $P(v, c_i)$ corresponds to the likelihood of the verb in the category (frequency divided by the size of the list of verbs) and $p(c_i)$ was the likelihood of the category (frequency divided by the number of documents belonging to the category). This conditional probability corresponds to the confidence of the rule. Finally, by varying the support threshold for creating the rules and combining the different features we obtain a set of seven classifiers: RB (verbs thr$\geq$2); RB (bigrams thr$\geq$1; thr$\geq$2; thr$\geq$3) and RB (trigrams thr$\geq$1; thr$\geq$2; thr$\geq$3). Each classifier was represented by a list of rules where each item was associated with its confidence and frequency.

The last learning approach consists of a combination of the Logit classifier and the RB classifier. The idea behind combining models was to improve the general decision boundary by applying a linear combination over the decision surfaces constructed by the individual classifiers. The expected error of the combined classifier will have a lower bound given by the minimum error of the classifiers involved. The Logit classifier throws out a vector of probabilities for each test document (one probability per category). The RB classifier also yields a vector of weights, where an element of the vector corresponds to the sum of confidences regarding one category. The two classifiers can be combined by summing the outputs per category of both models and taking the highest probability value of the resulting output vector. The category associated with the highest value will be the predicted category corresponding to a test document. Thus, two Logit classifiers from the first group (Logit and PCA Logit classifiers, both with words as features) were used for being combined with the seven RB classifiers from the second group. The result of the classifiers of both group yields a total of 14 combined models: Logit (words) + RB, PCA Logit (words) + RB.

In order to have a more robust analysis of the models, each one of them has been tested using a 10-fold cross validation schema. In this schema, nine folds were used for training and one for testing, repeating the process 10 times (until each fold has been used for training and testing the model).

The performance of the models was compared using as evaluation measures the accuracy, F1-measure and the AUC. The final values of these measures were averaged over the 10 folds. Furthermore, the parameter optimization was done using a global approach, meaning that 10 different parameters of the learners were given using the training data, and the parameter values with the best accuracy among the 10 learners were selected for training the final model.

In total, we train 30 classifiers: 9 linear, 7 RB and 14 combined. From these, the top five classifiers according to the accuracy, F1-measure and AUC were selected in order to develop a deeper performance and statistical analysis.

4   Experiments and Results

4.1   Experimental Evaluation Measures

Three groups of evaluations were conducted based on the three types of learning approaches. The first group corresponds to the linear classifiers which included the SVM and Logit classifiers together with the classifiers using PCA. The second group was composed of the RB classifiers using verbs, bigrams and trigrams as features. The third group included the combination of Logit classifiers with RB classifiers.

First Group: for this group, the SVM (words) classifier trained with the optimal parameter $C = 1$ reaches values of 0.6104, 0.3999 and 0.6815 for accuracy, F1-measure and AUC respectively. the Logit (words) classifier with optimal parameters ($C = 7$, L2 and $\varepsilon = $ 1e-8) reaches values of 0.6225, 0.4096 and 0.7023 for the same measures. When PCA is performed using the optimal number of components ($q=39$), the PCA Logit (words) classifier yields values of 0.6240, 0.4141 and 0.7122 for the measures, reaching a slightly better performance than the previous models. The performance of the others linear trained classifiers: SVM (words), Logit (verbs, bigrams, trigrams) and PCA Logit (verbs, bigrams, trigrams) is below the given results.

Second Group: for the second group, the accuracy, F1-measure, and AUC for the RB (trigrams thr$\geq$1) classifier are 0.6966, 0.6454 and 0.7838, respectively. The RB (bigrams thr$\geq$1) classifier reaches values of 0.6667, 0.5508 and 0.7494 for the same measures. The performance of the rest of the trained classifiers this group: RB (verbs thr$\geq$2), RB (bigrams thr$\geq$2; thr$\geq$3) and RB (trigrams thr$\geq$2; thr$\geq$3) is below the results of the two aforementioned classifiers.

Third Group: in this group, the Logit (words) + RB (trigrams thr$\geq$1) classifier reaches an accuracy, F1-measure and AUC of 0.7671, 0.6994 and 0.8230, respectively. The PCA Logit (words)+ RB (trigrams thr$\geq$1) obtains 0.7671, 0.7018 and 0.8246 for the same measures. The Logit (words) + RB (bigrams thr$\geq$1) classifier reaches a performance of 0.6838, 0.5703, 0.7542 for the measures. The results of the other combined classifiers: Logit (words) + RB (verbs thr$\geq$2), Logit (words) + RB (bigrams thr$\geq$2; thr$\geq$3), Logit (words) + RB (trigrams thr$\geq$2; thr$\geq$3), PCA Logit (words) + RB (verbs thr$\geq$2), PCA Logit (words) + RB (bigrams thr$\geq$2; thr$\geq$3) and PCA Logit (words) + RB (trigrams thr$\geq$2; thr$\geq$3) are lower than the ones obtained from the aforementioned classifiers.

4.2   Overall Results

After performing the three groups of experiments, the best five models were selected based on the highest values of accuracy, F1-measure and AUC: two of them correspond to linear classifiers (with and without PCA), one to the RB classifier and two to combined classifiers. These are shown in table 2.

Table 2. Summary of classifiers with best evaluation measures

Classifiers	Accuracy	F1-measure	AUC
Logit (words)	0.6225	0.4096	0.7023
PCA Logit (words q=39)	0.6240	0.4141	0.7122
RB (trigrams thr$\geq$1)	0.6966	0.6454	0.7838
Logit (words) + RB (trigrams thr$\geq$1)	0.7671	0.6994	0.8230
PCA Logit (words q=39) + RB (trigrams thr$\geq$1)	**0.7671**	**0.7018**	**0.8246**

Table 3. F1-measure and AUC per category

Category	Logit		PCA Logit		Trigrams		Logit + Trigrams		PCA Logit + Trigrams	
	F1-measure	AUC	F1-measure	AUC	F1-measure	AUC	F1-measure	AUC	F1-measure	AUC
Remembering	0.66	0.84	0.66	0.81	0.75	0.86	0.81	0.89	0.82	0.89
Understanding	0.65	0.65	0.62	0.66	0.73	0.78	0.79	0.82	0.78	0.82
Analyzing	0.00	0.00	0.00	0.00	0.60	0.72	0.61	0.72	0.61	0.72
Evaluating	0.00	0.00	0.00	0.00	0.53	0.68	0.53	0.68	0.53	0.68
Uncodable	0.73	0.90	0.76	0.86	0.62	0.77	0.76	0.83	0.78	0.83

Table 3 shows the results for F1-measure and AUC for the selected models regarding the individual categories. According to these results, the Logit (words) and PCA Logit (words) classifiers do not perform well in categories Analyzing and Evaluating. In contrast with the linear classifiers, the RB (trigrams thr$\geq$1) classifier reaches better performance for those categories. The results obtained from the combination of classifiers were better than the other ones. Here, the results of categories Remembering, Understanding and Uncodable increased considerably comparing with the results of individual classifiers; while the results for categories Analyzing and Evaluating remained the same as expected, mainly due to the misclassification coming from the linear classifiers.

In order to find the statistical significance of the results, the McNemar paired test was performed over the general results of the top five classifiers. In this test we applied the Bonferroni adjustment factor to calculate a new significance level α^* in order to avoid the multiplicity effect. This was calculated with the formula: $1 - (1 - \alpha^*)^n \leq \alpha$, where n was the number of learned models and α was the two tails significance level (0.05). Replacing the values and solving the equation, the new significance level for which the result of a paired test was compared is $\alpha^* = 0.0017083$. Thus, the paired test was performed between each pair of classifiers from table 2. Furthermore, The p-value resulting from the McNemar test was compared with the α^*. For this test, the null hypothesis $H_0 : p_a = p_b$ stands that there is not significant difference between two classifiers when $p \geq 0.00170832$. The p-value for each pair of classifiers is shown in table 4.

Table 4. Paired tests with their p-value

Classifier 1	Classifier 2	p-value
Logit (words)	PCA Logit (words q=39)	0.00005706
Logit (words)	RB (trigrams thr$\geq$1)	0.00000000
Logit (words)	Logit (words) + RB (trigrams thr$\geq$1)	0.00000024
Logit (words)	PCA Logit (words q=39) + RB (trigrams thr$\geq$1)	0.00000002
PCA Logit (words q=39)	RB (trigrams thr$\geq$1)	0.00000036
PCA Logit (words q=39)	Logit (words) + RB (trigrams thr$\geq$1)	0.00128306
PCA Logit (words q=39)	PCA Logit (words q=39) + RB (trigrams thr$\geq$1)	0.00060775
RB (trigrams thr$\geq$1)	Logit (words) + RB (trigrams thr$\geq$1)	0.00002019
RB (trigrams thr$\geq$1)	PCA Logit (words q=39) + RB (trigrams thr$\geq$1)	0.00030624
Logit (words) + RB (trigrams thr$\geq$1)	PCA Logit (words q=39) + RB (trigrams thr$\geq$1)	0.91383314

The McNemar test shows that the difference in performance between the Logit (words) + RB (trigrams thr$\geq$1) classifier and the PCA Logit (words) + RB (trigrams thr$\geq$1) is not significant (p=0.914). The test also shows that there is a statistical significance between the other results of the classifiers.

4.3 Analysis

Starting from the Logit classifier, the evaluation measures showed that while being a simple classifier it achieves a good performance, classifying correctly more than half of the test data. The F1-measure corroborates that there are many examples classified either as false positive or false negative. Moreover, the AUC measure shows a better value, but considering that the AUC is a weighted average per category, it could be the case that some categories with many examples helps to produce a high value for this measure.

In the same order of the Logit classifier is the PCA Logit classifier, which produces similar results. Eventually, if a choice of the learner needs to be done, it is important to consider the processing time of both classifiers. The training phase of the PCA Logit classifier needs to execute a grid search varying the number of components extracted from PCA in order to reach ist optimal performance, resulting in a higher computational cost. On the other hand, the test phase is performed faster because of the reduced number of features used. However, the general time spent for training and testing the Logit classifier is lower than the one of the PCA Logit classifier.

Equally important is the comparison of performance between the RB classifiers and Logit classifiers. The RB trigram learner has better performance than the Logit model. The reason of this behavior is given by the distribution of the data. Most of the data has many dependent features among categories. While the RB learner tries to manage the overlapping patterns in a better way allowing that a rule can belong to two or more classes, linear classifiers cannot deal with this type of data, and even the models using kernels did not succeed neither.

Finally, analyzing the results given by the combined classifiers, it can be easily observed how the performance increases when adding the output predictions of two classifiers. This clearly shows that the combination of classifiers based on probabilities helps to improve the accuracy by acting as a sum of weights where prior knowledge for categories has been modeled. The PCA Logit classifier combined with the RB trigram classifier achieves the best performance among all the classifiers. However, one would prefer the simplicity of creating a Logit classifier using words as features rather than searching for the best number of components in PCA. This selection could be motivated by the Occam's Razzor principle, which states that the hypothesis selected should be the simplest one.

5   Conclusions

This paper has shown that the labeling of forums using Bloom's taxonomy can be done using a Rule-Based (RB) classifier in combination with a linear classifier. The evaluation measures showed that taking a Logit and a RB classifier as basis for combining them could yield a better performance of classifiers than the individual classifiers. The combination of the models is given by summing the outputs of each classifier and taking the highest value of the resulting vector of probabilities as the predicted label.

In addition, the results of this study show that RB classifiers are better suited than models for this particular dataset. This is mainly due to the distribution of data over categories. Linear classifiers act as discriminant functions but when many data overlap, the accuracy for such overlapping categories decreases and consequently the overall accuracy decreases. In contrast, RB classifiers can discriminate a category with rules belonging to more than one category.

Although the current study is based on a small and unbalanced dataset, the findings suggest that a RB classifier is a good model solution. An idea to overcome the problem of the unbalanced data could be to fulfill a stratified sampling over the training data and have an homogeneous distribution, but certainly in real life all categories are not equally probable to appear. Therefore, it is better to treat the problem as was initially proposed without making any special treatment for this particularity on the data.

Also, it should be noted that, despite of the use of SVM as a primary classifier for TC tasks, the present study achieved a better performance with logistic regression (Logit) and Rule-Based (RB) classifiers.

Finally, several limitations need to be considered. First, the number of samples for training and testing was small and this affected directly the performance of classifiers in categories with a low distribution over the whole dataset. Another limitation is the origin of the data. The original data was collected in Spanish and translated into English by different people. At the moment of translation some misspellings or bad interpretations could have been carried out.

A further work could improve the selection of relevant features for the Rule-Based classifiers using a topic modeling and discriminate words that are off-topic, meaning that those off-topic words are relevant for the category. Another approach that could be implemented is the use of algorithms like sequential pattern mining using projection databases to find frequent sequential patterns.

Acknowledgments. We would like to thank to the Centro de Tecnologías de Información and to Professor Katherine Chiluiza G. for the data provided for this study. This research was supported partially by the KU Leuven project RADICAL (GOA 12/003).

References

1. Bloom, B.S., Engelhart, M., Furst, E.J., Hill, W.H., Krathwohl, D.R.: Taxonomy of Educational Objectives: The Classification of Educational Goals. In: Handbook I: Cognitive Domain, vol. 19, Longman, Green (1956)
2. Chang, W., Chung, M.: Automatic applying Bloom's taxonomy to classify and analysis the cognition level of English question items. In: 2009 Joint Conferences on Pervasive Computing (JCPC), pp. 727–734 (2009)
3. Chiluiza, K., Echeverria, V.: Cognitive and Meta-Cognitive Skills Measurement: What about the Task in Web 2.0 Environments? In: Proceedings of Society for Information Technology & Teacher Education International Conference, pp. 1685–1690 (2012)
4. Joachims, T.: Text categorization with support vector machines: learning with many relevant features. In: Nédellec, C., Rouveirol, C. (eds.) ECML 1998. LNCS, vol. 1398, pp. 137–142. Springer, Heidelberg (1998)
5. Jolliffe, I.: Principal Component Analysis. Wiley Online Library (2005)
6. Kittler, J.: Combining classifiers: A theoretical framework. Pattern Analysis and Applications 1(1), 18–27 (1998)
7. Krathwohl, D.: A revision of Blooms Taxonomy: An overview. Theory into Practice 41(4), 212–218 (2002)
8. Miyake, N.: Computer supported collaborative learning. In: The SAGE Handbook of E-Learning Research, pp. 248–267. SAGE Publications Ltd. (2007)
9. Omar, N., Haris, S.S., Hassan, R., Arshad, H., Rahmat, M., Zainal, N.F., Zulkifli, R.: Automated Analysis of Exam Questions According to Bloom's Taxonomy. Procedia - Social and Behavioral Sciences 59, 297–303 (2012)
10. Pincay, J., Ochoa, X.: Automatic Classification of Answers to Discussion Forums According to the Cognitive Domain of Blooms Taxonomy using Text Mining and a Bayesian Classifier. In: Proceedings of World Conference on Educational Multimedia, Hypermedia and Telecommunications, pp. 626–634 (2013)
11. Salton, G., Buckley, C.: Term-weighting approaches in automatic text retrieval. Information Processing & Management 24(5), 513–523 (1988)
12. Sebastiani, F.: Machine learning in automated text categorization. ACM Computing Surveys 34(1), 1–47 (2002)
13. Valcke, M., Wever, B.D., Zhu, C., Deed, C.: Supporting active cognitive processing in collaborative groups: The potential of Blooms taxonomy as a labeling tool. The Internet and Higher Education 12(3), 165–172 (2009)
14. Yusof, N., Hui, C.J.: Determination of Bloom's cognitive level of question items using artificial neural network. In: 10th International Conference on Intelligent Systems Design and Applications (ISDA), pp. 866–870 (2010)
15. Zurita, G., Nussbaum, M.: Computer supported collaborative learning using wirelessly interconnected handheld computers. Computers & Education 42(3), 289–314 (2004)

Classifying Papers from Different Computer Science Conferences

Yaakov HaCohen-Kerner[1], Avi Rosenfeld[2],
Maor Tzidkani[1], and Daniel Nisim Cohen[1]

[1] Dept. of Computer Science, Jerusalem College of Technology, 9116001 Jerusalem, Israel
`kerner@jct.ac.il`, `{maortz,sdanielco}@gmail.com`
[2] Department of Industrial Engineering,
Jerusalem College of Technology, 9116001 Jerusalem, Israel
`rosenfa@jct.ac.il`

Abstract. This paper analyzes what stylistic characteristics differentiate different styles of writing, and specifically types of different A-level computer science articles. To do so, we compared various full papers using stylistic feature sets and a supervised machine learning method. We report on the success of this approach in identifying papers from the last 6 years of the following three conferences: SIGIR, ACL, and AAMAS. This approach achieves high accuracy results of 95.86%, 97.04%, 93.22%, and 92.14% for the following four classification experiments: (1) SIGIR / ACL, (2) SIGIR / AAMAS, (3) ACL / AAMAS, and (4) SIGIR / ACL / AAMAS, respectively. The Part of Speech (PoS) and the Orthographic sets were superior to all others and have been found as key components in different types of writing.

Keywords: Classification and regression trees, Conference classification, Decision tree learning, Document classification, Feature sets, Text classification.

1   Introduction

Academic conference papers from different research domains are differ in their content and can be classified according to suitable content words. However, we intuitively feel that additional stylistic elements exist that differentiate writing styles in different research domains. The goal of this research is to quantify elements of stylistic characteristics that differentiate different types of different A-level computer science articles.

To create this model, this study uses text classification (TC) and supervised machine learning (ML) to create this automated classifier. TC is the supervised learning task of classifying natural language text documents to one or more pre-defined categories from training data [21]. The main difficulty with this learning task is that a large number of features exist in the training data, and some, if not many, are redundant and can be ignored. Additionally, having too many features can lead to the "curse of dimensionality" whereby the model's accuracy is reduced as

H. Motoda et al. (Eds.): ADMA 2013, Part I, LNAI 8346, pp. 529–541, 2013.

more features are considered [26]. Thus, an important question is identifying which features are necessary for building an accurate model.

To date, TC approaches typically focused on the content within a paper to identify these features in sentiment analysis [23, 24, 28], spam filtering [1, 39]; genre [3, 14, 29, 19] and gender [15, 2, 13]. For instance, economical texts use inherently different content words and phrases than political texts. In contrast, this work does not consider the actual content within the paper or any keywords, yet strives to build an accurate model nonetheless. Instead, we model writing differences based on four general feature sets: orthographic, quantitative, vocabulary richness, and part of speech.

Other works have taken aspects of this approach within other problems. These include use of linguistic features in classifying news stories [2], work on gender identification [15, 2, 13], work on classifying documents according to their author [29, 6, 22, 4, 32, 16, 17], and work on identifying the historical period and ethnic origin of different documents [10, 11].

The key contribution of this paper is its finding that a paper's vocabulary richness (variation) and use of certain parts of speech can accurately classify what type of writing it is. Specifically, we studied full papers from three leading computer science conferences from different research areas: SIGIR (Special Interest Group of Information Retrieval), ACL (Association for Computational Linguistics), and AAMAS (Autonomous Agents and Multi-Agent Systems) from the past 6 years. As each of these conferences studies different aspects of computer science, we wished to ascertain what characteristics differentiate the writing in these venues. We found that different conferences exhibited strong differences based on the average part of speech usage, suggesting that stylistic differences between venues can be quantified along the lines presented in this paper.

A general overview of systems that automatically evaluate and score written prose can be found in [7]. Dikli describes several Automated Essay Scoring (AES) systems. The systems check various aspects of the essay, e.g.: spelling, grammar, redundancy, ideas and content, organization of the essay, essay length, plagiarism, sentence structure, coherence, focus and unity (coherence), and development and elaboration. These systems use various techniques such as (1) Latent Semantic Analysis (LSA) [9]. Latent Semantic Analysis (LSA) is defined as "a statistical model of word usage that permits comparisons of the semantic similarity between pieces of textual information" [18, p. 2]; (2) empirical natural language methods that employ statistical or machine learning techniques, and (3) Bayesian text classification.

This paper is organized as follows: Section 2 describes four feature sets for classification used in related works. Section 3 presents the classification model. Section 4 describes the results of the experiments and analyzes them. Section 5 presents a summary and proposals for research directions.

2   Feature Sets

Various feature sets have been proposed and applied in many research domains, such as authorship attribution [21, 29, 31], genre-based text classification and retrieval [3, 14, 29, 19], and sentiment analysis [23, 28].

In this paper, we consider 103 features divided into four categories of feature sets: 10 Orthographic features (O), 3 Quantitative features (Q), 18 Vocabulary Richness features (V), and 72 Part of Speech (PoS) features.

(1) The O features focus on the special characters needed for correct language usage. These features typically focus on spelling and include: capitalization, word breaks and punctuation marks. We then normalized these features both by the number of characters and words (thus creating two normalized values per symbol) to account for differences in article length. For example, we checked the number of " ' " in the document and then normalized this value by the number of characters in the document. The implemented **O** set contains the frequencies for 10 common symbols used to write grammatically correct English. These 10 symbols are: " ' ", " " ", ":", ";", ".", ",", "!", "?", "-", and "/".

(2) The Q features present statistical measures that generally describe a document, e.g., average number of characters in a word/sentence/document; average number of word tokens in a sentence/document. Such features were first proposed by Yule [38]. The implemented **Q** set contains 3 normalized features: average number of characters in a word, average number of characters in a sentence, and average number of words in a sentence. We did not apply features such as the number of characters in the document, the number of word tokens in the document, and number of sentences in the document. This was done to prevent the model from focusing on trivial differences between different types of papers (e.g. their length).

(3) The V features capture the richness or the diversity of the vocabulary of a text. They have mainly been applied to authorship attribution research [35]. The most popular feature of this category is the type-token ratio S/N where S is the size of the vocabulary of the document, and N is the number of tokens of the document. Additional similar features are the hapax legomena/N where hapax legomena are words occurring once in the sample text and the dis legomena / N where dislegomena are words occurring twice in the sample text. Additionally, we considered the variations of this rule for words that appear only 3, 4, or 5 times. We considered two variations of the **V** features - those that first stemmed the words (**VRS**), and those that did not (**VRN**). To implement **VRS**, we used the established Porter stemmer [27]. Using a stemmer would reduce the number of unique words in **VRS** compared to **VRN**. For example, if previously "boy" and "boys" had appeared once each, a stemmer would identify these words as the same word, and thus are no longer unique. The **V** set was broken into the **VRN** and **VRS** variations. The implemented **VRN** contains 6 features: the number of different words in the tested document normalized by the number of words (Diff) (i.e., the type-token ratio), the number of unique words (One), words with two, three, four and five appearances. The implemented **VRS** contains 12 features, which are variations of the following 6 features: the number of different stems in the tested document normalized by the number of words (Diff), and the number of stems with only one, two, three, four and five appearances, respectively. Each one of these 6 features was normalized twice: one by the total number of words and one by the number of stems.

(4) The PoS features were created via the established Stanford Part-of-Speech tagger [34], which was used to identify the part of speech for every term in the documents. We again normalized each one of the 36 parts of speech by N the number of tokens of the document, and by the total parts of speech within the document. Hence, the implemented **PoS** set contains 72 normalized frequencies of tagged types.

3   The Classification Model

Current-day text classification presents challenges due to the large number of features present in the text set, their dependencies and the large number of training documents. Effective feature selection is essential to make the learning task efficient and more accurate.

Our general methodology is as follows: We first created a varied corpus with full articles from three A-level conferences: SIGIR, ACL, and AAMAS from the following 6 years: 2007-2012. We then used a commercial, off-the-shelf conversion program (www.abbyy.com) to convert the PDF files of the source articles into text files that could be analyzed for the four feature sets described above. In order to remove the potential influence the different lengths of these papers had on this analysis, we considered only the first 500 words for each paper. Our consideration is supported by the Heaps' law (also known as Herdan's law). This is an empirical linguistic law, discovered by Gustav Herdan in 1960 [8]. This law describes the number of distinct words in a document as a function of the document length in words. The Heaps' law means in general that as higher is the number of the words in the document as lower is the proportion of unique words in the document.

The V features are considered unreliable to be used alone; however, they can be useful in combination with other feature types [33, 20].

In the typical format of these conferences, 500 words, probably include the title, abstract, and part of the introduction section. While this may remove important content from body of the paper from our analysis, success with such a small number of words only strengthens the significance of the approach we present.

The feature sets from these text files were then used to create an input file for use with the recognized ML Weka Package [36, 12]. We then analyzed the output from the classification algorithms found within Weka applying the test mode of 10-fold cross-validation. The success of this classification approach is clearly based on the features entered into the model.

While many ML algorithms exist within the Weka package, we intentionally chose decision tree methods as the base of our classifier. Comparing other classification methods (e.g., Support Vector Machines, Bayesian learners and Neural Networks), decision trees have many advantages: (1) people are able to understand the produced trees quickly relatively to other ML methods, (2) they handle both numerical and categorical data, (3) having interpretable models instead of black boxes, in order to know which features constitute the stylistic difference, and (4) require little data preparation; i.e., no need for data normalization and production of dummy variables [37, 27].

Decision tree learning methods are used in various applications such as: classification, clustering, machine learning, pattern matching, and text mining [27]. These methods use a decision tree as a predictive model, which maps features' values of an item to decisions concerning the item's target value. In the produced decision trees, leaves represent class labels and branches represent conjunctions of features that lead to those class labels.

Specifically, we focus primarily on the results achieved by the Classification And Regression Tree (CART) machine learning (ML) method, implemented as SimpleCart in Weka [36, 12]. CART has been applied in our experiments since it is a ML method particularly suitable for classification tasks that use categorical or numeric features. CART refers to a non-parametric decision tree learning technique that produces either classification or regression trees, depending on whether the dependent variable is categorical or numeric, respectively. This ML method was first introduced by Breiman [5].

4   Experimental Results

We created a study of differences between accepted papers of three top rated (A) computer science conferences from the past 6 years (2007-2012). The three conferences are: SIGIR - Special Interest Group of Information Retrieval, ACL - Association of Computational Linguistics, and AAMAS - International Conference on Autonomous Agents and Multiagent Systems. These three conferences belong to three different research domains: autonomous agents and multiagent systems, computational linguistics (CL), and Information Retrieval (IR), respectively. We wished to ascertain if there are features that can consistently differentiate between different types of research papers, with even a limited fixed amount of 500 words from each conference. This corpus contains a total of 2082 full papers of which 565 are SIGIR papers, 714 are ACL papers, and 803 are AAMAS papers. As all of these conferences are A-level conferences, we assumed that with very few exceptions, all writing in all conferences was grammatically correct English.

Using the SimpleCart ML method, we performed four classification experiments representing all possible combinations between these conferences: (1) SIGIR / ACL, (2) SIGIR / AAMAS, (3) ACL / AAMAS, and (4) SIGIR / ACL / AAMAS. The measure of accuracy in all experiments is the fraction of the number of documents correctly classified to the total number of possible documents to be classified. The results of these four experiments are summarized in Tables 1-4 and Figures 1-4, respectively. The baseline classifier (row 1) in tables 1-4 naively assumes that the paper belongs to the largest category. Rows 2–6 present the classification results using each one of the **O, Q, VRN, VRS** and **PoS** feature sets, respectively. Row 7 (**All**) presents the accuracy using all features.

Table 1 presents the results of the SIGIR / ACL experiment, and Fig. 1 presents the produced CART decision tree for this classification task.

Table 1. Classifying full SIGIR / ACL papers, based on feature sets

Row		Accuracy in %
1	Majority	55.82
2	O	94.37
3	Q	63.57
4	VRN	73.89
5	VRS	74.28
6	PoS	90.15
7	ALL Sets	95.86
8	Weighted average F-Measure (ALL)	0.959
9	Mean absolute error (ALL)	0.0585
10	Root mean squared error (ALL)	0.1912

/_in_letters (O)< 0.000829: ACL (636/10)
/_in_letters (O)>= 0.000829
| PoS-normalized_by_ sentences_count (PoS)< 0.005435
| | /_in_words (O)< 0.0057915: SIGIR(16/0)
| | /_in_words (O)>= 0.0057915: ACL (48/5)
| PoS- normalized_by_ sentences_count (PoS) >= 0.005435: SIGIR (534/30)

Fig. 1. The decision tree developed for the SIGIR / ACL classification

Several general conclusions can be drawn from Table 1 and Fig. 1:

1. The improvement rates presented in Table 1 are significant improvements from the majority baseline (row 1) in all cases (rows 2-7). The best improvement rate (from the majority baseline to the combination of all sets) is 40.4%.
2. The Orthographic (row 2) and the PoS (row 6) sets were superior to all other sets.
3. The low values of the mean absolute error (row 9), and the root mean squared error (row 10) indicate that the classification results are stable.
4. The F-measure result was also almost optimal and indicates that this classification task was highly successful.
5. The produced CART decision tree presented in Fig. 1 is very simple and contains only three features represented in 7 nodes including 4 leaf nodes. On the right side of the tree's root we have "(636/10)", which means that 636 ACL papers were successfully recognized as ACL papers, while 10 SIGIR papers were wrongly recognized as ACL papers.

6. Much more ACL's full papers (than SIGIR's full papers) contain less than 0.000829 of appearances of '/' divided by the number of letters contained in the first 500 words.

7. Much more SIGIR's full papers contain more or equal to 0.005435 of appearances of PoS_normalized_by_sentences_count (the number of Possessive endings such as "system's" divided by the number of sentences that contain the first 500 words).

From the initial analysis of these results, it seems clear that the Orthographic and the PoS feature sets were the most important feature sets in differentiating between different types of research articles. One possible explanation for these findings is that different conferences have different styles. Certain elements of these styles can be quantified via analyzing that writing's frequency (normalized) for part of speech usage. It is important to note that the best result for the preformed experiments has been achieved using only three features: (1) "/_in_letters" (the number of '/' divided by the number of letters contained in the first 500 words), (2) PoS_normalized_by_sentences_count (the number of Possessive endings such as "system's" divided by the number of sentences that contain the first 500 words), and (3) "/_in_words" (the number of '/' divided by 500, which is the number of the words that are taken into account). This result confirms other ML results that a "curse of dimensionality" can exist when adding even seemingly important features [26].

Table 2 presents the results of the SIGIR / AAMAS experiment, and Fig. 2 presents the produced CART decision tree for this classification task.

Table 2. Classifying full SIGIR / AAMAS papers, based on feature sets

Row		Accuracy in %
1	Majority	59.21
2	O	95.67
3	Q	78.27
4	VRN	59.86
5	VRS	59.64
6	PoS	86.21
7	ALL Sets	97.04
8	Weighted average F-Measure (ALL)	0.97
9	Mean absolute error (ALL)	0.0483
10	Root mean squared error (ALL)	0.1634

/_in_letters (O)< 0.0008705 : AAMAS(770/10)
/_in_letters (O)>= 0.0008705
| NNPS_normalized (PoS)< 0.004654 : SIGIR(537/17)
| NNPS_normalized (PoS)>= 0.004654
| | CC_normalized (PoS)< 0.03256 : AAMAS (27/2)
| | CC_normalized (PoS)>= 0.032536 : SIGIR(16/6)

Fig. 2. The decision tree developed for the SIGIR / AAMAS classification

The general conclusions that can be drawn from Table 2 and Fig. 2 are rather similar to those concluded from Table 1 and Fig. 1. The main differences between the two experiments are that two features integrated in the decision tree have been changed. The "/_in_letters" feature remains in the root node. The two new features are taken from the PoS set: (1) "NNPS_normalized (PoS)", which means the number of plural proper nouns (e.g., Americans) divided by 500, the number of the words that are taken into account, and (2) "CC_normalized", which means the number of coordinating conjunctions (e.g., and, but, or, nor, for, yet, so) divided by 500. The AAMAS' full papers contain relatively less appearances of "/" than the SIGIR's full papers.

The results of the last two classification experiments: ACL / AAMAS, and SIGIR / ACL / AAMAS are shown in Tables 3-4 and Figures 3-4, respectively. The produced CART decision trees for these experiments are much larger and complicated than the previous two decision trees.

Due to space limitations, we do no present the analysis of these results in detail. The results again demonstrate the success of the presented approach, and the general conclusions are rather similar, except for minor differences, e.g., in the third experiment the best set was the PoS set and the feature at the root node was a PoS's feature (NNPS_normalized) and not an orthographic feature.

Table 3. Classifying full ACL / AAMAS papers, based on feature sets

Row		Accuracy in %
1	Majority	53.46
2	O	83.90
3	Q	79.47
4	VRN	76.34
5	VRS	69.95
6	PoS	92.96
7	ALL Sets	93.22
8	Weighted average F-Measure (ALL)	0.934
9	Mean absolute error (ALL)	0.0988
10	Root mean squared error (ALL)	0.2452

NNPS_normalized (PoS)< 0.002793
| LS_normalized_by_sentences_count (PoS)< 0.0066665
| | averageCharacterPerWords (Q)< 4.7439275
| | | averageCharacterPerSentences (Q)< 43.436955: ACL(6/0)
| | | averageCharacterPerSentences (Q)>= 43.436955: AAMAS(16/3)
| | averageCharacterPerWords (Q)>= 4.7439275
| | | FW_normalized (PoS)< 0.003861
| | | | averageCharacterPerWords (Q)< 5.187702
| | | | | averageCharacterPerSentences (Q)< 50.46679: ACL(19/3)
| | | | | averageCharacterPerSentences (Q)>= 50.46679: AAMAS(15/3)
| | | | averageCharacterPerWords (Q)>= 5.187702: ACL(89/4)
| | | FW_normalized (PoS)>= 0.003861: ACL(515/6)
| LS_normalized_by_sentences_count (PoS)>= 0.0066665
| | averageWordsPerSentences (Q)< 9.574204
| | | oneWordsNormalized (V)< 0.348445: ACL(28/1)
| | | oneWordsNormalized (V)>= 0.348445
| | | | VBN_normalized (PoS)< 0.018228: ACL(8/1)
| | | | VBN_normalized (PoS)>= 0.018228
| | | | | comma_in_letters (O)< 0.008367: ACL(4/0)
| | | | | comma_in_letters (O)>= 0.008367: AAMAS(15/2)
| | averageWordsPerSentences (Q)>= 9.574204
| | | FW_normalized_by_sentences_count (PoS)< 0.3156025
| | | | VBP_normalized_by_sentences_count (PoS)< 0.5138995: AAMAS(112/5)
| | | | VBP_normalized_by_sentences_count (PoS) >= 0.5138995: ACL(5/1)
| | | FW_normalized_by_sentences_count (PoS)>= 0.3156025: ACL(7/1)
NNPS_normalized (PoS)>= 0.002793
| JJ_normalized_by_sentences_count (PoS)< 1.646307: AAMAS(643/11)
| JJ_normalized_by_sentences_count (PoS)>= 1.646307: ACL(9/2)

Fig. 3. The decision tree developed for the ACL / AAMAS classification

Table 4. Classifying full SIGIR / ACL / AAMAS papers, based on feature sets

Row		Accuracy in %
1	Majority	39.07
2	O	84.04
3	Q	61.17
4	VRN	53.69
5	VRS	50.98
6	PoS	82.66
7	ALL Sets	92.14
8	Weighted average F-Measure (ALL)	0.921
9	Mean absolute error (ALL)	0.0812
10	Root mean squared error (ALL)	0.2159

/_in_letters (O)< 8.705E-4
| NNPS_normalized (PoS)< 0.002793
| | LS_normalized_by_sentences_count (PoS)< 0.0066665
| | | averageCharacterPerWords (Q)< 4.7439275
| | | | averageCharacterPerSentences (Q)< 43.436955: ACL(6/0)
| | | | averageCharacterPerSentences (Q)>= 43.436955: AAMAS(15/3)
| | | averageCharacterPerWords (Q)>= 4.7439275: ACL(553/29)
| | LS_normalized_by_sentences_count (PoS)>= 0.0066665
| | | averageWordsPerSentences (Q)< 9.574204
| | | | oneWordsNormalized (V)< 0.348445: ACL(27/2)
| | | | oneWordsNormalized (V)>= 0.348445
| | | | | LS_normalized (PoS)< 0.003799: AAMAS(14/5)
| | | | | LS_normalized (PoS)>= 0.0037990: ACL(9/1)
| | | averageWordsPerSentences (Q)>= 9.574204
| | | | FW_normalized_by_sentences_count (PoS)< 0.3156025: AAMAS(109/12)
| | | | FW_normalized_by_sentences_count (PoS)>= 0.3156025: ACL(6/0)
| NNPS_normalized (PoS)>= 0.002793
| | JJ_normalized_by_sentences_count (PoS)< 1.646307: AAMAS(604/10)
| | JJ_normalized_by_sentences_count (PoS)>= 1.646307: ACL(9/2)
/_in_letters (O)>= 8.705E-4
| POS_normalized_by_sentences_count (PoS)< 0.005435
| | NNPS_normalized_by_sentences_count (PoS)< 0.0370745
| | | /_in_words (O)< 0.0057915: SIGIR(15/2)
| | | /_in_words (O)>= 0.0057915: ACL(48/9)
| | NNPS_normalized_by_sentences_count (PoS)>= 0.0370745: AAMAS(22/1)
| POS_normalized_by_sentences_count (PoS)>= 0.005435
| | CC_normalized (PoS)< 0.0301605
| | | NNPS_normalized (PoS)< 0.004654
| | | | ;_in_words (O)< 0.0038135
| | | | | oneStemsNormalized (V)< 0.2607845: ACL(7/2)
| | | | | oneStemsNormalized (V)>= 0.2607845: SIGIR(24/2)
| | | | ;_in_words (O)>= 0.0038135: ACL(12/1)
| | | NNPS_normalized (PoS)>= 0.004654: AAMAS(9/2)
| | CC_normalized (PoS)>= 0.0301605: SIGIR(505/22)

Fig. 4. The decision tree developed for the SIGIR / ACL / AAMAS classification

5   Summary and Future Work

In this paper, we present a methodology for quantifying different types of writing based on orthographic, quantitative, vocabulary richness, and part of speech. This methodology achieves high accuracy results of 95.86%, 97.04%, 93.22%, and 92.14% for the following classification experiments: (1) SIGIR / ACL, (2) SIGIR / AAMAS, (3) ACL / AAMAS, and (4) SIGIR / ACL / AAMAS, respectively.

This paper represents a study of papers from six years of three A-level CS conferences, which belong to different research domains. We found that different conferences could be differentiated with more than 92% accuracy based on the feature sets, mentioned above. The PoS and the Orthographic sets were superior to all others. These initial results suggest that stylistic differences between venues can be quantified along the lines presented in this paper.

In ongoing work, we are considering several directions based on these results. We have begun to consider different types of writing – both within additional scientific conferences and within generalized (real-world) writing. As the methodology we present is general, we are confident that we will be find similar distinctions between other corpora. Our longer-term goal is to facilitate an application to automatically review papers: both scientific and general. We envision that this result will allow conference organizers and editors to automatically recommend rejecting, accepting, or other consideration (e.g. poster). An equally interesting application is for authors themselves to understand how to write better papers through our model. We believe that this research direction is exciting and presents a fascinating approach for quantifying writing styles.

References

1. Androutsopoulos, I., Koutsias, J., Chandrinos, K., Paliouras, G., Spyropoulos, C.D.: An Evaluation of Naive Bayesian Anti-spam Filtering. CoRR, cs.CL/0006013 (2000)
2. Argamon, S., Shimoni, A.R.: Automatically Categorizing Written Texts by Author Gender. Literary and Linguistic Computing 17, 401–412 (2003)
3. Argamon, S., Koppel, M., Avneri, G.: Style-based Text Categorization: What Newspaper am I Reading? In: AAAI Workshop on Learning for Text (1998)
4. Argamon, S., Koppel, M., Pennebaker, J.W., Schler, J.: Mining the Blogosphere: Age, Gender and the Varieties of Self-expression. First Monday 12(9) (2007)
5. Breiman, L., Friedman, J.H., Olshen, R.A., Stone, C.J.: Classification and Regression Trees. In: Monterey, C.A. (ed.) Wadsworth & Brooks/Cole Advanced Books & Software (1984) ISBN 978-0-412-04841-8
6. Diederich, J., Kindermann, J., Leopold, E., Paass, G.: Authorship Attribution with support vector machines. Applied Intelligence 19(1-2), 109–123 (2003)
7. Dikli, S.: An Overview of Automated Scoring of Essays. Journal of Technology, Learning, and Assessment 5(1), 1–35 (2006)
8. Egghe, L.: Untangling Herdan's Law and Heaps' Law: Mathematical and Informetric Arguments. Journal of the American Society for Information Science and Technology 58(5), 702–709 (2007)
9. Foltz, P.W.: Latent Semantic Analysis for Text-based Research. Behavior Research Methods, Instruments and Computers 28(2), 197–202 (1996)
10. HaCohen-Kerner, Y., Beck, H., Yehudai, E., Mughaz, D.: Stylistic Feature Sets as Classifiers of Documents According to their Historical Period and Ethnic Origin. Applied Artificial Intelligence 24(9), 847–862 (2010a)
11. HaCohen-Kerner, Y., Beck, H., Yehudai, E., Rosenstein, M., Mughaz, D.: Cuisine: Classification using Stylistic Feature Sets and/or Name-Based Feature Sets. JASIST 61(8), 1644–1657 (2010b)

12. Hall, M., Frank, E., Holmes, G., Pfahringer, B., Reutemann, P., Witten, I.H.: The WEKA Data Mining Software: an Update. ACM SIGKDD Explorations Newsletter 11(1), 10–18 (2009)

13. Hota, S.R., Argamon, S., Chung, R.: Gender in Shakespeare: Automatic Stylistics Gender Character Classification using Syntactic, Lexical and Lemma Features. In: Digital Humanties and Computer Science (DHCS) (2006)

14. Karlgren, J., Cutting, D.: Recognizing Text Genres with Simple Metrics using Discriminant Analysis. In: Proceedings of the 15th International Conference on Computational Linguistics, pp. 1071–1075 (1994)

15. Koppel, M., Argamon, S., Shimoni, A.R.: Automatically Categorizing Written Texts by Author Gender. Lit. Linguist Computing 17(4), 401–412 (2002)

16. Koppel, M., Schler, J., Argamon, S.: Computational Methods in Authorship Attribution. JASIST 60(1), 9–26 (2009)

17. Koppel, M., Schler, J., Argamon, S.: Authorship Attribution in the Wild. Language Resources and Evaluation 45(1), 83–94 (2011)

18. Lemaire, B., Dessus, P.: A System to Assess the Semantic Content of Student Essays. Educational Computing Research 24(3), 305–306 (2001)

19. Lim, C., Lee, K., Kim, G.: Multiple Sets of Features for Automatic Genre Classification of Web Documents. Information Processing Management 41(5), 1263–1276 (2005)

20. Luyckx, K.: Scalability Issues in Authorship Attribution. Ph.D. Dissertation, Universiteit Antwerpen. University Press, Brussels (2010)

21. Meretakis, D., Wüthrich, B.: Extending Naive Bayes Classifiers using Long Itemsets. In: Proceedings of the Fifth ACM SIGKDD International Conference on Knowledge Discovery and Data Mining (KDD), pp. 165–174. ACM (1999)

22. Novak, J., Raghavan, P., Tomkins, A.: Anti-aliasing on the Web. In: Proceedings of the 13th International Conference on World Wide Web (WWW), pp. 30–39. ACM (2004)

23. Pang, B., Lee, L.: Seeing Stars: Exploiting Class Relationships for Sentiment Categorization with Respect to Rating Scales. In: Proceedings of the 43rd Annual Meeting on Association for Computational Linguistics, pp. 115–124. Association for Computational Linguistics (2005)

24. Pang, B., Lee, L., Vaithyanathan, S.: Thumbs up?: Sentiment Classification using Machine Learning Techniques. In: Proceedings of the ACL 2002 Conference on Empirical Methods in Natural Language Processing (EMNLP 2002), vol. 10, pp. 79–86 (2002)

25. Porter, M.: An Algorithm for Suffix Stripping. Program 14(3), 130–137 (1980)

26. Rosenfeld, A., Zuckerman, I., Azaria, A., Kraus, S.: Combining Psychological Models with Machine Learning to Better Predict People's Decisions. Synthese 189, 81–93 (2012)

27. Rokach, L., Maimon, O.: Data Mining with Decision Trees: Theory and Applications. World Scientific Pub. Co. Inc. (2008) ISBN 978-9812771711

28. Snyder, B., Barzilay, R.: Multiple Aspect Ranking using the Good Grief Algorithm. In: Proceedings of the HLT-NAACL, pp. 300–307 (2007)

29. Stamatatos, E., Kokkinakis, G., Fakotakis, N.: Automatic Text Categorization in Terms of Genre and Author. Comput. Linguist. 26(4), 471–495 (2000)

30. Stamatatos, E., Fakotakis, N., Kokkinakis, G.: Computer-based Authorship Attribution without Lexical Measures. Computers and the Humanities 35(2), 193–214 (2001)

31. Stamatatos, E.: Authorship Attribution based on Feature Set Subspacing Ensembles. International Journal on Artificial Intelligence Tools 15(5), 823–838 (2006)

32. Stamatatos, E.: Author identification: Using Text Sampling to Handle the Class Imbalance Problem. Inf. Process. Manage. 44(2), 790–799 (2008)

33. Stamatatos, E.: A Survey of Modern Authorship Attribution Methods. Journal of the American Society for information Science and Technology 60(3), 538–556 (2009)
34. Toutanova, K., Klein, D., Manning, C.D., Singer, Y.: Feature-rich Part-of-speech Tagging with a Cyclic Dependency Network. In: Proceedings of the 2003 Conference of the North American Chapter of the Association for Computational Linguistics on Human Language Technology (NAACL 2003), vol. 1, pp. 173–180. Association for Computational Linguistics (2003)
35. Tweedie, F.J., Baayen, R.H.: How Variable a Constant Be? Measures of Lexical Richness in Perspective. Computers and the Humanities 32(5), 323–352 (1998)
36. Witten, I.H., Frank, E.: Data Mining: Practical Machine Learning Tools and Techniques, 2nd edn. Morgan Kaufmann Series in Data Management Systems. Morgan Kaufmann (2005)
37. Yuan, Y., Shaw, M.J.: Induction of Fuzzy Decision Trees. Fuzzy Sets and Systems 69, 125–139 (1995)
38. Yule, U.: On Sentence Length as a Statistical Characteristic of Style in Prose with Application to Two Cases of Disputed Authorship. Biometrika 30, 363–390 (1938)
39. Zhang, L., Zhu, J., Yao, T.: An Evaluation of Statistical Spam Filtering Techniques. ACM Transactions on Asian Language Information Processing (TALIP) 3(4), 243–269 (2004)

Vertex Unique Labelled Subgraph Mining for Vertex Label Classification

Wen Yu, Frans Coenen, Michele Zito, and Subhieh El Salhi

Department of Computer Science, University of Liverpool,
Ashton Building, Ashton Street, Liverpool, L69 3BX, UK
{yuwen,coenen,michele,hsselsal}@liverpool.ac.uk

Abstract. A mechanism is presented to classify (predict) the values associated with vertices in a given unlabelled graph or network. The proposed mechanism is founded on the concept of Vertex Unique Labelled Subgraphs (VULS). Two algorithms are presented. The first, the minimal Right-most Extension VULS Mining (minREVULSM) algorithm, is used to identify all minimal VULS in a given graph or nework. The second, the Match-Voting algorithm, is used to achieve the desired VULS based classification (prediction). The reported experimental evaluation demonstrates that by using the minimal VULS concept good results can be obtained in the context of a sheet metal forming application used for evaluation purposes.

Keywords: Data mining, Graph mining, Vertex unique labelled subgraph mining, Classification.

1 Introduction

This paper introduces a novel classification system founded on the concept of Vertex Unique Labelled Subgraphs (VULS) coupled with a Match-Voting algorithm to predict the unknown vertex labelling in a given graph or network G. The basic idea is that, given a suitably defined training set, if we can identify subgraphs that have unique vertex labelings associated with them we can use this knowledge to predict the vertex labelings for previously unencountered graphs. We refer to such subgraphs as Vertex Unique labelled Subgraphs (VULS). There are two elements to the proposed mechanism: (i) the process for identifying individual VULS in a vertex labelled training graph, and (ii) the process for matching the identified VULS to vertex-unlabelled testing graph so that vertex labels can be predicted. There are two challenges with respect to the first process:

1. The need to identify a sufficiently comprehensive set of VULS (a collection of VULS that will ensure good "coverage" with respect to unseen data).
2. The large number of potential VULS that can be contained in a reasonably sized graph (the identification process thus needs to be efficient).

With respect to the second challenge the issue is how to address the situations where, with respect to a specific vertex in unseen data, either: (i) several competing VULS can be used to label the vertex, or (ii) no appropriate VULS can

H. Motoda et al. (Eds.): ADMA 2013, Part I, LNAI 8346, pp. 542–553, 2013.
© Springer-Verlag Berlin Heidelberg 2013

be found. With respect to the first process the minimal Right-most Extension VULS Mining (minREVULSM) algorithm is presented to find all minimal VULS in a given graph or network. With respect to the second process we present the Match-Voting algorithm.

The proposed VULS based prediction mechanism has a variety of applications. For example given a social network we can use the VULS concept to identify unique structures in the network and consequently to classify the "type" of specific nodes in the network that are covered by VULS. However, to act as a focus for the work described in this paper, a sheet metal forming application is considered. More specifically Asymmetric Incremental Sheet Forming (AISF) [1–5]. An issue with sheet metal forming processes, such as AISF, is that distortions (referred to as "springback") are introduced as a result of the application of the process. These distortions are non-uniform across the "shape" but tend to be related to local geometries. The intuition is that by utilising a system, such as the proposed VULS based systems, to predict springback in a proposed shape (prior to its manufacture) knowledge of this springback can be used to formulate some form of mitigating error correction to the shapes definition.

The rest of this paper is organised as follows. In section 2 we provide a formalism for the concept of VULS together with some examples. The minREVULSM algorithm is then presented in section 3, and the Match-Voting algorithm in Section 4. An experimental analysis of the proposed algorithms, in the context of sheet metal forming, is presented in section 5. Section 6 then presents some conclusions and summarises the work and main findings.

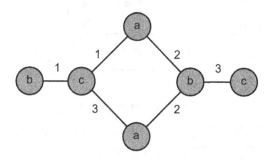

Fig. 1. Example input graph G_1

2   Formalism

A graph G is usually defined in terms of (V, E, L_V, L_E, F), where V is a set of n vertices, such that $V = \{v_1, v_2, \ldots, v_n\}$; E is a set of m edges, such that $E = \{e_1, e_2, \ldots, e_m\}$. The vertices are labelled according to a set of p vertex labels $L_V = \{l_{v_1}, l_{v_2}, \ldots, l_{v_p}\}$. The edges are labelled according to a set of q edge labels $L_E = \{l_{e_1}, l_{e_2}, \ldots, l_{e_q}\}$. F is some function that maps the vertices and edges onto the labels. A graph G can also be conceptualised as comprising k one-edge subgraphs: $G = \{P_1, P_2, \ldots, P_k\}$, where P_i is a pair of vertices linked by an edge, thus $P_i = \langle v_a, v_b \rangle$ (where $v_a, v_b \in V$). The size of a graph G ($|G|$) can thus be defined in terms of its one edge sub-graphs, we refer to 1-edge subgraphs, 2-edge subgraphs and so on up to k-edge subgraphs. We use the notation $P_i.v_a$ and $P_i.v_b$ to indicate the vertices v_a and v_b associated with a particular vertex pair P_i, and the notation $P_i.v_a.label$ and $P_i.v_b.label$ to indicate the labels associated with $P_i.v_a$ and $P_i.v_b$ respectively. We indicate the sets of labels which might be associated with $P_i.v_a$ and $P_i.v_b$ using the notation $L_{P_i.v_a}$ and $L_{P_i.v_b}$ ($L_{P_i.v_a}, L_{P_i.v_b} \in L_V$). We indicate the edge label associated with P_i using the notation $P_i.label$ ($P_i.label \in L_E$). We also define a function, $getVertexLabels$, that returns the potential list of labels S that can be assigned to the vertices in G_2 according to G: $getVertexLabels(G_2) \rightarrow S$, where $G_2 = \{P_1, P_2, \ldots, P_k\}$ and $S = [[L_{P_1.v_a}, L_{P_1.v_b}], [L_{P_2.v_a}, L_{P_2.v_b}], \ldots, [L_{P_k.v_a}, L_{P_k.v_b}]]$ (recall that $L_{P_i.v_a}$ and $L_{P_i.v_b}$ are the sets of potential vertex labels for vertex v_a and v_b associated with a one-edge subgraph P_i).

We typically define the size of G in terms of the number of edges $|E|$. Given a vertex labelled graph $G_1 = (V_1, E_1, L_{V_1}, L_{E_1}, F_1)$ and a vertex unlabelled

Table 1. 1-edge VULS

Table 2. 2-edge VULS

subgraph $G_2 = (V_2, E_2, L_{V_2}, L_{E_2}, F_2)$ such that $|E_2| \leq |E_1|$, $L_{E_2} \subseteq L_{E_1}$, $L_{V_1} \neq$ {} and $L_{V_2} = \{\}$, we can attempt to match G_2 to G_1 so as to populate L_{V_2}. There may be zero, one or more matches. If we have only one match then G_2 is a VULS. A VULS is a minimal VULS if none of its sub-graphs are also VULS.

Considering the example graph G_1 presented in Figure 1. We can identify a number of potential VULS in this graph. Considering only 1-edge VULS we can identify three potential VULS as shown in the left-hand column of Table 1. Populating these candidate VULS, using the vertex labels in G_1, we get the labelings shown in the right-hand column of Table 1. Note that the second 1-edge subgraph, featuring edge 2 is a VULS because it has only one possible vertex labelling associated with it, the remainder have more than one vertex labellings. If we now consider the 2-edge candidate VULS we can identify six such candidate VULS as listed in the left-hand column of Table 2. When we attempt to match these to G_1 we find that the first five exist in G_1 and the sixth does not. Note also that candidate VULS one, two, four and five are all VULS because they have unique vertex labeling, while candidate VULS three is not a VULS because it has more than one possible vertex labellings. No vertex labelling could be matched with the sixth candidate VULS so by default it is not a VULS.

Returning to Table 1 note that the identified one edge VULS (graph 2) is in fact a minimal VULS because it has no sub-graphs that are also VULS. If we only expand the non-minimal 1-edge VULS listed in the left-hand column of Table 1 we get the 2-edge minimal VULS candidates listed in the left-hand column of Table 3. The possible labelings are shown in the right-hand column of Table 3 (the first candidate is also a minimal VULS). Note that with respect to coverage, using only minimal VULS does not necessarily produce as good a coverage as in the case when all VULS are used (inspection of Figure 1 and Tables 1 and 3 indicates that the right hand edge of G_1 is not covered when only minimal VULS are considered, while when using all VULS a coverage of 100% is obtained). However, finding only minimal VULS clearly requires less computational resource. The effect of the trade off between coverage and efficiency is one of the issues that the work presented in this paper seeks to identify.

Table 3. 2-edge minimal VULS when 1-edge minimal VULS are not expanded

Num.	Candidate minimal VULS (G_2)	Populated Candidate minimal VULS
1	(?)—1—(?)—1—(?)	(b)—1—(c)—1—(a)
2	(?)—1—(?)—3—(?)	(a)—1—(c)—3—(a) (b)—1—(c)—3—(a)
3	(?)—3—(?)—3—(?)	

3   The minREVULSM Algorithm

The proposed minREVULSM algorithm is presented in this section. The algorithm is partially influenced by the well known gSpan algorithm [7] in that the graph representation, candidate generation process (right most extension and isomorphism testing) are similar. The pseudo code for minREVULSM is presented in Algorithms 1 and 2. Algorithm 1 presents the high level control structure while Algorithm 2 the detail for determining whether a specific subgraph is a VULS or not. Considering Algorithm 1 first, the algorithm comprises one main procedure ($main$) and a sub-procedure ($genMinVULS$). The algorithm commences with an input graph G_{input} and a parameter max that defines the maximum size for a desired minimal VULS. If we do not limit the size of the searched-for VULSs the entire input graph may ultimately be identified as a minimal VULS which in the context of the target application will not be useful. The output is a set of minimal VULS R. Note that all graphs are encoded using Minimal Depth First Search (DFS) lexicographical ordering (as also used in gSpan [7]). The global variable G (line 7 in Algorithm 1) is the part of the graph G_{input} which is not covered by any of the identified minimal VULS so far. The global variable $coverage$ (line 8) is employed to determine whether G_{input} is covered completely by the minimal VULS identified so far (if so the algorithm stops). The $coverage$ is the percentage of the number of vertices "covered" by the detected minimal VULS so far compared to the total number of vertices in the input graph G_{input} (Equation 1) (as will become apparent later in this paper, when using VULS for classification purposes, high coverage is desirable). The global variable T_k (line 9) is the set of k-edge non-VULS which will be extended further to produce $(k + 1)$-edge candidate VULS. Note that at the start of the procedure, G will be equal to G_{input} and coverage will be 0. We proceed in a breadth first manner starting with one-edge subgraphs ($k = 1$), then two edge sub-graphs ($k = 2$), and so on. We continue in this manner until either: (i) $k = max$ or (ii) the coverage is equal to 100%. On each iteration the $genMinVULS$ procedure is called (line 15).

$$coverage = \frac{num.\ vertices\ covered\ by\ VULS}{num.\ vertices\ in\ G_{input}} \times 100 \qquad (1)$$

The $genMinVULS$ procedure takes as input the current graph size k (where k is the number of edges) and the set of k-edge sub-graphs contained in the set G_k as pruned so far. The procedure returns the set of k-edge minimal VULS. On each call the procedure $genMinVULS$ loops through the input set of k-edge subgraphs and (line 23) for each sub-graph g determines whether it is a VULS or not by calling Algorithm 2 which is described in detail below. If g is a VULS it is added to the set R (line 24). We then (line 25) calculate the coverage so far, if this has reached 100% we have found the complete set of minimal VULS and we exit (line 27). Note that if coverage is equal to 100% the input set G will now be empty. Otherwise, if the coverage is not 100%, we continue processing and (line 29) remove g from the global set G. If g is not a VULS we add it to T_k (line 31), T_k is the set of k-edge sub-graphs which we will eventually be extended to form

Algorithm 1. minREVULSM

1: **Input:**
2: G_{input} = Training input graph
3: max = Max subgraph size
4: **Output:**
5: R = Set of minimal VULS
6: **Global variables:**
7: $G = G_{input}$ (Part of training input graph not covered by minimal VULS)
8: coverage = 0
9: T_k = the set of k-edge subgraphs which are not VULS

10: **procedure** $main(G_{input}, max)$
11:     $k = 1$
12:     G_k = the set of k-edge subgraphs in G
13:     $R = \emptyset$
14:     **while** $(k < max)$ **do**
15:         $R = R \cup genMinVULS(k, G_k)$
16:         G_{k+1} = Set of $(k+1)$-edge subgraphs in G (found by applying right most extension to each subgraph in T_k)
17:         $k = k + 1$
18:     **end while**
19: **end procedure**

20: **procedure** $genMinVULS(k, G_k)$
21:     $T_k = \emptyset$
22:     **for all** $g \in G_k$ **do**
23:         **if** $isaVULS(g, G_k) == true$ (Algorithm 2) **then**
24:             $R = R \cup g$
25:             $coverage$ = compute coverage using Equ. 1
26:             **if** $coverage == 100\%$ **then**
27:                 exit
28:             **end if**
29:             $G = G - g$
30:         **else**
31:             $T_k = T_k \cup \{g\}$
32:         **end if**
33:     **end for**
34:     **if** $T_k == \emptyset$ **then**
35:         exit
36:     **end if**
37:     return R
38: **end procedure**

G_{k+1}, the set of $(k + 1)$-edge sub-graphs, ready for the next level of processing. Eventually all g in G_k will have been processed. If, at this stage T_k is empty there will be no more sub-graphs that can be generated and the process will exit (line 35). Otherwise control will return to the *main* procedure and the set of $(k + 1)$-edge sub-graphs will be generated from T_k (the set of k-edge subgraphs that have not been found to be VULS) using a right most extension technique coupled with isomorphism checking to establish which $(k + 1)$-edge subgraphs are contained in G as processed so far (line 16). This part of the algorithm is not presented here because it is similar to that found in more traditional sub-graph mining algorithms (such as gSpan). The generated minimal VULS set R is then returned (line 37) back to the main process ready for the next iteration (unless the maximum value for k has been reached).

Algorithm 2 presents the pseudo code for identifying whether a given sub-graph g is a VULS or not with respect to the current set of k-edge sub-graphs G_k from which g has been removed. The algorithm returns *true* if g is a VULS and *false* otherwise. The process commences (line 8) by generating the potential list of vertex labels S that can be matched to g according to the content of G_k (see previous section for detail). The list S is then processed and tested. If there exists a vertex pair whose possible labelling is not unique (has more than one possible labelling that can be associated with it) g is not a VULS and the procedure returns *false*, otherwise g is a VULS and the procedure returns *true*. Thus, as the minREVULSM algorithm proceeds, the input graph G will be continuously pruned with respect to the identified VULS. As a result G can become disconnected, any disconnected sub graph of size less than the current value of k can not therefore contain any k-edge VULS. Although not shown in Algorithm 1 any disconnected sub-graphs of size less than k are discounted so as to speed up the overall process.

Algorithm 2. Identify VULS

1: **Input:**
2: g = a single k-edge subgraph (potential VULS)
3: G_k = a set of k-edge subgraphs to be compared with g
4: **Output:**
5: *true* if g is a VULS, *false* otherwise

6: **procedure** $isaVULS(g, G_k)$
7:     $isVULS = true$
8:     S = the list of vertex label pairs of edges that may be assigned to g
9:     **for all** vertex label pairs of each edge $[L_i, L_j] \in S$ **do**
10:         **if** either size of start vertex labels $|L_i| \neq 1$ or size of end vertex labels $|L_j| \neq 1$ **then**
11:             $isVULS = false$
12:             break
13:         **end if**
14:     **end for**
15:     return $isVULS$
16: **end procedure**

---

4   The Match-Voting Algorithm

Once we have identified a set of minimal VULS we can use the set of identified VULS to predict vertex labels with respect to further (unseen) graphs. The basic idea is to match subgraphs contained in the new graph with the collection of identified VULS. The matching is conducted using the Match-Voting Algorithm. The algorithm comprises one main procedure ($main$) and a sub-procedure ($matchVoteVULS$). The algorithm takes as input a collection of VULS and a new graph G_{new} which has edge labels but no known vertex labels. The algorithm also utilises a parameter max set to the same value as that used to generate the VULS so that the algorithm does not continue to try and find matches beyond the maximum size of the VULS presented as part of the input. The output is the graph G_{new} with vertices labelling. We match each VULS from VULSmodel to each of the potential edge lists in G_{new}. Whenever a match is found we attach the labels from the VULS to the vertices associated with the identified edge list. We continue in this manner for each VULS in our input set of VULSmodel. On completion at least some of vertices in G_{input} will have vertex labels associated with them (the more the better). In some cases an edge list in G_{new} may be matched to more than one VULS in which case the associated vertices will have a number of potential labelings. In this case a voting mechanism is used to select the "best" labelling (lines 15-18). We proceed in a breadth first manner starting with one-edge VULS ($k = 1$), then two edge VULS ($k = 2$), and so on. We continue in this manner until $k = max$ is reached. On each iteration the $matchVoteVULS$ procedure is called (lines 11 to 14).

5   Experiments and Performance Study

This section describes the evaluation of the proposed VULS classification system in terms of effectiveness and efficiency. Because of the novelty of the proposed VULS concept there are no alternative algorithms that can be used for comparison purposes. However, we were able to analyse the effectiveness of the prediction using standard classification performance measures (accuracy and AUC), and compare the effectiveness of using only minimal VULS with using all VULS. The rest of this section is organised as follows. Section 5.1 presents the data sets used for the evaluation and briefly describes the application domain. Section 5.2 reports on classification performance with respect to both minimal VULS and all VULS, and in Section 5.3 we consider the "runtime" of the proposed approach. Note the the REVULSM algorithms were implemented using the JAVA programming language; all experiments were conducted using a 2.7 GHz Intel Core i5 with 4 GB 1333 MHz DDR3 memory, running OS X 10.8.1 (12B19).

5.1   Data Sets

The data used for the reported experiments was taken from a real sheet metal forming application, namely the fabrication of flat topped pyramid shapes out

---

Algorithm 3. Match-Voting Algorithm

1: **Input:**
2: G_{new} = Edge labelled testing input graph
3: max = Max subgraph size (consistent with minREVULSM in training procedure)
4: $VULSmodel$ = a set of VULS or minimal VULS
5: **Output:**
6: G_{new} with vertices labelling

7: **procedure** $main(G_{new}, max, VULSmodel)$
8:     $G_{new} = \{P_1, P_2, \ldots, P_i\}$ (where i is the total number of edges in G_{new})
9:     $k = 1$
10:    $VULS_k$ = the set of k-edge VULS or minimal VULS in $VULSmodel$
11:    **while** ($k < max$) **do**
12:        matchVoteVULS($VULS_k$)
13:        $k = k + 1$
14:    **end while**
15:    **for all** $p \in G_{new}$ **do**
16:        $p.v_a.label$ =most frequently voted vertex label in Vote $(p.v_a)$
17:        $p.v_b.label$ =most frequently voted vertex label in Vote $(p.v_b)$
18:    **end for**
19: **end procedure**

20: **procedure** $matchVoteVULS(VULS_k)$
21:    **for all** $vuls \in VULS_k$ **do**
22:        $S \leftarrow getVertexLabels(vuls)$
23:        **if** $\exists\{P_x, \ldots, P_y\} \in G_{new}$ (where $y - x = k$) and $\{P_x.label = L_1.label, P_{(x+1)}.label = L_2.label, \ldots, P_y.label = L_k.label\}$ (where $[L_1, L_2, \ldots, L_k] \in S$) **then**
24:            Vote($P_x.v_a.label = L_{v_{a1}}, P_x.v_b.label = L_{v_{b1}}, \ldots, P_y.v_a.label = L_{v_{ak}}, P_y.v_b.label = L_{v_{bk}}$)
25:        **end if**
26:    **end for**
27: **end procedure**

---

of sheet steel. More precisely the application of Asymmetric Incremental Sheet Forming (AISF), a particular kind of sheet metal forming. A pyramid shape was chosen as it is frequently used as a benchmark shape for conducting experiments in the context of AISF [6]. The pyramid was manufactured twice so that we had one dataset to use as a training set and another to use as a test set. The required input labelled training and test graphs were extracted from the "before" and "after" grids describing the geometry of the piece to be manufactured and the resulting piece actually produced. Each grid square centre point was defined in terms of a Euclidean (X-Y-Z) coordinate scheme. Each before grid square centre point was then considered to represent a vertex and had a springback error value associated with it calculated by measuring the distance along the normal from the before surface at the grid centre to where the normal cut the after surface. Each vertex (except at the edges and corners) was then connected to each of its four neighbours by a set of four edges (one per neighbour) labelled with a "slope" value calculated from the absolute difference in Z value between each centre point and its neighbour. Finally the vertex and edge values were discretised, according to a set of labels L_V and L_E respectively, so that they were represented by nominal values (otherwise every edge pair was likely to be unique). A number of different grid sizes were also considered each related to a different grid size d.

Table 4. Number of VULS Comparison

max	$\|L_E\| \times \|L_V\|$	All VULS			min VULS		
		$d = 10$	$d = 12$	$d = 14$	$d = 10$	$d = 12$	$d = 14$
4	2×2	25	37	24	16	17	18
	3×2	137	135	112	41	35	25
	4×2	286	299	303	70	54	76
5	2×2	139	144	110	50	40	76
	3×2	611	689	573	53	113	37
	4×2	1466	1564	1628	112	113	84
6	2×2	665	696	526	118	46	204
	3×2	2732	3366	2886	87	277	42
	4×2	7137	7423	8392	205	173	89

Table 5. Accuracy Comparison

max	$\|L_E\| \times \|L_V\|$	All VULS			min VULS		
		$d = 10$	$d = 12$	$d = 14$	$d = 10$	$d = 12$	$d = 14$
4	2×2	29.00	9.03	43.37	38.00	26.39	83.67
	3×2	39.00	13.89	37.24	70.00	38.19	78.06
	4×2	37.00	22.92	42.35	68.00	72.22	90.82
5	2×2	30.00	11.81	51.02	70.00	65.28	88.78
	3×2	49.00	17.36	33.67	70.00	58.33	82.65
	4×2	57.00	23.61	15.31	67.00	93.75	91.33
6	2×2	24.00	15.97	47.45	70.00	72.92	92.86
	3×2	31.00	22.22	35.20	70.00	89.58	91.33
	4×2	52.00	20.83	12.76	67.00	77.78	91.33

5.2 Prediction Performance

For the prediction performance evaluation we used a number of different parameter settings for: $\|L_E\| \times \|L_V\|$, d and max. With respect to the evaluation results reported here the following were used: $\|L_E\| \times \|L_V\| = \{2 \times 2, 3 \times 2, 4 \times 2\}$, $d = \{10, 12, 14\}$ and $max = \{4, 5, 6\}$. Use of these parameters all gave a 100% coverage (in the context of the sheet metal forming application 100% coverage is desirable). The results obtained using the above parameters are presented in Tables 5 and 6[1]. From Table 5 it can be observed that using minimal VULS produces much better accuracy results than when using all VULS. This is because

---

[1] When $d = 10$ is used the number of vertices is $d \times d = 10 \times 10 = 100$, so the percentage coverage values are whole numbers.

Table 6. AUC Comparison

max	$\|L_E\| \times \|L_V\|$	All VULS			min VULS		
		$d = 10$	$d = 12$	$d = 14$	$d = 10$	$d = 12$	$d = 14$
4	2×2	44.90	52.88	58.35	34.92	61.87	75.09
	3×2	47.86	7.19	33.70	50.32	19.78	50.72
	4×2	44.41	21.51	31.17	49.93	37.41	49.72
5	2×2	44.58	54.32	59.88	50.32	33.81	72.56
	3×2	52.87	8.99	31.74	50.32	30.22	53.24
	4×2	53.40	12.23	19.03	48.20	48.56	50.00
6	2×2	31.18	56.47	60.58	50.32	37.77	72.13
	3×2	37.13	11.51	32.58	50.32	46.40	57.99
	4×2	44.78	10.79	20.29	48.20	40.29	50.00

when minimal VULS are used there are much fewer competing label predictions for each vertex and consequently the Match Voting algorithm is much more effective. Best results were obtained using $max = 6$ and $d = 14$. Inspection of Table 6 indicates that in terms of the Area Under the receiver operating Curve (AUC) $\|L_E\| \times \|L_V\| = 2 \times 2$ and $d = 14$ produced the best results. Overall the results indicate that the proposed VULS mechanism can be successfully applied in the context of sheet metal forming for the purpose of error (springback) prediction.

5.3 Runtime Analysis

The runtime results obtained, using the same parameters as above, are presented in Table 7. From the table it can be observed (as anticipated) that the runtime for mining minimal VULS is significantly less than when mining all VULS (gains with respect to memory usage are all realised). Thus we can conclude that finding only minimal VULS is more efficient than when finding all VULS while at the same time resulting in an improved accuracy.

Table 7. Runtime Comparison (seconds)

max	$\|L_E\| \times \|L_V\|$	All VULS			min VULS		
		$d = 10$	$d = 12$	$d = 14$	$d = 10$	$d = 12$	$d = 14$
4	2×2	0.34	0.52	0.50	0.27	0.32	0.37
	3×2	0.48	0.66	0.74	0.28	0.36	0.41
	4×2	0.59	0.82	0.87	0.40	0.38	0.41
5	2×2	0.7	1.07	0.93	0.35	0.45	0.55
	3×2	1.35	1.53	1.41	0.39	0.75	0.47
	4×2	1.50	1.69	1.64	0.42	0.60	0.46
	$4 \text{ v } 3$	1.66	1.69	2.38	0.76	0.99	1.04
6	2×2	1.62	1.81	1.54	0.53	0.63	0.95
	3×2	1.91	2.78	3.24	0.46	1.15	0.56
	4×2	4.03	4.88	5.84	0.42	0.80	0.68

6 Conclusions and Further Study

In this paper we have introduced the concept of minimal VULS mining and presented the minREVULSM algorithm for identifying all minimal VULSM in a given graph or network. One application for the VULS concept is classification and we have thus also presented the Match-Voting algorithm for applying the VULS concept to the vertex label classification problem. The work has been illustrated, and evaluated, using a sheet metal forming application (AISF) where we wish to predict the "springback" error as a result of metal forming. The results produced indicate that minimal VULS mining is both efficient and effective (at least in the context of the sheet metal forming application used for the evaluation). For future work the authors intend to investigate more sophisticated ways of conducting VULS based classification and to consider alternative application domains such as social network mining.

Acknowledgments. The research leading to the results presented in this paper has received funding from the European Union Seventh Framework Programme (FP7/2007-2013) under grant agreement number 266208.

References

1. Cafuta, G., Mole, N., Tok, B.: An enhanced displacement adjustment method: Springback and thinning compensation. Materials and Design 40, 476–487 (2012)
2. Firat, M., Kaftanoglu, B., Eser, O.: Sheet metal forming analyses with an emphasis on the springback deformation. Journal of Materials Processing Technology 196(1-3), 135–148 (2008)
3. Jeswiet, J., Micari, F., Hirt, G., Bramley, A., Allwood, J., Duflou, J.: Asymmetric single point incremental forming of sheet metal. CIRP Annals Manufacturing Technology 54(2), 88–114 (2005)
4. Liu, W., Liang, Z., Huang, T., Chen, Y., Lian, J.: Process optimal ccontrol of sheet metal forming springback based on evolutionary strategy. In: 7th World Congress on Intelligent Control and Automation, WCICA 2008, pp. 7940–7945 (June 2008)
5. Nasrollahi, V., Arezoo, B.: Prediction of springback in sheet metal components with holes on the bending area, using experiments, finite element and neural networks. Materials and Design 36, 331–336 (2012)
6. Salhi, S., Coenen, F., Dixon, C., Khan, M.: Identification of correlations between 3d surfaces using data mining techniques: Predicting springback in sheet metal forming. In: Proceedings Proc. AI 2012, pp. 391–404. Springer, Cambridge (2012)
7. Yan, X., Han, J.: gSpan: Graph-based substructure pattern mining. In: Proceedings of the 2002 International Conference on Data Mining, p. 721 (2002)

A Similarity-Based Grouping Method for Molecular Docking in Distributed System

Ruisheng Zhang[*], Guangcai Liu, Rongjing Hu, Jiaxuan Wei, and Juan Li

School of Information Science and Engineering, Lanzhou University, Lanzhou, 730000, China
zhangrs@lzu.edu.cn

Abstract. Molecular docking is one main technique in Virtual Screening. During a molecular docking process, the molecule docking time presents serious diversity because of different chemical structures. The time diversity can cause certain nodes to overload, thereby reducing the data processing ability of the whole distributed molecular docking system. Therefore, a reasonable and efficient data grouping strategy is essential in the molecular docking system. In this paper, molecular structural similarity is researched in depth, and a similarity-based data grouping method is proposed. On the basis of the work in Database Management System for Virtual Screening, the method takes advantage of the computational chemistry software Chemistry Development Kit and cluster analysis methods to process the chemical molecules data. Finally, we deploy and implement the data grouping method on the Hadoop distributed platform. The experimental results show that this data grouping method can improve the efficiency of molecular docking.

Keywords: Molecular Docking, Virtual Screening, Distributed System, Hadoop Platform.

1 Introduction

Virtual Screening (VS) is the primary method in computer-aided drug discovery process. It is also the promotion of drug design methods, the extension of research and development of new drugs. VS greatly shorten the cycle of drug development and save the cost of drug development. Currently, molecular docking is a key method to realize VS [1], which is based on the "Lock-Key" principle [2]. In addition, molecular docking is used to predict the binding mode and affinity between the ligand and receptor. However, due to the large number of chemical small molecules and complexity of molecular docking, the molecular docking technology has seriously hampered the development of VS. Therefore, a reasonable data pre-processing method is of great practical significance for the development of efficient molecular docking.

The distributed computing processing technology is conventionally used in the process of large data sets [3]. A large number of item data is divided into small pieces and calculated by multiple computers. The execution time of task depends on "Cannikin Law" in the distributed system [4], which is determined by the slowest

---

[*] Corresponding author.

H. Motoda et al. (Eds.): ADMA 2013, Part I, LNAI 8346, pp. 554–563, 2013.
© Springer-Verlag Berlin Heidelberg 2013

subtask. Therefore, it is particularly important to keep the load balance for distributed computing. Conventional load balancing scheme is emphasis on the balance of the amount of data, while ignoring the differences between the data [5]. In the Ref. [6], the author proposed a reasonable data grouping method based on molecular properties for molecular docking, however, in terms of VS, the completion of the molecular docking task not only depends on the quantity of sub-tasks data, but also is determined by the data structure which is the structure data of chemical small molecular, and the later is even more important. Therefore, for the load balancing of VS, we have to take into account not only balance data number, but also guaranteeing the similarity of molecular data structure in the sub-tasks.

The notion of molecular similarity is one of the most important concepts in the chemoinformatics. Molecular similarity provides a simple and popular method for VS and underlies the use of clustering methods on chemical databases [7]. As the basic of molecular diversity and cluster analysis, molecular similarity is widely used in drug design. And most studies of the molecular similarity are based on the similar property principle of Johnson and Maggiora, which states: similar compounds have similar properties [8]. To achieve high efficacy of similarity-based screening of databases containing millions of compounds, molecular structures are usually represented by molecular screens (structural keys) or by fixed-size or variable-size molecular fingerprints, whereas the processing of large databases performed with fingerprints has much higher information density. Fragment-based Daylight, BCI, and UNITY 2D fingerprints are some best known examples. Tanimoto (or Jaccard) coefficient T is the most popular similarity measure in the chemical structures represented by molecular fingerprints [9-11].

In this paper, we provide a similarity-based grouping method for molecular docking in distributed system and take advantage of Hadoop platform to achieve the purpose of rapid completion of the preliminary data for molecular docking mission [12]. Firstly, the small molecular chemical data information is collected and submitted to the Hadoop platform. Next, a molecular similarity calculation algorithm based on molecular fingerprints is proposed in detail. Then, a cluster analysis method based on molecular similarity is applied to get different clusters. Finally, a data grouping method to solve the problem of load unbalance caused by molecular structure is introduced.

2   Grouping Method

The grouping method is similar to a middleware between the chemical database Database Management System for Virtual Screening (DBMSVS) and the molecular docking platform, which mainly achieves the preprocessing of small molecule data and to solve the problem of load unbalance caused by molecular structure. Fig. 1 shows that the architecture of the data grouping method: brief exposition of the structure of the data stream. The DBMSVS is collected and submitted to the Hadoop platform. The Hadoop platform is mainly responsible for completing the following tasks: molecular similarity calculation, molecular similarity-based cluster analysis and data grouping. Finally the result data is submitted to molecular docking platform.

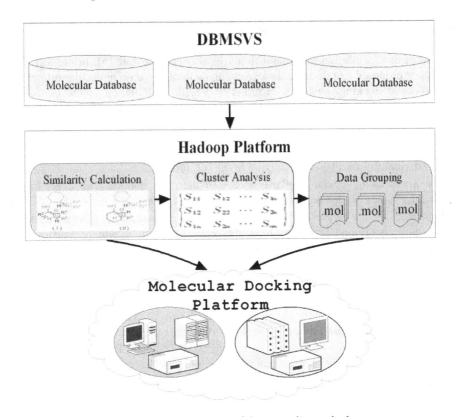

Fig. 1. The architecture of data grouping method

2.1   Data Collection and Submission

This step mainly completes the collection of small molecular chemical data information and submits these data to the Hadoop platform. The data information mainly includes the attribute and structure information of the small molecule. It comes from our chemical database which is a self-developed chemical database, and the database contains seven million small molecules which are from the ZINC databases and the PubChem databases [13, 14]. The structure file of chemical small molecules, which are in a form of the mol2 file, is used to conserve the structure information of small chemical molecules. The size of each mol2 file is about 6KB.

2.2   Molecular Similarity Calculation

In the chemistry discipline system, chemists generally agree with a view that the similar structure of chemical molecules have similar chemical properties or similar biological activities. Molecular similarity is on the basis of molecular diversity and molecular clustering analysis, and has been widely used in the field of drug discovery and development. Molecular similarity can be denoted with similarity or distance. According to the used descriptors, the algorithm is different. Molecular fingerprint is a usual method of determining the similarity of chemical structures. There are various definition methods of structure feature dataset including Daylight Chemical Information System Inc., MDL public keys, MACCS keys, etc. [15]. There are many methods of molecular similarity calculation based on molecular fingerprint, and the most famous of which is the Tanimoto coefficient [16, 17].

In this paper, Daylight fingerprint is adopted as chemical structural representation method. Similarity indexes are calculated according to Tanimoto Coefficient using the famous chemical library project: Chemical Development Kit (CDK) [18, 19]. Fig.2 shows the molecular similarity calculation process. The steps of molecular similarity calculation are as follows:

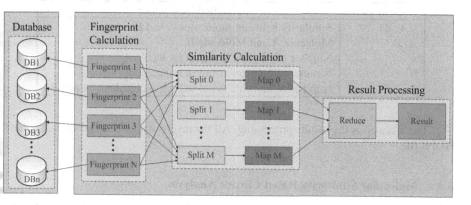

Fig. 2. Molecular similarity calculation flowchart

The first step is to preprocess the mol2 files of chemical molecular structure. In the paper, all the mol2 files are from the DBMSVS. The DBMSVS is a visualization Browser/Server database management tool and developed by our project team. Therefore, in this step, the attribute information of molecular mol2 file is collected from DBMSVS, and integrated into Data Definition Language (DDL) file for the Hadoop platform calls.

The second step is to calculate the fingerprint and store the calculation results in the Hadoop Distributed File System (HDFS) [20]. In this step, chemical molecular structure is represented as a molecular fingerprint by the finger of CDK, and the calculation results are stored in the HDFS. During the process of fingerprint calculation, the fingerprint calculation brings a lot of challenges for the computing platform like the huge number of chemical molecules and the high spatial complexity of molecular fingerprint algorithm. Therefore, the Hadoop platform is chosen as the calculation platform and the calculation results can be easily stored into the HDFS.

The third step is to calculate chemical similarity of the molecules and store the calculation results in matrix in the HDFS. There are many similarity coefficients which are commonly used in chemical information systems. In this paper, the widely used Tanimoto coefficient definition is adopted. In this method, there are two formulas for different variables: continuous variables and dichotomous variables. The range of the latter is 0 to 1, and the formula (1) for dichotomous variables is used to measure chemical similarity.

$$s\ (A,\ B) = c/(a + b - c) \tag{1}$$

The detailed description of the symbols is shown in Table 1:

Table 1. Formula symbols detailed description

Symbol	Detailed Description
s(A, B)	Similarity between molecules A and B
A, B	Molecule A and Molecule B
a	Number of bits "on"(true) in molecule A
b	Number of bits "on"(true) in molecule B
c	Number of bits "on"(true) in both molecules A and B

The last step is the results processing. All the results data are stored in the form of HDFS file.

2.3   Molecular Similarity-Based Cluster Analysis

The similarity matrix is a quantization matrix on the similarity of small molecules. In order to provide more efficient and reasonable grouping results for molecular docking, reasonable cluster analysis is needed. Conventional clustering algorithms include divided method, hierarchical method, density-based method, grid-based method, mode-based method and so on [21]. Here, we adopt the bottom-up hierarchical method. Hierarchical method decomposes a given data set hierarchically until it meets some conditions. Classical hierarchical algorithms mainly include BIRCH algorithm, CURE algorithm, CHAMELEON algorithm, etc.

In this paper, we mainly refer to the process of hierarchical clustering algorithm and combine the core algorithms in the literature to analyze the similarity matrix [22]. Firstly, we cluster every different object in the matrix respectively. Secondly, we merge these atomic clusters based on the cluster similarity. These merged clusters will become larger and larger, and this process willed repeat until it exceeds the cluster threshold. Finally, in order to guarantee the object similarity between different clusters to satisfy the threshold condition, the definition of the merged cluster is as follows:

$$s\ (M,\ N) = \min\{s(x,\ y)\ |\ x \in M,\ y \in N\} \tag{2}$$

The algorithm of molecular similarity-based cluster analysis is described as follows :

```
Algorithm: Molecular Similarity-Based Cluster Analysis
Input: Each row of the n*n similarity matrix Dn and
                    ⎧ S₁₁, S₁₂, ··· S₁ₙ ⎫
                    ⎪ S₂₁, S₂₂, ··· S₂ₙ ⎪
Thresholds r. Dₙ = ⎨ ..., ..., ... ... ⎬ , r∈ [0, 1]
                    ⎩ Sₙ₁, Sₙ₂, ··· Sₙₙ ⎭
Output:  The clustering results {Cᵢ}. i∈ {1, 2, ..., k}

STEP 1 FOR i =1 TO n DO
         Cᵢ = Dᵢ
         END FOR
STEP 2 s(Cₘ, Cₙ)
         =MAX {s(C₁, C₂), s(C₁, C₃), ..., s(C₁, Cₖ),
               s(C₂, C₃), s(C₂, C₄), ..., s(C₂, Cₖ),
               ..., s(Cₖ₋₁, Cₖ)
              }
STEP 3 IF s(Cₘ, Cₙ)>r
         Cₘ = Cₘ ∪ Cₙ
         DELETE Cₙ
         END IF
STEP 4 ITERATION STEP 2,STEP 3
         UNTIL s (Cₘ, Cₙ)<r
STEP 5 OUTPUT {C₁,C₂,...,Cₖ}
```

Fig. 3. The algorithm of molecular similarity-based cluster analysis

2.4   Data Grouping

Data grouping is of great practical significance for load balance of the molecular docking system. Reasonable data grouping can not only save the molecular docking time to improve its efficiency, but also solve the load balance problem of the distributed molecular docking system to improve the efficiency of VS.

As experience of chemists, taking into account the effects of the molecular structure in the molecular docking, the Division Method (DM) of Hash algorithm is selected to construct a grouping function. The calculation function of DM is as follows:

$$Hash (key) = key \bmod p \tag{3}$$

The remainder is the hash address of the keyword key which is divided by the integer p. In the process of using DM, we have not only to choose the right p-value, but also deal with the keyword key. Therefore, molecular is sequentially stored and has a prescribed sequence ID analyzed by cluster analysis. The sequence ID is the keyword key-value, and p-value is the number of sub-docking task.

The data grouping method and focus algorithm are described in detail on four parts: data collection and submission, molecular similarity calculation, molecular similarity-based cluster analysis and data grouping. The feasibility and efficiency of the data grouping method will be discussed in the next part of this article.

3   Result and Analysis

In this section, we actually implement some experiments using the proposed similarity-based grouping method for molecular docking on Hadoop platform. The entire data grouping method is realized from data preparation to molecular docking based on H5N1.

3.1   Experimental Environment

10 IBM servers are used as the management nodes to build the entire Hadoop cluster. The computing nodes are linked with Dolphin PCI/SCI interfaces with a bandwidth of 100MB/s. The configurations of each node are identically and detailed described in Table 2.

Table 2. Description of experimental environment

Environmental Attributes	Environmental Description
Hardware Environment	Intel® Xeon® CPU E5420 @ 2. 50GHz; 4096MB RAM.
Software Environment	Cent OS 5.5; Hadoop 0.20.2; CDK; DOCK 6.0; MySQL 5.1; Sun Java JDK 1. 6. 25.

3.2   Performance Analysis

1000 small molecular data is tested to verify the performance of the Hadoop platform. Molecular similarity is calculated respectively on conventional distributed platforms and the Hadoop platform. These data are uniformly distributed to the cluster and the cluster consisting of nodes 2, 4, 6, 8 and 10. The experimental results of nodes are shown in Table 3.

Table 3. Performance comparison on different platform

Number of Nodes	Conventional Platforms	Hadoop Platforms	Time Ratio
2	319. 72s	124. 64s	2. 57
4	198. 33s	67. 59s	2. 93
6	169. 57s	52. 56s	3. 23
8	114. 68s	49. 56s	2. 31
10	89. 37s	39. 73s	2. 25

There exists a linear proportional between the number of nodes and the docking-task execution time. The maximum value of time ratio is 3.23. Accordingly, Hadoop platform is more efficient and can handle the same task in a shorter time compared with conventional distributed platforms.

Table 4. The execution time with different method

Node ID	Execution Time (order-based)	Execution Time (factor-based)	Execution Time (similarity-based)
1	7179s	8160s	8038s
2	8336s	8185s	8017s
3	8096s	7655s	7988s
4	8518s	7805s	8059s
5	7661s	8155s	8101s
6	6964s	8026s	8253s
7	8631s	8363s	8022s
8	8714s	7842s	8132s
9	8259s	8081s	8119s
10	7615s	7959s	7931s

To evaluate the load performance of molecular docking, three trials are carried out with different grouping strategies, and they are molecular similarity-based, order-based and based on the molecule factor of rotate key. In this experiment, 1000 data are used to execute docking task in the molecular docking system. Monitoring the operating conditions of each node, the execution time of different methods is shown in Table 4 and the corresponding curve graph is shown in Fig.4.

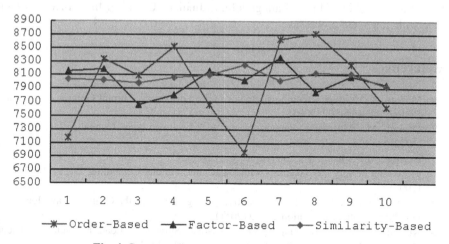

Fig. 4. Corresponding curve graph of execution time

According to the classic "Cannikin Law" and data shown in Table4, the execution time of entire docking task is 8714 seconds with order-based grouping method, the factor-based is 8363 seconds and the similarity-based is 8253 seconds. Although the execution time has a little difference between factor-based and similarity-based method, the running time of each node using the similarity-based grouping method is smoother and almost equal as shown in the Fig.4. Therefore, our grouping method is feasible and efficient, and can be used to solve the problem of the docking system load unbalance caused by molecular structure.

4   Conclusions and Future Work

This study proposes a similarity-based grouping method based on the work of DBMSVS to solve the load unbalance of the docking system fast and effectively caused by molecular structure. This method mainly includes the data preparation and submission, molecular similarity calculation, matrix analysis, and data grouping. All work is meaningful because the grouping data can be used multiple times after processing, although the computational complexity is increased. In addition, the Hadoop Distributed platform is realized in the process of molecular data preprocessing and entire data flow. Our data grouping method is effective and feasible through the validation experiments in real environment.

The future work will focus on improving the accuracy of molecular similarity and the integration of different databases [23-26], and improving the flexibility and efficiency of molecular docking system. Even more important is to improve and expand the existing molecular docking system to adapt higher demands of the entire VS.

Acknowledgment. This paper is supported by National Natural Science Foundation of China (No. 90912003 and 90812001) and ChunHui Plan of Ministry of Education (No. Z2012114). In addition, I am grateful to Jiuqiang Chen and Jingfei Hou for their help with document processes.

References

1. Mclnnes, C.: Virtual screening strategies in drug discovery. Current Opinion in Chemical Biology 11, 494–502 (2007)
2. Conrad, M.: Molecular computing: the lock-key paradigm. Computer 25(11), 11–20 (1992)
3. Beynon, M.D., Kurc, T., Catalyurek, U., Chang, C., Sussman, A., Saltz, J.: Distributed processing of very large datasets with DataCutter. Parallel Computing 27(11), 1457–1478 (2001)
4. Yi, Z.: The Rethinking of the Competitive Strategy Based on the Cannikin Law. Journal of Ningbo Institute of Education 2, 029 (2011)
5. Khetan, A., Vivek, B., Gupta, S.C.: A Novel Survey on Load Balancing in Cloud Computing. International Journal of Engineering 2(2) (2013)
6. Jingwei, L., Rongjing, H., Ruisheng, Z., Jiuqiang, C., Guangcai, L.: An Effective Data Management Solution for Distributed Virtual Screening. In: The 2012 IET International Conference on Frotier Computin., pp. 280–285 (2012)
7. Maldonado, A.G., Doucet, J.P., Petitjean, M., Fan, B.T.: Molecular similarity and diversity in chemoinformatics from theory to applications. Molecular Diversity 10(1), 39–79 (2006)
8. Johnson, M.A., Gerald, M.: Maggiora: Concepts and applications of molecular similarity, vol. 8. Wiley, New York (1990)
9. Daylight Chemical Information Systems Int., http://www.daylight.com/

10. Barnard Chemical Information Ltd., http://www.bci.gb.com/
11. Tripos Inc., http://www.tripos.com/
12. White, T.: Hadoop: The definitive guide. O'Reilly Media, Inc. (2012)
13. ZINC- A free database for virtural screening, http://zinc.docking.org/
14. PubChem, http://pubchem.ncbi.nlm.nih.gov/
15. Taylor, R.C.: An overview of the Hadoop/MapReduce/HBase framework and its current applications in bioinformatics. BMC Bioinformatics 11(suppl. 12) (2010)
16. Ellingson, S.R., Jerome, B.: High-throughput virtual molecular docking: Hadoop implementation of AutoDock4 on a private cloud. In: Proceedings of the Second International Workshop on Emerging Computational Methods for the life Sciences. ACM (2011)
17. Holliday, J.D., Hu, C.Y., Peter, W.: Grouping of coefficients for the calculation of intermolecular similarity and dissimilarity using 2D fragment bit-strings. Combinatorial Chemistry & High Throughput Screening 5(2), 155–166 (2002)
18. Steinbeck, C., Hoppe, C., Kuhn, S., Floris, M., Guha, R., Willighagen, E.L.: Recent developments of the chemistry development kit (CDK) – an open-source Java library for chemo- and bioinformatics. Curr. Pharm. Des. 12(17), 2111–2120 (2006)
19. Steinbeck, C., Han, Y., Kuhn, S., Horlacher, O., Luttman, E., Willighagen, E.: The Chemistry Development Kit (CDK): an open-source Java library for Chemo-and Bioinformatics. J. Chem. Inf. Comput. Sci. 43(2), 493–500 (2003)
20. Borthakur, D.: HDFS architecture guide. Hadoop Apache Project, http://hadoop.apache.org/common/docs/current/hdfs_design.pdf
21. Chen, X., Frank, K.B.: Asymmetry of chemical similarity. Chem. Med. Chem. 2(2), 180–182 (2007)
22. Kaufman, L., Peter, J.R.: Finding groups in data: an introduction to cluster analysis, vol. 344. Wiley-Interscience (2009)
23. Hai, M., Zhang, S., Zhu, L., Wang, Y.: A Survey of Distributed Clustering Algorithms. In: 2012 International Conference on Industrial Control and Electronics Engineering (ICICEE), pp. 1142–1145. IEEE (2012)
24. Yuan, D., et al.: A data dependency based strategy for intermediate data storage in scientific cloud workflow systems. Concurrency and Computation: Practice and Experience 24(9), 956–976 (2012)
25. Ping, S.H.E.N.: The Research on Mining High Dimensional Data. Computer Knowledge and Technology 6, 011 (2009)
26. Zhou, T., Caflisch, A.: Data management system for distributed virtual screening. Journal of Chemical Information and Modeling 49(1), 145–152 (2008)

A Bag-of-Tones Model with MFCC Features for Musical Genre Classification

Zengchang Qin[1], Wei Liu[1,2], and Tao Wan[3,4,*]

[1] Intelligent Computing and Machine Learning Lab
School of ASEE, Beihang University, Beijing, 100191, China
[2] School of Advanced Engineering
Beihang University, Beijing, 100191, China
[3] School of Biological Science and Medical Engineering
Beihang University, Beijing, 100191, China
[4] Department of Biomedical Engineering
Case Western Reserve University, Cleveland, OH 44106, USA
tao.wan.wan@gmail.com

Abstract. Musical genres are categorical labels created by humans to characterize pieces of music. These labels may be highly subjective but typically are related to the instrumentation, rhythmic structure, and harmonic content of the music. In this paper, we propose a model for music genre classification. The new model is referred to as the bag-of-tones (BOT) model which follows the conceptually similar idea of the bag-of-words (BOW) model in natural language processing and the bag-of-feature (BOF) model in image processing. The basic low-level music features such as Mel-frequency cepstral coefficients (MFCC) are clustered into a set of codewords referred to as "tones". By using such a model, each piece of music can be represented by a new feature vector of distribution on tones. Classical machine learning models such as support vector machines (SVM) can be applied for genre classification. The model is tested using two datasets. We found that the polynomial kernel function has the best performance in the SVM classification. By comparing to the previous work, we found the new proposed model outperform classical models on a given benchmark dataset. In general, this model can be used to structure the large collections of music available on the Web. It can play an important role in automatic digital music categorization and retrieval.

Keywords: bag-of-words, bag-of-tones, MFCC, musical genre classification.

1 Introduction

A musical genre is a conventional category that identifies pieces of music as belonging to a shared tradition or set of conventions. With the rapid growth of the Internet, it has become much easier to access digital music and songs. However, it is always difficult for users to find right music of their personal preferences.

---

* Corresponding author.

H. Motoda et al. (Eds.): ADMA 2013, Part I, LNAI 8346, pp. 564–575, 2013.
© Springer-Verlag Berlin Heidelberg 2013

Currently musical genre annotation is performed manually using meta-data. Automatic musical genre classification can assist or replace the human user in this process and would be a valuable addition to music information retrieval systems. So we need to find an efficient way to categorize and recommend music automatically to users. Because of the lack of standards in music classification -or the lack of well-posed standards -there is a huge amount of unclassified music in the world. The notion of similarity between music is complex because there are numerous dimensions of similarity including objective similarity based on musical features such as tempo, rhythm, timbre, but also less objective features such as personal history, social context and even emotions from the music.

Automatically extracting music information is gaining importance. The study of musical genre classification problems has been reported in literatures for years. For example, Dannenberg [1] extracted 13 low-level features from MIDI and used 3 different classifiers (Naive Bayes, linear classifier, and neural networks) to recognize music styles. Chai [2] proposes hidden Markov models to classify the folk music from Irish, German and Austrian based on their monophonic melodies. Shan [3] investigated the classification of music style by melody from a collection of MIDI music. However, MIDI data is a structured format, it is easy to extract features according to its structure. But sound such as .wav and MP3 files are different from MIDI, and the former two formats are more common, thus MIDI style classification is not practical in real applications.

For non-MIDI files, Matiyaho [4] uses a multi-layer neural network to recognize the music types. Han [5] extracts spectral and temporal properties from TV sound signals and use spectrogram principle to classify the genres. Pye [6] used Mel-frequency cepstral coefficients (MFCC) and Gaussian mixture model (GMM) to classify music into six types: blues, easy listening, classic, opera, dance and rock. Jiang [7] used octave-based spectral contrast feature and GMM to classify music into five types. Liu [8] uses features from both MIDI and MFCC to calculate the similarity of music clips.

In this paper, we proposed a new bag-of-tones model for musical genre classification. We deal with music clips in .wav format which is well used in the real-world. The structure of the remainder of the paper is as follows: in Section 2, we introduce the basic feature for music representation. In Section 3, we fully describe the bag-of-tones model in details. In Section 4, we designed experiments to test the model and the experimental results are analyzed. Finally, the conclusions are given in Section 5.

2   Music Features Extraction

The Mel frequency cepstral coefficients (MFCCs) are a set of perceptually motivated features that have been widely used in speech recognition [9]. They provide a compact representation of the spectral envelope, such that most of the signal energy is concentrated in coefficients. There is an increasing use of MFCC in music information retrieval as well as musicology [9]. Seven steps are needed to extract the coefficients, as shown in Fig. 1:

Fig. 1. Extracting a sequence of MFCC feature vectors

Preemphasis. The first step is to boost the energy in the high frequencies. Boosting the high frequency energy makes information in these frequencies more available and this improves the accuracy of the model. In this step, we need to use the first-order-high-pass filter:

$$y_n = x_n - \alpha x_{n-1} \tag{1}$$

where x_n is in time domain representing the input signal; α is a coefficient, $0.9 \leq \alpha \leq 1.0$. For example, Fig. 2 shows the signal waveform before and after the preemphasis step.

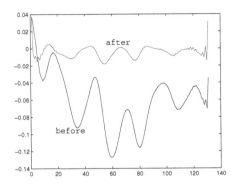

Fig. 2. An example of spectrum before and after preemphasis

Windowing. As the spectrum changes very quickly, we extract features from a small window so we can make the assumption that the signal is stationary. There are three parameters to characterize the window: the width of the window, the offset between successive windows and the shape of the window. The piece extracted from the window is a frame, the frame size is the number of milliseconds and the number of milliseconds between the left edges of successive windows the frame shift.

Discrete Fourier Transform (DFT). In order to calculate how much energy the signal contains at different frequency bands, we employ DFT to extract the spectral information. The sequence of N complex numbers $x_0, x_1, \ldots, x_{N-1}$ is transformed into an N-periodic sequence of complex numbers $X_0, X_1, \ldots, X_{N-1}$ according to the DFT formula:

$$X_K = \sum_{n=0}^{N-1} x_n (e^{-j\frac{2\pi}{N}kn}) \tag{2}$$

where x_n is a windowed signal which is the input of DFT; and the output, X_K is a complex number representing the magnitude and phase of that frequency component in the input signal; N is the sample length of analysis window; K is the length of DFT.

Mel Filter Bank and Log. The mapping between frequency in Hertz and the Mel scale is linear below 1000Hz and logarithmic above 1000Hz, and the Mel frequency can be computed from the sound wave as follows:

$$Mel(f) = 1127 \ln(1 + \frac{f}{700}) \tag{3}$$

During MFCC computation, we create a bank of filters that collects energy from each frequency band to implement this task. The filter bank has 10 filters spaced linearly below 1000Hz and the remaining filters spread logarithmically above 1000Hz. Fig. 3 shows the bank of triangular filters that implement this idea. Finally, we take the log of each of the Mel spectrum values.

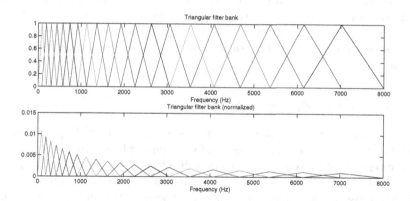

Fig. 3. An example of triangular filter bank

Cepstrum. The last step to extract the coefficients is to compute the cepstrum of the log of the Mel spectrum values. There are many different methods to calculate it, here we calculate the Discrete Cosine Transform (DCT):

$$X_k = 2 \sum_{n=0}^{N-1} x_n \cos(\frac{\pi k(2n + 1)}{2N}), \quad 0 \le k \le N - 1 \tag{4}$$

where N is the number of filters in the filter bank, k is the number of cepstral coefficients which are computed, and x_n is formulated as the "log-energy output of the n-th filter".

3 Bag-of-Tones Model

The bag-of-words model is a simplifying representation used in natural language processing and information retrieval (IR) [10]. In this model, a text is represented as an unordered collection of words, disregarding grammar and even word order. The bag-of-words model is commonly used in methods of document classification, where the (frequency of) occurrence of each word is used as a feature for training a classifier [11,12]. In computer vision, the bag-of-words model (BoW model), also referred to as the bag-of-features model, can be applied to image classification [13,14], by treating image features as "visual words". In document classification, a bag of words is a sparse vector of occurrence counts of words; that is, a sparse histogram over the vocabulary. In computer vision, a bag of visual words is a sparse vector of occurrence counts of a vocabulary of local image features.

Following the similar idea, we propose the bag-of-tones model. Similarly, we treat music as a document, and the "words" need to be well defined. To achieve this, all pieces of music are transformed into a high dimensional space of low-level features (i.e, MFCC in this paper) where they are clustered to obtain some significant basic units such as topics in text processing and visual words in image processing - here, we call them the "tones". Each piece of music then can be represented as a distribution of the tones. The details are described as the following.

3.1 Feature Description

Modeling music with appropriate features is a complex task. In [8], the author extract features from both MIDI files and wave format files. He extracted average pitch, pitch entropy, pitch density, average duration, duration entropy and others form MIDI files and MFCCs from wave format files. In [15], the author selected high zero-crossing rate ratio, low short-time energy ratio, and spectrum flux as features of MIDI files. And other meta data such as the singers, the lyric and some other information also can be used as features. However, it is hard for people to extract information about the pitch and the duration from a .wav file. For its good performance in speech recognition, the MFCC features are employed to transform each piece of music into a 13-dimensional matrix. It is like to have 13 channels where the length is proportional to the length of the original music piece L_M and the frame shift L_w. In this paper, the first and second derivatives of each channel are also used. In all, we have $13 \times 3 \times L$ dimensions where $L = L_M / L_w$.

3.2 Codebook Generation

The essential step of the Bag-of-Tones (BOT) model is to convert vector represented musical clips to "codewords", which also produces a "codebook". Given a piece of music, by calculating the MFCC, each point of sound is transformed into a vector with 39 dimensions. Therefore, each piece of music can be transformed into a feature matrix of 39 rows.

In clustering, all the matrices are mapped into a 39-dimensional space, where each sound is represented as a point in this high dimensional space. Given the size of the codebook (number of clusters), K-means are used to cluster all these sound in this space, where each cluster center can be regarded as a codeword or *tone*. Given a piece of music, each sound is classified to either of these basic tones based on nearest Euclidean distance in this space. The distribution of a piece of music on tones can be simply calculated using frequency counting.

In this work, the music clips are of the same length and what we meet more often is that songs are of different lengths. When this happens, bag-of-tones model can also be applied with only few changes in the final step. In such situations, normalization can be used. The sum of all numbers in a certain histogram is divided by each number in this histogram, thus we can get a new vector which is the same dimension with the original one, and which can be used as input vector of SVM.

3.3 Classification

Once descriptors have been assigned to clusters to form the feature vectors, we reduce the problem to that of multi-class supervised learning. The classifier performs two separate steps in order to predict the classes of the unlabeled music

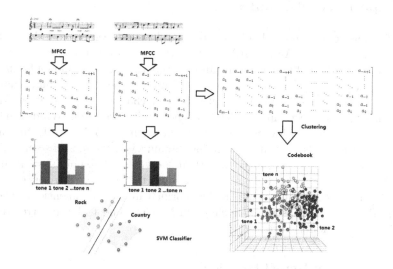

Fig. 4. A schematic illustration of the BOT model. The MFCC features of each music piece are extracted. The feature matrices are mapped into the feature space and clustered into basic "tones". Each feature matrix can be represented as a distribution on tones that serves as a new feature vector. The SVM is then used to classify these music pieces based on these new feature vectors.

clips: training and testing. During training, labeled data is sent to the classifier and used to adapt a statistical decision procedure for distinguishing categories. Among many classifiers, we use Support Vector Machine (SVM) and compare the performance of different kernel functions. The complete process of applying the BOT model is illustrated in Fig. 4.

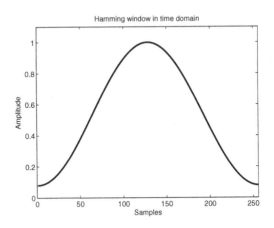

Fig. 5. The shape of Hamming window in time domain

4   Experimental Studies

To verify the performance of the new proposed model, we first build a test set of our own called BUAA5 (BUAA music dataset with 5 genres)[1], in which the music clips are dual-channel, in enframing, and the signal is monophonic. We extract signal from both channels separately and reshape them to single channel. In order not to lose information, we just join one channel after the other. When pre-emphasize the spectrum, $\alpha = 0.9375$. The window we use in Windowing is a 256-point Hamming window whose shape is shown in Fig. 5. It shrinks the values of the signal towards zero at the window boundaries, avoiding discontinuities. The Hamming window has 256-point with an increment of 80 points, so the frame size is 6ms, and the frame shift is 2ms. Based on the steps we introduced in the last section, a single musical clip is transformed into a 39-dimensional feature matrix. Because the cepstral coefficients in MFCCs tend to be uncorrelated, it helps make our model more accurate.

4.1   Influence of Kernel Functions

In our experiments, we use the dataset BUAA5 which contains 1543 music clips from 132 songs of five different genres: country, rock, jazz, punk, electronic. Each clip is saved at 44100Hz sampling rate, with dual channel and 32 bits per sample

---

[1] http://icmll.buaa.edu.cn/members/WeiLiu/liuwei.html

and each music clips is 10s in length. We divide the dataset into two parts: 1057 clips for training and the test set is consisting by the remaining 486 clips. To avoid any possible bias in the experiments, all the clips are randomly truncated from the songs, and the clips are randomly selected from the data base.

In order to study the influence caused by codewords, we set the size of code-book K to 30, 40, 50, 60, 80, 90, 100, 110, respectively, and calculate the accuracy with different kernels in SVM, the results are shown in Table 1 and 2. By compar-ing the above two tables, we can see that the kernel function with the polynomial function has the best average accuracy 75.93%. When the kernel function is poly-nomial, rock music can be best recognized at the rate about 69.61%, with 50 codewords; country music gets 90.43% in accuracy when with 110 codewords ; the accuracy of jazz music is about 84.62%; for the other two categories, elec-tronic and punk music, can be classified correctly with the rate of 71%.

Table 1. Accuracy on BUAA5 with linear kernel function where K is the number of codewords

Genre \ K	30	40	50	60	80	90	100	110
Rock	45.10%	54.90%	56.86%	64.71%	59.80%	64.71%	74.51%	62.75%
Country	54.26%	59.57%	67.02%	61.70%	65.96%	73.40%	68.09%	69.15%
Jazz	63.46%	57.69%	54.81%	54.81%	57.69%	59.62%	62.50%	63.46%
Electronic	50.93%	56.48%	59.26%	60.19%	64.81%	62.04%	60.19%	56.48%
Punk	44.87%	39.74%	47.44%	53.85%	51.28%	48.72%	41.03%	50.00%
Average	52.06%	54.32%	57.41%	59.26%	60.29%	62.14%	62.14%	60.70%

Table 2. Accuracy on BUAA5 with polynomial kernel function where K is the number of codewords

Genre \ K	30	40	50	60	80	90	100	110
Rock	62.75%	64.71%	69.61%	67.65%	60.78%	61.76%	65.69%	65.69%
Country	68.09%	79.79%	81.91%	84.04%	88.30%	85.11%	89.36%	90.43%
Jazz	65.38%	66.35%	71.15%	75.96%	76.92%	77.88%	83.65%	84.62%
Electronic	62.69%	65.74%	61.11%	62.96%	64.81%	71.30%	63.89%	69.44%
Punk	55.13%	48.72%	52.56%	61.54%	66.67%	71.79%	67.95%	69.23%
Average	63.17%	65.64%	67.70%	70.58%	71.40%	73.46%	74.07%	75.93%

4.2   GTZAN Data

As the BUAA5 has not yet been tested by other researchers, it is not fair to use it to verify performance of the new proposed model. In this experiment, we test the new model on the GTZAN Genre Collection, which was first introduced in [16]

Table 3. Average accuracy with different number of codewords

K	50	150	200	300	400	500	600
Average Accuracy	51.41%	56.22%	58.23%	59.83%	60.24%	61.04%	**61.85%**
K	610	650	660	670	680	690	700
Average Accuracy	60.64%	59.84%	58.63%	61.04%	57.83%	61.04%	60.64%

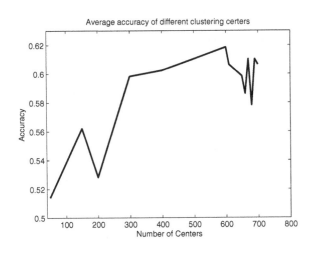

Fig. 6. Average accuracy of the BOT model with different number of codewords

and be available to public[2]. The dataset consists of 1000 audio tracks each has 30 seconds long. It contains 10 genres, each genre is with 100 tracks. The tracks are all 22050Hz Mono and 16-bit audio files in .wav format. In the experiments, 75% of the clips are used for training classifier and the rest 25% are used as test examples. The average accuracy against different number of codewords are listed in Table 3. As we can see from the table, the best accuracy is achieved when codebook size is 600. In order to get more precise results, we also test the number from 610 to 700. The change of accuracy is also illustrated in Fig. 6, the detailed accuracy for each genre are shown in Table 5 and Table 6, respectively.

Table 4. Result comparisons to Random, RT-GS, GS and GMM [16]

Model	Random	RT-GS	GS	GMM(2)	GMM(3)
Average Accuracy	10.00 %	44.00 %	59.00 %	60.00 %	61.00 %
Model	GMM(4)	KNN(1)	KNN(3)	KNN(5)	BOT(600)
Average Accuracy	61.00 %	59.00 %	60.00 %	56.00 %	**61.85 %**

---

[2] http://marsyas.info/download/data_sets/

Table 5. Accuracy of the BOT model with different number of codewords

Genre \ Size of Codebook	50	150	200	300	400	500	600	700
Blues	56%	44%	56%	60%	56%	44%	60%	52%
Classical	79%	79%	83%	75%	75%	79%	71%	71%
Country	32%	44%	56%	60%	60%	68%	60%	60%
Disco	56%	68%	56%	48%	60%	64%	60%	52%
Hiphop	60%	68%	64%	56%	64%	68%	60%	64%
Jazz	48%	44%	44%	56%	56%	44%	60%	56%
Metal	60%	64%	68%	80%	68%	72%	84%	84%
Pop	52%	68%	64%	76%	72%	76%	64%	72%
Reggae	36%	56%	48%	48%	44%	52%	48%	44%
Rock	36%	28%	44%	40%	48%	44%	52%	52%

Table 6. Accuracy of the BOT model with different number of codewords

Genre \ Size of Codebook	600	610	650	660	670	680	690	700
Blues	60%	52%	44%	44%	64%	48%	64%	52%
Classical	71%	67%	79%	71%	71%	67%	67%	71%
Country	60%	64%	56%	56%	68%	60%	60%	60%
Disco	60%	64%	60%	60%	52%	52%	60%	52%
Hiphop	60%	60%	60%	64%	60%	64%	64%	64%
Jazz	60%	60%	52%	60%	52%	52%	52%	56%
Metal	84%	68%	80%	68%	76%	76%	80%	84%
Pop	64%	72%	72%	68%	68%	68%	68%	72%
Reggae	48%	56%	48%	52%	52%	48%	48%	44%
Rock	52%	44%	48%	44%	48%	44%	48%	52%

Given the GTZAN dataset, the BOT model is compared to the following models: the Random classifier, RT [16], RT-GS [16], simple Gaussian Classifier (GS), Gaussian mixture model (GMM(k) where k is the number of Gaussian) and K-Nearest Neighbor (KNN). How these models can be used for musical genre classification are discussed in [16]. As we can see from Table 4, the new proposed BOT model (with K=600) outperform or at least equivalent (considering the case there is no statistical significance test due to the limited number of data) to the previous models. Another advantage of this model is its simplicity. By using the idea of the bag-of-words, the BOT model is basically based on MFCC transform, K-means clustering and support vector machines. All these techniques are classical and easy to be implemented.

However, the experimental results refer to single runs for each setting and this is not allow assuming that the results are significant and will be the same for other experiments. To solve this, more experiments are needed to be done, the average accuracy should be recorded and the variation of all those average accuracies under the same setting can be calculated, thus we can get a more convincing result.

Besides, the relationship of accuracy and the number of codewords is not very obvious and we can just see from the results that there may be a best number of codewords that has the highest accuracy; and when starting a new experiment, the number of codewords is not known and for now we can only try different numbers to find the best one. In the future work, we will try to find a method by which to estimate the range of the codeword numbers.

5 Conclusion and Future Work

In this paper, we have presented a simple but novel approach for musical genre classification. Each piece of music can be represented by MFCC feature matrices. These features are clustered into codewords called "tones". Following the idea of the bag-of-words model, we proposed the bag-of-tones model that each piece of music can be represented by a distribution on the codebook. Such distribution can be used as new feature vectors that can be classified using support vector machines. The approach has been evaluated on a five category dataset BUAA5 and a benchmark data GTZAN Genre Collection. The experimental results show that the polynomial kernel function outperform linear kernels in classification. The general performance is the same or better than some classical methods on given a benchmark.

However, this model has disadvantages that we have ignored the relations between tones. All the basic codewords are basically modeled on sound or tones. If we dynamically consider a duration of music, we can model the music on "tunes" - which is an ordered set of tones. This will be our future work to find what are the basic tunes and how they can contribute to musical genre classification. It is also the case that one clip is labeled with more than one genres. This phenomenon indicates that this clip has more than one styles. We should tell the users the likelihood of this clip belonging to a certain musical genre.

Acknowledgements. This work is partially funded by the National Natural Science Foundation of China (NSFC) under Grant No. 61305047.

References

1. Dannenberg, R.B., Thom, B., Watson, D.: A machine learning approach to musical style recognition. In: Proc. International Computer Music Conference (1997)
2. Chai, W., Barry, V.: Folk music classification using hidden Markov models. In: Proceedings of International Conference on Artificial Intelligence, vol. 6 (2001)

3. Shan, M.K., Kuo, F.-F.: Music style mining and classification by melody. IEICE Transactions on Information and Systems 86(3), 655–659 (2003)
4. Matityaho, B., Furst, M.: Neural network based model for classification of music type. In: Eighteenth Convention of Electrical and Electronics Engineers in Israel. IEEE (1995)
5. Han, K.-P., Park, Y.-S., Jeon, S.-G., Lee, G.-C.: Genre classification system of TV sound signals based on a spectrogram analysis. IEEE Transactions on Consumer Electronics 44(1), 33–42 (1998)
6. Pye, D.: Content-based methods for the management of digital music. In: Proceedings of the 2000 IEEE International Conference on Acoustics, Speech, and Signal Processing, vol. 6. IEEE (2000)
7. Jiang, D.N., Lu, L., Zhang, H.J., Tao, J.-H.: Music type classification by spectral contrast feature. In: Proceedings of the IEEE International Conference on Multimedia and Expo, ICME 2002, vol. 1. IEEE (2002)
8. Liu, N.H.: Comparison of content-based music recommendation using different distance estimation methods. Applied Intelligence 38(2), 160–174 (2013)
9. Logan, B.: Mel frequency cepstral coefficients for music modeling. In: MUSIC IR (2000)
10. Qin, Z., Thint, M., Huang, Z.: Ranking answers by hierarchical topic models. In: Chien, B.-C., Hong, T.-P., Chen, S.-M., Ali, M. (eds.) IEA/AIE 2009. LNCS (LNAI), vol. 5579, pp. 103–112. Springer, Heidelberg (2009)
11. Zhao, Q., Qin, Z., Wan, T.: What is the Basic Semantic Unit of Chinese Language? A Computational Approach Based on Topic Models. In: Kanazawa, M., Kornai, A., Kracht, M., Seki, H. (eds.) MOL 12. LNCS (LNAI), vol. 6878, pp. 143–157. Springer, Heidelberg (2011)
12. Zhao, Q., Qin, Z., Wan, T.: Topic modeling of Chinese language using character-word relations. In: Lu, B.-L., Zhang, L., Kwok, J. (eds.) ICONIP 2011, Part III. LNCS, vol. 7064, pp. 139–147. Springer, Heidelberg (2011)
13. Yuan, X., Yu, J., Qin, Z., Wan, T.: A bag-of-features model with integrated SIFT-LBP features for content-based image retrieval. In: Proceedings of the International Conference on Image Processing, pp. 1061–1064 (2011)
14. Yu, J., Qin, Z., Wan, T., Zhang, X.: Feature integration analysis of bag-of-features model for image retrieval. Neurocomputing 120, 355–364 (2013)
15. Lie, L., Jiang, H., Zhang, H.: A robust audio classification and segmentation method. In: Proceedings of the Ninth ACM International Conference on Multimedia (2001)
16. Tzanetakis, G., Cook, P.: Musical genre classification of audio signals. IEEE Transactions on Speech and Audio Processing 10, 293–302 (2002)

The GEPSO-Classification Algorithm

Weihong Wang[1,2], Dandan Jin[2], Qu Li[2], Zhaolin Fang[2], and Jie Yang[2]

[1] State Key Laboratory of Software Development Environment, Beihang University,
Bejing, China
[2] College of Computer Science, Zhejiang University of Technology, Hangzhou, China
{wwh,lq,fzl}@zjut.edu.cn, troubleking@yeah.net,
yangjie4699@163.com

Abstract. In order to solve the problem that the evolutionary algorithm based class center classification algorithm easily falls into a local optimum later in the process, this paper proposes a Gene Expression Programming (GEP) classification algorithm which is optimized by Particle Swarm Optimization(PSO). It's named after the GEPSO-Classification Algorithm, and the word GEPSO comes from the combination of the word GEP and PSO. This algorithm first finds a suboptimal solution on the merit that GEP can converge rapidly in the early stage, then with this suboptimal solution, the algorithm searches the optimal solution on the merit that PSO is more likely to converge to the optimal solution. The experimental result shows that this algorithm has a better performance on classification.

Keywords: GEP, PSO, Classification.

1 Relevant Knowledge

Classification [1] is important in data mining, it needs to construct a classifier to mark the unlabelled data points. In [2], the classification methods are divided into four types: distance-based classification, decision tree classification, Bayesian classification and rule induction classification. Distance-based classification methods have the simple idea and according to its idea, a class centers based classification method called N-centers algorithm has proposed[1]. The algorithm requires a lot of searching and computation work to construct a group of centers which can accurately represent the classes. When classifying, calculate the dissimilarity between the data point and each class, and the classification result is the class label which the minimum dissimilarity is corresponding to.

Evolutionary algorithm is a kind of stochastic search algorithm which is based on the theory of biological evolution, and it searches the optimal solution by simulating the biological evolution process. According to the various realizations of different parts in the algorithm, Andries [3] divides evolutionary algorithms into: Genetic Algorithms(GA), Genetic Programming(GP), Evolutionary Programming(EP), Evolutionary

---

[1] Wang Weihong, Jin Dandan, Li Qu. The GEP based N-centers classification algorithm.

H. Motoda et al. (Eds.): ADMA 2013, Part I, LNAI 8346, pp. 576–583, 2013.
© Springer-Verlag Berlin Heidelberg 2013

Strategies(ES), Differential Evolution(DE), Cultural Evolution(CE) and Co-evolution(CoE) and so on. Besides, Gene Expression Programming(GEP) [4] and Particle Swarm Optimization(PSO) [3]. The former is a novelty genetic algorithm which is proposed on the base of GA and GP by a Portuguese scholar, C.Ferreira. The latter is an evolutionary algorithm similar to Genetic Algorithms. Evolutionary algorithms has many important applications in various fields of data mining, in classification, evolutionary algorithms are often combined with existing classification methods in different forms, and thus bring up a kind of classification method which is based on evolutionary algorithms (e.g. [5-7]). As the combination of two algorithms is on the principle of giving play to the advantages of the two algorithms and avoiding their disadvantages, so the new algorithm often has better performance.

Evolutionary algorithm based classification method utilizes the strong optimization ability of evolutionary algorithm to search class centers, and the classifier is constructed by these centers. N-centers algorithm elaborated the idea how to classify with class centers, and then it was combined with GEP and proposed the NGEP-Classification. In [8], it didn't point out the concept of using class centers to classify, but it can be seen that the process to construct the classifier is namely the process of constructing a group of class centers. Evolutionary algorithm based classification method using evolutionary algorithm to searches class center, thus improves the efficiency, but it has the shortcoming of convergence to local optimum.

2   Summary

Compared with traditional search algorithms, evolutionary algorithms has the characteristic of being parallel and intelligent(cf. [9]), which makes the evolutionary algorithm achieve better search performance and be wider applied. In classification problems, evolutionary algorithms are often combined with traditional classification methods to form a kind of evolutionary algorithm based classification methods. Compared with original classification algorithms, the performance of evolutionary algorithm based classification methods has improved in different degrees. NGEP-Classification[1] the can quickly converge in the early stage, but as the best individual fitness of the population continues to improve, the convergence speed gradually slows down, and it's easy to converge to a local optimal value, and difficult to break through the local optimal value. In order to overcome the shortcoming of this algorithm, this paper introduces PSO into the algorithm on the purpose of optimizing the solution which is found by GEP, thus proposes the GEPSO-Classification algorithm.

GEP has a strong ability of global optimization, it converges fast in the early stage of the algorithm, later, the convergence speed slows down, and it becomes easy to converge to a local optimum value. When it comes to PSO, larger space it searches in, the lower convergence speed and accuracy it has. But while the search space becomes small, PSO can converge very fast with a high accuracy. Therefore this paper combines GEP with PSO in the below form: in the early stage, use GEP to find a suboptimal, then construct a suitable search space for PSO with the suboptimal solution found by GEP. This article assumes the optimal solution exists in this suitable space, GEP can't find it because GEP falls into a local optimal and it's difficult for GEP to break through the local optimal. PSO will be fast flying in the small space until the solution which satisfies the stop condition is found or the maximum generation number is reached. The experiment results suggest that GEPSO-Classification has a better performance.

3   GEPSO-Classification Algorithm

This algorithm composes of two complete processes of evolutionary algorithm and an interim: in the first stage, GEP finds suboptimal solutions, it is also possibly to find the optimal solution; in interim: the algorithm will step into PSO stage from GEP stage; in latter stage, PSO finds the optimal solution.

3.1   The GEP Stage

In this stage, NGEP-Classification algorithm is used to find a suboptimal solution, the detail realization of this algorithm is introduced in its article, so this paper doesn't introduce it any more.

3.2   The Interim

This stage sends the suboptimal solution found in GEP stage to PSO stage, and this suboptimal solution is named base particle X_{base} with which PSO search for the optimal solution.

3.3   The PSO Stage

In the PSO stage of GEPSO-Classification algorithm, this paper will improve PSO to make it suitable for searching a classifier which is constructed by n class centers. With the base particle X_{base} and a parameter called Enable Offset, PSO defines a small search space V_{small}. The optimal solution exists in V_{small} When the size of V_{small} is suitable. Before the first iteration, PSO lets X_{base} be the global optimal position P_{global}. The initial particles fly in V_{small}, and the algorithm compares the optimal position p'_{global} of each iteration with P_{global}. If the current p'_{global} is better than P_{global}, P_{global} will be replaced by p'_{global}. When the solution which is satisfies the stop condition or the count of iterations reaches the maximum, the search process stops, and P_{global} is namely the output solution. The flowchart of PSO stage is as follows.

Particle Coding. In general PSO algorithm, each particle has a current position and a current speed. Both of them are represented by space vectors. While in the PSO stage of GEPSO-Classification algorithm, the current position P_i composes of n component

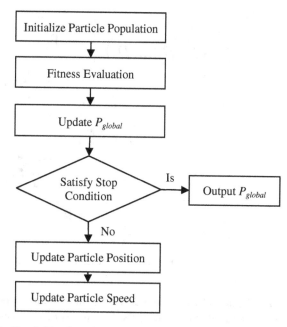

Fig. 1. The flowchart of PSO stage

positions $p_k(k=$ 1, 2, ... , n), a component position represents a class center c_i, so the current position P_i is namely a classifier. Corresponding to the current position, the current speed V_i is also constituted of n component speed v_k, and v_k only works on the corresponding p_k. The particle i is represented as follows:

$$X_i=\{ P_j(p_1, p_2, ... , p_n); V_j(v_1, v_2, ... , v_n), (p_i, v_i=(x_1, x_2, ... , x_m), j=1, 2, ... , n)\} \quad (1)$$

Here, n means the count of classes, m represents the dimension amount of P_j and V_j. The representation of Particle X_i in V_{small} is shown in Fig.2, here suppose it is a *3-class* classification problem.

As it is shown in Fig.2, the position of the particle in space is no longer a point, it is represented by several spatial component positions. A component position is changed by the corresponding component speed, and the update of speed is also caused by the component speeds. Formula(2) and Formula(3) show how the position and speed of particle X_i updates.

$$p_k(t+1)=v_k(t)+ p_k(t) \quad (2)$$

$$v_k(t+1)=v_k(t)+c_1 \cdot r_{1,k}(t) \cdot (p_{best}(t)-p_k(t))+ c_2 \cdot r_{2,k}(t) \cdot (p_{global}(t)-p_k(t)) \quad (3)$$

Here, $p_k(t)$ represents the component position of X_i in generation t, and $v_k(t)$ is namely the component speed of X_i in generation t. From Formula (2) and Formula (3), it can be seen that the position and speed update in the same way with the general PSO algorithm except that this algorithm updates on each component position and speed.

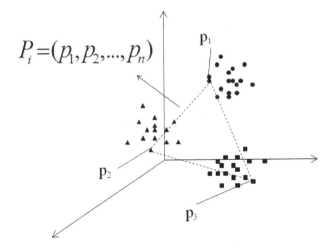

Fig. 2. The representation in V_{small}

The Search Space of Particles. Corresponding to current position and current speed, the search space in this algorithm also composes of n component space v_k. The component speed only works on the corresponding position, and the corresponding position only changes in corresponding space, the corresponding speed must make sure that the component position will not be out of its corresponding component space. That is to say, one corresponds to one among the three parameters. In this algorithm, particles' search space V_{small} is co-determined by X_{base} and *Enable Offset*. The *Enable Offset* is a real number in the interval [0, 1], and it is preset by the algorithm. $p_{base,k}$ represents the component position of X_{base}, each dimension value of $p_{base,k}$ multiplies 1-*offset*, thus get the v_k which is X_{base}'s the lower limit of the component space's dimension k. and the upper limit of dimension k is got by multiplying each dimension value of $p_{base,k}$ with 1+*offset*. Component spaces are constructed just in this way. As is referred above, when particles fly in these spaces, the component speed must make sure the component position will not exceed the component space.

Particle Swarm Initialization. Different with the basic particle swarm initialization, this algorithm needs X_{base} to do the initialization. Each particle is produced by multiplying a random r and X_{base}, and r is in interval [1-*offset*, 1+*offset*]. For example, a 3-*class* classification problem, in which data points have four attributes. X_{base} is represented by P_{base}=((4, 3, 2, 6),(3, 5, 3, 4), (2, 2, 1, 7)). Set *Enable Offset*= 0.2, thus r is in interval [0.8, 1.2]. While initializing a particle, assume the value of r is 0.9, 0.85, 0.95, 1.1, 1.2, 1.15, 0.82, 0.96, 0.99, 1.18, 1.05, 1.15, thus the initial position of this particle is P=((3.6, 2.25, 1.9, 6.6), (3.6, 5.75, 2.46, 3.84), (1.98, 2.36, 1.05, 8.05)). And the upper limit of each dimension value for each component search space is ((4.8, 3.6, 2.4, 7.2), (3.6, 6, 3.6, 4.8), (2.4, 2.4, 1.2, 8.4)). And the lower limit is ((3.2, 2.4, 1.6, 4.8), (2.4, 4.0, 2.4, 3.2), (1.6, 1.6, 0.8, 5.6)).

Fitness Computation. In this algorithm, the individuals in GEP stage and PSO stage are equivalence relation, so they have the same fitness computation, the details are shown

4   Experiment

4.1   Settings of Experimental Prameters

In order to verify the validity of this algorithm, this paper compares the experimental result of this algorithm with NGEP-Classification on the same data sets. Experimental data sets are from the UCI database. The information of these data sets is described in Table 1.

Table 1. The main information of data sets

Data set	Number of examples	Number of attributes	Number of classes
Breast Cancer W.	683	9	2
Glass	214	9	6
Iris	150	4	3
Lung Cancer	32	56	3
Pima Indian	768	8	2
Wine	178	13	3
Zoo	101	17	7

In this experiment the maximum generation number in GEP stage is 200, while it's 100 in PSO stage. And their sum is equal to NGEP-Classification's maximum generation number, and other parameters are also set the same. The detail settings of experimental parameters are shown in Table 2.

Table 2. Settings of parameters

Stage	Parameters	Value
GEP stage	Maximum generation	200
	Population size	100
	The count of elites	1
	Mutate probability	0.8
	Transfer probability	0.8
PSO stage	Maximum generation	100
	Population size	100
	C1, c2	2
	offset	0.25

4.2   The Comparison of Experimental Results

This paper applies 5-fold cross-validation, and the results of GEPSO-Classification and NGEP-Classification are shown in Table 3. Among these datasets, only dataset Wine are normalized because of the big disparity of its attributes' values. C4.5 is a kind of classical classification algorithm, it constructs reasonable classifier and its accuracy is higher than many other classification algorithms. In order to make the judgement more objective, the result of C4.5 is also shown in Table 3.

Table 3. The experimental results

Data set	C4.5	Rule Classification with GEP		NGEP-Classification	
		Average accuracy	Best accuracy	Average accuracy	Best accuracy
BreastCancerW	94.7	97.0±0.2	97.4±0.1	97.1±0.3	97.5±0.1
Glass	65.7	61.5±1.2	63.0±0.2	60.5±2.7	63.2±0.6
Iris	93.9	96.1±1.1	97.5±0.3	97.1±1.4	97.7±0.5
LungCancer	44.3	43.8±2.8	51.9±3.7	43.8±5.6	54.4±3.2
PimaIndian	74.8	75.7±0.7	77.0±0.7	76.6±0.5	77.2±0.6
Wine	91.6	95.7±0.5	96.9±0.6	96.0±0.7	97.0±0.6
Zoo	92.0	92.9±0.7	95.2±1.2	95.1±1.4	96.8±1.0

From Table 3, it can be seen that the accuracy of GEPSO-Classification is generally higher than C4.5, except that in Glass and LungCancer, GEPSO-Classification is lower. Compared with NGEP-Classification which is without PSO, the accuracy of GEPSO-Classification has also improved significantly. For the average accuracy, the accuracy of GEPSO-Classification is lower than NGEP-Classification only in Glass, and this algorithm do the same well in LungCancer. While in other 5 datasets Breast-CancerW, Iris, PimaIndian, Wine and Zoo, GEPSO-Classification is 0.1%, 1.0%, 0.9%, 0.3% and 2.2% higher than NGEP-Classification. For the best accuracy, this paper's algorithm does better in all datasets, it's respectively 0.1%, 0.2%, 0.2%, 2.5%, 0.2, 0.1 and 1.6% higher than NGEP-Classification. The above comparison shows that, the GEPSO-Classification algorithm has a higher accuracy and a better performance in Classification.

5   Conclusion

The NGEP-Classification easily falls into local optimum value, so this paper introduces PSO to this algorithm, proposes GEPSO-Classification. With the characteristic that GEP converge quickly in the early stage, this algorithm searches for a suboptimal solution, then searches for the optimal solution with the characteristic that PSO can search accurately in a small space. The experimental result on several standard data sets show that GEPSO-Classification algorithm outperforms C4.5 and NGEP-Classification algorithms. The performance of this algorithm depends on the choice of parameters to some extends, but this paper hasn't done some further study of it, that is where this paper needs to improve.

Acknowledgement. This paper is supported by the National Natural Science Foundation of China (60873033), the Natural Science Foundation of Zhejiang Province (R1090569 and LY12F02039) and the State Key Laboratory of Software Development Environment open Fund(SKLSDE-2012KF-05).

References

1. Jiawei, H., Micheline, K.: Data Mining: Concept and Techniques. China Machine Press, Beijing (2011)
2. Guojun, M., Lijuan, D., Wang, S.: The Principle and Algorithm of Data Ming. TsingHua University Press, Beijing (2007)
3. Andries, P.E.: Computer Intelligence An Introduction. TsingHua University Press, Beijing (2010)
4. Ferreira, C.: Gene Expression Programming: Mathematical Modeling by an Artificial Intelligence. SCI, vol. 21. Springer, Heidelberg (2006)
5. Zengwei, Z., Ping, W.: The Study of Naive Bayes algorithm based Genetic Algorithm Classification. Computer Engineering and Design (2012)
6. Chi, Z., Weimin, X., Tirpak: Evolving Accurate and Compact Classification Rules with Gene Expression Programming. IEEE Transactions on Evolutionary Computation (2003)
7. Weihong, W., Wei, R., Qu, L.: Decision Tree Algorithm by Gene Expression Programming Based on Differential Evolution. Computer Engineering (2011)
8. Rui, D., Hongbing, D., Xianbin, F.: Particle Swarm Optimization Genetic Algorithm Applied in Classification Question. Computer Engineering (2009)
9. Licheng, J., Jing, L., Weicai, Z.: Co-evolutionaryComputation and Multi-agent System. Science Press, Beijing (2006)

Author Index